Lecture Notes in Computer Science   10411

Commenced Publication in 1973
Founding and Former Series Editors:
Gerhard Goos, Juris Hartmanis, and Jan van Leeuwen

More information about this series at http://www.springer.com/series/7409

Michael Gertz · Matthias Renz
Xiaofang Zhou · Erik Hoel
Wei-Shinn Ku · Agnes Voisard
Chengyang Zhang · Haiquan Chen
Liang Tang · Yan Huang
Chang-Tien Lu · Siva Ravada (Eds.)

Advances in Spatial and Temporal Databases

15th International Symposium, SSTD 2017
Arlington, VA, USA, August 21–23, 2017
Proceedings

 Springer

Editors

Michael Gertz
Institute of Computer Science
Heidelberg University
Heidelberg
Germany

Matthias Renz
George Mason University
Fairfax, VA
USA

Xiaofang Zhou
University of Queensland
Brisbane, QLD
Australia

Erik Hoel
ESRI
University of Minnesota
Minneapolis, MN
USA

Wei-Shinn Ku
Auburn University
Auburn, AL
USA

Agnes Voisard
Free University of Berlin
Dahlem, Berlin
Germany

Chengyang Zhang
Microsoft
Redmond, WA
USA

Haiquan Chen
California State University
Sacramento, CA
USA

Liang Tang
LinkedIn
Sunnyvale, CA
USA

Yan Huang
University of North Texas
Denton, TX
USA

Chang-Tien Lu
Virginia Tech
Falls Church, VA
USA

Siva Ravada
Oracle
Redwood Shores, CA
USA

ISSN 0302-9743   ISSN 1611-3349   (electronic)
Lecture Notes in Computer Science
ISBN 978-3-319-64366-3   ISBN 978-3-319-64367-0   (eBook)
DOI 10.1007/978-3-319-64367-0

Library of Congress Control Number: 2017947509

LNCS Sublibrary: SL3 – Information Systems and Applications, incl. Internet/Web, and HCI

Printed on acid-free paper

This Springer imprint is published by Springer Nature
The registered company is Springer International Publishing AG
The registered company address is: Gewerbestrasse 11, 6330 Cham, Switzerland

Preface

This volume contains the proceedings of the 15th International Symposium on Spatial and Temporal Databases (SSTD 2017). Included are research contributions in the area of spatial and temporal data management and related computer science domains presented at SSTD 2017 in Arlington, VA, USA. The symposium brought together, for three days, researchers, practitioners, and developers for the presentation and discussion of current research on concepts, tools, and techniques related to spatial and temporal databases. SSTD 2017 was the 15th in a series of biannual events. Previous symposia were held in Santa Barbara (1989), Zurich (1991), Singapore (1993), Portland (1995), Berlin (1997), Hong Kong (1999), Los Angeles (2001), Santorini, Greece (2003), Angra dos Reis (2005), Boston (2007), Aalborg (2009), Minneapolis (2011), Munich (2013), and Hong Kong (2015).

Before 2001, the series was devoted solely to spatial database management, and was called The International Symposium on Spatial Databases. Starting in 2001, the scope was extended in order to also integrate the temporal dimension and accommodate spatial and temporal database management issues, owing to the increasing importance of research that considers spatial and temporal dimensions of data as complementary challenges.

This year the symposium received 90 submissions in total out of which 58 contributions were submitted as research papers including two industrial papers. All papers were reviewed by four of the 58 Program Committee members. At the end of a thorough process of reviews and discussions, 19 submissions were accepted as full research papers for presentation at the symposium. SSTD 2017 also continued several innovative topics that were successfully introduced in previous events. In addition to the research paper track, the conference hosted a demonstration track and a vision/challenge track. Demonstration and vision/challenge papers were solicited by separate calls for papers. While proposals for demonstrations had to illustrate running systems that showcase the applicability of interesting and solid research, the vision/challenge submissions had to discuss novel ideas that are likely to guide research in the near future and/or challenge prevailing assumptions. The Computing Research Association's Computing Community Consortium (CCC) sponsored awards in the form of travel grants for the top three papers under their CCC Blue Sky Ideas initiative. The submissions to the demo and vision/challenge track (19 demonstration submissions and 13 vision/challenge papers submissions) were evaluated by dedicated Program Committees (each paper was reviewed by at least three members). Finally, eight demo and five vision/challenge papers were selected for the conference program. We were also very fortunate to have had two accomplished researchers from academia and industry as keynote speakers opening the first two days of the conference.

This year SSTD also organized a panel where renowned members of academia, industry, and federal government organizations discussed current trending and promising future research topics on spatio-temporal database and related research fields. Vivid discussions were inspired on topics such as smart cities, big spatial data systems, spatial crowdsourcing, spatial privacy, as well as insights from behinds the scenes from a National Science Foundation program director.

In addition to the panel, SSTD 2017 initiated the Early Career Researcher Workshop as a new event under the scope of the conference aiming to help early-career researchers, e.g., junior faculty member and senior PhD students, in their professional development.

The success of SSTD 2017 was the result of a team effort. Special thanks go to many people for their dedication and hard work, in particular to the demo track chairs, Wei-Shinn Ku (Auburn University, USA), Agnes Voisard (Free University of Berlin), and Chengyang Zhang (Microsoft, USA), the industry track chair, Erik Hoel (ESRI, USA), and local arrangements chair, Nektaria Tryfona (College of Science, George Mason University, USA). Further special thanks go to our panel chairs, publicity chairs, sponsorship chairs, proceedings chairs, registration chairs, webmaster, and the organizers of the Junior Faculty Workshop. Naturally, we owe our gratitude to more people, and in particular we would like to thank the authors, irrespectively of whether their submissions were accepted or not, for supporting the symposium series and for sustaining the high quality of the submissions. We also want to express our thanks to the sponsors of SSTD 2017, HERE, Google, and the Computing Community Consortium (CCC) for their generous support of this event. Last but most definitely not least, we are very grateful to the members of the Program Committees (and the external reviewers) for their thorough and timely reviews.

Finally, these proceedings reflect the state of the art in the domain of spatiotemporal data management, and as such we believe they form a strong contribution to the related body of research and literature.

June 2017

Michael Gertz
Matthias Renz
Xiaofang Zhou
Yan Huang
Chang-Tien Lu
Siva Ravada

Organization

Steering Committee

The SSTD Endowment

General Chairs

Yan Huang	University of North Texas, USA
Chang-Tien Lu	Virginia Tech, USA
Siva Ravada	Oracle, USA

Program Chairs

Michael Gertz	Heidelberg University, Germany
Matthias Renz	George Mason University, USA
Xiaofang Zhou	University of Queensland, Australia

Industry Chair

Erik Hoel	ESRI, USA

Demo Chairs

Wei-Shinn Ku	Auburn University, USA
Agnes Voisard	Free University of Berlin, Germany
Chengyang Zhang	Microsoft, USA

Panel Chairs

Liang Zhao	George Mason University, USA
Andreas Züfle	George Mason University, USA

Publicity Chairs

Guoliang Li	Tsinghua University, China
Jin Soung Yoo	Indiana University – Purdue University, Fort Wayne, USA

Sponsorship Chairs

Hua Lu	Aalborg University, Denmark
Raimundo Dos Santos	Army Geospatial Center, USA

Local Arrangements Chair

Nektaria Tryfona   College of Science, George Mason University, USA

Proceedings Chairs

Haiquan Chen   California State University, Sacramento, USA
Liang Tang   LinkedIn, USA

Registration Chairs

Yao-Yi Chiang   University of Southern California, USA
Kerone Jones Wetter   George Mason University, USA

Webmaster

Taoran Ji   Virginia Tech, USA

Junior Faculty Workshop

Zhe Jiang   University of Alabama, USA
Xun Zhou   University of Iowa, USA

Research/Industry Program Committee

Walid Aref
Masatoshi Arikawa
Nikolaos Armenatzoglou
Spiridon Bakiras
Michela Bertolotto
Claudio Bettini
Haiquan Chen
Lei Chen
Reynold Cheng
Chi-Yin Chow
Christophe Claramunt
Gao Cong
Maria Luisa Damiani
Ke Deng
Johann Gamper
Ralf Hartmut Güting
Yan Huang
Sergio Ilarri
Fang Jin
Kyoung-Sook Kim

Peer Kröger
Bart Kuijpers
Lars Kulik
Jae-Gil Lee
Dan Lin
Cheng Long
Nikos Mamoulis
Michael Mcguire
Claudia Medeiros
Mohamed Mokbel
Kyriakos Mouratidis
Mirco Nanni
Reza Nourjou
Dimitris Papadias
Spiros Papadimitriou
Torben Bach Pedersen
Dieter Pfoser
Chiara Renso
Dimitris Sacharidis
Jagan Sankaranarayanan

Mohamed Sarwat
Markus Schneider
Erich Schubert
Matthias Schubert
Bernhard Seeger
Shashi Shekhar
Yufei Tao
Yannis Theodoridis
Ranga Raju Vatsavai
Carola Wenk
Martin Werner
Ouri Wolfson
Raymond Chi-Wing
   Wong
Xiaokui Xiao
Jiang Zhe
Yu Zheng
Andreas Züfle

Vision Program Committee

Christian S. Jensen	Timos Sellis	Shashi Shekhar
Mario Nascimento	Cyrus Shahabi	Vassilis Tsotras

Demo Program Committee

Haiquan Chen	Yaron Kanza	Kristian Torp
Rui Chen	Hua Lu	Wendy Hui Wang
Feng Chen	Apostolos	Ting Wang
Jing Dai	N. Papadopoulos	Fusheng Wang
Cedric Du Mouza	Cyril Ray	Karine Zeitouni
Shen-Shyang Ho	Marcos Salles	Ji Zhang
Xunfei Jiang	Dimitris Skoutas	

External Reviewer

Furqan Baig	Zhi Liu	Dimitris Tsakalidis
Leilani Battle	Somayeh Naderivesal	Fabio Valdés
Xin Chen	Elham Naghizade	Oleksii Vedernikov
Theodoros	Phuc Nguyen	Jianqiu Xu
Chondrogiannis	Lefteris Ntaflos	Siyuan Zhang
Yixiang Fang	Márcio C. Saraiva	Yudian Zheng
Jiafeng Hu	Haiqi Sun	Zizhan Zheng
Christos Koutras	Panagiotis Tampakis	Yang Zhou

Gold Sponsor

HERE International B.V.

Silver Sponsors

Google
The Computing Community Consortium (CCC)

Contents

Routing and Trajectories

Big Spatial Data

Indexing and Aggregation

Demonstrations

Routing and Trajectories

Multi-user Itinerary Planning for Optimal Group Preference

Liyue Fan[1(✉)], Luca Bonomi[2], Cyrus Shahabi[3], and Li Xiong[4]

[1] University at Albany SUNY, Albany, NY, USA
liyuefan@albany.edu
[2] University of California - San Diego, La Jolla, CA, USA
lbonomi@ucsd.edu
[3] University of Southern California, Los Angeles, CA, USA
shahabi@usc.edu
[4] Emory University, Atlanta, GA, USA
lxiong@emory.edu

Abstract. The increasing popularity of location-based applications creates new opportunities for users to travel together. In this paper, we study a novel spatio-social optimization problem, i.e., Optimal Group Route, for multi-user itinerary planning. With our problem formulation, users can individually specify sources and destinations, preferences on the Point-of-interest (POI) categories, as well as the distance constraints. The goal is to find a itinerary that can be traversed by all the users while maximizing the group's preference of POI categories in the itinerary. Our work advances existing group trip planning studies by maximizing the group's social experience. To this end, individual preferences of POI categories are aggregated by considering the *agreement* and *disagreement* among group members. Furthermore, planning a multi-user itinerary on large real-world networks is computationally challenging. We propose one approximate solution with bounded approximation ratio and one exact solution which computes the optimal itinerary by exploring a limited number of paths in the road network. In addition, an effective compression algorithm is developed to reduce the size of the network, providing a significant acceleration in our exact solution. We conduct extensive empirical evaluations on the road network and POI datasets of Los Angeles and our results confirm the effectiveness and efficiency of our solutions.

1 Introduction

The prosperity of social networks and location-based services has created exciting opportunities to enable social interactions between users online as well as in the real world. Especially, we are interested in the problem where multiple users want to plan a trip together, jointly visiting one or more locations that maximize the entire group's experience. For instance, two friends would like to meet up on

---

L. Bonomi—Joint first author.

© Springer International Publishing AG 2017
M. Gertz et al. (Eds.): SSTD 2017, LNCS 10411, pp. 3–23, 2017.
DOI: 10.1007/978-3-319-64367-0_1

their way home after work. As in Fig. 1, they leave from their work locations, s_1 and s_2, have dinner (1), tour a museum exhibit (2), have coffee near the Trevi's Fountain (3), and then head back home separately, i.e., t_1 and t_2. Realistically, each of them would have certain constraints, e.g., how far he/she is willing to detour, and different personal preferences for the types of location to visit, e.g., restaurants are preferable over food trucks.

While route planning has been largely studied in literature, the majority of the studies focused on finding an optimal itinerary for a single user [1–3]. A few recent papers [4–7] studied the problem of *Group Trip Planning* for multiple users to travel together. The goal of those studies is to minimize the trip distance for the group in order to cover the specified categories from the input. Despite the effort in including multiple users, those studies present several limitations in modeling the social aspect in a *group* trip. Firstly, existing studies do not consider the possible differences in individual preferences, which occur often in real scenarios. For example, among a group of friends, some may want to dine at a sports bar while watching a game, while others may prefer fine-dining restaurants. Therefore, it is crucial to model the group's global preferences to maximize the social experience of the multi-user trip. In addition, existing studies do not consider the limited mobility or time each user has for traveling. Due to such constraints, the group trip should have the flexibility to dismiss certain POI categories. For instance, even though one user would like to visit a "museum", the best group itinerary may not include such a POI if other members are not willing to travel that far.

To this end, we formulate a novel spatio-social optimization problem, named *Optimal Group Route* (OGR), for multi-user itinerary planning. Our goal is to maximize the *group preference* on the visited locations within *individual distance constraints*. Each user can specify source and destination locations as well as the maximum distance he/she is willing to travel. In addition, each user can specify individual preference scores for the POI categories to visit. The itinerary will be planned on a road network graph where each edge is a directed road segment and each node can have multiple category labels.

There are two key challenges in addressing the OGR problem. Firstly, the individual preferences on each category may vary greatly across the multi-user group. This requires a robust scoring method that considers both the agreement and disagreement among group members to optimize the group experience. Secondly, the OGR problem is computationally expensive. We will demonstrate that it can be considered as a generalization of the Hamiltonian Path problem [8], which is known to be NP-complete. To enable real-world applications, the solutions should demonstrate efficiency despite the large size of the road network.

Contributions. Our specific contributions in this paper are as follows:
1. We formally define the OGR problem to optimize the group preference within individual distance constraints. We show that the problem is NP-complete.
2. We model the social experience of the group by aggregating individual category preference scores. Specifically, the group's global preference considers both *Group Relevance* and *Group Disagreement* measures.

3. We design an approximate solution that greedily constructs the itinerary and achieves a $1/k$-approximation, where k is the number of categories specified by the group. Furthermore, we propose an exact solution which constructs the optimal itinerary by exploring a limited number of paths in the graph. Even though the exact algorithm runs exponentially with k, our algorithm is shown to be very efficient as k is much smaller than the graph size in practice.

4. We further propose an effective algorithm that reduces the search space via a compression step. In our empirical evaluation, the compression technique is shown to be very effective and can accelerate the exact solution up to one order of magnitude.

5. We conduct extensive empirical evaluation using real road network and POIs of Los Angeles, which confirm the practicability of our proposed solutions.

The rest of the paper is organized as follows. In Sect. 2, we define the OGR problem and show its hardness. In Sect. 3, we describe our solutions in detail. In Sect. 4, we summarize the related works on route planning queries. In Sect. 5, we provide our empirical results. Finally, we conclude the paper in Sect. 6.

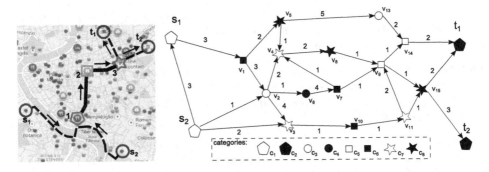

Fig. 1. Example itinerary for two users for 3 categories   **Fig. 2.** Location graph: each category is a different shape

2   Problem Definition and Preliminaries

A location graph $G = (V, E)$ consists of a set of nodes V and a set of edges E. Each node $v \in V$ represents a location, and $e = (v_i, v_j)$ denotes a directed edge between v_i and v_j. Each edge $e = (v_i, v_j)$ is associated with a positive value $\delta(e)$ representing the cost of traveling from v_i to v_j, e.g., distance. We consider G as a directed graph to represent a real road network where the traveling cost between two locations may not be symmetric. Note that our solutions are applicable on undirected graphs as well.

Some nodes in the location graph represent Point-of-Interests (POIs). Specifically, each node v_i is associated with a set of categories $C(v_i)$ that provides the nature of the POI. We adopt this formulation because in real road networks

POIs may be associated with multiple categories. For example, a shopping mall may have several restaurants, a post office, a theater, etc. We denote $\mathcal{C} = \{c_1, c_2, \ldots, c_N\}$ the universe of categories for nodes in the location graph. An example location graph with eight categories is depicted in Fig. 2. Only one category is shown for each node to simplify the diagram.

Definition 1 (Route). *A route $r = \langle v_0, v_1, \ldots, v_n \rangle$ is a path that starts at v_0 and ends at v_n, through the edges (v_i, v_{i+1}) where $i = 0, \ldots, n-1$.*

Definition 2 (Route Cost). *Given a route $r = \langle v_0, v_1, \ldots, v_n \rangle$, the cost of r is the sum of the edge costs in r: $\delta(r) = \sum_{i=0}^{n-1} \delta(v_i, v_{i+1})$.*

Our formulation is quite general and in principle the function δ may be used to model different costs associated with the route. For example, the time required for traveling from a point to another or just their distance. In the rest, of the paper, we will focus on considering the road network distance as a cost function.

Definition 3 (Route Coverage). *The route coverage $\mathcal{C}(r)$ for $r = \langle v_0, v_1, \ldots, v_n \rangle$ is the union of node categories in r: $\mathcal{C}(r) = \bigcup_{i=0}^{n} \{C(v_i)\}$.*

In our formulation, we assume that each user traveling in a group has own start and destination locations, a maximum distance he/she is willing to travel, and a set of preferred categories to visit.

Definition 4 (User). *A user u_i is a 4-tuple vector $\{\mathcal{P}(u_i), \Delta_i, s_i, t_i\}$, where s_i is the starting location, t_i is the destination, and Δ_i is the distance constraint. $\mathcal{P}(u_i) = \{\alpha_1^i, \alpha_2^i, \ldots, \alpha_N^i\}$ is the preference set, where $\alpha_j^i \in [0, \alpha_{max}]$ is u_i's preference score for category c_j.*

The parameter Δ_i indicates how much u_i is willing to detour to accommodate the group and can be specified as $\theta \geq 1$ times the shortest path distance from s_i to t_i. The preference set $\mathcal{P}(u_i)$ provides u_i's preference on the specified categories. Note that α_{max} upper-bounds the preference score and is a pre-defined parameter in our problem setting. In practice, we can assume that u_i specifies only the categories with non-zero preference in $\mathcal{P}(u_i)$. Alternatively, it may be possible to extract the preference set automatically from user profile or location history.

Given a group of m users, i.e., $\mathcal{U} = \{u_1, u_2, \ldots, u_m\}$, we can find $\mathcal{C}_{\mathcal{U}}$ as the union of the categories with non-zero individual preferences. We are interested in finding an *itinerary* which allows for a feasible route for each user in the group, such that they can jointly visit a series of POIs belonging to categories in $\mathcal{C}_{\mathcal{U}}$.

Definition 5 (Itinerary). *Given a user group $\mathcal{U}$, $I = \langle v_0, v_1, \ldots, v_l \rangle$ is a candidate itinerary, if for each $u_i \in \mathcal{U}$ there exists a route $r_i = \langle s_i, \ldots, I, \ldots, t_i \rangle$ and $\delta(r_i) \leq \Delta_i$.*

Example 1. Consider the graph in Fig. 2, where two users u_1 and u_2 specify their starting points s_1, s_2 and ending points t_1, t_2 respectively. Let $\Delta_1 = 15$ and $\Delta_2 = 20$ be their distance constraints. Then, $I = \langle v_2, v_4, v_8, v_9, v_{15} \rangle$ is a candidate itinerary, as we can construct $r_1 = \langle s_1, v_1, I, t_1 \rangle$ for u_1 and $r_2 = \langle s_2, I, t_2 \rangle$ for u_2 such that $\delta(r_1) \leq \Delta_1$ and $\delta(r_2) \leq \Delta_2$.

Among all the candidate itineraries, we are interested in those that maximize the group social experience by visiting the categories in $\mathcal{C_U}$. Since a category covered by an itinerary may have different individual preference scores, we propose to aggregate the individual preferences and compute a single score for each category, named *global preference score*. To this end, we consider possible agreement and disagreement among the group members. Specifically, inspired by the work of [9], we introduce *group relevance* and *group disagreement* for the categories from the input.

Definition 6 (Group Relevance). *The relevance of a category c_j for a group $\mathcal{U}$, denoted $rel(\mathcal{U}, c_j)$, is the aggregation of individual preferences:*

$$rel(\mathcal{U}, c_j) = \frac{1}{m} \sum_{i=1}^{m} \alpha_j^i \tag{1}$$

Definition 7 (Group Disagreement). *The disagreement of a category c_j for a group $\mathcal{U}$, denoted $dis(\mathcal{U}, c_j)$, is the average pairwise disagreement among individual preferences:*

$$dis(\mathcal{U}, c_j) = \frac{2}{m(m-1)} \sum_{i,i'=1, i \neq i'}^{m} |\alpha_j^i - \alpha_j^{i'}| \tag{2}$$

Definition 8 (Global Preference). *Given $w \in [0,1]$, the global preference λ_j of a category c_j for a group $\mathcal{U}$ is defined as a weighted combination of the group relevance and group disagreement as follows:*

$$\lambda_j = w \cdot rel(\mathcal{U}, c_j) + (1-w) \cdot (\alpha_{max} - dis(\mathcal{U}, c_j)) \tag{3}$$

In this definition, categories with high relevance and low disagreement would have high global preference, as illustrated in the following example.

Example 2. Let $\alpha_5^1 = 4$, $\alpha_7^1 = 3$ and $\alpha_5^2 = 2$, $\alpha_7^2 = 3$ be the preferences for the categories c_5 and c_7 specified by users u_1 and u_2. In this setting, we observe that the relevance for c_5 and c_7 is 3 while their disagreements are 2 and 0 respectively. Assuming for example, $\alpha_{max} = 5$ and $w = 0.5$, we have that $\lambda_5 = 3$ and $\lambda_7 = 4$. Hence, the category c_7 leads to higher value of global preference compared to c_5.

The global preference combines both group relevance and disagreement providing a balanced and robust aggregation of individual preferences, e.g., preventing the scores from being dominated by a single user. Using these notions, we aim to represent the social experience of the group in visiting the categories covered by the itinerary as the *profit* of the itinerary.

Definition 9 (Itinerary Profit). *The profit $Ps(I)$ of an itinerary I, is the sum of the global preference for the user-specified categories covered by I. Formally,*

$$Ps(I) = \sum_{c_j \in \mathcal{C}(I) \cap \mathcal{C_U}} \lambda_j. \tag{4}$$

Problem 1 (Optimal Group Route). Given a location graph G, we define a group trip query $\mathcal{Q} = \langle \mathcal{U}, \mathcal{C}_{\mathcal{U}} \rangle$, where $|\mathcal{U}| = m$ and $|\mathcal{C}_{\mathcal{U}}| = k$. The goal is to answer $\mathcal{Q}$ by providing an itinerary I that maximizes $Ps(I)$.

The Optimal Group Route (OGR) problem requires to explore the location graph in order to search for the most profitable itinerary. While there may exist more than one itineraries achieving the maximum profit, we are interested in returning only one to the group of users. In the following we demonstrate the hardness of this problem.

Theorem 1 (Hardness). *Determining the existence of an itinerary for a user group that achieves profit p on a location graph G is NP-complete.*

Proof. Clearly OGR is in NP. Then, we show the hardness of OGR by reducing the Hamiltonian path problem (HP) in directed graphs to it. Given a directed input graph $G = (V, E)$, a starting node s and an ending node t, the HP problem requires one to decide if there exists a Hamiltonian path from s to t in G. Given an instance for HP, we can create a new instance for OGR as follows. We consider a graph $G' = G$, where each node $v_i \in V$ is labeled with a unique category $c(v_i) = c_i$, and the cost for each edge is set to 1. Then, we consider the input query $\mathcal{Q} = \langle \mathcal{U}, \mathcal{C}_{\mathcal{U}} \rangle$ formed by only one user u with start and end points s and t respectively, with cost constrain $\Delta = |V| - 1$, and with preference set $\mathcal{P}(u) = \{\alpha_1, \alpha_2, \ldots, \alpha_{|V|}\}$, where $\alpha_i = 1$ for $i = 1, \ldots, |V|$. Therefore, an itinerary I with preference score $p = |V|$ for the group, i.e., user u, exists in G' if and only if the original graph G has a Hamiltonian path from s to t. Since the reduction is polynomial with the size of G and given the fact that HP is NP-complete, it follows that OGR is NP-complete.

3   Proposed Solutions

In this section, we describe the cornerstones for solving the OGR problem and present our two solutions.

3.1   Meeting Graph and Node Profit

In principle, one can examine all the possible paths in the location graph, finding the optimal itinerary in a brute-force manner. However, it incurs exponential running time with respect to the number of nodes in the graph, which is impractical for real road networks. Given our problem formulation, we can take advantage of users' distance constraints and preferences to reduce the search space for the optimal itinerary. Specifically, we will use the geospatial information of the nodes and their POI categories to significantly reduce the computational cost as we will demonstrate in this section.

Meeting Graph. In our problem formulation, a candidate itinerary can be traversed by each group member without violating their distance constraints.

In practice, only a small portions of nodes in G are reachable by all the users via feasible routes. Therefore, only those special nodes, named *meeting points*, need to be considered when constructing the optimal itinerary. We formally define this concept as follows.

Definition 10 (Meeting Point). *A node $v \in V$ is a meeting point for group $\mathcal{U}$, if $\delta(\pi(s_i, v)) + \delta(\pi(v, t_i)) \leq \Delta_i$ for every $u_i \in \mathcal{U}$, where $\pi(s,t)$ is a shortest path from s to t in G.*

Continuing from Example 1, the meeting points of u_1 and u_2 are $\{v_2, v_3, v_4, v_6, v_7, v_8, v_9, v_{10}, v_{11}, v_{15}\}$. With this definition, we construct a subgraph of G, i.e., $G_M = (V_M, E_M)$, induced by the set of meeting points $V_M \subseteq V$, as illustrated in Fig. 3a. We call G_M the *meeting graph* for users in $\mathcal{U}$. The search space for the optimal itinerary can now be restricted to G_M. Although in the worst case $G_M = G$, in real scenarios group members may not willing to detour much (i.e., restrictive distance constraints). Therefore, the meeting graph G_M is considerably smaller than the original graph G. As a preprocessing step, the meeting graph can be derived in an efficient way by utilizing the geospatial information of the graph G, i.e., *latitude* and *longitude* of each node in V. Specifically, since the geospatial information of the road network is known a priori, we can construct a R*-tree offline to index the nodes. When a OGR query $\mathcal{Q}$ is received in input, the index structure allows us to efficiently obtain the meeting graph without expensive computations. Given the start/end locations and the distance constraints of the group members, we utilize the R*-tree together with the geometric properties of ellipses as in [6] to quickly discard those nodes not reachable to the entire group. We apply further pruning to the remaining nodes using the road network distance to derive the set of meeting points V_M. In our empirical evaluation, the meeting graph derivation is very effective and efficient.

Node Profit in Itinerary. Intuitively, when adding nodes to an itinerary, we would like to select those that increase the itinerary profit. When a new node v is added to an itinerary I, its contribution to the profit of the current itinerary I depends on its categories and those already covered by I. Such a profit increment, denoted as $\sigma(v, I)$, is defined as follows.

$$\sigma(v, I) = \sum_{c_i \in (C(v) \setminus C(I)) \cap C_{\mathcal{U}}} \lambda_i \tag{5}$$

The computation of the profit increment for each node is dynamically performed during the itinerary construction process. We observe that this operation can be performed efficiently in $O(k)$ times where k is the number of categories specified in query $\mathcal{Q}$.

3.2 Greedy Itinerary Construction

We first propose a baseline solution that greedily constructs an itinerary by selecting the best node in each iteration, i.e., the node with the highest profit

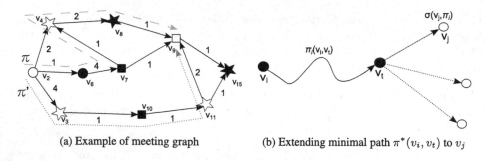

(a) Example of meeting graph   (b) Extending minimal path $\pi^*(v_i, v_t)$ to v_j

Fig. 3. Example of meeting graph and minimal paths

increment as defined in Eq. 5. Among the nodes with the same profit, the algorithm selects the closest node to the end of the current itinerary. Initially, the algorithm starts by selecting the node with the highest profit and the minimum average distance from all the user starting locations. Each time a new node is selected, the algorithm verifies whether the new node can be appended to the itinerary without violating the distance constraint of any user. If infeasible, the algorithm moves on to consider the best among the remaining nodes.

Greedy Algorithm. Our greedy approach receives as input the original graph G, the set of users $\mathcal{U}$ and the meeting graph G_M pre-computed. Starting from an empty itinerary I, it iteratively extends the itinerary with one node at the time. For each iteration, the greedy algorithm computes: (i) the score for each node v in the meeting graph according to its categories and those already covered by the current itinerary, i.e., $\mathcal{C}(I)$ as in Eq. 5, and (ii) the distance from the end node of current I to node v. The computation of (i) is performed to ensure the the the node with the highest profit increment is found in each iteration. On the other hand, the computation of (ii) is performed to ensure the feasibility of the itinerary, i.e., every user in the group can traverse the itinerary from start to destination within the individual distance constraint. As every $v \in V_M$ is reachable to the group, both $\delta(\pi(s_i, v))$ and $\delta(\pi(v, t_i))$ have been computed for each user u_i during the meeting graph derivation. To efficiently compute (ii), we adopt the all-shortest paths procedure for the nodes in V_M at the beginning of the algorithm.

Utility. Due to varying (local) graph connectivity, it is likely that greedily selecting a node to add to the itinerary leads to a sub-optimal solution. In fact, in any iteration by selecting the best current node, the greedy approach may leave out those nodes covering other categories specified in input. The following theorem illustrates such a situation and quantifies the approximation ratio of the greedy algorithm.

Theorem 2 (Greedy Approximation). *The solution returned by the Greedy approach is a $(1/k)$-approximation to the optimal solution, where k is the number of categories with non-zero preference scores specified in input.*

Proof. Let I_G be the itinerary computed by the greedy algorithm and I^* be optimal solution. It is clear that $Ps(I_G) \leq Ps(I^*)$, since the algorithm selects each node in a greedy way. To show the underestimate error, we consider the case where I_G contains only one node v that covers one category with the maximum preference score $\sigma = \max_{i=1,...,k}\{\lambda_i\}$, while the optimal itinerary I^* covers all k categories in the input query. Therefore, we have: $Ps(I^*) = \sum_{j=1}^{k} \lambda_j \leq k \cdot \sigma = k \cdot Ps(I_G)$. Hence, the following inequalities hold: $\frac{Ps(I^*)}{k} \leq Ps(I_G) \leq Ps(I^*)$.

Time Complexity. The greedy algorithm requires all-shortest paths within the meeting graph to ensure the feasibility of the itinerary. Therefore, using the Floyd-Warshall algorithm this operation takes $O(|V_M|^3)$ time. Let k be the number of categories specified by all users in input (i.e. non-zero preferences), then the scoring process for each node takes $O(k|V_M|)$ time. Overall, each node is tested at most once for each user; therefore, the total complexity for this algorithm is $O(km|V_M|^2 + |V_M|^3)$, where $|V_M|$ denotes the number of meeting points. In practice, since the size of the meeting graph is much smaller than the original graph, the greedy algorithm is very efficient as we will demonstrate in the experiments section.

3.3   Optimal Itinerary Construction

In this section, we develop an optimal solution to the OGR problem which does not enumerate all possible paths in the meeting graph. The construction of the itinerary is performed by considering only a subset of paths, named *minimal paths*, which have the property of being the shortest paths achieving a given profit.

Definition 11 (Minimal Path). *Given a profit score p, a path from v_i to v_j is called minimal path of profit p, denoted as $\pi_p^*(v_i, v_j)$, if it has minimal distance cost among all the paths from v_i to v_j with profit exactly p.*

$$\pi_p^*(v_i, v_j) = \underset{\pi(v_i,v_j)|Ps(\pi(v_i,v_j))=p}{\arg\min} \delta(\pi(v_i,v_j)) \qquad (6)$$

Example 3. Continuing from Example 1, let $\mathcal{P}(u_1) = \{\alpha_3^1 = 2, \alpha_5^1 = 4, \alpha_6^1 = 2, \alpha_7^1 = 3\}$ and $\mathcal{P}(u_2) = \{\alpha_5^2 = 1, \alpha_6^2 = 2, \alpha_7^2 = 3\}$ be the preference sets for u_1 and u_2 on the location graph illustrated in Fig. 2. In the induced meeting graph in Fig. 3a, there are only two paths $\pi = \langle v_2, v_6, v_7, v_4, v_8, v_9 \rangle$ and $\pi' = \langle v_2, v_3, v_{10}, v_{11}, v_9 \rangle$ covering the categories c_3, c_5, c_6 and c_7. Let $p = \lambda_3 + \lambda_5 + \lambda_6 + \lambda_7$. Since $\delta(\pi') < \delta(\pi)$, π' is a minimal path from v_2 to v_9 achieving profit p.

The existence of minimal paths has important implications on the construction of an optimal itinerary I^*. Given the profit value $p = Ps(I^*)$, a start node v_i and end node v_j, if there exists a minimal path of profit p between v_i and v_j that all the users can traverse, this path is equivalent to an optimal itinerary I^*. On the other hand, if such a minimal path does not exist, there is no path achieving

---

Algorithm 1. One-Source Minimal Path

```
1: procedure ONE-SOURCE MINIMAL PATH(G_M, v_i)
      Input: meeting graph G_M = (V_M, E_M), starting node v_i
      Output: Table of minimal paths

2:    for (all v and p) do
3:        C_i[p, v] =< P = null, δ = ∞ >
4:    end for
5:    σ(v_i) ← Σ_{j∈C(v_i)} λ_j
6:    C_i[σ(v_i), i] = {P = {π = ⟨v_i⟩}, δ = 0}
7:    Q.insert(key = δ, (v_i, C_i[σ(v_i), i]))
8:    while (Q is not empty) do
9:        (δ, (v_t, C_i[p, t])) = Q.remove_head()
10:       Let P be the current set of minimal paths in C_i[p, t]
11:       for (edges (v_t, v_j) in E_M) do
12:           Extend each path π_i ∈ P to reach v_j and update their profit as p' = p + σ(v_j, π_i)
13:           Update their cost as δ_1 = δ + δ(v_t, v_j)
14:           for (each new profit p' obtained extending π_i ∈ P ) do
15:               Let P' be the current set of minimal paths in C[p', j] with current cost δ_2
16:               if δ_1 ≤ δ_2 then
17:                   Update the entry C[p', j] with the newly extended paths π'_i = ⟨π_i, v_j⟩
18:                   Q.set_key(key = δ_1, C_i[p', j])
19:               end if
20:           end for
21:       end for
22:   end while
23:   return C_i
24: end procedure
```

---

profit p that from v_i to v_j. These observations give us a way to test and find an optimal itinerary between a pair of nodes in the meeting graph. In general, for a pair of nodes (v_i, v_j) and profit value p, there may exist multiple minimal paths from v_i to v_j achieving profit p. We can further classify them based on the categories covered by each path. In practice, minimal paths covering the same set of categories are equivalent to the algorithm. Hence, we can just select and store one representative path among them. The following Lemma bounds the number of minimal paths starting from a given node.

Lemma 1 (Number of Minimal Paths). *Given a starting node v_i, and the number of non-zero categories k specified in the query $\mathcal{Q} = \langle \mathcal{U}, \mathcal{C_U} \rangle$, the number of distinct minimal paths from v_i to any v_j is at most 2^k.*

Proof. For any given node pair, the maximum number of distinct minimal paths achieving profit p is equal to the number of subsets with the sum of category preference p, i.e., $\{S | \sum_{c_i \in S} \lambda_i = p, S \subseteq \mathcal{C_U}\}$. As $k = |\mathcal{C_U}|$, there are at most 2^k possible subsets of $\mathcal{C_U}$. We use that as a upper bound for the number of minimal paths.

Given a starting node v_i, by visiting the nodes in the increasing order of their distance from v_i, we can iteratively construct the minimal paths by profit. Similarly to the Dijkstra's algorithm for one-source shortest path [10], we name this procedure *one-source minimal path*.

One-Source Minimal Path. Starting from a node v_i, our algorithm fills a table C_i, where each entry $C_i[p, t] = \{P = \{\pi_1, \ldots, \pi_l\}, \delta\}$ represents the set P

of minimal paths from v_i to v_t with profit p covering different category subsets, and where δ is the cost of the minimal paths. The pseudo-code of our procedure is outlined in Algorithm 1. To guide our construction process, we keep an heap Q for all the possible entries in C_i, where the key for entry $C_i[p,t]$ is the cost of the minimal paths from v_i to v_t of profit p computed so far. Given a current node v_t, we use $C_i[p,t]$ to extend the minimal paths to each adjacent node v_j (lines 12-13). Specifically, for each minimal path π_i of profit p, we evaluate the profit increment $\sigma(v_j, \pi_i)$ for each v_j adjacent to v_t as in Eq. 5. Then the path π_i is extended to form a new path of profit $p + \sigma(v_j, \pi_i)$ from v_i to v_j (as in Fig. 3b). Among all these resulting paths from v_i to v_j, we update the entry $C_i[p + \sigma(v_j, \pi_i), j]$ by keeping only the shortest paths computed so far (lines 14-21). A running example for our One-Source Minimal Path is provided below.

Example 4. Continuing from Example 3, let $\lambda_3, \lambda_5, \lambda_6, \lambda_7$ be the global preference scores for the categories c_3, c_5, c_6, and c_7 specified in input. The construction process for the minimal paths from v_2 to v_9 is illustrated in Fig. 4. In step (a), the closest node to v_2, i.e., v_6, is processed. At this moment the only possible minimal path is π_1 with profit λ_3. In step (b), the minimal path π_2 from v_2 to v_4 is constructed where the profit increment of the node v_4 is λ_7. In the next step (c), path π_2 is further extended to reach v_8. In this case, the overall profit does not change since v_8 does not cover any categories in input. When the algorithm finally reaches v_9, we obtain three minimal paths from v_2 to v_9: π_4, π_5, and π_6 with different values of profit as reported in step (d). Notice that, before the node v_9 is processed, our algorithm computes the path $\pi = \langle v_2, v_6, v_7, v_4, v_8 \rangle$, which is the minimal from v_2 to v_8 with profit $\lambda_3 + \lambda_6 + \lambda_7$. However, such a path when extended to v_9 is not minimal since π_6 achieves the same profit but has a lower cost (see Example 3).

From the example above, we observe that the minimal paths are constructed iteratively by exploring the closest node to the source node. The following Lemma proves the correctness of our procedure.

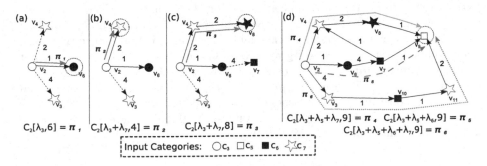

Fig. 4. Running example of one-source minimal path algorithm with v_2 as a source node.

Lemma 2 (Correctness of One-Source Minimal Path). *Given a starting node v_i, when all the reachable nodes from v_i have been visited by our procedure, the entry $C_i[p, j]$ contains the minimal paths of profit p from v_i to v_j.*

Proof. A complete proof would proceed by induction on the number of remove operations from the heap Q. For brevity we only present a sketch of the proof. At each iteration, the entries in Q are removed according to the distance cost from v_i. The paths $C_i[p, j]$ are the shortest reaching v_j with profit p, hence are minimal.

Computing the Optimal Itinerary. To compute the final optimal itinerary, we run the one-source minimal path procedure for each node v_i in the meeting points V_M. Among all the existing minimal paths returned, we start with the one with the highest profit and test whether it can be traversed by all the users without violating individual distance constraints. Specifically, we use the shortest path distance $\delta(\pi(s_i, v))$ and $\delta(\pi(v, t_i))$ for every i and every $v \in V_M$ to test the feasibility of the itinerary. Since the shortest paths of any node in V_M to user start s_i and destination t_i have been computed in the meeting graph construction process, this feasibility test can be performed in $O(m)$ time, where m is the number of users in the group. Let $\pi_{p^*}^*(v_i, v_j)$ be a minimal path and feasible for all the users. Then we return a route r_i for every user u_i, where $r_i = \langle \pi(s_i, v_i), \pi_{p^*}^*(v_i, v_j), \pi(v_j, t_i) \rangle$ is obtained by concatenating the shortest path from s_i to v_i with $\pi_{p^*}^*(v_i, v_j)$ and the shortest path from v_j to t_i.

Time Complexity. The overall complexity of this solution is dominated by the One-Source Minimal Path procedure. Therefore, we start by analyzing the time complexity of Algorithm 1. In the most inner loop (lines 14-20), the current set of minimal paths is extended. Since the number of minimal paths is at most $O(2^k)$ as shown in Lemma 1, we have that updating the paths requires $O(k2^k)$ due to the computation of node profit. We observe, that there are at most $|V_M|p_\Sigma$ insert and delete operations in Q, where $p_\Sigma = \sum_{c_i \in C_u} \lambda_i$ denotes the maximum itinerary profit. Therefore, using an implementation of Q based on a binary heap, the time complexity for one-source minimal path is $O(p_\Sigma(|V_M| + |E_M|) \log(|V_M|p_\Sigma) + k2^k|E_M|p_\Sigma)$. In the final step, the returned paths for each users are obtained concatenating the minimal paths on the meeting graph G_M with the shortest paths from the start and to the end of each user. Again, the shortest paths are available to use at this point since they have been computed in the meeting graph construction process. Hence, we do not incur additional computational cost. Therefore, the overall time complexity is $O(p_\Sigma|V_M|(((|V_M| + |E_M|) \log(|V_M|p_\Sigma) + k2^k|E_M|)))$. Although it is exponential in k, k is a constant and independent of the location graph G. Recall that k represents the number of categories with non-zero global preference. In practice k is a small number and $k \ll |V_M| \ll |V|$.

3.4   Acceleration via Graph Compression

In both of our solutions, the size of the meeting graph, i.e., $|V_M|$, has important effects on the time complexity. We are thus motivated to further reduce $|V_M|$.

The meeting graph G_M, already pruned by applying individual distance constraints, is much smaller than the location graph G. We observe that only of a small fraction of the meeting points V_M are associated with the user-specified categories. Hence, the majority of the nodes in the meeting graph can be considered as points of transit without contributing to the itinerary profit. Therefore, we propose compressing the meeting graph to preserve only the nodes labeled with user-specified categories along with their connectivity, while the other nodes are removed.

In short, our compression algorithm examines one node of $G_M = (V_M, E_M)$ at a time and outputs a new graph $G_C = (V_C, E_C)$, where $V_C \subseteq V_M$ contains the nodes that cover some categories in input, i.e., $\mathcal{C}_\mathcal{U}$. Any node $v \in V_M$ will be removed along with all its edges, if v does not cover any of the categories in $\mathcal{C}_\mathcal{U}$. To preserve the graph connectivity, for each incoming edge (v^-, v) and outgoing edge (v, v^+), a new edge between v^- and v^+ is constructed with cost equal to $\delta(v^-, v) + \delta(v, v^+)$. It will be added to E_C if it is the edge with the lowest cost between v^- to v^+.

Time Complexity. For each node in V_M without covering any category in $\mathcal{C}_\mathcal{U}$, all its incoming and outgoing edges are replaced with new edges. Let d_{max} be the maximum degree (for both in/out-degree) for the nodes in V_M; then, the overall complexity of our compression approach is $O(|V_M| d_{max}^2)$. Road networks tend to have small degrees as we will demonstrate in the experiment section. Hence, the compression process is very efficient in practice.

Lemma 3 (Paths on Compressed Graph). *For each pair of nodes (v_i, v_j) in $G_C = (V_C, E_C)$, the following inequality holds: $\delta(\pi_C(v_i, v_j)) \leq \delta(\pi_M(v_i, v_j))$, where $\pi_M(v_i, v_j)$ is a path between (v_i, v_j) in G_M and $\pi_C(v_i, v_j)$ is its compressed version in G_C.*

Proof. Here, we provide a sketch of the proof, where we distinguish two cases: (i) no nodes are removed on the path between v_i and v_j during the compression process, and (ii) at least one node is removed. In case (i), since no nodes are removed, we have that $\delta(\pi_C(v_i, v_j)) = \delta(\pi_M(v_i, v_j))$. In case (ii), the path $\pi_M(v_i, v_j) = \langle v_i, \ldots, v_x, v_t, v_y, \ldots v_j \rangle$ uses at least one node v_t that does not cover labels in $\mathcal{C}_\mathcal{U}$. Then, our compression algorithm removes v_t and it adds an edge to connect v_x to v_y of cost $\delta'(v_x, v_y) = \min\{\delta(v_x, v_t) + \delta(v_t, v_y), \delta(v_x, v_y)\}$, where $\delta(v_x, v_y)$ denoted the cost of a current edge (v_x, v_y). Hence, $\delta(\pi_C(v_i, v_j)) \leq \delta(\pi_M(v_i, v_j))$.

From the result in Lemma 3, it follows that the minimal paths in G_M are preserved during the compression process. Therefore, this compression process allows us to reduce the size of the meeting graph while preserving the minimal paths. In the experiments section, we will demonstrate the benefit of the compression procedure for the optimal solution which leads to a $\sim$10 times speed-up.

4   Related Work

Below we briefly review the most recent, relevant studies to our work.

Single User Route Planning. Li et al. [11] are the first to study the route planning by introducing the Trip Planning Query (TPQ) problem. A user specifies a set of categories that wishes to visits, a starting node, and a destination node as input. The TPQ problem aims to find the shortest route from the start to the end that passes through at least one location for each specified category. Many variations of this problem formulation have been proposed in recent years [1,12,13]. Among them, the formulation in [12] considers a fixed distance constraint for the user while the goal is to maximize the entities covered by the route. In the work of Sharifzadeh et al. [14], the search of the optimal route is performed by considering user defined order constraints on the sequence of categories visited in the route.

Group Trip Planning. Recently a few studies looked into the problem of finding a common route for a group of users. In [4,6], the authors considered the Group Trip Planning (GTP) problem which aims to find the shortest route such that all the users in a group can traverse while covering a set of input categories. Samrose et al. [7] proposed a variation of GTP, called Group Optimal Sequenced Route (GOSR), where the route traveled by the group follows a preferred sequence of point-of-interests (POIs). Zhang et al. [15] studied the problem of computing mutually beneficial routes for a group of users. Specifically, the goal is to find those paths that minimize the effective distance traveled by the users, where the distance for the shared portion of the route is discounted. Recently, the authors in [16] investigated the Collective Travel Planning (CTP) problem, which aims to find the shortest route connecting multiple sources to a single destination using at most a fixed number of intermediate locations where the users can meet. A similar study has been conducted by Ahmadi and Nascimento in [17]. Jahan et al. [18], studied a different problem, namely Group Trip Scheduling (GTS), with the objective of scheduling trips covering a set of required POIs for a group of users where the average traveled distance of the group members is minimized. The GTS problem differs from GTP as each POI type can be visited by a single group member rather than by all the users.

Our problem formulation presents major differences from those studies. First of all, our goal is to maximize the profit of the POI categories visited by the group while the traveled distance is considered as a constraint. Secondly, we only consider actual road network distance between locations rather than the approximation by Euclidean distance as in [4,6]. Lastly, we are the first, to the best of our knowledge, to model the social aspect of the group trip by computing global preference scores for the group from individual user preferences.

5 Experiments

This section presents the performance study of our proposed algorithms, i.e., the greedy solution, the optimal solution based on minimal path, and the optimal solution with graph compression, denoted by `Greedy`, `MP`, and `MPC`.

5.1 Experiment Design

Datasets. We evaluated our solutions using real-world datasets from Los Angeles. To compute shortest paths, we utilized a real *road network* dataset [19] containing $111,532$ nodes and $183,945$ directed edges, which represent the road network in the LA metro area. It can be shown as in Fig. 5a that both the maximum in-degree and the maximum out-degree of this road network are 6, while the majority nodes have 1 and 2 in/out-degree measures. In addition, we collected a *POI* dataset using Foursquare API, which contains $72,883$ businesses in the LA metro area, representing 30 categories[1], such as restaurants, shops, zoos, etc. We plot the number of POIs in each category in Fig. 5b, the top 3 popular categories are Food, Mexican Restaurant, and Pizza Place. The two dimensional dataspace is normalized into 1000×1000 sq. units as in [6].

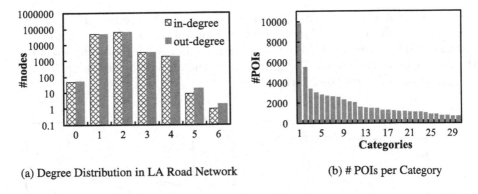

(a) Degree Distribution in LA Road Network   (b) # POIs per Category

Fig. 5. Dataset characteristics

Settings. To test our solutions in a wide range of settings, we varied the following parameters, such as the group size m, the number of categories k, and the distance constraint θ. As a user's source and destination, i.e., s_i and t_i, are randomly generated for each query, we simplified the user's distance constraint as the shortest distance from s_i to t_i times a constant θ, i.e., $\Delta_i = \theta \cdot \delta(\pi(s_i, t_i))$. For each query, the user-specified categories were randomly selected, and the preference scores were randomly sampled from the range $[\lambda_{min}, \lambda_{max}]$. In addition, we varied the query *area* as in [6], which is the minimum bounding rectangle covering the source and destination locations. The parameter descriptions and the default values are summarized as in Table 1.

---

[1] After data cleaning.

Table 1. Parameter settings

Parameter	Description	Default value
m	Number of Users	2
k	Number of Categories in Query	4
θ	Mobility Constraint Factor	1.5
$[\lambda_{min}, \lambda_{max}]$	Profit Range	$[1,100]$
Area (in sq. units)	MBR for Query	100

Metrics. To evaluate our solutions, we measure utility and runtime in our experiments, where utility is defines as $\frac{P_s(I)}{p_\Sigma}$. Since p_Σ is the sum of global preferences of all user-specified categories, when $\frac{P_s(I)}{p_\Sigma} = 100\%$, all user-specified categories are covered by itinerary I. Our algorithms are implemented in Python and run on a 2.5 GHz Intel i7 core with 16 GB of RAM. For each experiment, the results are averaged across 200 runs[2].

5.2   Experimental Results

We first evaluated the effectiveness and efficiency of our meeting graph generation and compression procedures. Using the real-world Los Angeles dataset, the meeting graph generation takes ~0.1 s and the compression procedure takes less than 0.01 s in most settings. The graph size reduction is greater than 80% after our graph compression procedure.

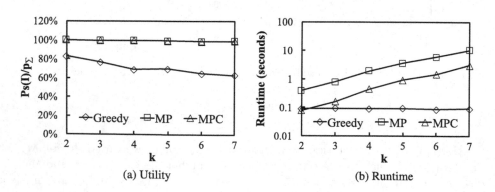

(a) Utility   (b) Runtime

Fig. 6. Performance vs categories k

Varying k. We plot the utility and the runtime of each algorithm with respect to k in Fig. 6. By increasing k, the OGR solutions need to consider a larger set

---

[2] We discarded the trivial instances, which result in empty or single-category meeting graphs.

of POI categories and p_Σ is expected to grow. In Fig. 6a, we observe that the optimal solution MP yields 100% utility most of time, finding itineraries that cover all the user-specified categories. The same utility performance can be observed for MPC, which is optimal in nature. On the other hand, the utility of Greedy is 20% to 40% less than optimal. By comparison, it is clear that the Greedy solution has a disadvantage: it fails to cover as many categories as possible due to its greedy nature. We further observe the utility gap between Greedy and MP widens as the number of categories k increases, illustrating the drawback of Greedy as the problem space becomes larger. From Fig. 6b, the runtime for the optimal solutions, i.e., MP and MPC, increases with k, since a larger number of minimal paths are constructed. Greedy does not show a significant increase in runtime due to the lower complexity in evaluating the node scores. The MPC solution is consistently faster than MP, thanks to graph compression, providing one order of magnitude speedup without compromising utility.

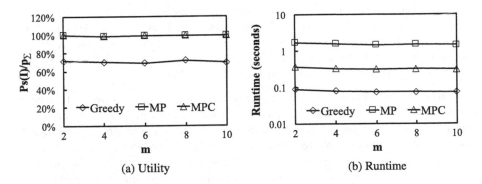

(a) Utility   (b) Runtime

Fig. 7. Performance vs users m

Varying m. Figure 7 depicts the utility and the runtime of each algorithm when varying the number of users in the group, i.e., m. We observe in Fig. 7a that our optimal solutions MP and MPC can cover all user-specified categories, despite the distance constraints imposed by more users, which is attributed to the prevalence of POIs in Los Angeles road network. On the other hand, Greedy is shown to be 25% worse than optimal consistently. As for runtime, there is no significant increase in computation cost for any method to plan for a larger group, shown in Fig. 7b. MPC is superior to MP, taking only 0.3 s to plan a 10-user itinerary.

Varying θ. The distance constraint θ is another important parameter: by increasing θ, users are able to detour more and reach a larger number of nodes. As expected, all the algorithms in Fig. 8a show higher utility as we relax the distance constraint. Greedy is around 30% lower than optimal initially $\theta = 1.25$ and catches up as θ increases. Interestingly, the optimal solutions, i.e., MP and MPC, are able to achieve 90% profit at a small cost, i.e., $\theta = 1.25$, which is 25% detour

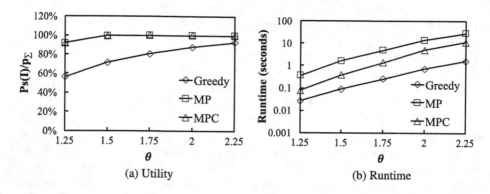

Fig. 8. Performance vs distance constraint θ

from the user's shortest path between source and destination; with $\theta = 1.5$, the optimal solutions can cover all user-specified categories, achieving 100% profit. Again, this can be attributed to the prevalence of POIs in Los Angeles. The runtime for all the algorithms increases in Fig. 8b, as θ increases. The reason is the meeting graph becomes larger as the distance constraint is relaxed. We still observe ~10 times speedup achieved by MPC over MP.

Varying Query Area. Recall that the query area is the minimum bounding rectangle for user source and destination locations. Enlarging the query area may also increase the problem space as sources and destinations are further apart. Consequently, users are able to reach a larger number of nodes in the query area. The effect the increasing query area is quite similar to that of θ in the previous experiment. As a matter of fact, we conduct another experiment to examine the presence of POI categories in the query area and discover that on average more than 80% categories can be found in an area of 50 sq. unit, and all the categories in an area of 100 sq. unit and above. This confirms the feasibility of the OGR problem in a large metropolitan environment. We observe in Fig. 9a that all the

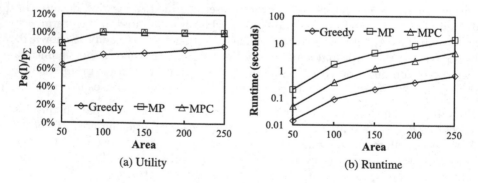

Fig. 9. Performance vs query area

algorithms achieve higher utility when query area is larger and their runtime increases along with query area as shown in Fig. 9b.

Additional Experiments. To showcase the efficacy and efficiency of our meeting graph generation and compression steps, we control the distance constraint θ to vary the size of the meeting graph V_M. As can be seen in Fig. 10, both procedures incur little overhead, e.g., under 0.2 s and 0.04 s, respectively. We further study the performance of our solutions when the user-specified categories are chosen from the most popular ones, as in Fig. 11. We observe that the performance of Greedy degenerates as users specify less popular categories, since the algorithm tends to have fewer choices at each iteration.

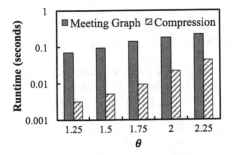

Fig. 10. Graph compression runtime

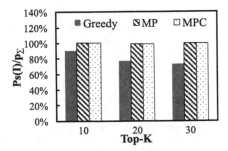

Fig. 11. Varying category popularity

6 Discussions and Conclusion

We defined the Optimal Group Route problem, to find the itinerary feasible for multiple users that maximizes the group preference. This problem was shown to be NP-complete. We proposed an exact solution that computes the optimal itinerary without enumerating all possible paths. We also designed an approximation algorithm with bounded worst case approximation ratio that reduces the computational cost. Our empirical evaluation with large-scale real-world datasets confirmed the applicability of both solutions, showing Greedy is more efficient at the cost of 20% utility.

Future work may include the extension of our solutions to address other user-specified constraints, such as *order*, i.e., visit restaurant first and cinema afterwards, which will reduce the search space. In addition, we plan to enable users to specify traveling time as the cost constraint, as opposed to distance used in this paper. To that end, real-time traffic data [19] can be incorporated to model the edge cost dynamically. Many challenges will arise for time-dependent shortest paths.

Acknowledgement. The authors wish to thank the anonymous reviewers for their valuable comments.

References

1. Cao, X., Chen, L., Cong, G., Xiao, X.: Keyword-aware optimal route search. VLDB Endow. **5**(11), 1136–1147 (2012)
2. Chen, H., Ku, W.S., Sun, M.T., Zimmermann, R.: The multi-rule partial sequenced route query. In: GIS 2008, pp. 10:1–10:10 (2008)
3. Kanza, Y., Levin, R., Safra, E., Sagiv, Y.: Interactive route search in the presence of order constraints. VLDB Endow. **3**(1–2), 117–128 (2010)
4. Hashem, T., Hashem, T., Ali, M.E., Kulik, L.: Group trip planning queries in spatial databases. In: Nascimento, M.A., Sellis, T., Cheng, R., Sander, J., Zheng, Y., Kriegel, H.-P., Renz, M., Sengstock, C. (eds.) SSTD 2013. LNCS, vol. 8098, pp. 259–276. Springer, Heidelberg (2013). doi:10.1007/978-3-642-40235-7_15
5. Chen, G., Wu, S., Zhou, J., Tung, A.: Automatic itinerary planning for traveling services. TKDE **26**(3), 514–527 (2014)
6. Hashem, T., Barua, S., Ali, M.E., Kulik, L., Tanin, E.: Efficient computation of trips with friends and families. In: Proceedings of the 24th ACM International on Conference on Information and Knowledge Management, CIKM 2015, pp. 931–940. ACM, New York (2015)
7. Samrose, S., Hashem, T., Barua, S., Ali, M.E., Uddin, M.H., Mahmud, M.I.: Efficient computation of group optimal sequenced routes in road networks. In: 2015 16th IEEE International Conference on Mobile Data Management, vol. 1, pp. 122–127, June 2015
8. Cormen, T.H., Leiserson, C.E., Rivest, R.L., Stein, C.: Introduction to Algorithms, 3rd edn. MIT Press, Cambridge (2009)
9. Amer-Yahia, S., Roy, S.B., Chawlat, A., Das, G., Yu, C.: Group recommendation: semantics and efficiency. VLDB Endow. **2**(1), 754–765 (2009)
10. Dijkstra, E.: A note on two problems in connexion with graphs. Numer. Math. **1**(1), 269–271 (1959)
11. Li, F., Cheng, D., Hadjieleftheriou, M., Kollios, G., Teng, S.-H.: On trip planning queries in spatial databases. In: Bauzer Medeiros, C., Egenhofer, M.J., Bertino, E. (eds.) SSTD 2005. LNCS, vol. 3633, pp. 273–290. Springer, Heidelberg (2005). doi:10.1007/11535331_16
12. Kanza, Y., Safra, E., Sagiv, Y., Doytsher, Y.: Heuristic algorithms for route-search queries over geographical data. In: GIS 2008, pp. 11:1–11:10 (2008)
13. Roy, S.B., Das, G., Amer-Yahia, S., Yu, C.: Interactive itinerary planning. In: ICDE, pp. 15–26 (2011)
14. Sharifzadeh, M., Kolahdouzan, M., Shahabi, C.: The optimal sequenced route query. VLDB J. **17**(4), 765–787 (2008)
15. Zhang, X., Asano, Y., Yoshikawa, M.: Mutually beneficial confluent routing. IEEE Transactions on Knowledge and Data Engineering (2016). Preprint
16. Shang, S., Chen, L., Wei, Z., Jensen, C.S., Wen, J.R., Kalnis, P.: Collective travel planning in spatial networks. IEEE Trans. Knowl. Data Eng. **28**(5), 1132–1146 (2016)
17. Ahmadi, E., Nascimento, M.A.: k-optimal meeting points based on preferred paths. In: Proceedings of the 24th ACM SIGSPATIAL International Conference on Advances in Geographic Information Systems, GIS 2016, pp. 47:1–47:4. ACM, New York (2016)

18. Jahan, R., Hashem, T., Barua, S.: Group trip scheduling (GTS) queries in spatial databases. In: Proceedings of the 20th International Conference on Extending Database Technology, EDBT 2017, Venice, Italy, 21–24 March 2017, pp. 390–401 (2017)
19. Demiryurek, U., Banaei-Kashani, F., Shahabi, C.: TransDec: a spatiotemporal query processing framework for transportation systems. In: 2010 IEEE 26th International Conference on Data Engineering (ICDE 2010), pp. 1197–1200, March 2010

Hybrid Best-First Greedy Search
for Orienteering with Category Constraints

Paolo Bolzoni and Sven Helmer[✉]

Free University of Bozen-Bolzano, Bolzano, Italy
{paolo.bolzoni,sven.helmer}@unibz.it

Abstract. We develop an approach for solving rooted orienteering problems with category constraints as found in tourist trip planning and logistics. It is based on expanding partial solutions in a systematic way, prioritizing promising ones, which reduces the search space we have to traverse during the search. The category constraints help in reducing the space we have to explore even further. We implement an algorithm that computes the optimal solution and also illustrate how our approach can be turned into an anytime approximation algorithm, yielding much faster run times and guaranteeing lower bounds on the quality of the solution found. We demonstrate the effectiveness of our algorithms by comparing them to the state-of-the-art approach and an optimal algorithm based on dynamic programming, showing that our technique clearly outperforms these methods.

1   Introduction

Imagine a user arriving at the train station of a city, depicted on the left-hand side of Fig. 1(a), and needing to be at the airport, located on the right-hand side of the figure, five hours later[1]. They do not want to immediately go to the airport, but have a look at the city first. If the user takes the shortest route from the train station to the airport, represented by the solid line, they are only able to see a basilica on the way and still have a lot of spare time when arriving at the airport. According to the ratings of a tourist guide, the basilica, a pagoda, and a cathedral are the top points of interest (POIs) that can be visited en route to the airport within five hours (see dashed line). While this makes much better use of the time, the user may not be in the mood to visit these POIs, as they may be tired from traveling and would like a more relaxing route. In this instance, the dotted line, connecting a park, a ferris wheel, and a statue, is much more appropriate. As this example shows, the goal is to find a trip with the best points of interest that makes good use of the available time while at the same time considering the preferences of a user.

Finding itineraries for domains such as tourist trip planning and logistics often involves solving an orienteering problem. This is because those tasks are

---

[1] The icons are from icons8.com, used under Creative Commons License CC BY-ND 3.0. To view a copy, visit https://creativecommons.org/licenses/by-nd/3.0/.

© Springer International Publishing AG 2017
M. Gertz et al. (Eds.): SSTD 2017, LNCS 10411, pp. 24–42, 2017.
DOI: 10.1007/978-3-319-64367-0_2

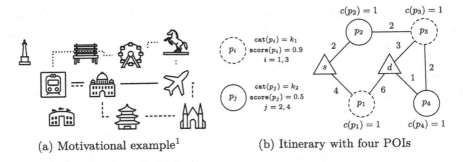

(a) Motivational example[1]   (b) Itinerary with four POIs

Fig. 1. Example scenarios

not about determining the shortest path, but the most attractive or the one covering the most needy customers while satisfying a strict time constraint. We focus on a variant that assumes that every point of interest (POI) or customer has a category. This categorization helps a user in expressing preferences, e.g. a tourist may only be interested in certain types of venues, such as museums, galleries, and cafes, while certain vehicles may only be able to serve particular customers.

In general, orienteering is an NP-hard problem, and adding categories does not change this fact [2]. We propose an approach based on a best-first strategy to explore the search space, meaning that we first expand the partial solutions that show the greatest potential. We do so with the help of a function approximating the attractiveness of POIs that can still be added. Similar to admissible heuristics in an A*-search, this function needs to satisfy certain properties. Additionally, we are able to prune partial solutions that cannot possibly result in an optimal route.

Even though this technique will speed up the search for an optimal solution by pruning unpromising partial solutions, in the worst case it still has an exponential run time. Therefore, we describe how to transform our method into much more efficient approximation algorithms, creating different variants with important properties concerning the quality of the generated solution and the run time. In summary, we make the following contributions: we show how to apply a best-first strategy to the problem of orienteering with category constraints; we turn the optimal algorithm into different approximation algorithms proving lower bounds for the quality of a solution or upper bounds for the run time; in the experimental evaluation we compare our approach to state-of-the-art algorithms, demonstrating its effectiveness. For some scenarios we improved the quality of the solutions by about 10% with a run time of just five seconds (with the main competitor taking up to 50 s and producing worse solutions).

2   Related Work

Introduced by Tsiligrides in [16], the orienteering problem (OP) is about determining a path from a starting node to an ending node in an edge-weighted graph with a score for each node, maximizing the total score while staying within a certain time budget. Orienteering is an NP-hard problem and algorithms computing exact solutions using branch and bound [6,10] as well as dynamic programming techniques [9,12] are of limited use, as they can only solve small problem instances. Consequently, there is a body of work on approximation algorithms and heuristics, most of them employing a two-step approach of partial path construction [7,16] and (partial) path improvement [1,3,13]. Meta-heuristics, such as genetic algorithms [15], neural networks [17], and ant colony optimization [8] have also been tested. For a recent overview on orienteering algorithms, see [5]. However, none of the approaches investigate OP generalized with categories.

There is also work on planning and optimizing errands, e.g., someone wants to drop by an ATM, a gas station, and a pharmacy on the way home. The generalized traveling salesman version minimizes the time spent on this trip [11], while the generalized orienteering version maximizes the number of visited points of interest (POIs) given a fixed time budget. However, as there are no scores, no trade-offs between scores and distances are considered.

Adapting an existing algorithm for OP would be a natural starting point for developing an approximation algorithm considering categories. However, many of the existing algorithms have a high-order polynomial complexity or no implementation exists, due to their very complicated structure. Two of the most promising approaches we found were the segment-partition-based technique by Blum et al. [1] and the method by Chekuri and Pál, exploiting properties of submodular functions [4]. The latter approach, a quasi-polynomial algorithm, is still too slow for practical purposes. Nevertheless, Singh et al. modified the algorithm by introducing spatial decomposition for Euclidean spaces in the form of a grid, making it more efficient [14]. It has been adapted by us for OPs on road networks with category constraints in the form of the CLuster Itinerary Planning algorithm (CLIP) [2], where we use agglomerative hierarchical clustering and multidimensional knapsack to separate routing from POIs selection.

3   Problem Formalization

We assume a set $\mathbf{P}$ of points of interest (POIs) $p_i, 1 \leq i \leq n$. The POIs, together with a starting and a destination node, denoted by s and d, respectively, are connected by a complete, metric, weighted, undirected graph $G = (\mathbf{P} \cup \{s, d\}, \mathbf{E})$, whose edges, $e_l \in \mathbf{E} = \{(x, y) \mid x, y \in \mathbf{P} \cup \{s, d\}\}$ connect them. Each edge e_l has a cost $c(p_i, p_j)$ that signifies the duration of the trip from p_i to p_j, while every node $p_i \in \mathbf{P}$ has a cost $c(p_i)$ that denotes its visiting time. Each POI belongs to a certain category, such as *museums, restaurants*, or *galleries*. The set of m categories is denoted by $\mathbf{K}$ and each POI p_i belongs to exactly one category $k_j, 1 \leq j \leq m$. Given a p_i, $\mathtt{cat}(p_i)$ denotes the category p_i belongs to and $\mathtt{score}(p_i)$ its score or reward, with higher values indicating higher interest to the user. Finally, users have a certain maximum time in their budget to complete the itinerary, denoted by $t_{\max}$.

Definition 1 (Itinerary). *An itinerary $\mathcal{I}$ starts from a starting point s and finishes at a destination point d (s and d can be identical). It includes an ordered sequence of connected nodes $\mathcal{I} = \langle s, p_{i_1}, p_{i_2}, \ldots, p_{i_q}, d \rangle$, each of which is visited once. We define the* cost *of itinerary $\mathcal{I}$ to be the total duration of the path from s to d passing through and visiting the POIs in $\mathcal{I}$, $\mathtt{cost}(\mathcal{I}) = c(s, p_{i_1}) + c(p_{i_1}) + \sum_{j=2}^{q}(c(p_{i_{j-1}}, p_{i_j}) + c(p_{i_j})) + c(p_{i_q}, d)$, and its score to be the sum of the scores of the individual POIs visited, $\mathtt{score}(\mathcal{I}) = \sum_{j=1}^{q} score(p_{i_j})$.*

Example 1. Figure 1(b) shows an example with four POIs, p_1, p_2, p_3, and p_4, along with their distances, visiting times, scores, and categories. We simplify the graph slightly to keep it readable: all POIs of the same category have the same score and we also omit some edges. Three example itineraries with one, two, and three POIs, respectively, are: $\mathcal{I}_1 = \langle s, p_1, d \rangle$, $\mathcal{I}_2 = \langle s, p_2, p_3, d \rangle$, and $\mathcal{I}_3 = \langle s, p_2, p_3, p_4, d \rangle$. Their costs and scores are as follows:

- $\mathcal{I}_1 = \langle s, p_1, d \rangle$: $\mathtt{cost}(\mathcal{I}_1) = 4 + 1 + 6 = 11$, $\mathtt{score}(\mathcal{I}_1) = 0.9$;
- $\mathcal{I}_2 = \langle s, p_2, p_3, d \rangle$: $\mathtt{cost}(\mathcal{I}_2) = 2 + 1 + 2 + 1 + 3 = 9$, $\mathtt{score}(\mathcal{I}_2) = 0.5 + 0.9 = 1.4$;
- $\mathcal{I}_3 = \langle s, p_2, p_3, p_4, d \rangle$: $\mathtt{cost}(\mathcal{I}_3) = 2 + 1 + 2 + 1 + 2 + 1 + 1 = 10$, $\mathtt{score}(\mathcal{I}_3) = 0.5 + 0.9 + 0.5 = 1.9$.

Given traveling and visiting times as well as scores (obtained from sources such as tourist offices or web sites and possibly combined with user preferences), we need to build an itinerary starting at s and ending at d from a subset P of $\mathbf{P}$ with duration smaller than $t_{\max}$ and maximum cumulative score. We introduce an additional constraint specifying the number of POIs per category that can be included in the final itinerary. More precisely, we introduce a parameter $\max_{k_j}$ for each category k_j that is set by the user to the maximum number of POIs in a category that he or she prefers to visit during the trip. The category constraints are directly determined via user preferences. We are now ready define the *Orienteering Problem with Maximum Point Categories* (*OPMPC*).

Definition 2 (OPMPC). *Given a starting point s, a destination point d, n points of interest $p_i \in \mathbf{P}$, each with $\mathtt{score}(p_i)$, visiting times $c(p_i), 1 \leq i \leq n$, traveling times $c(x, y)$ for $x, y \in \mathbf{P} \cup \{s, d\}$, categories $k_j \in \mathbf{K}, 1 \leq j \leq m$, and the following two parameters: (a) the maximum total time t_{max} a user can spend on the itinerary and, (b) the maximum number of POIs $\max_{k_j}$ that can be used for the category k_j ($1 \leq j \leq m$), a solution to the OPMPC is an itinerary $\mathcal{I} = \langle s, p_{i_1}, p_{i_2}, \ldots, p_{i_q}, d \rangle, 1 \leq q \leq n$, such that*

- *the total score of the points, $\mathtt{score}(\mathcal{I})$, is maximized;*
- *no more than $\max_{k_j}$ POIs are used for category k_j;*
- *the time constraint is met, i.e., $\mathtt{cost}(\mathcal{I}) \leq t_{max}$.*

Example 2. In the presence of categories k_1 with $\max_{k_1} = 1$ and k_2 with $\max_{k_2} = 1$, and assuming that $t_{\max} = 10$, we can observe the following about the itineraries in Example 1: Itinerary $\mathcal{I}_1$ is infeasible since its cost is greater than $t_{\max}$.

Comparing $\mathcal{I}_2$ and $\mathcal{I}_3$, we can see that $\mathcal{I}_3$ is of higher benefit to the user, even though it takes more time to travel between s and d. However, it cannot be chosen since it contains two POIs from k_2. Itinerary $\mathcal{I}_2$ contains two POIs, each from a different category and it could be one recommended to the user.

4   Best-First Search Strategy

We roughly follow a best-first strategy, meaning that we keep all solutions generated so far in a priority queue sorted by their potential score. A solution is represented by the set of POIs it contains. In each step, we take the solution with the highest potential score from the queue, expand it with POIs that have not been visited yet, and re-insert the expanded solutions back into the queue. An important difference to the classic best-first search is that in our case it is not straightforward to identify the goal state, as we do not know the score of the optimal itinerary a priori. Consequently, we have to run the algorithm until there are no solutions left in the queue that have a higher potential score than the best found so far (which then becomes the result we return). Applying this approach in a straightforward fashion is not practical as OPMPC is an intractable problem. Therefore, after presenting an algorithm that computes the optimal solution, we turn it into an approximation algorithm.

4.1   Potential Score

An important aspect of our search strategy is the computation of the potential score of a partial solution, which has to be an upper bound of the score of the fully expanded solution based on this partial solution. Basically, this follows the principle of admissible heuristics found in the A*-algorithm, with the difference that we never underestimate the value, since we consider scores and not costs.

Definition 3 (Potential score). *Given a (partial) itinerary $\mathcal{I} = \langle s, p_1, p_2, \ldots, p_i, d \rangle$, its potential score $pot(\mathcal{I})$ is defined as follows:*

$$pot(\mathcal{I}) = score(\mathcal{I}) + extra(\mathcal{I})$$

where $score(\mathcal{I}) = \sum_{j=1}^{i} score(p_j)$ and $extra(\mathcal{I})$ is a heuristic never underestimating the additional score that is still possible for $\mathcal{I}$.

For efficiency reasons, we need to find an approximation for $extra(\mathcal{I})$ that is as small as possible without underestimating it. We do so by getting a rough estimate of the top-scoring POIs that still fit into the partial itinerary $\mathcal{I}$. In a first step, we discard all POIs in the set of remaining POIs, $p_k \in \mathbf{P} \setminus \mathcal{I}$, that are too costly for traveling from p_i via p_k to d, as $cost(\mathcal{I}) - c(p_i, d) + c(p_i, p_k) + c(p_k) + c(p_k, d) > t_{\max}$. We call this set of remaining POIs $\mathbf{P}_{rem(\mathcal{I})}$. Having reduced the set of POIs, we turn to category constraints for a more accurate approximation. For each category we know how many POIs still fit into the itinerary without violating the category constraints. As previously defined,

$\max_{k_j}$ is the maximum value of POIs that can be chosen for category k_j. Let $u_{k_j} = |\{p_i \in \mathcal{I} | \mathtt{cat}(p_i) = k_j\}|$ be the number of POIs of category k_j currently found in $\mathcal{I}$, then we know that we can only fit $\max_{k_j} - u_{k_j}$ more POIs of category k_j into $\mathcal{I}$. For each category k_j, we sort the POIs in $\mathbf{P}_{rem(\mathcal{I})}$ according to their score in descending order and add up the scores of the top $\max_{k_j} - u_{k_j}$. This comprises $\mathtt{extra}(\mathcal{I})$.

4.2   Our Algorithm

Algorithm 1 shows our approach in pseudocode. For the moment disregard the parts marked in gray: these are optimizations that will be discussed in the following section. Partial solutions are kept in a double-ended priority queue (also called deque). The deque is initialized with the empty solution e just containing s and d and a score of 0. We take the solution with the highest potential score out of the deque and expand it, one by one, with the POIs that have not been visited yet. In order to figure out whether a solution violates the time constraint $t_{\max}$, we have to compute the (shortest) length of the itinerary of a solution. As every partial solution contains only a limited number of POIs, we compute this length in a brute-force way. We discard any expanded solutions that violate the time constraint or any of the category constraints and put all valid expansions back into the deque. If an expanded solution is better than the current best solution, we update it. We continue until the potential score of the solution taken from the deque is smaller than the best found so far (which then becomes the answer we return).

4.3   Further Optimizations

The score of the current best solution helps us to prune earlier and more aggressively, because only partial solutions with a better potential score have to be kept. We initialize the best solution found so far with an itinerary found quickly using a greedy algorithm, which works as follows. Given a partial itinerary $\mathcal{I} = \langle s, p_1, p_2, \ldots, p_i, d \rangle$ (starting with the empty itinerary), the greedy strategy adds to this path a POI $p \in \mathbf{P}_{rem(\mathcal{I})}$ such that its $\mathtt{utility}(p) = \frac{\mathtt{score}(p)}{c(p_i,p)+c(p)+c(p,d)}$ is maximal and no constraints are violated. We repeat this until no further POIs can be added to the itinerary.

We can also use a greedy approach to improve the current best solution while the algorithm is running (see lines 15 to 17). When expanding a partial solution taken from the deque, we also complete it with the greedy strategy. This improves the current best solution more rapidly, leading to more aggressive pruning, and is especially useful in earlier stages of the algorithm, as there is still a lot of unexplored search space. We may want to stop the greedy expansion in later stages, due to it being less effective. The threshold $g \in [0, 1]$ defines when the algorithm is in an *early stage*, it checks each expanded solution z: whenever the ratio $z.\mathtt{cost}/t_{\max} \leq g$, it does a greedy expansion. We investigate this parameter in more detail in Sect. 7 on the experimental evaluation.

Algorithm 1. $b \leftarrow$ MainLoop(q, G)

Input: query q (consisting of s, d, $t_{\max}$, and category constraints), graph G, set **P** of all POIs

Output: best solution b

1   Q priority deque for partial solutions
2   e.cost $\leftarrow c(q.s, q.d)$
3   e.extra $\leftarrow$ Extra$(q, e, G, \mathbf{P})$
4   $b \leftarrow$ ExtendGreedily(q, e)
5   Q.push(e)
6   **while not** Q.empty() **do**
7      $s \leftarrow Q$.popMaximum()
8   **foreach** $z \in$ Expand(s, G) **do**
9   **if** z *violates any category constraint* **then continue**
10         $z \leftarrow$ ComputePathAndScore(z)
11   **if** z.cost $> q.t_{max}$ **then continue**
12   z.extra $\leftarrow$ Extra$(q, z, G, \mathbf{P})$
13   **if** z.score $> b$.score **then** $b \leftarrow z$
14   Q.push(z)
15   **if** *at an early stage* **then**
16          $y \leftarrow$ ExtendGreedily(q, z)
17   **if** y.score $> b$.score **then** $b \leftarrow y$
18   **while** b.score $> Q$.minimum().score $+ Q$.minimum().extra **do**
19   Q.popMinimum()
20 **return** b

---

5   Approximation Algorithms

Even though pruning and further optimizations speed up the search, in the worst case we still face exponential run time. After all, OPMPC is an NP-hard problem. Here we present variations of Algorithm 1 that turn it into an approximation algorithm. In principle, we can guarantee an upper bound on the run time or a lower bound on the score.

5.1   Bounding the Score

In line 18 of Algorithm 1 we remove (partial) solutions from the end of the deque that have a potential score smaller than the best solution b found so far. Since solutions are sorted by an overestimation of their score, we never accidentally get rid of a partial solution that could be expanded into the optimal one. However, we could prune more aggressively by introducing a *cut factor* c $(c \geq 1)$ and checking whether $c \cdot b$.score $> Q$.minimum().score $+ Q$.minimum().extra (setting c to 1 results in the original algorithm). The larger c, the more partial solutions will be ignored. However, by doing so we also risk losing the optimal and other good solutions.

Nevertheless, we are sure to get a solution that guarantees at least $\alpha = 1/c$ of the score of the optimal solution.

5.2 Bounding the Run Time

While the approach in the previous section makes sure that we always get a certain amount of the optimal score, we are not able to bound its run time. All we know is that it is faster, since we prune more partial solutions. By modifying Algorithm 1 in a different way, we can obtain a run time bound. Instead of allowing the queue to become arbitrarily long, we limit its length to a maximum of l_{max} entries. In this way, we put a limit on the number of partial solutions that can be expanded. Before pushing an expanded solution z back into the deque in line 14, we check whether there is still space. If there is not enough space, we remove the partial solution with the lowest potential score (this could also be z). While we cannot determine the quality of the answer a priori, we know the value for α upon completion of the algorithm if we keep track of the largest potential score of all the partial solutions we discarded. Assume that z_{max} has the largest potential score among the discarded solutions. Then we know that $\alpha = b/\text{pot}(z_{max})$, where b is the optimal solution.

Another, more direct, way of controlling the run time is to set an explicit limit r for it. After the algorithm uses up the allocated time, it returns the best solution found so far. For example, we can check between line 7 and 8 whether there is any time left. If not, we jump out of the while-loop. Again, it is not possible to prescribe a value for α beforehand, but we can determine its value when the algorithm finishes: in this case z_{max} is the partial solution with the largest potential score that is still in the deque when we stop.

6 Properties and Bounds

In the following we prove that conservative pruning (i.e. cut factor $= 1$) allows us to find the optimal solution. Additionally, we prove the bounds for the score and the run time of the approximation algorithms.

6.1 Correctness of Pruning

Let $\mathcal{I}_c$ be a *complete itinerary*, i.e., no further POIs can be added to $\mathcal{I}_c$ without violating a constraint. We have to prove that for every complete itinerary $\mathcal{I}_c$, $\text{pot}(\mathcal{I}_c) = \text{score}(\mathcal{I}_c) \leq \text{pot}(\mathcal{I}_p)$, where $\mathcal{I}_p$ is a partial itinerary of $\mathcal{I}_c$ ($\mathcal{I}_p \subset \mathcal{I}_c$). $\mathcal{I}_p \subset \mathcal{I}_c$ iff the POIs $p_1, p_2, \ldots, p_p$ in $\mathcal{I}_p$ are a subset of the POIs $p_1, p_2, \ldots, p_c$ in $\mathcal{I}_c$. The potential score of $\mathcal{I}_c$ is equal to its score, since $\text{extra}(\mathcal{I}_c) = 0$ (no more POIs can be added). We can actually show a stronger statement that includes the one above.

Lemma 1. *As we expand solutions, their potential score decreases: given two itineraries $\mathcal{I}_1$ and $\mathcal{I}_2$ with $\mathcal{I}_1 \subset \mathcal{I}_2$, it follows that $pot(\mathcal{I}_2) \leq pot(\mathcal{I}_1)$.*

Proof. We prove the lemma by structural induction.

Induction step $(n \to n+1)$**:** Given two itineraries $\mathcal{I}_n$ with POIs $p_1, p_2, \ldots, p_n$ and $\mathcal{I}_{n+1}$ with POIs $p_1, p_2, \ldots, p_{n+1}$, we have to show that $\mathrm{pot}(\mathcal{I}_{n+1}) \leq \mathrm{pot}(\mathcal{I}_n)$

$$\Leftrightarrow \mathrm{score}(\mathcal{I}_{n+1}) + \mathrm{extra}(\mathcal{I}_{n+1}) \leq \mathrm{score}(\mathcal{I}_n) + \mathrm{extra}(\mathcal{I}_n)$$

$$\Leftrightarrow \mathrm{score}(\mathcal{I}_n) + \mathrm{score}(p_{n+1}) + \mathrm{extra}(\mathcal{I}_{n+1}) \leq \mathrm{score}(\mathcal{I}_n) + \mathrm{extra}(\mathcal{I}_n)$$

$$\Leftrightarrow \mathrm{score}(p_{n+1}) + \mathrm{extra}(\mathcal{I}_{n+1}) \leq \mathrm{extra}(\mathcal{I}_n)$$

Assuming that we can still add (up to) k POIs to $\mathcal{I}_n$, $\mathrm{extra}(\mathcal{I}_n)$ is computed by taking the top-k POIs from $\mathbf{P}_{rem(\mathcal{I}_n)}$, whereas $\mathrm{extra}(\mathcal{I}_{n+1})$ is computed by taking the top-(k-1) POIs from $\mathbf{P}_{rem(\mathcal{I}_{n+1})}$. We have just expanded $\mathcal{I}_n$ by adding p_{n+1} to it, so we know that $\mathbf{P}_{rem(\mathcal{I}_{n+1})} \cup \{p_{n+1}\} \subseteq \mathbf{P}_{rem(\mathcal{I}_n)}$, as p_{n+1} is missing from $\mathbf{P}_{rem(\mathcal{I}_{n+1})}$ (compared to $\mathbf{P}_{rem(\mathcal{I}_n)}$) and $\mathbf{P}_{rem(\mathcal{I}_{n+1})} \cup \{p_{n+1}\}$ can at most reach $\mathbf{P}_{rem(\mathcal{I}_n)}$, because the remaining time in itinerary $\mathcal{I}_{n+1}$ is less than the remaining time in itinerary $\mathcal{I}_n$. Therefore, the choice we get when picking top-scoring POIs we pick for the left-hand side of the inequality is more restricted compared to the right-hand side.

Base case (empty itinerary): The empty itinerary $\mathcal{I}_0$ contains no POIs, therefore $\mathrm{pot}(\mathcal{I}_0) = \mathrm{extra}(\mathcal{I}_0)$ and we choose the top-scoring POIs from the largest possible set $\mathbf{P}_{rem(\mathcal{I}_0)}$, which means that $\mathcal{I}_0$ has the largest possible potential score.

6.2   Lower Bounding the Score

Using the approximation technique from Sect. 5.1, we can guarantee a lower bound for the score. At any given time, when discarding a partial solution $\mathcal{I}_p$, we know that its potential score $\mathrm{pot}(\mathcal{I}_p)$ is at most a factor c greater than the best solution b found so far: $c \cdot b \geq \mathrm{pot}(\mathcal{I}_p)$. If there is no further change for b, the ratio of achieving $\alpha = 1/c$ of the best score holds, as $\mathrm{pot}(\mathcal{I}_p)$ never underestimates the score of a full expansion of $\mathcal{I}_p$. If b is replaced by a new best score b', we then know that the previously discarded partial solutions are actually closer to the optimal solution, as b' can only be greater than b. In fact, the cut factor of the previously discarded partial solutions improves to $c' = b/b' \cdot c$. Additionally, the lemma in the previous section states that the potential score of an expanded solution can never go up, which means that we are also on the safe side here: we cannot lose a complete solution that is less than a factor α of the optimal solution by discarding $\mathcal{I}_p$.

Even when using the approximation techniques bounding the run time shown in Sect. 5.2, which do not allow us to set a lower bound for the score a priori, we can draw conclusions about the factor α after the algorithm finishes. When running our algorithm with a limited queue length, we keep track of the solution with the largest potential score $\mathrm{pot}(\mathcal{I}_{\max})$ that we have discarded. When execution stops, we can now compare $\mathrm{pot}(\mathcal{I}_{\max})$ to the best solution b that is returned, knowing that α has to be at least $b/\mathrm{pot}(\mathcal{I}_{\max})$. Using an explicit time limit r, the largest (implicitly) discarded $\mathrm{pot}(\mathcal{I}_{\max})$ can be found at the head of the queue and we can compute α as indicated above. We can run our algorithm in an iterative fashion: once the time limit r has been reached, but a user is not satisfied with the current level of quality, they can run the algorithm a while longer, checking α again. In this way, the quality of the solution and the run time can be balanced on the fly.

6.3   Upper Bounding the Run Time

When setting an explicit time limit r, bounding the run time is straightforward. In case of a limited queue length, drawing conclusions about the run time is more complicated. The worst case for a best-first search algorithm is a scenario in which all the POIs have the same score, the same cost, and a similar distance from s, d and each other. In this case we first expand the empty partial solution generating all itineraries of length one. In the next step, we expand all these itineraries to those of length two, then all of those to those of length three, and so on, meaning that no pruning is taking place. We now give an upper bound for the number of generated itineraries and then illustrate the effect of a bounded queue length on this number.

We have m categories k_j, $1 \le j \le m$, each with the constraint max_{k_j} and the set $K_{k_j} = \{p_i | p_i \in P, cat(p_i) = k_j\}$ containing the POIs belonging to this category. The set of relevant POIs R is equal to $\bigcup_{j=1,max_{k_j}>0}^{m} K_{k_j}$, we denote its cardinality by $|R|$ (assuming that R only contains POIs actually reachable from s and d in time $t_{\max}$). For the first POI we have $|R|$ choices, for the next one $|R|-1$, and so on until we have created paths of length $\lambda = \sum_{j=1,max_{k_j}>0}^{m} max_{k_j}$. We cannot possibly have itineraries containing more than λ POIs, as this would mean violating at least one of the category constraints. So, in a first step this gives us $\prod_{i=0}^{\lambda-1}(|R|-i)$ different itineraries. This is still an overestimation, though, as it does not consider the t_{max} constraint, nor does it consider that once an itinerary has reached max_{k_j} POIs for a category k_j, the set K_{k_j} can be discarded completely when extending the itinerary further. Introducing a maximum queue length of $l_{\max}$, we arrive at the final number $\prod_{i=0}^{\lambda-1} \min((|R| - i), l_{\max})$.

7   Experimental Evaluation

We conducted our experiments on a Linux machine (Arch Linux, kernel version 4.8.4) with an i7-4800MQ CPU running at 2.7 GHz and 24 GB of RAM (of which 15 GB were actually used). The algorithms were written in C++, compiled with gcc 6.2 using -O2. The priority deque was implemented with an interval heap, the graph representing the city map with an adjacency list. For comparison with the CLIP we used black-box testing, configuring it with the default settings for cluster size and relaxed linear programming solver as suggested in [2]. The greedy strategy expands an empty itinerary containing just s and d.

We averaged 25 different test runs for every data point in the following plots. For each test run we selected a start and end node randomly, such that $c(s, d) <$ $t_{\max}$. When comparing different algorithms, we generated one set of queries and re-used it for every algorithm. Unless specified otherwise, as default values for the parameters we use two hours for $t_{\max}$, the largest number of categories possible for a graph (four or nine), a category constraint of two for each category, a cut ratio of 1.2, a greedy expansion threshold of 1, an unbounded queue length, and unlimited run time.

7.1 Data Sets

We used several different data sets in our evaluation. The first one was an artificial one consisting of a grid with 10,000 nodes (100×100) with a total of 3,000 POIs in four different categories. The distance between neighboring nodes in the grid was 60 s. The POIs were randomly scattered in the grid, had a visiting time between three minutes and one hour and a score between 1 and 100.

The second artificial network is a spider network having the same number of nodes and POIs as the grid, but the edges are placed as a 100-sided polygon with 100 levels. The edges between levels are 100 s, and the central, and smallest, polygon has sides of 8 s. The other levels become larger and larger as in a Euclidean plane.

The two real-world data sets were a map of the Italian city of Bolzano with a total of 1,830 POIs in nine categories and a map of San Francisco with a total of 1,983 POIs in four categories. Here the length of the edges is computed assuming a walking speed of $1\frac{m}{s}$.

7.2 Effects of Parameters

First of all, we illustrate the effects of the different parameters on the performance of our approximation algorithms. Namely, these are the values for the greedy expansion threshold, the cut ratio, the queue length limit, and the run time limit. We also show how the number of categories selected by a user influences our performance.

Greedy Expansion Threshold. Figure 2 shows the impact of the greedy expansion threshold on the performance of our algorithm (here on a grid network; the results for the other networks look very similar).

As can be seen in Fig. 2 (left), the effect on the score is minimal (for comparison a pure greedy strategy is also shown). However, the improvements in terms of run time are significant (Fig. 2, right). Running a greedy expansion during the execution of the algorithm updates the best solution found so far much faster, resulting in better pruning. On average, it was always worth running a greedy expansion right up to the point when the algorithm finishes, since it can be done very quickly. In fact, toward the end of the execution this can be done even faster, as only one or two more POIs can be added to an itinerary.

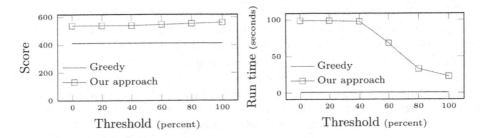

Fig. 2. Varying the greedy expansion threshold

Cut Ratio. Next, we investigate the impact of the cut ratio. Increasing the cut ratio allows us to prune more aggressively, albeit at the price of sometimes losing the optimal solution. In Fig. 3 we depict the results for varying the cut ratio from 1 (optimal case) to 2.5 (guaranteeing at least 0.4 of the optimal score). As expected, the score of the found solutions goes down when increasing the cut ratio. This is the case for all data sets. Nevertheless, it does so much slower than the guaranteed lower bound for the score and it also stays well above the score for the solutions found by the simple greedy algorithm. This is especially true for the real-world data sets (Figs. 3(c) and (d)). We found that our algorithm finds good solutions early (more on this in Sect. 7.2 about setting a fixed run time limit).

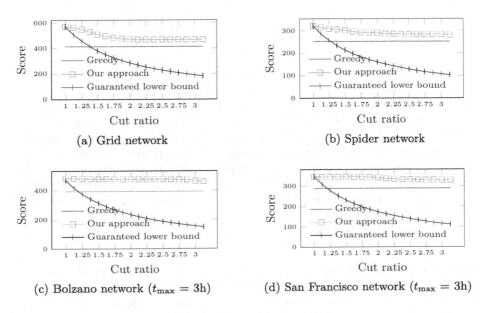

Fig. 3. Increasing the cut ratio

The impact of the cut ratio on the run time is clearly noticeable (see Fig. 4). For some data sets, namely the grid and San Francisco networks, increasing the cut ratio brings down the run time significantly. Grid networks, including the artificial grid and San Francisco, whose map is composed of many grid-like structures, have a very long run time for small cut ratios, while spider networks do not exhibit this behavior. (Bolzano, which has grown organically over centuries around an old city center, shares some of the features of a spider network.) The different edge lengths make the central part of the spider network easier to explore, whereas the external parts are more isolated. As a consequence, the computation of extra($\mathcal{I}$) is more precise, meaning that we get a more accurate picture about the potential scores of the yet unexpanded solutions that are still in the queue, making it easier to prune partial solutions. In a grid-like network, there are lots of alternatives, making it more difficult to compute extra($\mathcal{I}$) accurately, so the potential scores of some unexpanded solutions may still be high, and we need more time to check them.

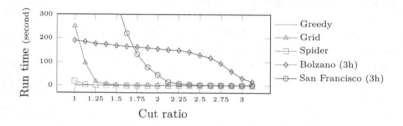

Fig. 4. Effect of the cut ratio on run time

Queue Length. Now we look at the impact of bounding the queue length to shorten the run time of our algorithm. The left-hand column of Fig. 5, labeled (a), shows the results for the grid network, the right-hand column, labeled (b), those for the spider network. The cut ratio is set to 1 to measure only the impact of the queue length limit.[2] The size of the queue on the x-axis is measured relatively to the number of POIs in the graph. Having a very short queue lowers the score slightly. Please note that the score is expressed as a ratio of the optimal one and that the y-axis does not start at 0 (greedy scored 0.73 for the grid and 0.78 for the spider network). Unsurprisingly, the run time goes up with increasing queue length and it will eventually reach the run time of the unbounded queue (represented by horizontal lines).

Run Time Limit. In Fig. 6 we show results for setting an explicit run time limit. The orange line at the top of of each plot represents the optimal score. We can see that even for very short run time limits we get results for every

---

[2] Due to the cut ratio, we did not run this experiment for the real-world data sets, as it would have taken too much time.

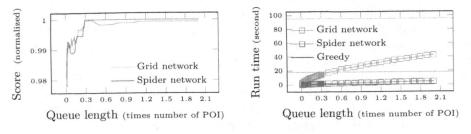

Fig. 5. Limiting the queue length

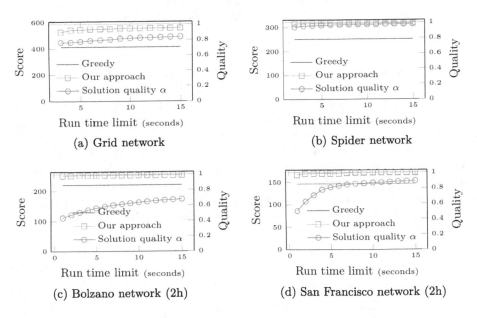

Fig. 6. Limiting the run time (Color figure online)

data set that are very close to the optimal score. What this means is that our algorithm finds very good solutions early in the search process and then, if we let it continue, spends the remaining time basically verifying that there are no significantly better solutions. We also clearly outperform the greedy heuristic in terms of score (the greedy algorithm takes between 0.2 and 1.5 s to find a solution).

When limiting the run time, we cannot guarantee the quality of the attained score a priori. Nevertheless, when the algorithm stops running, together with the score it returns the factor α, telling us the ratio of the optimal score we have reached at least (for details, see Sect. 6.2). We have plotted the lower bound α in Fig. 6, please note that for this measure the scale on the right-hand side of the y-axis is used. Looking at Fig. 6, we see that the results for α for two of the data sets, the spider network in Fig. 6(b) and the Bolzano network in Fig. 6(c) are somewhat different, which comes as a surprise, as we mentioned that the Bolzano network shares some of the properties of a spider network in Sect. 7.2. However, there are other effects at work here, too. For the Bolzano network the larger number of categories is responsible for the lower guarantee. This makes it more difficult to compute $\texttt{extra}(\mathcal{I})$ accurately, making it more difficult to discard potential unexpanded solutions. (We go into more details about the number of categories in the following section.) We would like to point out that the found solutions are close to the optimal for every data set, only the lower bounds for the score are affected differently.

In summary, we can obtain a solution and its quality from our algorithm at any time, enabling us to balance the quality of the solution and the run time on the fly.

Impact of Categories. For the final parameter, we illustrate the impact of categories on the run time. While our approximation algorithms scale well in terms of the itinerary length ($t_{\max}$), as we will see in Sect. 7.3, this is not independent of the number of categories. Figure 7(a), left column, shows the effects of increasing the number of categories. By putting a limit on the run time, we can keep the execution time of the algorithm low, while still generating very good solutions. In the upper diagram, we only show the plots for two and eight categories to keep it uncluttered. In the lower diagram, the run time for two categories actually goes down for longer itineraries. At the end of visiting two POIs of two categories each, we actually start having more and more spare time, which makes the problem easier to solve.

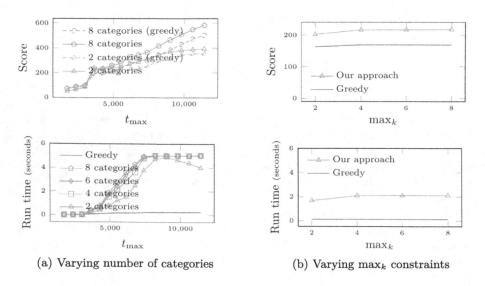

(a) Varying number of categories   (b) Varying $\max_k$ constraints

Fig. 7. Impact of categories

Figure 7(b), right column, shows the effect of increasing the maximum constraint of categories. Again, we quickly generate good solutions in a scalable way, due to the explicit limit on the run time.

7.3 Comparison with Competitors

Here we compare our algorithm to the state-of-the-art. We distinguish two different cases: (a) the artificial networks (grid and spider) and (b) the real-world data sets (Bolzano and San Francisco). For the real-world data sets we also run more realistic queries with a larger time constraint t_{max} and for that reason do not compare it to the optimal algorithm (as its run time explodes for long itineraries).

Grid and Spider Networks. Figure 8 illustrates the results of running our algorithm with a cut factor of 1.2 against the optimal algorithm (the left-hand column, labeled (a), shows the results for the grid network, the right-hand column, labeled (b), shows those for the spider network). In terms of the score our approach outperforms the greedy approach and comes close to the optimal solution, while at the same time having a much better run-time performance than the optimal algorithm. The final measurements for large values of t_{max} are missing, since we aborted runs taking longer than ten minutes to complete.

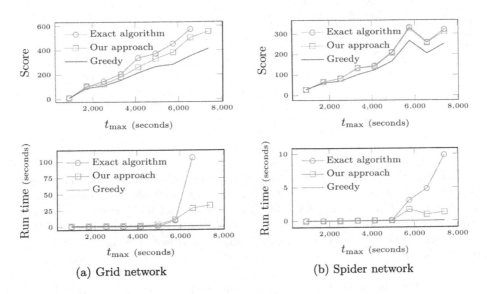

(a) Grid network   (b) Spider network

Fig. 8. Comparison with the exact algorithm

Real-World Data Sets. We now move to the real-world data sets. Figure 9(a), found in the left-hand column, compares three different variants of our algorithm (cut factor, bounded queue, and run time limit) with the state-of-the-art algorithm for orienteering with categories, CLIP on a map of Bolzano. In terms of the score (upper part of Fig. 9(a)) the three variants of our algorithm are almost identical: merely one measurement resulted in a difference two digits after the decimal point (thus we only depict one curve). Our algorithm outperforms both, CLIP and the greedy heuristic, the latter by a large margin. When we look at the run time (lower part of Fig. 9(a)), we see huge differences, though. The only competitive algorithms are the ones limiting the run time in some form, either directly or by limiting the queue length.

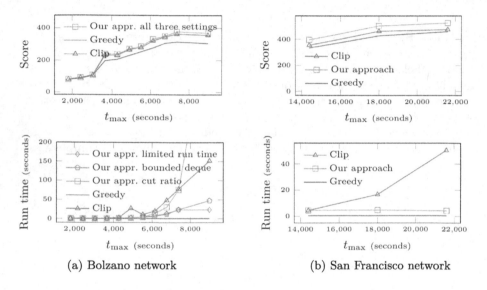

(a) Bolzano network   (b) San Francisco network

Fig. 9. Comparison with CLIP

In the final experiment we show the full strength of our approach for running queries with a large time constraint t_{max}. Here we run a blended version of our algorithm, combining a run time limit of five seconds with a queue length of half the number of total POIs in the graph. At first glance, combining a run time limit with a fixed queue length seems redundant. However, from a certain run time limit onwards, the queue length may become quite large and we may want to restrict it.

Figure 9(b) shows results for the San Francisco data set, comparing CLIP to our blended variant. In terms of the score, our approach outperforms both, CLIP and the greedy heuristic. We even get a better score than CLIP with a shorter run time, demonstrating that our technique scales much better than CLIP. In fact, as we were running the experiments on weaker hardware compared to the findings in [2], we could not replicate the ten-hour itineraries for CLIP shown in [2] in reasonable time. This clearly shows that our algorithm is much more suitable for deployment on mobile devices, which have limited computational resources. We also illustrate that limiting the queue length (to a certain extent) does not have adverse effects on the run-time-limited variant.

8   Conclusion and Future Work

We have developed an effective and efficient approximation algorithm for solving the orienteering problem with category constraints (OPMPC) by applying a best-first strategy, blending it with greedy search, and then limiting its run time. One major advantage of our technique over CLIP, the state-of-the-art approach for OPMPC, is the fact that our technique is an anytime algorithm, which immediately generates solutions with quality guarantees, as it keeps track of potential scores. Consequently, we can run our algorithm for a fixed time or until a certain quality level has been reached, whichever comes first.

Nevertheless, we still see some room for improvement. Profiling our algorithm, we noticed that we spend a considerable amount of time (about 40%) calculating the potential score. If we were able to do this more efficiently, maybe by parallelizing the task, we would be able to create an even more efficient algorithm. For longer itineraries it may also be interesting to move to more efficient techniques for computing itineraries from sets of POIs. On a more general level, our approach could also be viable for other orienteering variants, such as the team orienteering problem or orienteering with time windows.

References

1. Blum, A., Chawla, S., Karger, D.R., Lane, T., Meyerson, A., Minkoff, M.: Approximation algorithms for orienteering and discounted-reward TSP. SIAM J. Comput. **37**(2), 653–670 (2007)
2. Bolzoni, P., Helmer, S., Wellenzohn, K., Gamper, J., Andritsos, P.: Efficient itinerary planning with category constraints. In: SIGSPATIAL/GIS 2014, Dallas, Texas, pp. 203–212 (2014)
3. Chekuri, C., Korula, N., Pál, M.: Improved algorithms for orienteering and related problems. In: SODA 2008, pp. 661–670 (2008)
4. Chekuri, C., Pál, M.: A recursive greedy algorithm for walks in directed graphs. In: FOCS 2005, pp. 245–253 (2005)
5. Gavalas, D., Konstantopoulos, C., Mastakas, K., Pantziou, G.: A survey on algorithmic approaches for solving tourist trip design problems. J. Heuristics **20**(3), 291–328 (2014)
6. Gendreau, M., Laporte, G., Semet, F.: A branch-and-cut algorithm for the undirected selective traveling salesman problem. Networks **32**(4), 263–273 (1998)
7. Keller, C.: Algorithms to solve the orienteering problem: a comparison. Eur. J. OR **41**, 224–231 (1989)
8. Liang, Y.-C., Kulturel-Konak, S., Smith, A.: Meta heuristics for the orienteering problem. In: CEC 2002, pp. 384–389 (2002)
9. Lu, E.H.-C., Lin, C.-Y., Tseng, V.S.: Trip-mine: an efficient trip planning approach with travel time constraints. In: MDM 2011, pp. 152–161 (2011)

10. Ramesh, R., Yoon, Y.-S., Karwan, M.H.: An optimal algorithm for the orienteering tour problem. Inf. J. Comput. **4**(2), 155–165 (1992)
11. Rice, M.N., Tsotras, V.J.: Parameterized algorithms for generalized traveling salesman problems in road networks. In: SIGSPATIAL/GIS 2013, Orlando, Florida, pp. 114–123 (2013)
12. Righini, G., Salani, M.: Decremental state space relaxation strategies and initialization heuristics for solving the orienteering problem with time windows with dynamic programming. Comput. OR **36**(4), 1191–1203 (2009)
13. Sevkli, Z., Sevilgen, F.E.: Variable neighborhood search for the orienteering problem. In: Levi, A., Savaş, E., Yenigün, H., Balcısoy, S., Saygın, Y. (eds.) ISCIS 2006. LNCS, vol. 4263, pp. 134–143. Springer, Heidelberg (2006). doi:10.1007/11902140_16
14. Singh, A., Krause, A., Guestrin, C., Kaiser, W.J., Batalin, M.A.: Efficient planning of informative paths for multiple robots. In: IJCAI 2007, pp. 2204–2211 (2007)
15. Tasgetiren, F., Smith, A.: A genetic algorithm for the orienteering problem. In: IEEE Congress on Evolutionary Computation (2000)
16. Tsiligrides, T.A.: Heuristic methods applied to orienteering. J. Oper. Res. Soc. **35**(9), 797–809 (1984)
17. Wang, Q., Sun, X., Golden, B.L., Jia, J.: Using artificial neural networks to solve the orienteering problem. Ann. OR **61**, 111–120 (1995)

On Privacy in Spatio-Temporal Data: User Identification Using Microblog Data

Erik Seglem[1], Andreas Züfle[1(✉)], Jan Stutzki[2], Felix Borutta[2], Evgheniy Faerman[2], and Matthias Schubert[2]

[1] George Mason University, Fairfax, USA
{eseglem,azufle}@gmu.edu
[2] Ludwig-Maximilians-Universität München, Munich, Germany
{stutzki,borutta,faerman,schubert}@dbs.ifi.lmu.de

Abstract. Location data is among the most sensitive data regarding the privacy of the observed users. To collect location data, mobile phones and other mobile devices constantly track their positions. This work examines the question whether publicly available spatio-temporal user data can be used to link newly observed location data to known user profiles. For this study, publicly available location information about Twitter users is used to construct spatio-temporal user profiles describing a user's movement in space and time. It shows how to use these profiles to match a new location trace to their user with high accuracy. Furthermore, it shows how to link users of two different trace data sets. For this case study, 15,989 of the most prolific Twitter users in London in 2014 are considered. The experimental results show that the classification approach allows to correctly identify 98% of the most prolific 500 of these users. Furthermore, it can correctly identify more than 50% of any users by using three observations of these users, rather than their whole location trace. This alarming result shows that spatio-temporal data is highly discriminative, thus putting the privacy of hundreds of millions of geo-social network users at a risk. It further shows that it can correctly match most users of Instagram to users of Twitter.

1 Introduction

It is estimated that a third of the 130 billion copies of applications distributed by Apple's App Store® access a user's geographic location [1,3]. As an example, the recently launched augmented reality game "Pokémon Go", which has been downloaded more than 100 million times on Android devices alone [4], constantly synchronizes the GPS location of users with a company server. While users trust that their location data will be used in sensitive fashion, Apple® recently updated its privacy policy to allow sharing the spatio-temporal location of their users with "partners and licensees" [2].

The mobility behavior of a person often reveals a large variety of sensitive information, which they may not be aware of. A list of potentially sensitive professional and personal information that could be inferred about an individual, knowing only their mobility trace, was published recently by the Electronic

© Springer International Publishing AG 2017
M. Gertz et al. (Eds.): SSTD 2017, LNCS 10411, pp. 43–61, 2017.
DOI: 10.1007/978-3-319-64367-0_3

Frontier Foundation [8]. Such personal information could simply be marketing information, obtained from a user's choice of restaurants, or a user's religious beliefs, inferred through the proximity to a particular church. It can also indicate other, much more sensitive, information about an individual based on their presence in a motel or at a medical clinic.

In this work, the severity of privacy risks through publishing individual spatio-temporal data on the use case of Twitter data is investigated. In particular, it is shown that geotagged tweets might yield enough location information for building user specific trace profiles. Based on these profiles, Twitter accounts can be linked to additional trace data being observed from unknown users. Other location based services or mobile devices are also potential sources for traces. Additionally, face detection methods tag known persons in images in social networks. Thus, geotagged images can reveal a user's whereabouts at certain points in time. Given that there are multiple such images, it might be possible to build a trace and link it to a known user. To conclude, freely available location data might be used to link accounts and devices for the same user. Thus, the user reveals more of their movements and actions than might be intended.

To derive trace profiles for a given Twitter account, geotagged tweets containing an exact geolocation, a time, and a user ID were collected. Since this work focuses on the location aspect the content of the Tweet is completely ignored, even though it might add even more useful information to user profile. Using the Twitter API, or similar micro-blogging applications, users can publish a short text message, called a Tweet, together with their current geolocation, a current time-stamp, and their user ID.

The sequence of Tweets of a user is interpreted as a trace. For each user, all available Twitter data is used to build a trace profile to capture each user's specific mobility patterns. Using these profiles, new trace, for which the originating user is unknown, can be linked to a known user with an alarmingly high accuracy. To illustrate this classification problem, a typical Twitter trace of a single user is depicted in Fig. 1(a). The figure shows a twelve week trace of a user's tweets, in color-coded one-week intervals. For comparison, Fig. 1(b) shows the same twelve week traces for ten users, using a different color per user. Note that the tweets of this user are voluntarily published by the user, such that Fig. 1(a) and (b) do not raise any privacy concerns.

The challenge of this work is to match a new trace, such as a one week trace corresponding to a single color in Fig. 1(a), to the correct user corresponding to one of the colors in Fig. 1(b). Note that the ten selected user profiles in the example are located in relatively distinct activity regions. Thus, finding the right profile is relatively simple. In a more realistic setting, distinguishing thousands of users in the same area, and user identification is significantly more challenging. In these experiments up to 15,989 users, within the same bounding box of London, are used leading to a much more challenging classification task.

Twitter data is comparatively sparse to other location tracking applications, as tweets are typically published at a frequency of less than one per hour. Despite this data sparsity, it is shown that a large quantity of low-quality location data

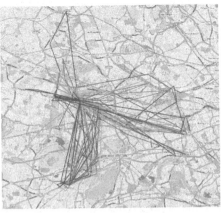
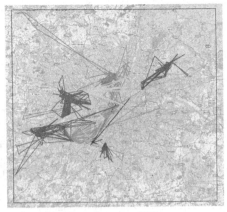

(a) Weekly history of a single user.   (b) 12-week trace of 10 users.

Fig. 1. Illustration of Twitter traces

can still be used to construct highly discriminative user models. To summarize the contributions of this work are as follows:

– Trace models to capture user-specific movement profiles from sparse traces obtained from Twitter.
– Methods for mapping a newly observed traces of an unknown user to the most likely user in the database.
– An experimental evaluation showing that individual patterns are highly unique and allow for a user classification accuracy of up to 98%.
– A case study of linking users of Twitter to users of Instagram, with an accuracy of up to 81%.

The remainder of this paper is organized as follows. Section 2 describes related work of analyzing trace data and user identification. In Sect. 3, problem setting is formalized and the task of linking new trace to users is defined. Section 4 describes the trace models and the approach to user identification. The results of the experimental evaluation are described in Sect. 5. Scalability of this solution is address in Sect. 5.5 and further user linkage experiments are address in Sect. 5.4. It is concluded in Sect. 6.

2   Related Work

This section provides a survey of the state-of-the-art in spatio-temporal user identification, user linkage, and spatial privacy. User identification is focused on identifying the same user again within the same database, while user linkage is focused on linking two users together across multiple sources of data. This work assumes that user trace data is fully available, without any notion of privacy preservation. This assumption is appropriate in the experimental evaluation, using publicly available Twitter data. However, other datasets may employ some form of privacy, thus it is important to understand privacy methods, which might be used on this data.

2.1   User Identification

A problem similar to the problem of trace based user-identification was considered in [11]. This work estimates the number of points needed to uniquely identify an individual trace. The user-identification method in [11] assumes that a trace of the user to be identified is already in the database. Thus, a new trace, which has not been seen before, cannot be classified. The work presented in [7] investigates the problem of how to prevent the identification of actual persons behind the users of location based services. Thus de-anonymizing the user. However, the stated problem is different from this work, as they make use of external sources to finally get the real names behind the pseudonames.

2.2   User Linkage

There are a variety of publications considering the problem of user linkage or more general record linkage. In the database community, record linkage generally aims at detecting duplicate records within one or several databases. Records describing the same entity may not share a common key or contain faulty attribute values, which makes the detection of such duplicates non-trivial. A survey on the proposed approaches can be found in [12].

Considering networks, record linkage is widely understood as user linkage and is stated as the problem of linking corresponding identities from different communities appearing within one or many networks [23]. An important area of user linkage is social networks where the user linking problem aims at connecting user profiles from different platforms that are used by the same persons. [17] differentiate between three types of user linkage across social networks: user-profile-based methods, which use information provided by the profiles to connect corresponding profiles [19], user-generated-content-based approaches, which analyze the content published by the users to link profiles [17] and user-behavior-model-based methods that generate models based on the (temporal) user behaviors and finally link user profiles based on the similarity of these models [18]. Most related to our approach is the recent work of [10]. In this work, the authors use various sources for data for the trajectories and propose a MapReduce-based framework called Automatic User Identification (AUI). Signal Based Similarity is introduced as a measurement of the similarity of two trajectories. In contrast to that approach, this work assumes much sparser trajectories. While the authors of [10] do consider sparse social media data, they accumulate these trajectories during a long time interval of at least multiple months. In this work, a long term mobility history of user is not assumed to be available. Instead, it aims at identifying users with the fewest observations possible.

2.3   Spatial Privacy

The predominantly used measurement for privacy is k-anonymity [22], which works with a closed world assumption and assures that, for each query that could be used to identify the identity of a user, at least $k - 1$ other users are returned as possible results.

Common approaches to guarantee a defined degree of anonymity are suppression, obfuscation and generalization [13]. To achieve k-anonymity by suppression, every element that does not fit into an anonymity set is removed [9, 16]. For trajectories, suppression would require discarding observations in discriminative locations such as a user's home. While this method is effective, the use of suppression alone can lead to a significant loss of information. Perturbation is another method used to obfuscate the data [5]. The goal is to generate a synthetic dataset with the same properties of the original dataset using a generative model. For generalization, k-groups of users could simply be unified into a single entity.

This work does not try to maintain privacy of users, and can be seen as an adversary approach of trying to breach the privacy of users. A highly relevant future piece of work is to investigate how existing privacy preservation methods for trajectories can be employed to suppress, obfuscate and generalize trajectories to minimize the user identification accuracy of this solutions, while further minimizing the loss of information in the data.

3   Problem Definition

In this work, the question of to what extent a set of spatio-temporal observations, such as geotagged tweets, are sufficient to derive spatial user profiles for the observed users and reliably link location traces of unknown users to one of the known user profiles is answered. Therefore, this section will define terms and notations, and formally define the problem of user identification using trace data.

In this paper, spatio-temporal data is considered. That is data of users annotated with a geolocation and a timestamp, such as obtained from Twitter.

Definition 1 (Spatio-Temporal Database). *Let $\mathcal{U}$ denote a set of unique user identifiers, let $\mathcal{S}$ be a set of spatial regions, and let $\mathcal{T}$ denote a time domain. A* spatio-temporal database *$\mathcal{DB} \subseteq \mathcal{U} \times \mathcal{S} \times \mathcal{T}$ is a collection of triples ($id \in \mathcal{U}, s \in \mathcal{S}, t \in \mathcal{T}$). Each triple $(u, s, t) \in \mathcal{DB}$ is called an* observation.

Furthermore, a trajectory is defined as a sequence of location and time pairs.

Definition 2 (Trajectory). *A* trajectory *$tr \subseteq \mathcal{S} \times \mathcal{T}$ is a collection of pairs $(s \in \mathcal{S}, t \in \mathcal{T})$.*

These trajectories are then partitioned temporally and spatially to distill the information down to a minimum set of components.

To build a user specific mobility pattern, the data is temporally partitioned into equal sized time intervals called epochs. Within an epoch the set of observations of a specific user is called a location trace, formally defined as follows.

Definition 3 (Location Trace). *Let $\mathcal{DB}$ be a spatio-temporal database. Let $\mathcal{E} = \{e_1, ...e_n\}$ be a partitioning of $\mathcal{T}$ into n temporal intervals denoted as epochs. For each epoch $e \in \mathcal{E}$, and each user u, the trace*

$$\mathcal{DB}(u, e) := \{(u', s, t) \in \mathcal{DB} | u' = u, t \in e\}, \tag{1}$$

is called the location trace of user id during epoch e.

In the remainder of this paper, location trace models are introduced to capture the motion of a user in space and time. The models are derived from the set of all trace of a user, called their trace profile formally defined as follows.

Definition 4 (Trace Profile). *Let $\mathcal{DB}$ be a spatio-temporal database, let $u \in \mathcal{U}$ be a user and let $\mathcal{E}$ be a temporal partitioning of $\mathcal{DB}$ into n epochs. The trace profile $\mathcal{P}(u)$ is the set of all traces of u, i.e.,*

$$\mathcal{P}(u) = \{D(u', e) | u' = u, e \in \mathcal{E}\}. \tag{2}$$

This trace profile is used to establish a pattern between multiple traces over discrete epochs, in order to allow for more accurate user identification.

The main challenge of this paper is to map the trace of an unknown user to a user already in the database, thus identifying them.

Definition 5 (User Identification). *Let $\mathcal{DB}$ be a spatio-temporal database and let $Q \subseteq \mathcal{S} \times \mathcal{T}$ be a trace of an unknown user u. The task of user identification is to predict the identity of user u of Q given $\mathcal{DB}$. The function*

$$\mathcal{I} : \mathcal{P}(\mathcal{S} \times \mathcal{T}) \mapsto \mathcal{U}, \tag{3}$$

maps a trace Q to user x as a user identification function.

Thus, user identification is a classification task mapping a trace to its unknown user. This is not to be confused with a de-anoymization attack, such as in [7] where a user's real world identity is uncovered. This task is only attempting to identify which user's Trace Profile is most similar to the new trace. To train a user identification function $I(Q)$, the next chapter presents the classification approach, which uses the traces in $\mathcal{DB}$ as a training set, in order to predict the user of a new trace $Q \notin \mathcal{DB}$.

User linkage, takes this task further and attempts to map two users, from separate databases, together and is formally defined as follows.

Definition 6 (User Linkage). *Let $\mathcal{DB}_1$ and $\mathcal{DB}_2$ be two trace feature databases, and let $\mathcal{U}_{DB_1}$ and $\mathcal{U}_{DB_1}$ be two user databases. Such that each entry $(T, u) \in \mathcal{DB}_1$ corresponds to a trace feature vector T and a user $u \in \mathcal{U}_{DB_1}$, and each $(T', u') \in \mathcal{DB}_2$ corresponds to a trace feature vector T' and a user $u' \in \mathcal{U}_{DB_2}$. The task of user linkage is to map a user in $\mathcal{U}_{DB_1}$ to a user in $\mathcal{U}_{DB_2}$.*

This is achieved by using the user identification techniques discussed in the next chapter and will be discussed in Sect. 5.4.

4   Trajectory Based User Identification

Trace models are introduced to capture the motion of a user $u \in \mathcal{U}$ in space and time by learning from their trace profile $\mathcal{P}(u)$ in Subsect. 4.1. Note that this first approach does not consider the time component of observations of a user within an epoch. The time component is only used to divide the whole trajectory of a user into different epochs that can be used for learning and testing. For each model, a similarity measure to quantify similarity between different trace models is proposed. Based on these similarity measures, the user identification approach is presented in Subsect. 4.4. As mentioned before, the prediction is based on the assumption that there exists a profile $P(u_i)$ for each user $u_i \in U$.

4.1   Trace Profile Modeling

Each trace $\mathcal{DB}(u, e)$ of user u during epoch e is a sequence of observations, i.e., time-stamped geo-locations. A spatial grid to partition geo-space into equal sized regions $\mathcal{S} = \{S_1, S_{|S|}\}$ is used, thus reducing a trace to a sequence of time-stamped grid-cells. To model such a sequence, two kinds of approaches are proposed:

- The first approach using *set descriptors* treats a trace as a *set* of grid-cell observations, thus ignoring the sequence, ordering, and time-stamps of these observations.
- The second approach using *frequent transitions* considers the transitions of users from one spatial region to another, thus explicitly modeling the order of observations.

4.2   Set Descriptors

Ignoring the temporal aspect, a trace $\mathcal{DB}(u, e)$ of user u during epoch e can be described by a vector $v(u, e)$ of all spatial regions in S. In other words, each spatial region is represented by a dimension of $v(u, e)$.

Note that $v(u, e)$ contains zero values in the majority of dimensions as each user usually only traverses a small fraction of space during an epoch. In other words, $v(u, e)$ is sparse. Modeling trace using frequency descriptions has a strong resemblance to handling bag of words vectors known in text mining. To describe, if and how often a domain was visited within trace $\mathcal{DB}(u, e)$, the following two approaches are examined.

Binary Descriptor. In this rather simple method, a trace $\mathcal{DB}(u, e)$ is represented as a set of visited spatial regions. Thus, each feature value v^{bit} equals one if user u visited region S_i (at least once) during epoch e, formally:

$$v_i^{\text{bit}}(u, e) := \begin{cases} 1, & \text{if } \exists (u', s, t) \in \mathcal{DB} : u' = u \wedge s \in S_i \wedge t \in e, \\ 0, & \text{otherwise} \end{cases} \tag{4}$$

To compare binary vectors $v, v' \in \{0, 1\}^n$, the Jaccard coefficient is employed [15], which is a standard similarity measure for sets:

Definition 7 (Jaccard Coefficient). *Let* $v, v' \in \{0,1\}^n$ *be two bit vectors, then the Jaccard coefficient is defined as follows:*

$$Jac(v, v') = \frac{\sum_{i=1}^{n} v_i \wedge v_i'}{\sum_{i=1}^{n} v_i \vee v_i'} \tag{5}$$

Frequency Descriptors. A frequency, or term weighted, vector [21] v^{freq} contains the number of visits of each spatial region of user u in epoch e. This allows to distinguish between users visiting a particular region more or less often than other users.

$$v^{\text{freq}}(u, e)_i = |\{(u', s, t) \in \mathcal{DB}|u' = u \wedge s \in S_i \wedge t \in e\}|. \tag{6}$$

A common way to compute the similarity in sparse numerical vectors is the cosine coefficient:

Definition 8 (Cosine Coefficient). *Let* $v, v' \in \mathbb{N}^n$ *be two vectors, then the Cosine coefficient is defined as follows:*

$$Cos(v, v') = \frac{v \cdot v'}{||v|| \cdot ||v'||} \tag{7}$$

Since the cosine coefficient can be strongly dominated by dimensions having high average frequency values, spatial regions are normalized by their total number of observations [21].

4.3   Transition Descriptors

All of the previous trace descriptors had in common that they treat a trace as an unordered set of locations, without considering any notion of sequence or time. In this section, a trace is treated as a sequence of regions. As a baseline to compute the similarity between two sequences, dynamic time-warping [6] (DTW), a state-of-the-art method for similarity search on sequences, is used. Since the experimental evaluation shows that using DTW without any adaption as a similarity measure yields a fairly low classification accuracy, this section presents two approaches to directly model the transitions of a trace. A transition is a pair (s, s') of regions where s is called source and s' is called destination. Using a descriptor for each pair of spatial regions s_i, s_j, describing the number of times the specific sequence (s_i, s_j) has been observed in a trace $\mathcal{DB}(u, e)$, is proposed.

Definition 9 (Trace Transitions). *Let* $\mathcal{DB}(u, e) = \{(s_1, t_1), ..., (s_n, t_n))\}$ *be a trace, the set of n transitions* $\uparrow \mathcal{DB}(u, e)$ *is defined as the multi-set (thus allowing duplicates)*

$$\uparrow \mathcal{DB}(u, e) := \bigvee_{1 \leq i < n} (s_i, s_{i+1}). \tag{8}$$

The number of occurrences of (s, s') *in trace* $\mathcal{DB}(s, e)$ *is denoted as* $\uparrow$ $\mathcal{DB}(u, e)(s, s')$.

Since modeling all observed transitions blows up the feature space quadratically, using only the k globally most frequent transitions as features is proposed.

Definition 10 (Top-k Most Frequent Transitions). *Let k be a positive integer, then the set FT is a set of pairs of spatial regions defined as*

$$FT^k(\mathcal{DB}) = argmax^k_{s_i,s_j \in \mathcal{S}} |\{ \sum_{u \in \mathcal{U}, e \in \mathcal{E}} \uparrow \mathcal{DB}(u,e)(s_i,s_j) \}|, \qquad (9)$$

where $argmax^k_X(\varphi)$ returns the set of k arguments $x \in X$ yielding the maximum value substituted in term φ.

Now the k most frequent transitions $FT^k(\mathcal{DB})$ can be used as additional features. Similar to the set descriptors presented in Subsect. 4.2, the features are described using

– Bit vectors, using the feature vector

$$v_i^{\uparrow \text{bit}(u,e)} = \begin{cases} 1 & \text{if } FT^k(\mathcal{DB})_i \in \uparrow \mathcal{DB}(u,e) \\ 0 & \text{otherwise} \end{cases} \qquad (10)$$

– Frequency vectors, using the binary feature vector

$$v^{\uparrow \text{freq}}(u,e)_i = \uparrow \mathcal{DB}(u,e)(FT^k(\mathcal{DB})_i) \qquad (11)$$

For these vectors, the same similarity functions defined in Sect. 4.2 can be used.

4.4 Classification

Regardless of which of the modeling approaches presented in this section is employed, the result is a high-dimensional feature vector. To classify a new trace of an unknown user, the next section proposes the classification procedure, using the previously proposed user-specific trace models. To classify the user of a new trace, a k-nearest neighbor classification approach is employed. This choice is made due to the extremely high dimensional feature space, having one dimension per spatial grid-cell. Therefore, given a trace database $\mathcal{DB}$, traces $\mathcal{DB}(u,e)$ are extracted for each user u in each epoch e. Since the user is known for each of these traces, the result is a labeled dataset P_{train} of feature vectors. Given a new trace Q, map Q to its feature description v_{new} and search the k-nearest neighbors of v_{new} in P_{train} w.r.t. a corresponding similarity measure. To decide the final class decision, each queried neighbor is weighted by its similarity value and the class is predicted as the one having the largest cumulated similarity.

Formally, the k-nearest neighbors classification can be defined as follows. Let $P_{train} = \{(v_i, y_i) \mid v_i \in \{0,1\}^n \wedge y_i \in \mathcal{L}\}$ be the set of training instances consisting of pairs (v_i, y_i) with v_i being the feature description of the user trace i and y_i being the label, i.e., identity of the user, assigned to trace i. $\mathcal{L}$ denotes the set of labels. Given the feature description v_{new} of a query trace, the identity, resp.

label, y_{new} of v_{new} is determined by cumulating the similarities, i.e., $d(.,.)$, for each label $l \in \mathcal{L}$ represented among the k-nearest neighbors of v_{new} and taking the most representative label.

$$y_{new} = argmax_{l \in \mathcal{L}}\{\sum d(v_{new}, v_k^l) \mid v_k^l \in kNN(v_{new})\} \qquad (12)$$

Note that no index structure is used to support the kNN-search due to the high dimensionality of the feature space.

4.5 User Linkage

In addition to the identification of individual users, another application of the user trace profiling is to link users between two trace datasets. Therefore, let $\mathcal{DB}$ and $\mathcal{DB}'$ be two trace databases having the set of users $\mathcal{U}$ and $\mathcal{U}'$, respectively. The task of user linkage is to find pairs of database users $(u \in \mathcal{U}, u' \in \mathcal{U}')$ that correspond to the same individual in the real world, i.e., having $u = u'$. As an example, the two datasets may correspond to Twitter and Instagram. The same individual may have different user names in both social networks. The task of user linkage is to find such individuals.

Clearly, using the approach presented in Sect. 4.4, the trace of each user are classified in $\mathcal{DB}$, and the most similar user in $\mathcal{DB}'$ is classified. The drawback of such approach is that multiple users in $\mathcal{DB}$ may be matched to the same user in $\mathcal{DB}'$, and some users in $\mathcal{DB}\prime$ might not have any match. To avoid this drawback, the matching problem is formalized as a bipartite graph, containing for each $(u \in \mathcal{U}, u' \in \mathcal{U}')$ a weight of similarity. This similarity is chosen by performing a kNN search of each trace in $\mathcal{DB}$ on the database $\mathcal{DB}'$. Then, the score of (u, u') corresponds to the number of occurrences of $u\prime$ in kNN sets of all traces of user u.

Given this bipartite graph, the Hopcroft-Karp algorithm [14] is used to find an optimal matching, i.e., mapping of each user in the smaller database to exactly one user in the other that maximizes the total score.

5   Experimental Evaluation

The proposed approach is initially evaluated on a dataset mined from Twitter using their public API, feeding from a global 1%-sample using a $(51.25, 51.75)$ degrees longitude to $(-0.55, 0.30)$ degrees latitude window covering the London region shown in Fig. 1(b). London was chosen as a starting location for having a high population, while still being predominantly an English speaking location. Furthermore, London shows a high density of users and tweets, thus increasing the size of the trace database $\mathcal{DB}$, and allowing more significant conclusions to be made.

It was also decided to use one-week periods for the temporal epoch. This choice was meant to minimize the daily variability in a users locations. For example: a user may work Monday through Friday, but only go to the gym Monday, Wednesday and Friday, or night classes on Tuesday and Thursday.

They may also go grocery shopping, or to a religious institution only on the weekends, thus going locations significantly different than were they would be on a week day. Additionally, Twitter data is extremely sparse for most users, thus more than a day was necessary to a reasonable number of tweets to create location traces from. This lead to the choice of a twelve week time interval from December 30, 2013 to March 24, 2014 being used. The choice of twelve weeks was to allow for multiple weeks of traces for each user, while not having too large of a dataset to use for initial testing purposes.

Out of these London-Tweets, the 500 users with the most Tweets during the study period were selected, excluding obvious spammer or bot users. This dataset was then split into temporal epochs of one-week. Thus, the database contains a total of $|\mathcal{U}| = 500$ users, and a total of $|\mathcal{E}| = 12$ epochs. Consequently, the database $\mathcal{DB}$ contains a total of $\mathcal{U} \times \mathcal{E} = 6000$ location traces.

To discretize space, a spatial grid is applied on the aforementioned rectangle covering the London region, having an extent ext in longitude and latitude ranging from $0.01'$ to $0.001'$. The set of all resulting grid cells constitutes the set of spatial regions $\mathcal{S}$, having $|\mathcal{S}| = 4,250$ cells for $ext = 0.01'$ and $425,000$ cells for $ext = 0.001'$.

Consequently, for a user $u \in \mathcal{U}$ and an epoch $e \in \mathcal{E}$ a trace $\mathcal{DB}(u, e)$ is a sequence of cells in $\mathcal{S}$. To give a more detailed intuition of the characteristics of the dataset, Fig. 2 shows statistics about the traces of these 500 users. Figure 2(a) shows the number of traces having at least one observation in the corresponding epoch.

Of users, 42% have an observation have at least one observation in each of the twelve epochs, and 75% of the users have at least one observation in at least eight epochs. In addition, Fig. 2(b) shows the number of observed cells for each trace. Most users only visited a small number of space cells each week, as half of the trace contain six or less cells. Note that any trace having zero observations were removed from the dataset.

The classification experiments in this work were performed using an eight-fold cross validation. Eight folds for optimal parallelization on an eight core processor. Thus, in each experiment a test set of tracestrace $Q(u, e) \subset \mathcal{DB}(u, e)$ is selected, and user mobility profiles are built using the techniques of Sect. 4.1,

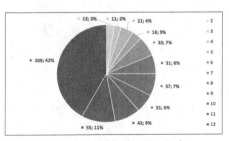

(a) Traces within the 12 epochs

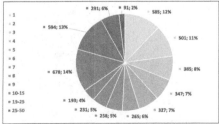

(b) Observations per one-week Trace

Fig. 2. Distribution of the Top 500 most prolific users in our London-Twitter Dataset.

without using the test traces, i.e. $\mathcal{DB}(u, e)\backslash Q(u, e)$, in the training step to avoid over-fitting.

Note that this important avoidance of over-fitting is a main differentiation to the trace identification approach proposed in [11]. By having the query trace in the training data, a $k = 1$-NN classification would always return a 100% classification accuracy, but defeating the purpose of user identification. Consequently, since the related work in [11], solves a different problem, a comparison would be unfair and non-explanatory. See Sect. 2 for more details on [11].

As a classifier, k-nearest neighbor classification was utilized, using a distance-weighting in case of ties, which is able to perform well despite an extremely large number of $|\mathcal{S}|$ features. Classifications are performed using scikit-learn, a Python machine learning framework [20].

5.1   Accuracy Using Set Descriptors

In the first set of experiments, the accuracy of the user identification is evaluated for different grid-resolutions ext, using binary descriptors for the Jaccard similarity measure (c.f. Definition 4). The results of this evaluation are shown in Fig. 3(a). In the basic setting having a relatively coarse spatial grid of $ext = 0.01'$, a simple distance weighted kNN classification is able to correctly identify (c.f. Definition 5) up to 85% of individuals for $k = 5$. This result improves even further as the grid-resolution ext is increased. In the case of the most detailed grid having $ext = 0.001'$, the solution is able to break the 97% classification accuracy line. This result is quite concerning, as it shows that the motion of individual real-persons is quite characteristic, and that the motion model allows to capture this individuality and allows to discriminate different users very well.

The classification result are worse for $k = 1$ and $k = 3$. This result is contributed to chance, as another user may, by chance, have a trace very similar to the query trace $q_u^e \in Q(u, e)$ of user u. However, by using more neighbors, it is likely that the correct user u appears at least twice in the $k = 3$ or $k = 5$ set, thus out-weighting the erroneous user in the first rank. Yet, for $k > 5$ there is a drop in accuracy. This is contributed that the query user only has at most 11 traces in the training set. This number might be less than 11 if a user was

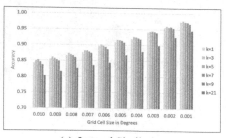
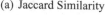

(a) Jaccard Similarity

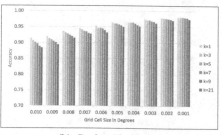

(b) Cosine Similarity

Fig. 3. Classification accuracy for varying grid-cell size and varying k.

not active in all epochs. This is the case for many users, shown by Fig. 2. In the extreme case having $k = 21$, at least 10 trace of wrong users must be in the kNN result, allowing noise have a much greater effect, especially in the case where u has few trace.

Furthermore, Fig. 3(b) shows the results using frequency vectors as descriptors, and using the cosine coefficient as a similarity measure (c.f. Definition 4). The improvement in classification accuracy is relatively minor, but are able to hit the 98% accuracy mark. This result can be contributed to the fact that binary descriptors already perform so well. Summarizing, knowing the set of places that a user visited is descriptive enough, such that the frequency of visits does not yield much additional descriptiveness.

5.2 Accuracy Using Frequent Transitions

In the next set of experiments, how the usage of transition descriptors (c.f. Sect. 4.3) instead of set descriptors affects the classification accuracy is evaluated. The results depicted in Fig. 4 indicate that using from-to-transitions, as opposed to just using sets of cells, further allows to improve the classification quality. An increase in classification accuracy of around 10% (absolute) is observed using transitions, achieving an classification accuracy of nearly 95%. This result indicates that the sequence, and thus the motion in space and time is more descriptive than just sets of regions, and thus the motion in space-only.

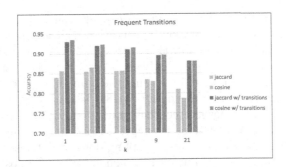

Fig. 4. Classification accuracy using frequent transitions.

While this was method did allow for a slight increase in accuracy, this increase came at a cost. It causes the dimensionality of the data, and greatly increase complexity. And thus, it greatly increases the processing requirements. Because there was only a small increase in accuracy for this increase in complexity, transitions were not use used in the remainder of the experiments. Though, further research in the subject could be worthwhile.

5.3   Accuracy for Different Observation Counts

Next, the number of observations required to identify (c.f. Definition 5) a user accurately is evaluated. Therefore counts are created according to the observation distribution in Fig. 2. Then tests are for each count. If a trace does not have the minimum number of observations for the corresponding group, it is not tested, and if a trace has more observations than the allowed maximum for the corresponding group, a random sample is taken and tested instead. Thus, instead of testing the accuracy on the original traces this tests the accuracy on controlled observation counts.

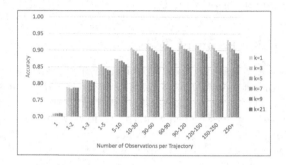

Fig. 5. User identification accuracy for different observation counts.

The classification results for each group can be seen in Fig. 5. Surprisingly, in the case of having only one random observation for each trace, it is possible to identify over 70% of the users in this dataset. This is likely due to the fact that a random location from a trace is likely to pick a users most frequent grid cell, which is most discriminative. Increasing the number of observation samples to two, a significant increase in accuracy to 78% is seen, and a steady growth in accuracy from there is shown. Accuracy starts to level off after having 30 or more obervations from a user. This is surprising, as the vast majority of trace has more than 60 observations. Thus, sampling down to 30 observations, yields a significant reduction in data, but as Fig. 5 shows, yields almost no reduction of discriminative information.

The leveled accuracy level is above 90%, which is extremely high for a classification task having 500 different classes. This positive result is also a consequence of large trace (i.e., traces having a large number of observations) generally having larger trace in the training set, as the frequency distribution of tweets among these 500 Twitter users in London is very skewed. Finally, the classification performs the best, if the parameter of the kNN classification is set to $k = 1$. This result is in line with Fig. 3(b), as Cosine-Similarity is used per default in this experiment.

Summarizing this experiment, very short trace having 10 or less observations in space and time are enough to unveil the identity of a user. This is a concerning result.

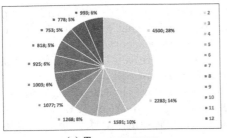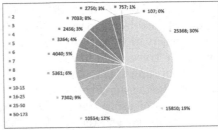

(a) Traces per user.   (b) Observations per one-week Trace

Fig. 6. Distribution of all 15,989 users in our London-Twitter Dataset.

5.4   User Linkage Between Different Social Networks

In all the previous experiments, a single user had to be identified based on a new trace. In this section, the next step is evaluated. Linking whole sets of users of two different social networks, based on their traces, as described in Sect. 4.5 and defined in Definition 6. For this purpose, two new datasets are employed, one generated synthetically by splitting the scalability (c.f. Sect. 5.5) dataset randomly, and one splitting the same dataset based on links between Twitter and Instagram.

Synthetic Database Split: For the synthetic database, a fraction of p Tweets is uniformly sampled from the Twitter dataset $\mathcal{DB}$, and pretend that this set belongs to a different social network $\mathcal{DB}'$. In this sampled database $\mathcal{DB}'$, the user-labels as ground-truth, which the algorithm tries to predict given the data in $\mathcal{DB}$ can be used. For this experiment, only traces having at least 10 tweets to sample from are considered. If uniform sampling of a trace yields an empty set, it is re-sampled.

Instagram Data: Out of the 2.7 million tweets in the dataset, a significant portion of 204 thousand tweets is labelled as coming from the Instagram network. These Tweets were cross posted by the user, on both Instagram and Twitter. Thus, the Instagram database $\mathcal{DB}^I$ consists of all these cross-linked posts. For the Twitter database, two cases are evaluated. In the first case, the full dataset $\mathcal{DB}$ can simply be used, thus assuming that the Instagram observations were made in both datasets. In the second case, the database $\mathcal{DB}_T = \mathcal{DB} \setminus \mathcal{DB}_I$ is used, thus assuming that the Instagram observations were made in the Instagram network only.

The results on the synthetic database split are shown in Fig. 7(a). For each value of p, 10 random samples of the database $\mathcal{DB}$ are obtained, and results from each are averaged in order to avoid effects generated due to random sampled. In all ten runs, the depicted values showed almost no deviation, all being in a $\pm 0.5\%$ interval. An even 50/50 split yields a correct linkage rate of almost 85%. Yet, this split becomes biased towards a smaller value p. This can be explained by having a larger sample in the training database $\mathcal{DB}$, on which the traces of

$\mathcal{DB}'$ are queried on. However, for $p = 0.1$, this accuracy drops significantly. This can be explained by the previous experiments, showing that a sample of as little as three observations suffices for a high classification accuracy. However, since many of the traces only have 10–20 observations, there is a high chance that a 10% random sample may only have one or two observations.

For the Instagram-Twitter matching, the results are shown in Fig. 7(b), for the two cases of using the data as is, thus having all Instagram observations also present in the Twitter database, and the case of splitting the dataset, thus removing the Instagram observations from the Twitter traces. Using the raw dataset a prediction accuracy of roughly 80% using $k = 1$ nearest neighbor classification to build the bipartite graph is observed.

In contrast, the case of splitting Instagram off of Twitter, the accuracy drops to about 10%. These disappointing results can be explained by making the hypothesis that users use Instagram and Twitter in different ways, such as using Instagram when on a far-away vacation, while also using Twitter in locations where you don't usually take a picture, such as work and home. Also, some of the users had all their tweets linked to Instagram, such that the algorithm had no training data left in the Twitter database, thus having to random guess the user. Thus, it appears that Twitter and Instagram are used differently by users, making the Instagram sample much harder to match than a uniform random sample taken from Twitter.

5.5   Scalability

In all of the previous experiments, only the top 500 Twitter users in London were used. In the final experiment, this number of users is scaled up, by using 15,989 users that have a least two trace containing at least two observations each. This larger dataset contains over 2.7 million Tweets, including the original dataset. Statistics for this dataset are shown in Fig. 6. The quality of the observed traces is much worse compared to the earlier 500 users explored in Fig. 2: In Fig. 6(a) more than half of the users have less than five traces within the twelve epochs, and only a small fraction of 6% of the users have maximum number of twelve traces. In addition, Fig. 6(b) shows the quality of these trace is much lower, as nearly 50% of the traces have three or less observations. Due to the quality of this data a eight-fold split was no longer possible. A stratified shuffle split was used instead, taking 10 iterations of 20% samples.

The results on this dataset, in terms of classification accuracy as well as run-times are shown in Fig. 8. In terms of accuracy, there is a vast decrease in accuracy observed, even for the default setting of 500 users. This is because the experiments are no longer using the top users, but just a random sample of users, and the data quality, in terms of number of observations per trace, as well as the number of trace per user, is much lower for these users.

Clearly, less frequent users are harder to classify, since there is less information. As the experiments are scaled up the number of users, there is a decrease in classification accuracy, as the classification problem becomes harder having

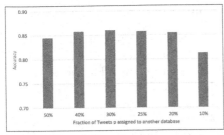

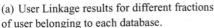

(a) User Linkage results for different fractions of user belonging to each database.

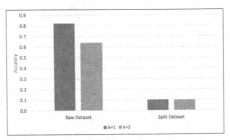

(b) User Linkage results for linking Twitter and Instagram.

Fig. 7. Classification accuracy for different Social Networks.

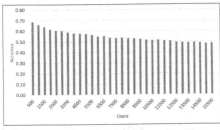

(a) Classification Accuracy

(b) Run-time (in seconds)

Fig. 8. Scalability: scaling the number of Twitter users.

more users. Still, the classification accuracy remains at almost 50%, despite the large number of 15,989 users, and the much lower trace quality.

Since a kNN classification is employed, and thus a lazy learning method is used, there is no model learning phase. The run-time results for the classification is shown in Fig. 8(b).[1] a linear run-time is observed, which is attributed to the extreme high dimensionality of the feature vectors, which cannot be beneficially supported by an index structure for the kNN search. But even at the full 15,989 users, the time to classify each trace is less than 1 ms.

6   Conclusions

In this work, the challenge of identifying users in a spatio-temporal database was approached. This approach uses historic traces of a user to learn their motion in space and time, by proposing various feature extraction and similarity search methods. Using a 12-week dataset of Tweets in the London region, the experimental results show that it is possible to map a trace to a ground-truth user with extremely high accuracy.

---

[1] Run-time tests were performed on AWS using a m4.2xlarge EC2 instance running Amazon Linux. This instance type has 8 CPU cores and 32 GB of RAM.

This raises various concerns and opportunities:

- **The Threat** of loss of privacy: Traces of real people are publicly available. Given only few observations of an individual. For example, one person inadvertently appearing in the background of another person's Facebook images, then identifying this user in a trace database, and linking them to additional data, such as username or real name.
- **The Potential** through record linkage of trace databases. For example, joining the personal interests in locations (such as restaurants, bars, cafes) from a LBSN with textual thoughts of a user from a micro-blog. Thus, a micro-blog tweet from user u such as **"W00T, I'm going to George Mason University!"**, might be used to recommend restaurants to u by mining their restaurant preferences using their check-ins in the LBSN.
- **The Challenge** of privacy preservation by trace obfuscation and other means. By learning the characteristics that make a trace matchable to its user, techniques to hide particularly descriptive and discriminative observations from the public trace can be developed.

References

1. App genome report. www.mylookout.com/resources/reports/appgenome. Accessed 12 Aug 2016
2. Apple privacy policy. www.apple.com/legal/privacy/. Accessed 12 Aug 2016
3. Apples app store downloads top 130 billion. http://www.gsmarena.com/apple_app_store_now_has_2000000_apps_50_billion_paid_to_devs-blog-18798.php. Accessed 12 Aug 2016
4. Google Play: Pokémon Go Download. https://play.google.com/store/apps/details? id=com.nianticlabs.pokemong. Accessed 22 Aug 2016
5. Aggarwal, C.C., Yu, P.S.: A condensation approach to privacy preserving data mining. In: Bertino, E., Christodoulakis, S., Plexousakis, D., Christophides, V., Koubarakis, M., Böhm, K., Ferrari, E. (eds.) EDBT 2004. LNCS, vol. 2992, pp. 183–199. Springer, Heidelberg (2004). doi:10.1007/978-3-540-24741-8_12
6. Berndt, D.J., Clifford, J.: Using dynamic time warping to find patterns in time series. In: AAAI 1994 Workshop on Knowledge Discovery in Databases (KDD-1994), vol. 398, pp. 359–370 (1994)
7. Bettini, C., Wang, X.S., Jajodia, S.: Protecting privacy against location-based personal identification. In: Jonker, W., Petković, M. (eds.) SDM 2005. LNCS, vol. 3674, pp. 185–199. Springer, Heidelberg (2005). doi:10.1007/11552338_13
8. Blumberg, A.J., Eckersley, P.: On locational privacy, and how to avoid losing it forever. Electronic Frontier Foundation, Technical report, pp. 1–7, August 2009
9. Byun, J.-W., Kamra, A., Bertino, E., Li, N.: Efficient k-anonymization using clustering techniques. In: Kotagiri, R., Krishna, P.R., Mohania, M., Nantajeewarawat, E. (eds.) DASFAA 2007. LNCS, vol. 4443, pp. 188–200. Springer, Heidelberg (2007). doi:10.1007/978-3-540-71703-4_18
10. Cao, W., Wu, Z., Wang, D., Li, J., Wu, H.: Automatic user identification method across heterogeneous mobility data sources. In: 2016 IEEE 32nd International Conference on Data Engineering (ICDE), pp. 978–989. IEEE, May 2016

11. de Montjoye, Y.-A., Hidalgo, C.A., Verleysen, M., Blondel, V.D.: Unique in the Crowd: the privacy bounds of human mobility. Sci. Rep. **3**, 1376 (2013)
12. Elmagarmid, A.K., Ipeirotis, P.G., Verykios, V.S.: Duplicate record detection: a survey. IEEE Trans. Knowl. Data Eng. **19**(1), 1–16 (2007)
13. Hashem, T., Kulik, L.: Safeguarding location privacy in wireless ad-hoc networks. In: Krumm, J., Abowd, G.D., Seneviratne, A., Strang, T. (eds.) UbiComp 2007. LNCS, vol. 4717, pp. 372–390. Springer, Heidelberg (2007). doi:10.1007/978-3-540-74853-3_22
14. Hopcroft, J.E., Karp, R.M.: An $n^{5/2}$ algorithm for maximum matchings in bipartite graphs. SIAM J. Comput. **2**(4), 225–231 (1973)
15. Jaccard, P.: The distribution of the flora in the alphine zone. New Phytol. **11**(2), 37–50 (1912)
16. LeFevre, K., DeWitt, D., Ramakrishnan, R.: Mondrian multidimensional K-anonymity. In: 22nd International Conference on Data Engineering (ICDE 2006), vol. 2006, p. 25. IEEE (2006)
17. Liu, J., Zhang, F., Song, X., Song, Y.-I., Lin, C.-Y., Hon, H.-W.: What's in a name? In: Proceedings of the Sixth ACM International Conference on Web Search and Data Mining - WSDM 2013, p. 495. ACM Press, New York (2013)
18. Liu, S., Wang, S., Zhu, F., Zhang, J., Krishnan, R.: HYDRA: large-scale social identity linkage via heterogeneous behavior modeling. In: Proceedings of the 2014 ACM SIGMOD International Conference on Management of Data - SIGMOD 2014, pp. 51–62. ACM Press, New York (2014)
19. Malhotra, A., Totti, L., Meira, W., Kumaraguru, P., Almeida, V.: Studying user footprints in different online social networks. In: 2012 IEEE/ACM International Conference on Advances in Social Networks Analysis and Mining, pp. 1065–1070. IEEE, August 2012
20. Pedregosa, F., Varoquaux, G., Gramfort, A., Michel, V., Thirion, B., Grisel, O., Blondel, M., Prettenhofer, P., Weiss, R., Dubourg, V., Vanderplas, J., Passos, A., Cournapeau, D., Brucher, M., Perrot, M., Duchesnay, É.: Scikit-learn: machine learning in python. J. Mach. Learn. Res. **12**, 2825–2830 (2012)
21. Salton, G., Buckley, C.: Term-weighting approaches in automatic text retrieval. Inf. Process. Manage. **24**(5), 513–523 (1988)
22. Sweeney, L.: k-anonymity: a model for protecting privacy. Int. J. Uncertainty Fuzziness Knowl. Based Syst. **10**(05), 557–570 (2002)
23. Zafarani, R., Liu, H.: Connecting corresponding identities across communities. In: Proceedings of the Third International Conference on Weblogs and Social Media - ICWSM 2009, pp. 354–357, November 2009

Big Spatial Data

Sphinx: Empowering Impala for Efficient Execution of SQL Queries on Big Spatial Data

Ahmed Eldawy[1]([✉]), Ibrahim Sabek[2], Mostafa Elganainy[3], Ammar Bakeer[3], Ahmed Abdelmotaleb[3], and Mohamed F. Mokbel[2]

[1] University of California, Riverside, USA
eldawy@cs.ucr.edu
[2] University of Minnesota, Twin Cities, USA
{sabek,mokbel}@cs.umn.edu
[3] KACST GIS Technology Innovation Center, Mecca, Saudi Arabia
{melganainy,abakeer,aothman}@gistic.org

Abstract. This paper presents Sphinx, a full-fledged open-source system for big spatial data which overcomes the limitations of existing systems by adopting a standard SQL interface, and by providing a high efficient core built inside the core of the Apache Impala system. Sphinx is composed of four main layers, namely, *query parser*, *indexer*, *query planner*, and *query executor*. The *query parser* injects spatial data types and functions in the SQL interface of Sphinx. The *indexer* creates spatial indexes in Sphinx by adopting a two-layered index design. The *query planner* utilizes these indexes to construct efficient query plans for range query and spatial join operations. Finally, the *query executor* carries out these plans on big spatial datasets in a distributed cluster. A system prototype of Sphinx running on real datasets shows up-to three orders of magnitude performance improvement over plain-vanilla Impala, SpatialHadoop, and PostGIS.

1 Introduction

There has been a recent marked increase in the amount of spatial data produced by several devices including smart phones, space telescopes, medical devices, among others. For example, space telescopes generate up to 150 GB weekly spatial data, medical devices produce spatial images (X-rays) at 50 PB per year, NASA satellite data has more than 1 PB, while there are 10 Million geo-tagged tweets issued from Twitter every day as 2% of the whole Twitter firehose. Meanwhile, various applications and agencies need to process an unprecedented amount of spatial data including brain simulation [1], the track of infectious disease [2], climate studies [3], and geo-tagged advertising [4].

As a result of that rise of big spatial data and its applications, several attempts have been made to extend big data systems to support big *spatial* data. This includes HadoopGIS [5], SpatialHadoop [6], $\mathcal{M}D$-HBase [7], Distributed Secondo [8], GeoMesa [9], GeoSpark [10], Simba [11], and ESRI Tools for Hadoop [12]. However, these systems suffer from at least one of these limitations: (1) The lack of the ANSI-standard SQL interface, and (2) performance

© Springer International Publishing AG 2017
M. Gertz et al. (Eds.): SSTD 2017, LNCS 10411, pp. 65–83, 2017.
DOI: 10.1007/978-3-319-64367-0_4

limitations for interactive SQL-like queries due to significant startup time and disk IO overhead.

This paper introduces Sphinx[1], a highly-scalable distributed spatial database with a standard SQL interface. Sphinx is an open source project[2] which extends Impala [14] to efficiently support big spatial data. Impala is a distributed system designed from the ground-up to efficiently answer SQL queries on disk-resident data. It achieves orders of magnitude speedup [14–16] on standard TPC-H and TPC-DS benchmarks as compared to its competitors such as Hive [17] and Spark-SQL [18]. To achieve this impressive performance, Impala is built *from scratch* with several key design points including: (1) ANSI-standard SQL interface, (2) DBMS-like query optimization, (3) C++ runtime code generation, and (4) direct disk access. While Sphinx utilizes this design to achieve orders of magnitude speedup over competitors, it also makes it very challenging due to the fundamental differences over the existing research in this area such as SpatialHadoop [6], GeoSpark [10], and Simba [11].

Even though there has been a body of research in implementing parallel spatial indexes and query processing in traditional DBMS [19,20], this work is not directly applicable in Impala due to three fundamental differences in their underlying architectures. First, Impala relies on the Hadoop Distributed File System (HDFS) which, unlike traditional file systems, only supports sequential file writing and cannot modify existing files. Second, Impala adopts standard formats of big data frameworks, such as CSV, RCFile, and Parquet, which are more suitable to the big data environment where the underlying data should be accessible to other systems, e.g., Hadoop and Spark. This is totally different from DBMS which has exclusive access to its data giving it more space for optimizing the underlying format. Third, the query processing of Impala relies on runtime code generation where the master node generates an optimized code which is then run by the worker nodes. However, traditional DBMS relies on a precompiled fixed code that cannot be changed at runtime. As a collateral result of these differences, Impala misses key features in traditional databases including indexes which is one of the features we introduce in Sphinx.

Sphinx consists of four layers that are implemented inside the core of Impala, namely, *query parser*, *indexing*, *query planner*, and *query executor*. In the *query parser*, Sphinx introduces spatial data types (e.g., Point and Polygon), functions (e.g., Overlap and Touch), and spatial indexing commands. In the *indexing* layer, Sphinx constructs two-layer spatial indexes, based on grid, R-tree and Quad tree, in HDFS. Sphinx also extends the *query planner* by adding new query plans for two spatial operations, *range* and *spatial join* queries. Finally, the *query executor* provides two core components which utilize *runtime code generation* to efficiently execute the *range* and *spatial join* queries on both indexed and non-indexed tables. Extensive experiments on real datasets of up to 2.7 billion points, show that Sphinx achieves up-to three orders of magnitude speedup over traditional

---

[1] The idea of Sphinx was first introduced as a poster here [13].

[2] Project home page is http://www.spatialworx.com/sphinx/ and source code is available at https://github.com/gistic/SpatialImpala.

Impala [14] and PostGIS and three times faster than SpatialHadoop [6]. We believe that Sphinx will open new research directions for implementing more spatial queries using Impala.

The rest of this paper is organized as follows. Section 2 gives a background on Impala to make the paper self contained. Section 3 provides an overview of Sphinx. The details of the *query parser, indexing, query planner*, and *query executor* layers are described in Sects. 4, 5, 6 and 7. Section 8 gives an experimental evaluation of Sphinx using real data. Section 9 reviews related work. Finally, Sect. 10 concludes the paper.

2   Background on Impala

Impala [14,15] is an open-source engine for distributed execution of SQL queries in the Hadoop ecosystem. It achieves real-time response for analytic queries on gigabytes of data and runs much faster than other comparable SQL engines, including Hive [17] and SparkSQL [18]. In this section, we describe the key features of Impala that are needed to understand the contributions of Sphinx.

Similar to other big data engines (e.g., Hadoop and Spark), Impala adopts the Hadoop Distributed File System (HDFS) as its main storage engine. Impala provides an ANSI-standard SQL interface which deals with input files as tables and enforces the schema only as the file is read to reduce the loading time of new data. This *schema-on-read* feature also allows Impala to use a wide range of storage methods including text files in HDFS, HBase tables, RC Files, and Parquet compressed column store. In Impala, users can provide user-defined functions (UDF) and user-defined aggregate functions (UDAF), which Impala integrates with SQL queries.

SQL queries in Impala are executed through the following four steps: (1) The *query parser* decodes the input query and checks for any syntax errors. (2) The *query planner* applies simple optimization rules (e.g., pushing the selection and projection down) to generate a *single-node* query plan as an initial *logical* plan that is never executed. (3) The *query planner* continues its job by converting the initial *logical* plan to a *distributed physical* plan, which has enough details to be executed on cluster nodes. The physical plan consists of *query fragments* where each fragment can run independently on one machine, which minimizes the movement of data across machines. (4) The *query executor* completes the job by running the physical plan on cluster nodes. The query executor is implemented mainly in C++ and uses runtime code generation, which makes it faster and more memory efficient compared to other engines, e.g., Java [15].

3   Architecture

Sphinx is implemented in the core of Impala to provide orders of magnitude speedup with spatial queries while maintaining backward compatibility with non-spatial queries. Figure 1 provides a high level overview of Sphinx which shows the

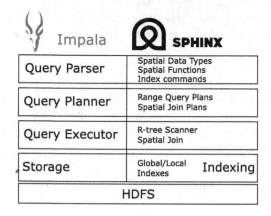

Fig. 1. Overview of Sphinx

four layers that have been extended from Impala, namely, *query parser*, *query planner*, *query executor*, and *storage/indexing* layers, described briefly below.

Query parser. (Sect. 4) Sphinx modifies the *query parser* layer by adding spatial data types (e.g., `Point` and `Polygon`), spatial predicates (e.g., `Overlap` and `Touch`), and spatial functions (e.g., `Intersect` and `Union`). It also adds a new `Create Index` command to construct a spatial index. This *syntactic sugar* makes the system user-friendly, especially, for non-technical users.

Storage. (Sect. 5) In the *storage/indexing* layer, Sphinx employs a two-layer spatial index of one *global index* that partitions the data into blocks and *local indexes* that organize records in each block. Users can build this index using the `Create Index` command in Sphinx or import an existing index from SpatialHadoop. This allows Sphinx to inherit the various indexes [21] supported by SpatialHadoop.

Query planner. (Sect. 6) In the *query planner* layer, Sphinx adds two new query plans for range query and three plans for spatial join. Sphinx automatically chooses the most efficient plan depending on the existence of spatial indexes.

Query executor. (Sect. 7) In the *query executor* layer, the query plan, created by the planner, is physically executed on the worker nodes of the cluster. Sphinx introduces two new components, *R-tree scanner* and *spatial join*, which are both implemented in C++ for efficiency. These new components use runtime code generation to optimize the generated machine code based on query selectivity and the types of constructed indexes, if any.

4   Query Parser

Sphinx modifies the *query parser* of Impala to allow users to express spatial queries in an easy way. In particular, Sphinx adds a new `Geometry` data type that expresses spatial objects, and four sets of spatial functions that manipulate geometry objects. In addition, Sphinx introduces the `CREATE INDEX` command to build spatial indexes. Finally, we extend the `CREATE EXTERNAL TABLE` command to import SpatialHadoop indexes into Sphinx.

4.1   Geometry Data Type

Impala supports only the primitive relational data types, such as numbers, Boolean, and string. In addition, it does not provide a way to add user-defined abstract data types (ADT). Therefore, Sphinx modifies the query parser to introduce the new primitive datatype `Geometry`, which represents shapes of various types including `Point`, `Linestring`, `Polygon`, and `MultiPolygon`, as defined by the Open Geospatial Consortium (OGC). Furthermore, we adopt the Well-Known Text (WKT) format as a default textual representation of shapes for better integration with existing systems, such as PostGIS and Oracle Spatial.

4.2   Spatial Functions

Sphinx adds OGC-compliant spatial functions which are categorized into four groups, namely, *basic functions*, *spatial predicates*, *spatial analysis*, and *spatial aggregates*. Functions in the first three categories are implemented as user-defined functions (UDF), while those in the last one are implemented as user-defined aggregate functions (UDAF). It is imperative to mention that all those functions only work in Sphinx as the input and/or the output of each function is of the `Geometry` datatype, which is supported only in Sphinx.

Basic Spatial Functions which are used to either construct shapes, e.g., `MakePoint`, or retrieve basic information out of them, e.g., `Area`.

Spatial Predicates test the spatial relationship of two `Geometry` objects, such as, `Touch`, `Overlap`, and `Contain`. The return value is always `Boolean`, which allows these functions to be used in the `WHERE` clause of the SQL query.

Spatial Analysis functions are used to manipulate `Geometry` objects, such as `Centroid`, `Intersection`, and `Union`. These functions are usually combined together to perform spatial analysis on a large dataset.

Spatial Aggregate Functions take as input a set of geometries and return one value that summarizes all of the input shapes, e.g., `Envelope` returns the minimum-bounding rectangle (MBR) of a set of objects.

4.3   Spatial Operations

Users can use the above spatial data types and functions to express several spatial operations such as the range query and spatial join operations. The following example show how the range query finds all points in the table P that lie in the rectangle defined by two corner points (x_1, x_2) and (y_1, y_2).

```
SELECT COUNT(*) FROM Points AS P
WHERE Contains(MakeBox(x1, y1, x2, y2), P.coords)
```

The following example uses spatial join to associate points with ZIP code boundaries using the point-in-polygon predicate.

```
SELECT * FROM Points JOIN ZIPCodes
  ON Contains(ZIPCodes.geom, Points.coords)
```

4.4 Spatial Indexing

Sphinx introduces the new CREATE INDEX command which the user can use to build spatial indexes. The following example builds an R-tree index on the the field coordinates in the table Points.

```
CREATE INDEX PointsIndex
  ON Points USING RTREE (coordinates)
```

In addition, we extend the CREATE EXTERNAL TABLE command in Impala to allow users to import existing indexes built using SpatialHadoop [6] as shown in the example below.

```
CREATE EXTERNAL TABLE OSM_Points
(
... /* Columns definitions */
)
INDEXED ON coords AS RTREE
LOCATION('/osm_points')
```

5 Spatial Indexing

In this section, we describe how Sphinx constructs spatial indexes on HDFS-resident tables. The main goal of spatial indexing is to store the records in a spatial-aware manner by grouping nearby records and storing them physically together in the same HDFS block. Sphinx employs the two-layered index previously employed by other major big spatial data systems including Spatial-Hadoop [21] and Simba [11] where one global index partitions records into blocks and several local indexes organize records in each block. Figure 2 shows an example of an R+-tree global index build in Sphinx for a 130 GB of geotagged tweets. Each rectangle in the figure represents a block of 128 MB of records.

This section shows two methods of using indexes in Sphinx. The first method constructs the index within Sphinx while the second method imports imports an existing index constructed by SpatialHadoop. The next two sections show how these indexes are used with two fundamental spatial queries, range query and spatial join.

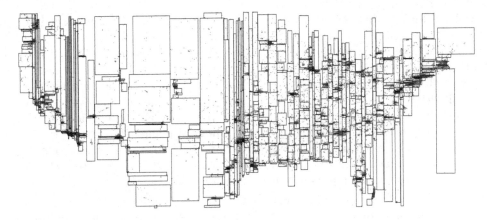

Fig. 2. An R+-tree index in Sphinx built on a table which stores geotagged tweets in the US with a total size of 130 GB

5.1   Index Construction in Sphinx

Figure 3 shows the index construction plan in Sphinx which consists of four steps, *sample*, *divide*, *partition*, and *local index*. This method is similar to the index method used in SpatialHadoop [21] and other systems [11,22]. Implementing the construction plan in Sphinx allows it to be more efficient by supporting the optimized file formats in Impala including RCFile and Parquet. The four steps are described below.

(1) Sample: This step reads a uniform random sample which acts as a summary of the data that can fit on a single machine. Previous work [21] shows that a 1% sample is enough to produce a high quality index. Distributed random sampling is typically implemented by scanning the entire file and selecting each record in the sample with a probability of 1%. However, due to a limitation in Impala, we cannot apply any random functions in the SQL query. To work around this limitation, we apply a filter function that applies a hash function to the record, uses the hash code to initialize a random number generator, and generates one number from that generator. This way, the function becomes deterministic which satisfies the requirements of Impala.

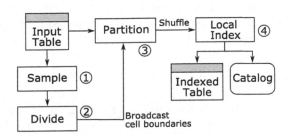

Fig. 3. Indexing plan in Sphinx

(2) Divide: The divide step reads the sample and divides the input space into n partitions such that each partition contains roughly the same number of sample points. The number of partitions n is equal to the number of HDFS blocks in the input table. By default, Sphinx applies the STR-based partitioning algorithm [23] which runs in two passes. In the first pass it sorts all points by x and divides the space into $\lceil\sqrt{n}\rceil$ vertical strips each containing roughly the same number of points. In the second pass, it sorts each strip by y and divides it into $\lceil\sqrt{n}\rceil$ horizontal strips with roughly equal number of points. Previous work has shown that this step can be customized to support a wide range of partitioning techniques, e.g., Quad-tree-based and Hilbert-curve-based [21]. This step returns a set of rectangles that represent the boundaries of the partitions.

(3) Partition: In the partition step, the records of the input table are assigned to the index partitions based on their locations. First, the partition boundaries computed in the divide step are broadcast to all cluster nodes. Then, the input table is scanned again and each record is either assigned to one partition or replicated to all overlapping partitioning, depending on the partitioning technique being used. For example, STR-based partitioning assigns a record to exactly one partition while Quad-tree-based partitioning might replicate a record that overlaps multiple partitions. The records from all the machines are then shuffled based on their partition IDs so that all records in each partition are physically colocated in the same machine.

(4) Local Index: In this last step, each partition is locally indexed and then written to HDFS. Each machine processes the partitions, one-by-one, and for each partition is bulk loads the records in main memory and writes the index to the output as a single file. Since each partition fits in one HDFS block, typically 128 MB, it is feasible to construct and write the index using any traditional technique. After all partitions from all machines are written to HDFS, the *catalog serve* in Impala is updated with the index information which includes the MBR of each partition and the corresponding file name.

5.2   Importing SpatialHadoop Indexes

For a better user experience, Sphinx can import the spatial indexes constructed by SpatialHadoop [21]. This allows it to seamlessly work with various indexes based on Quad-tree, K-d tree and others. In this case, the user issues a **CREATE EXTERNAL TABLE** command, and provides the path to the index in HDFS. In addition to creating the table that maps to these files, Sphinx loads the global index into the catalog server.

6   Query Planner

The *query planner* in Sphinx is responsible of generating the *query plan* for a user query, which is later executed by the *query executor*. Sphinx introduces new query plans for both the *range query* and *spatial join* operations. Since Impala does not support any indexes, Sphinx is the first system that shows how to integrate indexes into the query planner of Impala. This can open new research directions of integrating indexes into more complex queries in Impala.

In general, the *query planner* in Impala runs in two phases, namely, *single node planning* and *plan parallelization and fragmentation*. The *single node planning* phase generates a non-executable *logical* plan where each operator is expressed as a node in the plan tree. In the *plan parallelization and fragmentation*, the *logical* plan is translated into a *physical* distributed plan, which can be executed by the cluster nodes. This entire process takes a fraction of a second and runs on a single machine. To conform with this design, Sphinx only utilizes the global index in the query planning phase.

6.1 Range Query Plans

In range query, users want to retrieve all the records that overlap a given query range. Sphinx applies two query plans, a full table scan if the input table is not indexed, and an R-tree search if it is indexed.

Full table scan. In this plan, Sphinx simply scans the entire input table and tests each record according to the query range. This plan is the only one supported by traditional Impala as it does not require indexes and Sphinx applies it if the table is not indexed on the query column. This plan is a direct translation of the following simple SQL query:

```
SELECT * FROM Points AS P
WHERE Overlap(A, P.x);
```

R-tree search. If the query table is spatially indexed, Sphinx produces this efficient plan that utilizes the index. This plan improves over the full table by employing three new features: (1) *Early pruning*, in which Sphinx utilizes the global index to prune partitions that are completely outside the query range. (2) The *R-tree scanner* replaces the regular scanner in Impala and utilizes the local R-tree indexes to avoid reading records that are outside the query range. (3) The *push down* feature pushes the predicate down to the scanner so that it can utilize the local index. We had to introduce this feature as Impala isolates the query predicate from the scanners. The details of the execution of the *R-tree scanner* are described in Sect. 7.1.

6.2 Spatial Join Plans

The spatial join query finds all overlapping pairs of records in two input tables, R and S. The only supported plan in Impala is a cross (Cartesian) product followed by a filter which is extremely inefficient for big tables. Instead, Sphinx supports three efficient plans depending on the whether the two input tables have two, one, or zero spatial indexes. Notice that in all these plans, only two new operators are introduced in these plans, namely, spatial join and spatial multicast, which are further described below.

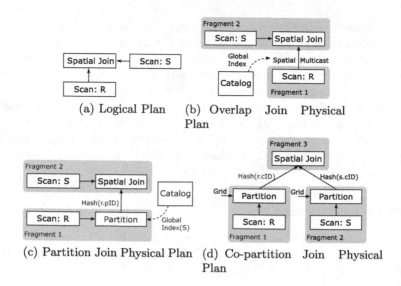

(a) Logical Plan   (b) Overlap Join Physical Plan

(c) Partition Join Physical Plan   (d) Co-partition Join Physical Plan

Fig. 4. Spatial Join Query Plans

(1) Two Indexes (Overlap Join): If the two input tables are indexed on the join columns, Sphinx provides the *overlap join* algorithm as shown in Fig. 4(b). Without loss of generality, we assume that R is the smaller table. The plan consists of two fragments assigned to tables R and S. Each machine in fragment 1 reads one partition of the R index while each machine in fragment 2 reads one partition in the S index. Then, each machine in fragment 1 replicates the entire partition to all machines in fragment 2 with an overlapping partition. This replication happens over network connections created between machines based on the partition MBRs extracted from the two global indexes. After that, each machine in fragment 2 performs a spatial join between each partition received from fragment 1 and the assigned partition from S. Sphinx assigns the larger table to fragment 2 to increase the level of parallelism in the join phase as the larger table contains more partition. In order to implement the above plan, Sphinx introduces two new components, the *spatial multicast* communication pattern, and the *spatial join* operator, described below.

Spatial multicast: Communication 'patterns in Impala control the flow of records from one fragment to another. Sphinx introduces the new *spatial multicast* pattern which can be also helpful to various spatial operations including spatial join. This component first extracts the global indexes for the two tables from the catalog server and uses it to assign an MBR to each partition in the two fragments. Then, it creates a communication link from each partition in fragment 1 to all overlapping partitions in fragment 2.

Spatial join: The *spatial join* operator in Sphinx takes the contents of two partitions and performs a spatial join between the two partitions and generates the answer. The details how the *spatial join* operator is implemented will be described in Sect. 7.

(2) One Index (Partition Join): The *partition join* plan shown in Fig. 4(c) is used when only one table is indexed. This plan resembles the bulk-index-join algorithm [24] and its MapReduce implementation [6]. Without loss of generality, we assume that S is the indexed table. The plan consists of two fragments. Each machine in fragment 1 extracts the global index of S and scans the assigned partition in R. It compares each record $r \in R$ against the partitions of S and outputs the pair $\langle pID, r \rangle$ for each partition pID it overlaps. Then, we use the *hash* communication pattern to send each record r to all overlapping partitions based on the computed attribute pID. Each machine in fragment 2 receives a set of records from R and is assigned a partition in S. It locally computes the spatial join between these two sets of records and output the final answer.

(3) No Indexes (Co-partition Join): If none of the input files is indexed, Sphinx employs the *co-partition join* which is an Impala port of the traditional *partition-based spatial-merge* (PBSM) join algorithm [25]. This plan consists of three fragments. In the query planning phase, Sphinx defines a uniform grid based on the MBR of the two files and the number of machines. The two fragments 1&2 scan record in R and S, respectively, and compares each record to the grid. An output record $\langle cID, r \rangle$ or $\langle cID, s$ is written for each overlap between a record and a grid cell. These hash communication pattern is used to group these records by cID and sent to machines in fragment 3. Each machine in fragment 3 is assigned a grid cell and it locally joins all received records from the two tables. It also applies the reference point technique [26] to detect and remove duplicate answers caused by replication.

7 Query Executor

The *query executor* is the component that executes the physical query plan, created by the *query planner*. Sphinx introduces two new components in the query executor, *R-tree scanner* for range queries, and *spatial join* operator. These components are written in C++ for higher performance and less memory overhead.

7.1 R-tree Scanner

The *R-tree scanner* takes as input one table partition P, typically 128 MB, and a rectangular query range A, and returns all records in the partition that overlaps A. The R-tree scanner is implemented as an input scanner which gives it direct access to the disk. To make this possible in Sphinx, we introduced the *predicate push down* feature which pushes the predicate down to the input scanner. This allows the scanner to skip chunks of disk that do not match the query as shown below. A post processing *duplicate avoidance* step might be needed if the input table is indexed with a replicated index, e.g., R+-tree.

The R-tree scanner has three modes of execution depending on the estimated selectivity of the range query. The estimated selectivity is computed as $\sigma = \frac{Area(A \cap P)}{Area(P)}$, where P and A are the MBRs of the partition and the query range, respectively. Depending on the value of σ the three modes of executions are described below. Notice that unlike traditional databases where σ is computed for the entire table, Sphinx computes the value of σ for each partition allowing mixed modes of executions for the same query.

1. Match All ($\sigma = 1.0$): This case applies when the MBR of the partition is completely contained in the query range. In this case, all records in the partition match the query predicate and all of them can be returned without even testing the query predicate.

2. R-tree Search (P is R-tree indexed and $\sigma < \delta$): This case applies under two conditions: (1) the partition P is locally indexed with an R-tree and (2) the estimated selectivity σ is below a threshold δ. In this case, we apply an on-disk R-tree search which can skip big chunks of disk and quickly get the result. Based on our empirical results, we set δ to a default value of 6%.

3. Full Scan (P is not indexed or $\delta \leq \sigma < 1.0$): This case applies if the partition P is not R-tree indexes or the selectivity σ is above the threshold δ. In this case, even if the partition P is indexed, it would be too much overhead to use the R-tree search algorithm as compared to the little saving of disk IO.

Post Processing - Duplicate Avoidance: If the input table is indexed with a replicated index, e.g., R+-tree, this duplicate avoidance step applies the reference point technique to ensure the answer does not contain any duplicates [26].

7.2   Spatial Join Operator

The *spatial join* operator joins two partitions P_1 and P_2 retrieved from the two input files and returns every pair of overlapping records in the two partitions. Similar to the *R-tree scanner*, this operator runs in two steps, *selection* and *duplicate avoidance*. The *selection* step selects pairs of overlapping records in the two partitions. It has three modes of execution depending on the types of local indexes on the two partitions. Unlike the range query, all partitions in one spatial join query follow the same mode of execution as all partitions in one table have the same type of index.

1. R-tree Join: If both partitions are locally indexed with R-trees, i.e., in the *overlap join* algorithm, this operator applies the synchronized traversal algorithm [27] which concurrently traverses both trees while pruning disjoint tree nodes.

2. Bulk Index Join: If only one partition is indexed, i.e., in the *partition join* algorithm, this operator applies the *bulk index join* algorithm [24] which partitions the non-indexed partition according to the R-tree of the indexed one and then joins each partition with the corresponding R-tree node.

3. Plane-sweep Join: If none of the partitions are indexed, i.e., in the *copartition join* algorithm, we apply the traditional planesweep join algorithm [28].

Post Processing - Duplicate Avoidance: If both partitions are indexed with a replicated index, we apply the reference point [26] duplicate avoidance technique to ensure the answer has not duplicates.

8 Experiments

This section provides an extensive experimental study of the initial prototype of Sphinx. The goal is to show the performance gain of Sphinx geared with the spatial indexes and query execution. We run the experiments on four different systems: (1) our proposed prototype of Sphinx, (2) plain-vanilla Impala without any spatial indexes or spatial operators, (3) SpatialHadoop [6] with its spatial indexes and operators, and (4) PostGIS a popular open-source spatial DBMS. Sphinx is built in the code-base of Impala 2.2.0 and is compared to Impala 2.2.0, SpatialHadoop 2.3, and PostGIS 9.3.0. All cluster experiments are conducted on an Amazon EC2 cluster of up-to 20 nodes of type 'm3.xlarge' with quad-core processor and 15GB of RAM. PostGIS is run on a single machine with 64GB of RAM. We use three real datasets extracted from OpenStreetMap and available at (http://spatialhadoop.cs.umn.edu/datasets.html#osm2). (1) *Nodes* is a set of 2.7 Billion points with a total size of 100 GB. (2) *Squares* is derived from the *nodes* dataset by generating $100\,m^2$ squares centered at each point. (3) *Cities* is a set of 170 K polygons extracted from OpenStreetMap with a total size of 500 MB. In our experiments, we use the end-to-end query processing time to measure the overall performance.

8.1 Index Construction

Figures 5(a) and (b) show the indexing time in PostGIS, SpatialHadoop, and Sphinx. In general, Sphinx is two orders of magnitude faster than PostGIS and three times faster than SpatialHadoop. We also found that Sphinx running on Parquet is much faster than text files. This finding will urge researchers to consider Parquet format which is not yet the default option in Impala.

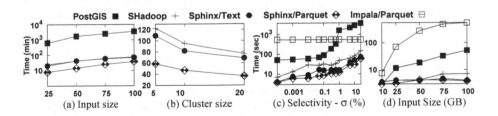

Fig. 5. Index construction and range query performance

Figure 5(a), compares the overall indexing time as the input size increases from 25 to 100 GB. PostGIS is out of the competition with almost two orders of magnitude slowdown as compared to Sphinx on Parquet. SpatialHadoop is on-par with Sphinx on text files due to the similarity of their indexing plans. However, Sphinx on Parquet is three times faster than SpatialHadoop due to the optimizations which are accessible to Sphinx as it is built inside Impala.

Figure 5(b) shows the indexing time for the 100 GB dataset as the cluster size is increased from 5 to 20 machines. By taking PostGIS out and using a regular (non-log) scale, we can see that Sphinx on text files is slightly faster than SpatialHadoop due to the more efficient core of Sphinx on Impala.

8.2   Range Query

Figures 5(c) and (d) give the performance of the range query on the *nodes* dataset on Sphinx, Impala, SpatialHadoop, and PostGIS. For each query, we select a random point p from the input and create a query range A as a square centered around p with an area of $\sigma Area(R)$, where $Area(R)$ is the overall area of the input domain. In these experiments, we use COUNT(*) to return the total number of results rather than the individual records as this is a typical use-case for querying big spatial data. We rely on the query optimizer of PostGIS to decide whether or not to use the spatial index depending on the selectivity. In general, we found that Sphinx is significantly faster than all other techniques with both text files and Parquet with the latter being the fastest.

In Fig. 5(c), we increase the selectivity ratio σ from 0.001% to 25% and measure the query response time. As shown, Sphinx is consistently faster than all other techniques due to the spatial index coupled with the efficient query processing. The R-tree index in PostGIS is only good when the selectivity is low and its performance degrades as the query range increases. The performance of Impala is not affected by the selectivity as it always scans the whole file due to the lack of indexes. While SpatialHadoop uses the same index of Sphinx, the latter is significantly faster even with text files due to the more efficient C++ execution layer in Sphinx.

Figure 5(d) shows the performance of the range query with $\sigma = 0.001\%$ as the input size increases from 10 to 100 GB. While all the techniques scale well with the input size, Sphinx is much faster and the performance gap increases with larger input sizes. It is interesting that the performance of Sphinx and SpatialHadoop remains almost constant as the input size increases as they both utilize the global index to limit the number of processed partitions.

8.3   Spatial Join

Figure 6 gives the performance of the spatial join operation, where we compare the nested loop join in Impala with the overlap join approaches. In Fig. 6(a), we compare the performance of the nested loop join in Impala and the spatial join in PostGIS to the *overlap join* approach in SpatialHadoop and Sphinx. Both Impala and Sphinx run on Parquet files to give their best performance.

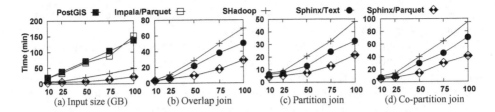

Fig. 6. Spatial Join Performance

In this experiment, the two inputs are equally sized and obtained using sampling from the *squares* dataset. This experiment clearly shows the superiority of Sphinx over Impala and PostGIS in the spatial join query where it achieves up-to an order of magnitude speedup. As with other experiments, Sphinx is also significantly faster than SpatialHadoop.

In the remaining experiments, we rule out both traditional Impala and PostGIS as they do not support the various spatial join algorithms. In addition, we use two larger datasets as input, *cities* is always used as the first datasets, and samples of the **squares** dataset of different sizes are used. We compare the performance of the three spatial join algorithms in Sphinx, running on both HDFS files and Parquet, to the performance of the same algorithm running in Spatial-Hadoop [6,29]. Figures 6(b), (c), and (d) show the performance of the *overlap join*, *partition join* and *co-partition* join, respectively, as described in Sect. 6.2. Sphinx consistently outperforms SpatialHadoop when they both run the same algorithm. Using Parquet in Sphinx gives an additional boost as compared to using text files in HDFS. Overall, Sphinx provides up-to 3x and 1.7x speedup over SpatialHadoop when it runs on text files and Parquet, respectively.

Figure 7(a) gives the scalability experiments of the three join algorithms in Sphinx as the cluster size is increased from 5 to 20 machines. The results indicate that all of the three algorithms scale out nicely on the cluster as they parallelize the work over all of the available processing cores in the cluster. Notice that this experiment is not intended to compare the relative performance of the three

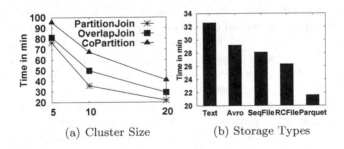

Fig. 7. Spatial join tuning in Sphinx

algorithms as each one runs in a difference input configuration, i.e., two indexes, one index, or no indexes, as described in Sect. 6.2.

Figure 7(b) shows how the storage type affects the performance of the *overlap join* algorithm in Sphinx. In this experiment, we build two indexes on the *squares* and *cities* datasets and then run the overlap join algorithm on them. First, this figure shows the advantage of building Sphinx, which supports a wide range of storage types, as opposed to other systems, e.g., SpatialHadoop, which only supports text files. Second, it shows that a raw text storage gives the worst performance, even though it is the default option in Impala. On the other hand, Parquet provides the best performance and it is the recommended storage by Cloudera. While other workloads might produce a different behavior, a detailed comparison of these storage types is out of the scope of this paper.

9   Related Work

MapReduce [30] and Hadoop were released as powerful alternatives to traditional DBMS to support Big Data. Hadoop was followed by a series of systems that either build on top of it (e.g., Hive [17] and Pig [31]) or use a renovated system design suitable for other application domains, such as Spark [32] for iterative processing and Impala [14] for running interactive SQL queries. Meanwhile, there has been an ongoing research in supporting spatial queries on these systems to process Big Spatial Data. Such research can be classified into three categories. (1) The *on-top* approach uses an existing system as a black box and provides spatial support through user-defined functions. Techniques in this category are easy to implement but they provide a sub-par performance due to the limitation of the underlying system. Yet, the *on-top* approach was used to express a wide range of spatial queries including, R-tree construction [33], range query [34,35], k-nearest neighbor (kNN) [34,36], and spatial join [29,37]. (2) The *from-scratch* approach which is the other extreme where a system is constructed from scratch to support spatial data. While these systems achieve higher performance, they are very complex to build and maintain. Systems in this category include SciDB [38], an array database for scientific applications, and BRACE [39], an in-memory MapReduce engine for behavioral simulations. (3) The *built-in* approach in which an existing system is extended to support spatial data by injecting spatial data types, primitive spatial functions, spatial indexes, and spatial operations in the core of an existing system to transform it from a spatial-agnostic to a spatial-aware system. Techniques in this category reach a balance between complexity and performance. Systems in this category include HadoopGIS [5]; an extended version of Hive [17] for spatial data, SpatialHadoop [6]; a MapReduce framework for spatial data, $\mathcal{MD}$-HBase [7]; an extension to HBase for supporting spatial data, Parallel Secondo [40]; a parallel spatial DBMS, GeoMesa [9]; a key-value store for spatio-temporal datasets built on Accumulo, ESRI Tools for Hadoop [12], which integrates Hadoop with ArcGIS, GeoSpark [10] a Spark-based system for spatial data, and Simba [11] an in-memory spatial analytics framework built in Spark.

Sphinx belongs to the category of *built-in* approach as it extends the core of Impala with spatial data types, functions, indexes, and operations. Sphinx distinguishes itself from other *built-in* systems as: (1) It adopts a standard SQL interface which makes it easy for existing DBMS users to adopt. (2) It is the only system that constructs spatial indexes inside Impala, which provides up to an order of magnitude speedup compared to other SQL-based systems. (3) It is the only system that uses the runtime code generation feature in Impala to produce an optimized machine code running natively on worker nodes. (4) It executes queries in real-time, where the answer is returned within seconds, by avoiding the huge overhead of the Hadoop runtime environment.

10   Conclusion

In this paper, we introduced Sphinx, the first and only system that extends the core of Impala to provide real-time SQL query processing on big spatial data. Sphinx reuses the spatial HDFS indexing introduced by earlier systems like SpatialHadoop. However, it introduces a completely renovated query processing engine based on the efficient design of Impala. Sphinx introduces primitive spatial data types and functions in the *query parser*. In the *storage/indexing* layer, it either links to SpatialHadoop indexes or construct its own spatial indexes. In the *query planner* layer, Sphinx introduces two new query plans for range query and three new query plans for spatial join. In the *query executor* layer, Sphinx adds two new components for *range query* and *spatial join* which are written in C++. Finally, the comprehensive experimental evaluation showed that Sphinx is much faster than plain-vanilla Impala, SpatialHadoop, and PostGIS in all queries.

Acknowledgement. This work is supported in part by the National Science Foundation under Grants IIS-1525953, CNS-1512877, IIS-0952977, and IIS-1218168.

References

1. Markram, H.: The blue brain project. Nat. Rev. Neurosci. **7**(2), 153–160 (2006)
2. Auchincloss, A., et al.: A review of spatial methods in epidemiology: 2000–2010. Annu. Rev. Public Health **33**, 107–122 (2012)
3. Faghmous, J., Kumar, V.: Spatio-temporal data mining for climate data: advances, challenges, and opportunities. In: Chu, W. (ed.) Data Mining and Knowledge Discovery for Big Data. Studies in Big Data, vol. 1, pp. 83–116. Springer, Heidelberg (2014). doi:10.1007/978-3-642-40837-3_3
4. Sankaranarayanan, J., Samet, H., Teitler, B.E., Sperling, M.: TwitterStand: news in tweets. In: SIGSPATIAL (2009)
5. Aji, A., et al.: Hadoop-GIS: a high performance spatial data warehousing system over MapReduce. In: VLDB (2013)
6. Eldawy, A., Mokbel, M.F.: SpatialHadoop: a MapReduce framework for spatial data. In: ICDE (2015)
7. Nishimura, S., et al.: $\mathcal{MD}$-HBase: design and implementation of an elastic data infrastructure for cloud-scale location services. DAPD **31**(2), 289–319 (2013)

8. Nidzwetzki, J.K., Güting, R.H.: Distributed SECONDO: a highly available and scalable system for spatial data processing. In: Claramunt, C., Schneider, M., Wong, R.C.-W., Xiong, L., Loh, W.-K., Shahabi, C., Li, K.-J. (eds.) SSTD 2015. LNCS, vol. 9239, pp. 491–496. Springer, Cham (2015). doi:10.1007/978-3-319-22363-6_28

9. Fox, A., et al.: Spatio-temporal indexing in non-relational distributed databases. In: International Conference on Big Data (2013)

10. Yu, J., et al.: A demonstration of GeoSpark: a cluster computing framework for processing big spatial data. In: ICDE (2016)

11. Xie, D., et al.: Simba: efficient in-memory spatial analytics. In: SIGMOD, San Francisco, CA, June 2016

12. Whitman, R.T., et al.: Spatial indexing and analytics on hadoop. In: SIGSPATIAL (2014)

13. Eldawy, A., et al.: Sphinx: distributed execution of interactive SQL queries on big spatial data (Poster). In: SIGSPATIAL (2015)

14. Kornacker, M., et al.: Impala: A Modern. CIDR, Open-Source SQL Engine for Hadoop (2015)

15. Wanderman-Milne, S., Li, N.: Runtime code generation in cloudera impala. IEEE Data Eng. Bull. **37**(1), 31–37 (2014)

16. Floratou, A., et al.: SQL-on-hadoop: full circle back to shared-nothing database architectures. In: PVLDB (2014)

17. Thusoo, A., et al.: Hive: a warehousing solution over a map-reduce framework. In: PVLDB (2009)

18. Armbrust, M., et al.: Spark SQL: relational data processing in spark. In: SIGMOD (2015)

19. Schnitzer, B., Leutenegger, S.T.: Master-client r-trees: a new parallel r-tree architecture. In: SSDBM (1999)

20. DeWitt, D., Gray, J.: Parallel database systems: the future of high performance database systems. In: CACM (1992)

21. Eldawy, A., Alarabi, L., Mokbel, M.F.: Spatial partitioning techniques in Spatial-Hadoop. In: PVLDB (2015)

22. Yu, J., et al.: GeoSpark: a cluster computing framework for processing large-scale spatial data. In: SIGSPATIAL (2015)

23. Leutenegger, S., et al.: STR: a simple and efficient algorithm for R-tree packing. In: ICDE (1997)

24. den Bercken, J.V., et al.: The bulk index join: a generic approach to processing non-equijoins. In: ICDE (1999)

25. Patel, J., DeWitt, D.: Partition based spatial-merge join. In: SIGMOD (1996)

26. Dittrich, J.P., Seeger, B.: Data redundancy and duplicate detection in spatial join processing. In: ICDE (2000)

27. Brinkhoff, T., Kriegel, H., Seeger, B.: Efficient processing of spatial joins using R-trees. In: SIGMOD, pp. 237–246 (1993)

28. Arge, L., et al.: Scalable sweeping-based spatial join. In: VLDB (1998)

29. Zhang, S., et al.: SJMR: parallelizing spatial join with MapReduce on clusters. In: CLUSTER, pp. 1–8 (2009)

30. Dean, J., Ghemawat, S.: MapReduce: simplified data processing on large clusters. Commun. ACM **51**, 107–113 (2008)

31. Olston, C., et al.: Pig latin: a not-so-foreign language for data processing. In: SIGMOD (2008)

32. Zaharia, M., et al.: Spark: cluster computing with working sets. In: HotCloud (2010)

33. Cary, A., Sun, Z., Hristidis, V., Rishe, N.: Experiences on processing spatial data with MapReduce. In: Winslett, M. (ed.) SSDBM 2009. LNCS, vol. 5566, pp. 302–319. Springer, Heidelberg (2009). doi:10.1007/978-3-642-02279-1_24
34. Zhang, S., et al.: Spatial queries evaluation with MapReduce. In: GCC, pp. 287–292 (2009)
35. Ma, Q., Yang, B., Qian, W., Zhou, A.: Query processing of massive trajectory data based on MapReduce. In: CLOUDDB (2009)
36. Akdogan, A., et al.: Voronoi-based geospatial query processing with MapReduce. In: CLOUDCOM (2010)
37. You, S., et al.: Large-scale spatial join query processing in cloud. In: CLOUDDM (2015)
38. Stonebraker, M., et al.: SciDB: a database management system for applications with complex analytics. Comput. Sci. Eng. $15(3)$, 54–62 (2013)
39. Wang, G., et al.: Behavioral Simulations in MapReduce. In: PVLDB (2010)
40. Lu, J., Guting, R.H.: Parallel secondo: boosting database engines with Hadoop. In: ICPADS (2012)

ST-Hadoop: A MapReduce Framework for Spatio-Temporal Data

Louai Alarabi$^{(\boxtimes)}$, Mohamed F. Mokbel$^{(\boxtimes)}$, and Mashaal Musleh

Department of Computer Science and Engineering,
University of Minnesota, Minneapolis, MN, USA
{louai,mokbel,musle005}@cs.umn.edu

Abstract. This paper presents ST-Hadoop; the first full-fledged open-source MapReduce framework with a native support for spatio-temporal data. ST-Hadoop is a comprehensive extension to Hadoop and Spatial-Hadoop that injects spatio-temporal data awareness inside each of their layers, mainly, language, indexing, and operations layers. In the language layer, ST-Hadoop provides built in spatio-temporal data types and operations. In the indexing layer, ST-Hadoop spatiotemporally loads and divides data across computation nodes in Hadoop Distributed File System in a way that mimics spatio-temporal index structures, which result in achieving orders of magnitude better performance than Hadoop and SpatialHadoop when dealing with spatio-temporal data and queries. In the operations layer, ST-Hadoop shipped with support for two fundamental spatio-temporal queries, namely, spatio-temporal range and join queries. Extensibility of ST-Hadoop allows others to expand features and operations easily using similar approach described in the paper. Extensive experiments conducted on large-scale dataset of size 10 TB that contains over 1 Billion spatio-temporal records, to show that ST-Hadoop achieves orders of magnitude better performance than Hadoop and SpaitalHadoop when dealing with spatio-temporal data and operations. The key idea behind the performance gained in ST-Hadoop is its ability in indexing spatio-temporal data within Hadoop Distributed File System.

1 Introduction

The importance of processing spatio-temporal data has gained much interest in the last few years, especially with the emergence and popularity of applications that create them in large-scale. For example, Taxi trajectory of New York city archive over 1.1 Billion trajectories [1], social network data (e.g., Twitter has over 500 Million new tweets every day) [2], NASA Satellite daily produces 4 TB of data [3,4], and European X-Ray Free-Electron Laser Facility produce large collection of spatio-temporal series at a rate of 40 GB per second, that collectively

This work is partially supported by the National Science Foundation, USA, under Grants IIS-1525953, CNS-1512877, IIS-1218168, and by a scholarship from the College of Computers & Information Systems, Umm Al-Qura University, Makkah, Saudi Arabia.

M. Gertz et al. (Eds.): SSTD 2017, LNCS 10411, pp. 84–104, 2017.
DOI: 10.1007/978-3-319-64367-0_5

Objects = **LOAD** 'points' **AS** (id:int, Location:**POINT**, Time:t);
Result   = **FILTER** Objects **BY**
    Overlaps (Location, **Rectangle**(x1, y1, x2, y2))
    AND t < t2 **AND** t > t1;

(a) Range query in SpatialHadoop

Objects = **LOAD** 'points' **AS** (id:int, **STPoint**:(Location,Time));
Result   = **FILTER** Objects **BY**
    Overlaps (**STPoint**, **Rectangle**(x1, y1, x2, y2), **Interval** (t1, t2));

(b) Range query in ST-Hadoop

Fig. 1. Range query in SpatialHadoop vs. ST-Hadoop

form 50 PB of data yearly [5]. Beside the huge achieved volume of the data, space and time are two fundamental characteristics that raise the demand for processing spatio-temporal data.

The current efforts to process big spatio-temporal data on MapReduce environment either use: (a) *General purpose* distributed frameworks such as Hadoop [6] or Spark [7], or (b) *Big spatial data systems* such as ESRI tools on Hadoop [8], Parallel-Secondo [9], $\mathcal{MD}$-HBase [10], Hadoop-GIS [11], GeoTrellis [12], GeoSpark [13], or SpatialHadoop [14]. The former has been acceptable for typical analysis tasks as they organize data as non-indexed heap files. However, using these systems as-is will result in sub-performance for spatio-temporal applications that need indexing [15–17]. The latter reveal their inefficiency for supporting time-varying of spatial objects because their indexes are mainly geared toward processing spatial queries, e.g., SHAHED system [18] is built on top of SpatialHadoop [14].

Even though existing big spatial systems are efficient for spatial operations, nonetheless, they suffer when they are processing spatio-temporal queries, e.g., *find geo-tagged news in California area during the last three months*. Adopting any big spatial systems to execute common types of spatio-temporal queries, e.g., *range query*, will suffer from the following: (1) The spatial index is still ill-suited to efficiently support time-varying of spatial objects, mainly because the index are geared toward supporting spatial queries, in which result in scanning through irrelevant data to the query answer. (2) The system internal is unaware of the spatio-temporal properties of the objects, especially when they are routinely achieved in large-scale. Such aspect enforces the spatial index to be reconstructed from scratch with every batch update to accommodate new data, and thus the space division of regions in the spatial-index will be jammed, in which require more processing time for spatio-temporal queries. One possible way to recognize spatio-temporal data is to add one more dimension to the spatial index. Yet, such choice is incapable of accommodating new batch update without reconstruction.

This paper introduces ST-Hadoop; the first full-fledged open-source MapReduce framework with a native support for spatio-temporal data, available to download from [19]. ST-Hadoop is a comprehensive extension to Hadoop and

SpatialHadoop that injects spatio-temporal data awareness inside each of their layers, mainly, indexing, operations, and language layers. ST-Hadoop is compatible with SpatialHadoop and Hadoop, where programs are coded as *map* and *reduce* functions. However, running a program that deals with spatio-temporal data using ST-Hadoop will have orders of magnitude better performance than Hadoop and SpatialHadoop. Figures 1(a) and (b) show how to express a spatio-temporal range query in SpatialHadoop and ST-Hadoop, respectively. The query finds all points within a certain rectangular area represented by two corner points $\langle x1, y1 \rangle, \langle x2, y2 \rangle$, and a within a time interval $\langle t1, t2 \rangle$. Running this query on a dataset of 10 TB and a cluster of 24 nodes takes 200 s on SpatialHadoop as opposed to only one second on ST-Hadoop. The main reason of the sub-performance of SpatialHadoop is that it needs to scan all the entries in its spatial index that overlap with the spatial predicate, and then check the temporal predicate of each entry individually. Meanwhile, ST-Hadoop exploits its built-in spatio-temporal index to only retrieve the data entries that overlap with *both* the spatial and temporal predicates, and hence achieves two orders of magnitude improvement over SpatialHadoop.

ST-Hadoop is a comprehensive extension of Hadoop that injects spatio-temporal awareness inside each layers of SpatialHadoop, mainly, *language, indexing, MapReduce*, and *operations* layers. In the *language* layer, ST-Hadoop extends Pigeon language [20] to supports spatio-temporal data types and operations. The *indexing* layer, ST-Hadoop spatiotemporally loads and divides data across computation nodes in the Hadoop distributed file system. In this layer ST-Hadoop scans a random sample obtained from the whole dataset, bulk loads its spatio-temporal index in-memory, and then uses the spatio-temporal boundaries of its index structure to assign data records with its overlap partitions. ST-Hadoop sacrifices storage to achieve more efficient performance in supporting spatio-temporal operations, by replicating its index into *temporal hierarchy index structure* that consists of two-layer indexing of temporal and then spatial. The *MapReduce* layer introduces two new components of *SpatioTemporal-FileSplitter*, and *SpatioTemporalRecordReader*, that exploit the spatio-temporal index structures to speed up spatio-temporal operations. Finally, the *operations* layer encapsulates the spatio-temporal operations that take advantage of the ST-Hadoop *temporal hierarchy index structure* in the indexing layer, such as spatio-temporal range and join queries.

The key idea behind the performance gain of ST-Hadoop is its ability to load the data in Hadoop Distributed File System (HDFS) in a way that mimics spatio-temporal index structures. Hence, incoming spatio-temporal queries can have minimal data access to retrieve the query answer. ST-Hadoop is shipped with support for two fundamental spatio-temporal queries, namely, spatio-temporal range and join queries. However, ST-Hadoop is extensible to support a myriad of other spatio-temporal operations. We envision that ST-Hadoop will act as a research vehicle where developers, practitioners, and researchers worldwide, can either use it directly or enrich the system by contributing their operations and analysis techniques.

The rest of this paper is organized as follows: Sect. 2 highlights related work. Section 3 gives the architecture of ST-Hadoop. Details of the *language, spatio-temporal indexing,* and *operations* are given in Sects. 4, 5 and 6, followed by extensive experiments conducted in Sect. 7. Section 8 concludes the paper.

2 Related Work

Triggered by the needs to process large-scale spatio-temporal data, there is an increasing recent interest in using Hadoop to support spatio-temporal operations. The existing work in this area can be classified and described briefly as following:

On-Top of MapReduce Framework. Existing work in this category has mainly focused on addressing a specific spatio-temporal operation. The idea is to develop map and reduce functions for the required operation, which will be executed on-top of existing Hadoop cluster. Examples of these operations includes spatio-temporal range query [15–17], spatio-temporal join [21–23]. However, using Hadoop as-is results in a poor performance for spatio-temporal applications that need indexing.

Ad-hoc on Big Spatial System. Several big spatial systems in this category are still ill-suited to perform spatio-temporal operations, mainly because their indexes are only geared toward processing spatial operations, and their internals are unaware of the spatio-temporal data properties [8–11,13,14,24–27]. For example, SHAHED runs spatio-temporal operations as an ad-hoc using Spatial-Hadoop [14].

Spatio-Temporal System. Existing works in this category has mainly focused on combining the three spatio-temporal dimensions (i.e., x, y, and time) into a single-dimensional lexicographic key. For example, GeoMesa [28] and GeoWave [29] both are built upon Accumulo platform [30] and implemented a space filling curve to combine the three dimensions of geometry and time. Yet, these systems do not attempt to enhance the spatial locality of data; instead they rely on time load balancing inherited by Accumulo. Hence, they will have a sup-performance for spatio-temporal operations on highly skewed data.

ST-Hadoop is designed as a generic MapReduce system to support spatio-temporal queries, and assist developers in implementing a wide selection of spatio-temporal operations. In particular, ST-Hadoop leverages the design of Hadoop and SpatialHadoop to loads and partitions data records according to their time and spatial dimension across computations nodes, which allow the parallelism of processing spatio-temporal queries when accessing its index. In this paper, we present two case study of operations that utilize the ST-Hadoop indexing, namely, spatio-temporal range and join queries. ST-Hadoop operations achieve two or more orders of magnitude better performance, mainly because ST-Hadoop is sufficiently aware of both temporal and spatial locality of data records.

3   ST-Hadoop Architecture

Figure 2 gives the high level architecture of our ST-Hadoop system; as the first full-fledged open-source MapReduce framework with a built-in support for spatio-temporal data. ST-Hadoop cluster contains one master node that breaks a map-reduce job into smaller tasks, carried out by slave nodes. Three types of users interact with ST-Hadoop: (1) *Casual users* who access ST-Hadoop through its spatio-temporal language to process their datasets. (2) *Developers*, who have a deeper understanding of the system internals and can implement new spatio-temporal operations, and (3) *Administrators*, who can tune up the system through adjusting system parameters in the configuration files provided with the ST-Hadoop installation. ST-Hadoop adopts a layered design of four main layers, namely, *language*, *Indexing*, *MapReduce*, and *operations* layers, described briefly below:

Language Layer: This layer extends Pigeon language [20] to supports spatio-temporal data types (i.e., STPOINT, TIME and INTERVAL) and spatio-temporal operations (e.g., OVERLAP, and JOIN). Details are given in Sect. 4.

Indexing Layer: ST-Hadoop spatiotemporally loads and partitions data across computation nodes. In this layer ST-Hadoop scans a random sample obtained from the input dataset, bulk-loads its spatio-temporal index that consists of two-layer indexing of temporal and then spatial. Finally ST-Hadoop replicates its index into *temporal hierarchy index structure* to achieve more efficient

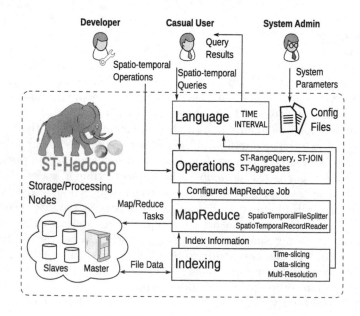

Fig. 2. ST-Hadoop system architecture

performance for processing spatio-temporal queries. Details of the index layer are given in Sect. 5.

MapReduce Layer: In this layer, new implementations added inside Spatial-Hadoop MapReduce layer to enables ST-Hadoop to exploits its spatio-temporal indexes and realizes spatio-temporal predicates. We are not going to discuss this layer any further, mainly because few changes were made to inject time awareness in this layer. The implementation of MapReduce layer was already discussed in great details [14].

Operations Layer: This layer encapsulates the implementation of two common spatio-temporal operations, namely, spatio-temporal range, and spatio-temporal join queries. More operations can be added to this layer by ST-Hadoop *developers*. Details of the operations layer are discussed in Sect. 6.

4 Language Layer

ST-Hadoop does not provide a completely new language. Instead, it extends Pigeon language [20] by adding spatio-temporal data types, functions, and operations. Spatio-temporal data types (STPoint, Time and Interval) are used to define the schema of input files upon their loading process. In particular, ST-Hadoop adds the following:

Data types. ST-Hadoop extends `STPoint`, `TIME`, and `INTERVAL`. The `TIME` instance is used to identify the temporal dimension of the data, while the time `INTERVAL` mainly provided to equip the query predicates. The following code snippet loads NYC taxi trajectories from 'NYC' file with a column of type `STPoint`.

```
trajectory = LOAD 'NYC' as
(id:int, STPoint(loc:point, time:timestamp));
```

NYC and `trajectory` are the paths to the non-indexed heap file and the destination indexed file, respectively. `loc` and `time` are the columns that specify both spatial and temporal attributes.

Functions and Operations. Pigeon already equipped with several basic spatial predicates. ST-Hadoop changes the `overlap` function to support spatio-temporal operations. The other predicates and their possible variation for supporting spatio-temporal data are discussed in great details in [31]. ST-Hadoop encapsulates the implementation of two commonly used spatio-temporal operations, i.e., range and Join queries, that take the advantages of the spatio-temporal index. The following example *"retrieves all cars in State Fair area represented by its minimum boundary rectangle during the time interval of August 25th and September 6th"* from trajectory indexed file.

```
cars = FILTER trajectory
 BY overlap( STPoint,
    RECTANGLE(x1,y1,x2,y2),
    INTERVAL(08-25-2016, 09-6-2016));
```

ST-Hadoop extended the JOIN to take two spatio-temporal indexes as an input. The processing of the join invokes the corresponding spatio-temporal procedure. For example, one might need to understand the relationship between the birds death and the existence of humans around them, which can be described as *"find every pairs from birds and human trajectories that are close to each other within a distance of 1 mile during the last year"*.

```
human_bird_pairs = JOIN human_trajectory, bird_trajectory
  PREDICATE =  overlap( RECTANGLE(x1,y1,x2,y2),
               INTERVAL(01-01-2016, 12-31-2016),
               WITHIN_DISTANCE(1) );
```

5   Indexing Layer

Input files in Hadoop Distributed File System (HDFS) are organized as a heap structure, where the input is partitioned into chunks, each of size 64 MB. Given a file, the first 64 MB is loaded to one partition, then the second 64 MB is loaded in a second partition, and so on. While that was acceptable for typical Hadoop applications (e.g., analysis tasks), it will not support spatio-temporal applications where there is always a need to filter input data with spatial and temporal predicates. Meanwhile, spatially indexed HDFSs, as in SpatialHadoop [14] and ScalaGiST [27], are geared towards queries with spatial predicates only. This means that a temporal query to these systems will need to scan the whole dataset. Also, a spatio-temporal query with a small temporal predicate may end up scanning large amounts of data. For example, consider an input file that includes all social media contents in the whole world for the last five years or so. A query that asks about contents in the USA in a certain hour may end up in scanning all the five years contents of USA to find out the answer.

ST-Hadoop HDFS organizes input files as spatio-temporal partitions that satisfy one main goal of supporting spatio-temporal queries. ST-Hadoop imposes temporal slicing, where input files are spatiotemporally loaded into intervals of a specific time granularity, e.g., days, weeks, or months. Each granularity is represented as a level in ST-Hadoop index. Data records in each level are spatiotemporally partitioned, such that the boundary of a partition is defined by a spatial region and time interval.

Figures 3(a) and (b) show the HDFS organization in SpatialHadoop and ST-Hadoop frameworks, respectively. Rectangular shapes represent boundaries of the HDFS partitions within their framework, where each partition maintains a 64 MB of nearby objects. The dotted square is an example of a spatio-temporal range query. For simplicity, let's consider a one year of spatio-temporal records loaded to both frameworks. As shown in Fig. 3(a), SpatialHadoop is unaware of the temporal locality of the data, and thus, all records will be loaded once and partitioned according to their existence in the space. Meanwhile in Fig. 3(b), ST-Hadoop loads and partitions data records for each day of the year individually, such that each partition maintains a 64 MB of objects that are close to each other

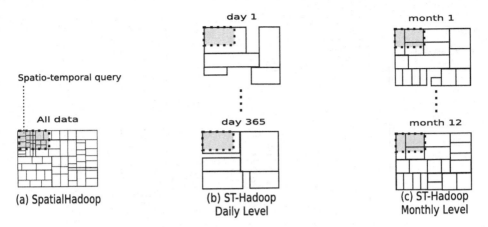

Fig. 3. HDFSs in ST-Hadoop vs. SpatialHadoop

in both space and time. Note that HDFS partitions in both frameworks vary in their boundaries, mainly because spatial and temporal locality of objects are not the same over time. Let's assume the spatio-temporal query in the dotted square *"find objects in a certain spatial region during a specific month"* in Figs. 3(a), and (b). SpatialHadoop needs to access all partitions overlapped with query region, and hence SpatialHadoop is required to scan one year of records to get the final answer. In the meantime, ST-Hadoop reports the query answer by accessing few partitions from its daily level without the need to scan a huge number of records.

5.1 Concept of Hierarchy

ST-Hadoop imposes a replication of data to support spatio-temporal queries with different granularities. The data replication is reasonable as the storage in ST-Hadoop cluster is inexpensive, and thus, sacrificing storage to gain more efficient performance is not a drawback. Updates are not a problem with replication, mainly because ST-Hadoop extends MapReduce framework that is essentially designed for batch processing, thereby ST-Hadoop utilizes incremental batch accommodation for new updates.

The key idea behind the performance gain of ST-Hadoop is its ability to load the data in Hadoop Distributed File System (HDFS) in a way that mimics spatio-temporal index structures. To support all spatio-temporal operations including more sophisticated queries over time, ST-Hadoop replicates spatio-temporal data into a *Temporal Hierarchy Index*. Figures 3(b) and (c) depict two levels of days and months in ST-Hadoop index structure. The same data is replicated on both levels, but with different spatio-temporal granularities. For example, a spatio-temporal query asks for objects in one month could be reported from any level in ST-Hadoop index. However, rather than hitting 30 days' partitions from the daily-level, it will be much faster to access less number of partitions by obtaining the answer from one month in the monthly-level.

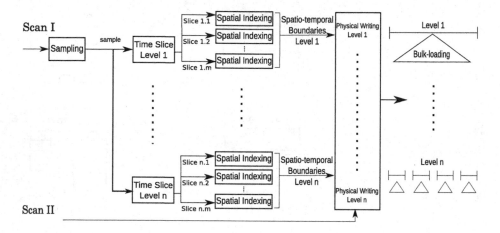

Fig. 4. Indexing in ST-Hadoop

A system parameter can be tuned by ST-Hadoop administrator to choose the number of levels in the *Temporal Hierarchy index*. By default, ST-Hadoop set its index structure to four levels of days, weeks, months and years granularities. However, ST-Hadoop users can easily change the granularity of any level. For example, the following code loads taxi trajectory dataset from "NYC" file using one-hour granularity, Where the `Level` and `Granularity` are two parameters that indicate which level and the desired granularity, respectively.

```
trajectory = LOAD 'NYC' as
            (id:int, STPoint(loc:point, time:timestamp))
            Level:1  Granularity:1-hour;
```

5.2 Index Construction

Figure 4 illustrates the indexing construction in ST-Hadoop, which involves two scanning processes. The first process starts by scanning input files to get a random sample, and this is essential because the size of input files is beyond memory capacity, and thus, ST-Hadoop obtains a set of records to a sample that can fit in memory. Next, ST-Hadoop processes the sample n times, where n is the number of levels in ST-Hadoop index structure. The temporal slicing in each level splits the sample into m number of slice (e.g., $slice_{1.m}$). ST-Hadoop finds the spatio-temporal boundaries by applying a spatial indexing on each temporal slice individually. As a result, outputs from temporal slicing and spatial indexing collectively represent the spatio-temporal boundaries of ST-Hadoop index structure. These boundaries will be stored as meta-data on the master node to guide the next process. The second scanning process physically assigns data records in the input files with its overlapping spatio-temporal boundaries. Note that each record in the dataset will be assigned n times, according to the number of levels.

ST-Hadoop index consists of two-layer indexing of a temporal and spatial. The conceptual visualization of the index is shown in the right of Fig. 4, where lines signify how the temporal index divided the sample into a set of disjoint time intervals, and triangles symbolize the spatial indexing. This two-layer indexing is replicated in all levels, where in each level the sample is partitioned using different granularity. ST-Hadoop trade-off storage to achieve more efficient performance through its index replication. In general, the index creation of a single level in the *Temporal Hierarchy* goes through four consecutive phases, namely sampling, temporal slicing, spatial indexing, and physical writing.

5.3   Phase I: Sampling

The objective of this phase is to approximate the spatial distribution of objects and how that distribution evolves over time, to ensure the quality of indexing; and thus, enhance the query performance. This phase is necessary, mainly because the input files are too large to fit in memory. ST-Hadoop employs a map-reduce job to efficiently read a sample through scanning all data records. We fit the sample into an in-memory simple data structure of a length (L), that is an equal to the number of HDFS blocks, which can be directly calculated from the equation $L = (Z/B)$, where Z is the total size of input files, and B is the HDFS block capacity (e.g., 64 MB). The size of the random sample is set to a default ratio of 1% of input files, with a maximum size that fits in the memory of the master node. This simple data structure represented as a collection of elements; each element consist of a time instance and a space sampling that describe the time interval and the spatial distribution of spatio-temporal objects, respectively. Once the sample is scanned, we sort the sample elements in chronological order to their time instance, and thus the sample approximates the spatio-temporal distribution of input files.

5.4   Phase II: Temporal Slicing

In this phase ST-Hadoop determines the temporal boundaries by slicing the in-memory sample into multiple time intervals, to efficiently support a fast random access to a sequence of objects bounded by the same time interval. ST-Hadoop employs two temporal slicing techniques, where each manipulates the sample according to specific slicing characteristics: (1) *Time-partition*, slices the sample into multiple splits that are uniformly on their time intervals, and (2) *Data-partition* where the sample is sliced to the degree that all sub-splits are uniformly in their data size. The output of this phase finds the temporal boundary of each split, that collectively cover the whole time domain.

The rational reason behind ST-Hadoop two temporal slicing techniques is that for some spatio-temporal archive the data spans a long time-interval such as decades, but their size is moderated compared to other archives that are daily collect terabytes or petabytes of spatio-temporal records. ST-Hadoop proposed the two techniques to slice the time dimension of input files based on either time-partition or data-partition, to improve the indexing quality, and thus gain

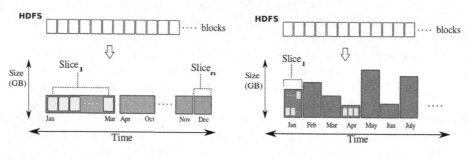

Fig. 5. Data-Slice   **Fig. 6.** Time-Slice

efficient query performance. The time-partition slicing technique serves best in a situation where data records are uniformly distributed in time. Meanwhile, data-partition slicing best suited with data that are sparse in their time dimension.

- *Data-partition Slicing.* The goal of this approach is to slice the sample to the degree that all sub-splits are equally in their size. Figure 5 depicts the key concept of this slicing technique, such that a $slice_1$ and $slice_n$ are equally in size, while they differ in their interval coverage. In particular, the temporal boundary of $slice_1$ spans more time interval than $slice_n$. For example, consider 128 MB as the size of HDFS block and input files of 1 TB. Typically, the data will be loaded into 8 thousand blocks. To load these blocks into ten equally balanced slices, ST-Hadoop first reads a sample, then sort the sample, and apply Data-partition technique that slices data into multiple splits. Each split contains around 800 blocks, which hold roughly a 100 GB of spatio-temporal records. There might be a small variance in size between slices, which is expectable. Similarly, another level in ST-Hadoop temporal hierarchy index could loads the 1 TB into 20 equally balanced slices, where each slice contains around 400 HDFS blocks. ST-Hadoop users are allowed to specify the granularity of data slicing by tuning α parameter. By default four ratios of α is set to 1%, 10%, 25%, and 50% that create the four levels in ST-Hadoop index structure.
- *Time-partition Slicing.* The ultimate goal of this approach is to slices the input files into multiple HDFS chunks with a specified interval. Figure 6 shows the general idea, where ST-Hadoop splits the input files into an interval of one-month granularity. While the time interval of the slices is fixed, the size of data within slices might vary. For example, as shown in Fig. 6 Jan slice has more HDFS blocks than April.

ST-Hadoop users are allowed to specify the granularity of this slicing technique, which specified the time boundaries of all splits. By default, ST-Hadoop finer granularity level is set to one-day. Since the granularity of the slicing is known, then a straightforward solution is to find the minimum and maximum time instance of the sample, and then based on the intervals between the both times ST-Hadoop hashes elements in the sample to the desired granularity.

The number of slices generated by the time-partition technique will highly depend on the intervals between the minimum and the maximum times obtained from the sample. By default, ST-Hadoop set its index structure to four levels of days, weeks, months and years granularities.

5.5   Phase III: Spatial Indexing

This phase ST-Hadoop determines the spatial boundaries of the data records within each temporal slice. ST-Hadoop spatially index each temporal slice independently; such decision handles a case where there is a significant disparity in the spatial distribution between slices, and also to preserve the spatial locality of data records. Using the same sample from the previous phase, ST-Hadoop takes the advantages of applying different types of spatial bulk loading techniques in HDFS that are already implemented in SpatialHadoop such as Grid, R-tree, Quad-tree, and Kd-tree. The output of this phase is the spatio-temporal boundaries of each temporal slice. These boundaries stored as a meta-data in a file on the master node of ST-Hadoop cluster. Each entry in the meta-data represents a partition, such as $<id, MBR, interval, level>$. Where id is a unique identifier number of a partition on the HDFS, MBR is the spatial minimum boundary rectangle, $interval$ is the time boundary, and the level is the number that indicates which level in ST-Hadoop temporal hierarchy index.

5.6   Phase IV: Physical Writing

Given the spatio-temporal boundaries that represent all HDFS partitions, we initiate a map-reduce job that scans through the input files and physically partitions HDFS block, by assign data records to overlapping partitions according to the spatio-temporal boundaries in the meta-data stored on the master node of ST-Hadoop cluster. For each record r assigned to a partition p, the map function writes an intermediate pair $\langle p, r \rangle$ Such pairs are then grouped by p and sent to the reduce function to write the physical partition to the HDFS. Note that for a record r will be assigned n times, depends on the number of levels in ST-Hadoop index.

6   Operations Layer

The combination of the spatiotemporally load balancing with the temporal hierarchy index structure gives the core of ST-Hadoop, that enables the possibility of efficient and practical realization of spatio-temporal operations, and hence provides orders of magnitude better performance over Hadoop and SpatialHadoop. In this section, we only focus on two fundamental spatio-temporal operations, namely, range (Sect. 6.1) and join queries (Sects. 6.2), as case studies of how to exploit the spatio-temporal indexing in ST-Hadoop. Other operations can also be realized following a similar approach.

6.1 Spatio-Temporal Range Query

A range query is specified by two predicates of a spatial area and a temporal interval, A and T, respectively. The query finds a set of records R that overlap with both a region A and a time interval T, such as *"finding geotagged news in California area during the last three months"*. ST-Hadoop employs its spatio-temporal index described in Sect. 5 to provide an efficient algorithm that runs in three steps, *temporal filtering, spatial search,* and *spatio-temporal refinement,* described below.

In the **temporal filtering** step, the hierarchy index is examined to select a subset of partitions that cover the temporal interval T. The main challenge in this step is that the partitions in each granularity cover the whole time and space, which means the query can be answered from any level individually or we can mix and match partitions from different level to cover the query interval T. Depending on which granularities are used to cover T, there is a tradeoff between the number of matched partitions and the amount of processing needed to process each partition. To decide whether a partition P is selected or not, the algorithm computes its *coverage ratio* r, which is defined as the ratio of the time interval of P that overlaps T. A partition is selected only if its *coverage ratio* is above a specific threshold $\mathcal{M}$. To balance this tradeoff, ST-Hadoop employs a top-down approach that starts with the top level and selects partitions that covers query interval T, If the query interval T is not covered at that granularity, then the algorithm continues to the next level. If the bottom level is reached, then all partitions overlap with T will be selected.

In the **spatial search** step, Once the temporal partitions are selected, the *spatial search* step applies the spatial range query against each matched partition to select records that spatially match the query range A. Keep in mind that each partition is spatiotemporally indexed which makes queries run very efficiently. Since these partitions are indexed independently, they can all be processed simultaneously across computation nodes in ST-Hadoop, and thus maximizes the computing utilization of the machines.

Finally in the **spatio-temporal refinement** step, compares individual records returned by the *spatial search* step against the query interval T, to select the exact matching records. This step is required as some of the selected temporal partitions might partially overlap the query interval T and they need to be refined to remove records that are outside T. Similarly, there is a chance that selected partitions might partially overlap with the query area A, and thus records outside the A need to be excluded from the final answer.

6.2 Spatio-Temporal Join

Given two indexed dataset R and S of spatio-temporal records, and a spatio-temporal predicate θ. The join operation retrieves all pairs of records $\langle r, s \rangle$ that are similar to each other based on θ. For example, one might need to understand the relationship between the birds death and the existence of humans around

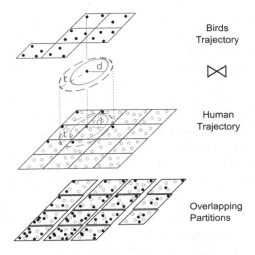

Fig. 7. Spatio-temporal join

them, which can be described as *"find every pairs from bird and human trajectories that are close to each other within a distance of 1 mile during the last week"*. The join algorithm runs in two steps as shown in Fig. 7, *hash* and *join*.

In the **hashing** step, the map function scans the two input files and hashes each record to candidate buckets. The buckets are defined by partitioning the spatio-temporal space using the two-layer indexing of temporal and spatial, respectively. The granularity of the partitioning controls the tradeoff between partitioning overhead and load balance, where a more granular-partitioning increases the replication overhead, but improves the load balance due to the huge number of partitions, while a less granular-partitioning minimizes the replication overhead, but can result in a huge imbalance especially with highly skewed data. The hash function assigns each point in the left dataset, $r \in R$, to all buckets within an Euclidean distance d and temporal distance t, and assigns each point in the right dataset, $s \in S$, to the one bucket which encloses the point s. This ensures that a pair of matching records $\langle r, s \rangle$ are assigned to at least one common bucket. Replication of only one dataset (R) along with the use of single assignment, ensure that the answer contains no replicas.

In the **joining** step, each bucket is assigned to one reducer that performs a traditional in-memory spatio-temporal join of the two assigned sets of records from R and S. We use the plane-sweep algorithm which can be generalized to multidimensional space. The set S is not replicated, as each pair is generated by exactly one reducer, and thus no *duplicate avoidance* step is necessary.

7   Experiments

This section provides an extensive experimental performance study of ST-Hadoop compared to SpatialHadoop and Hadoop. We decided to compare with this two frameworks and not other spatio-temporal DBMSs for two reasons. First, as our contributions are all about spatio-temporal data support in Hadoop. Second, the different architectures of spatio-temporal DBMSs have great influence on their respective performance, which is out of the scope of this paper. Interested readers can refer to a previous study [32] which has been established to compare different large-scale data analysis architectures. In other words, ST-Hadoop is targeted for Hadoop users who would like to process large-scale spatio-temporal data but are not satisfied with its performance. The experiments are designed to show the effect of ST-Hadoop indexing and the overhead imposed by its new features compared to SpatialHadoop. However, ST-Hadoop achieves two orders of magnitude improvement over SpatialHadoop and Hadoop.

Experimental Settings. All experiments are conducted on a dedicated internal cluster of 24 nodes. Each has 64 GB memory, 2 TB storage, and Intel(R) Xeon(R) CPU 3 GHz of 8 core processor. We use Hadoop 2.7.2 running on Java 1.7 and Ubuntu 14.04.5 LTS. Figure 8(b) summarizes the configuration parameters used in our experiments. Default parameters (in parentheses) are used unless mentioned.

Datasets. To test the performance of ST-Hadoop we use the Twitter archived dataset [2]. The dataset collected using the public Twitter API for more than three years, which contains over 1 Billion spatio-temporal records with a total size of 10 TB. To scale out time in our experiments we divided the dataset into different time intervals and sizes, respectively as shown in Fig. 8(a). The default size used is 1 TB which is big enough for our extensive experiments unless mentioned.

In our experiments, we compare the performance of a ST-Hadoop spatio-temporal range and join query proposed in Sect. 6 to their spatial-temporal implementations on-top of SpatialHadoop and Hadoop. For range query, we use system throughput as the performance metric, which indicates the number of MapReduce jobs finished per minute. To calculate the throughput, a batch of 20

Twitter Data	Size	Num-Records	Time window
Large	10TB	> 1 Billion	> 3 years
Average-Large	6.7TB	692 Million	1 years
Medium-Large	3TB	152 Million	9 months
Moderate-Large	(1TB)	115 Million	3 months

(a) Datasets

Parameter	Values (default)
HDFS block capacity (B)	32, 64, (128), 256 MB
Cluster size (N)	5, 10, 15, 20, (23)
Selection ratio (ρ)	(0.01), 0.02, 0.05, 0.1, 0.2, 0.5, 1.0
Data-pratition slicing ratio(α)	0.01, 0.02, 0.025, 0.05, (0.1), 1
Time-partition Slicing granularity(σ)	(days), weeks, months, years

(b) Parameters

Fig. 8. Experimental settings and Dataset

queries is submitted to the system, and the throughput is calculated by dividing 20 by the total time of all queries. The 20 queries are randomly selected with a spatial area ratio of 0.001% and a temporal window of 24 h unless stated. This experimental design ensures that all machines get busy and the cluster stays fully utilized. For spatio-temporal join, we use the processing time of one query as the performance metric as one query is usually enough to keep all machines busy. The experimental results for range and join queries are reported in Sects. 7.1, and 7.3, respectively. Meanwhile, Sect. 7.2 analyzes ST-Hadoop indexing.

7.1 Spatiotemporal Range Query

In Fig. 9(a), we increase the size of input from 1 TB to 10 TB, while measuring the job throughput. ST-Hadoop achieves more than two orders of magnitude higher throughput, due to the temporal load balancing of its spatio-temporal index. As for SpatialHadoop, it needs to scan more partitions, which explain why the throughput of SpatialHadoop decreases with the increase of data records in spatial space. Meanwhile, ST-Hadoop throughput remains stable as it processes only partition(s) that intersect with both space and time. Note that it is always the case that Hadoop needs to scan all HDFS blocks, which gives the worst throughput compared to SpatialHadoop and ST-Hadoop.

Figure 9(b) shows the effect of configuring the HDFS block size on the job throughput. ST-Hadoop manages to keep its performance within orders of magnitude higher throughput even with different block sizes. Extensive experiments are shown in Fig. 9(c), analyzed how slicing ratio (α) can affect the performance of range queries. ST-Hadoop keeps its higher throughput around the default HDFS block size, as it maintains the load balance of data records in its two-layer indexing. As expected expanding the block size from its default value will reduce the performance on SpatialHadoop and ST-Hadoop, mainly because blocks will carry more data records.

Experiments in Fig. 10 examines the performance of the temporal hierarchy index in ST-Hadoop using both slicing techniques. We evaluate different granularities of time-partition slicing (e.g., daily, weekly, and monthly) with various data-partition slicing ratio. In these two figures, we fix the spatial query range

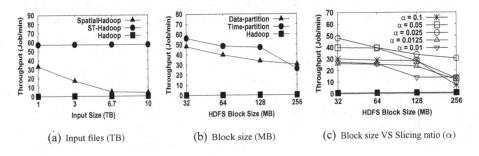

(a) Input files (TB)   (b) Block size (MB)   (c) Block size VS Slicing ratio (α)

Fig. 9. Spatio-temporal range query

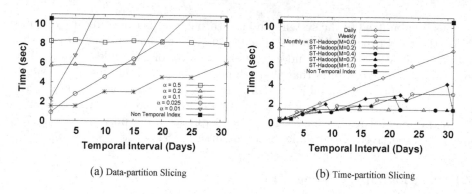

(a) Data-partition Slicing   (b) Time-partition Slicing

Fig. 10. Spatio-temporal range query interval window

and increase the temporal range from 1 day to 31 days, while measuring the total running time. As shown in the Figs. 10(a) and (b), ST-Hadoop utilizes its temporal hierarchy index to achieve the best performance as it mixes and matches the partitions from different levels to minimize the running time, as described in Sect. 6.1. ST-Hadoop provides good performance for both small and large query intervals as it selects partitions from any level. When the query interval is very narrow, it uses only the lowest level (e.g., daily level), but as the query interval expand it starts to process the above level. The value of the parameter $\mathcal{M}$ controls when it starts to process the next level. At $\mathcal{M} = 0$, it always selects the up level, e.g., monthly. If $\mathcal{M}$ increases, it starts to match with lower levels in the hierarchy index to achieve better performance. At the extreme value of $\mathcal{M} = 1$, the algorithm only matches partitions that are completely contained in the query interval, e.g., at 18 days it matches two weeks and four days while at 30 days it matches the whole month. The optimal value in this experiment is $\mathcal{M} = 0.4$ which means it only selects partitions that are at least 40% covered by the query temporal interval.

In Fig. 11 we study the effect of the spatio-temporal query range (σ) on the choice of $\mathcal{M}$. To measure the quality of $\mathcal{M}$, we define an optimal running time for a query Q as the minimum of all running times for all values of $\mathcal{M} \in [0, 1]$. Then, we determine the quality of a specific value of $\mathcal{M}$ on a query workload as the mean squared error (MSE) between the running time at this value of $\mathcal{M}$ and the optimal running time. This means, if a value of $\mathcal{M}$ always provides the optimal value, it will yield a quality measure of zero. As this value increases, it indicates a poor quality as the running times deviates from the optimal. In Fig. 11(a), We repeat the experiment with three values of spatial query ranges $\sigma \in \{1E - 6, 1E - 4, 0.1\}$. As shown in the figure, $\mathcal{M} = 0.4$ provides the best performance for all the experimented spatial ranges. This is expected as $\mathcal{M}$ is only used to select temporal partitions while the spatial range (σ) is used to perform the spatial query inside each of the selected partitions. Figure 11(b), shows the quality measures with a workload of 71 queries with time intervals

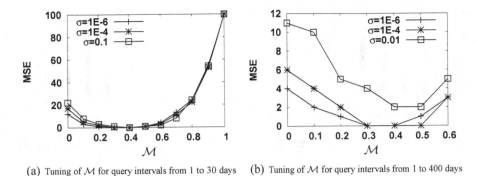

(a) Tuning of $\mathcal{M}$ for query intervals from 1 to 30 days   (b) Tuning of $\mathcal{M}$ for query intervals from 1 to 400 days

Fig. 11. The effect of the spatio-temporal query ranges on the optimal value of $\mathcal{M}$

that range from 1 day to 421 days. This experiment also provides a very similar result where the optimal value of $\mathcal{M}$ is around 0.4.

7.2 Index Construction

Figure 12 gives the total time for building the spatio-temporal index in ST-Hadoop. This is a one time job done for input files. In general, the figure shows excellent scalability of the index creation algorithm, where it builds its index using data-partition slicing for a 1 TB file with more than 115 Million records in less than 15 min. The data-partition technique turns out to be the fastest as it contains fewer slices than time-partition. Meanwhile, the time-partition technique takes more time, mainly because the number of partitions are increased, and thus increases the time in physical writing phase.

In Fig. 13, we configure the temporal hierarchy indexing in ST-Hadoop to construct five levels of the two-layer indexing. The temporal indexing uses Data-partition slicing technique with different slicing ratio α. We evaluate the indexing time of each level individually. Because the input files are sliced into splits according to the slicing ratio, which directly effects on the number of partitions. In general with stretching the slicing ratio, the indexing time decreases, mainly because the number of partitions will be much less. However, note that in some cases the spatial distribution of the slice might produce more partitions as in shown with 0.25% ratio.

7.3 Spatiotemporal Join

Figure 14 gives the results of the spatio-temporal join experiments, where we compare our join algorithm for ST-Hadoop with MapReduce implementation of the spatial hash join algorithm [33]. Typically, in this join algorithm we perform the following query, *"find every pairs that are close within an Euclidean distance of 1mile and a temporal distance of 2 days"*, this join query is executed on both ST-Hadoop and Hadoop and the response times are compared. The y-axis in the

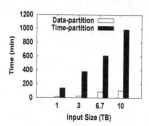

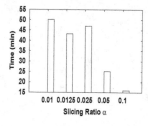

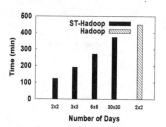

Fig. 12. Input files   **Fig. 13.** Data-partition   **Fig. 14.** Spatio-temporal join

figure represents the total processing time, while the x-axis represents the join query on numbers of days × days in ascending order. With the increase of joining number of days, the performance of ST-Hadoops join increases, because it needs to join more indexes from the temporal hierarchy. In general, ST-Hadoop gives the best results as ST-Hadoop index replicates data in several layers, and thus ST-Hadoop significantly decreases the processing of non-overlapping partitions, as only partitions that overlap with both space and time are considered in the join algorithm. Meanwhile, the same joining algorithm without using ST-Hadoop index gives the worst performance for joining spatio-temporal data, mainly because the algorithm takes into its consideration all data records from one dataset. However, ST-Hadoop only joins the indexes that are within the temporal range, which significantly outperforms the join algorithm with double to triple performance.

8   Conclusion

In this paper, we introduced ST-Hadoop [19] as a novel system that acknowledges the fact that space and time play a crucial role in query processing. ST-Hadoop is an extension of a Hadoop framework that injects spatio-temporal awareness inside SpatialHadoop layers. The key idea behind the performance gain of ST-Hadoop is its ability to load the data in Hadoop Distributed File System (HDFS) in a way that mimics spatio-temporal index structures. Hence, incoming spatio-temporal queries can have minimal data access to retrieve the query answer. ST-Hadoop is shipped with support for two fundamental spatio-temporal queries, namely, spatio-temporal range and join queries. However, ST-Hadoop is extensible to support a myriad of other spatio-temporal operations. We envision that ST-Hadoop will act as a research vehicle where developers, practitioners, and researchers worldwide, can either use directly or enrich the system by contributing their operations and analysis techniques.

References

1. NYC Taxi and Limousine Commission (2017). http://www.nyc.gov/html/tlc/html/about/trip_record_data.shtml
2. (2017). https://about.twitter.com/company
3. Land Process Distributed Active Archive Center, March 2017. https://lpdaac.usgs.gov/about
4. Data from NASA's Missions, Research, and Activities (2017). http://www.nasa.gov/open/data.html
5. European XFEL: The Data Challenge, September 2012. http://www.xfel.eu/news/2012/the_data_challenge
6. Apache. Hadoop. http://hadoop.apache.org/
7. Apache. Spark. http://spark.apache.org/
8. Whitman, R.T., Park, M.B., Ambrose, S.A., Hoel, E.G.: Spatial indexing and analytics on hadoop. In: SIGSPATIAL (2014)
9. Lu, J., Guting, R.H.: Parallel secondo: boosting database engines with hadoop. In: ICPADS (2012)
10. Nishimura, S., Das, S., Agrawal, D., El Abbadi, A.: $\mathcal{MD}$-HBase: design and implementation of an elastic data infrastructure for cloud-scale location services. DAPD **31**, 289–319 (2013)
11. Aji, A., Wang, F., Vo, H., Lee, R., Liu, Q., Zhang, X., Saltz, J.: Hadoop-GIS: a high performance spatial data warehousing system over mapreduce. In: VLDB (2013)
12. Kini, A., Emanuele, R.: Geotrellis: adding geospatial capabilities to spark (2014). http://spark-summit.org/2014/talk/geotrellis-adding-geospatial-capabilities-to-spark
13. Yu, J., Wu, J., Sarwat, M.: GeoSpark: a cluster computing framework for processing large-scale spatial data. In: SIGSPATIAL (2015)
14. Eldawy, A., Mokbel, M.F.: SpatialHadoop: a MapReduce framework for spatial data. In: ICDE (2015)
15. Ma, Q., Yang, B., Qian, W., Zhou, A.: Query processing of massive trajectory data based on MapReduce. In: CLOUDDB (2009)
16. Tan, H., Luo, W., Ni, L.M.: Clost: a hadoop-based storage system for big spatio-temporal data analytics. In: CIKM (2012)
17. Li, Z., Hu, F., Schnase, J.L., Duffy, D.Q., Lee, T., Bowen, M.K., Yang, C.: A spatiotemporal indexing approach for efficient processing of big array-based climate data with mapreduce. Int. J. Geograph. Inf. Sci. IJGIS **31**, 17–35 (2017)
18. Eldawy, A., Mokbel, M.F., Alharthi, S., Alzaidy, A., Tarek, K., Ghani, S.: SHAHED: a MapReduce-based system for querying and visualizing Spatio-temporal satellite data. In: ICDE (2015)
19. ST-Hadoop website. http://st-hadoop.cs.umn.edu/
20. Eldawy, A., Mokbel, M.F.: Pigeon: a spatial mapreduce language. In: ICDE (2014)
21. Han, W., Kim, J., Lee, B.S., Tao, Y., Rantzau, R., Markl, V.: Cost-based predictive spatiotemporal join. TKDE **21**, 220–233 (2009)
22. Al-Naami, K.M., Seker, S.E., Khan, L.: GISQF: an efficient spatial query processing system. In: CLOUDCOM (2014)
23. Fries, S., Boden, B., Stepien, G., Seidl, T.: PHiDJ: parallel similarity self-join for high-dimensional vector data with mapreduce. In: ICDE (2014)
24. Stonebraker, M., Brown, P., Zhang, D., Becla, J.: SciDB: a database management system for applications with complex analytics. Comput. Sci. Eng. **15**, 54–62 (2013)

25. Zhang, X., Ai, J., Wang, Z., Lu, J., Meng, X.: An efficient multi-dimensional index for cloud data management. In: CIKM (2009)
26. Wang, G., Salles, M., Sowell, B., Wang, X., Cao, T., Demers, A., Gehrke, J., White, W.: Behavioral simulations in MapReduce. PVLDB **3**, 952–963 (2010)
27. Lu, P., Chen, G., Ooi, B.C., Vo, H.T., Wu, S.: ScalaGiST: scalable generalized search trees for MapReduce systems. PVLDB **7**, 1797–1808 (2014)
28. Fox, A.D., Eichelberger, C.N., Hughes, J.N., Lyon, S.: Spatio-temporal indexing in non-relational distributed databases. In: BIGDATA (2013)
29. GeoWave. https://ngageoint.github.io/geowave/
30. Accumulo. https://accumulo.apache.org/
31. Erwig, M., Schneider, M.: Spatio-temporal predicates. In: TKDE (2002)
32. Pavlo, A., Paulson, E., Rasin, A., Abadi, D., DeWitt, D., Madden, S., Stonebraker, M.: A comparison of approaches to large-scale data analysis. In: SIGMOD (2009)
33. Lo, M.L., Ravishankar, C.V.: Spatial hash-joins. In: SIGMODR (1996)

GeoWave: Utilizing Distributed Key-Value Stores for Multidimensional Data

Michael A. Whitby[(⊠)], Rich Fecher, and Chris Bennight

DigitalGlobe, Herndon, VA, USA
{Michael.Whitby,Richard.Fecher}@digitalglobe.com

Abstract. To date, it has been difficult for modern geospatial software projects to take advantage of the benefits provided by distributed computing frameworks due to the implicit challenges of spatial and spatiotemporal data. Chief among these issues is preserving locality between multidimensional objects and the single dimensional sort order imposed by key-value stores. We will use the open source framework GeoWave to harness the scalability of various distributed frameworks and integrate them with geospatial queries, analytics, and map rendering. GeoWave performs dimensionality reduction by utilizing space–filling curves to convert n-dimensional data into a single dimension. This ensures that values close in multidimensional space are highly contiguous in the single dimensional keys of the datastore. By using various forms of geospatial data, we show that preserving locality in this way reduces the time needed to query, analyze, and render large amounts of data by multiple orders of magnitude.

1 Introduction

Distributed computing frameworks allow users to parallelize work across many cores to significantly reduce data access and processing time. By using data replication and storage across a cluster, these frameworks provide a powerful and scalable solution for handling massive and complex datasets. While the benefits of these frameworks are numerous, there are also inherent limitations that make them challenging to utilize when working with geospatial data. This paper will discuss these limitations and describe how we implemented efficient and effective solutions to these challenges within the open source framework GeoWave (https://github.com/locationtech/geowave). At its core, GeoWave is a software library that connects the scalability of various distributed computing frameworks and key-value stores with modern geospatial software to store, retrieve, and analyze massive geospatial datasets. GeoWave also contains intuitive command line tools, web services, and package installers. These features are designed to help minimize the learning curve for geospatial developers looking to utilize distributed computing, and for Big Data and distributed computing users looking to harness the power of Geographic Information Systems (GIS). The challenges we discuss in this paper are:

- Storing multidimensional data in a single dimensional key-value store leading to locality degradation
- Limiting the effectiveness of a distributed system due to downstream bottlenecks

© Springer International Publishing AG 2017
M. Gertz et al. (Eds.): SSTD 2017, LNCS 10411, pp. 105–122, 2017.
DOI: 10.1007/978-3-319-64367-0_6

– Working with multiple data types in a single table
– Creating GIS tools that work with multiple distributed key-value stores.

The greatest challenge in integrating geospatial systems into distributed clusters is the inability to preserve locality in multidimensional data types. Often when working with geospatial data, an analyst or developer does not want or need to use the entirety of a multi-billion point dataset for a spatially or temporally localized use case. It is beneficial to search only for qualified criteria within a bounding box around a city instead of searching a table that includes data of an entire country. Likewise, consider a user who wants to perform a k-means clustering analysis of data within a single city during the last month of a global, multi-year dataset. This analysis is considerably more efficient if the data is stored in a common, indexed location within the datastore. Unfortunately, distributed frameworks often store data in single dimensional keys that are then stored in the cluster using random hash functions [6, 14]. This feature, common to most distributed key-value stores, normally makes the previously mentioned use cases extremely challenging at scale. Naively, the entire dataset will need to be searched regardless of whether or not each data point is applicable for the intended use. Since geospatial data is naturally multidimensional, storing it primarily in a single dimension necessitates a tradeoff in the locality of every other dimension.

We solve this challenge and preserve locality in the GeoWave framework by using Space Filling Curves (SFCs) to decompose our bounding hyperrectangles into a series of single dimensional ranges. An SFC is a continuous, surjective mapping from R to R^d, where d is the number of dimensions being filled [9]. The complexities of the decomposition process are further discussed in Sect. 3.1 and related work specific to SFCs is part of Sect. 2.

A properly indexed distributed cluster can reduce the query and analytic times by multiple orders of magnitude, but these efficiency gains may still be masked by inefficiencies further downstream. One major bottleneck will be reached when attempting to render data in the context of a typical Web Map Service (WMS) request. Even a powerful rendering engine will cause serious delays when attempting to render billions of data points on a map. To display a WMS layer, a traditional rendering engine will first need to query and download all of the available data from the datastore. For smaller datasets this may not be a significant issue; however this will quickly become an unacceptable performance bottleneck when querying the massive datasets GeoWave was designed to work with. The fundamental issue here is the need to process all the data in order to visualize it. In this case that requires reading the data off of disk, but the same issue and scaling would take place if the data was stored in memory (system, GPU texture, etc.) – the constant factors and costs would simply change. In most cases this visualization of data ceases to convey additional information past a certain saturation point. A single pixel can only represent a finite amount of information at one time, so a high-level view of a WMS layer may have pixels representing thousands of individual data points. In this situation, there is a significant resource allocation directed at retrieving data that an end user will never actually see. We are able to make this process much more efficient by using spatial subsampling to identify and retrieve only the amount of data that will be visible in the layer. At a high level, GeoWave transforms the pixel space on the map into the context of the underlying SFC that is used to store the data. GeoWave then requests only the first data point that would be associated with that pixel. An in-depth description of this process is given in Sect. 3.2 of this paper. This approach turns what was a choke point into an interactive and responsive experience, and can be enabled by adding a simple pre-defined map stylization rule for a layer.

With continued advancements in data science practices it is often the case that a single type or source of data is not enough to answer a question or prove a hypothesis. The amount and variety of available data is growing at an ever-increasing rate. This should be to the benefit of an analyst. However, this data is often stored in different data structures and tables, adding a further level of complexity to its analysis. It may be beneficial to have these heterogeneous datasets in a commonly accessible location. Furthermore, for processing efficiency, it can be beneficial to have them indexed together and in fact, commingled in the same tables to prevent unnecessary client-server roundtrips when correlating these hybrid datasets. Mixing various data types may seem trivial but in spatiotemporal use cases, careful attention must be placed in storing attributes of the data that contribute to the SFC index in a common format and location. This will be discussed in more detail, but of particular note is that the SFC is intentionally an overly inclusive representation of the spatiotemporal bounds of the data. Careful attention is paid to eliminate false negatives, but false positives are possible using the SFC alone and it is only by explicitly reading and testing the actual attribute values that all false positives can be eliminated. The software design challenge is doing a fine-grained spatiotemporal evaluation on the server-side with many different types of data within the scope of the same scan. Regardless of the individual data type, an agnostic common model is used to store the indexed data. This common index model, along with the ability to add extended fields with a serialized adapter, allows the software to seamlessly interact with all varieties of data in a spatiotemporal context, as well as parse the extended information for each individual type. We will expand on this process and its architecture in Sect. 3.3.

As is the case with many modern technologies, there is not a single standard distributed computing framework that is used ubiquitously. While many function similarly, there are numerous options and in practice, there are generally performance tradeoffs that can make a certain framework better for a given task. A particular development team's experience and preference is another contributing factor – the significant differences in implementation details between these technologies means that a sizable investment in time and resources may be necessary up front. A development team will need to become familiar with a new distributed framework before they can begin to benefit from a library designed to work solely with that framework. Therefore, GeoWave has been designed to be key-value store agnostic. Functionality and performance parity between supported key-value stores is achieved by drawing on as many commonalities to keep the functionality as generic and abstract as possible. This happens, while also paying careful attention to microbenchmarks in the cases that tradeoffs and underlying expected behavior differs. While the full breadth of this task is beyond the scope of this paper, however, some relevant examples are discussed in Sects. 3.4 and 4.3.

2 Related Work

The concept of multidimensional indexing, or hashing, has been well researched and there is a significant amount of related work in this field. Efficient multidimensional hashing algorithms, as well as the limits to n-dimensional locality preservation, have been proposed by authors such as Indyk [11]. More recently, projects like Spatial Hadoop [7] have been developed to use grid files and R-trees to index two dimensional data into the Hadoop ecosystem. We have chosen to use SFCs for the multidimensional indexing component within our GeoWave project. SFCs have been discussed and studied frequently for many years. The research that has most directly impacted our design is from the work of Haverkort [10] in measuring the effectiveness of various SFCs and the introduction of the Compact Hilbert SFC by Hamilton [9].

In the Haverkort paper [10], a series of measures (Worst-case Dilation/Locality [WL], Worst-case Bounding Box Area Ratio [WBA], and Average Total Bounding Box Area [ABA]) were devised to test the effectiveness of various types of SFCs. The results from these tests are summarized in Table 1.

Table 1. Summary of results from Haverkort comparison [10]

	Z-Order	Hilbert	H-Order	Peano	AR^2W^2	$\beta\Omega$
WL_∞	∞	6.00	4.00	8.00	5.40	5.00
WL_2	∞	6.00	4.00	8.00	6.04	5.00
WL_1	∞	9.00	8.00	10.66	12.00	9.00
WBA	∞	2.40	3.00	2.00	3.05	2.22
ABA	2.86	1.41	1.69	1.42	1.47	1.40

The Z-Order curve has been successfully used for multidimensional indexing in products like MD-HBase [13], but was by far the worst in these experiments. The $\beta\Omega$ and Hilbert curves performed extremely well across all of these locality-based measures, but there are drawbacks to them as well. A major limitation of these SFCs is that the cardinality, or grid size, must be the same in each dimension. This can pose a significant issue in the ability to use the curve in spatiotemporal applications as well as when we consider other common dimensions such as elevation. These use cases can benefit from dissimilar cardinalities among dimensions to enable trading off dimensional precision when common access patterns are likewise unbalanced in dimensional precision. Hamilton [9] solves this issue by defining a new version of the Hilbert SFC called the Compact Hilbert SFC. The Compact Hilbert SFC maintains all of the locality preservation advantages of a standard Hilbert SFC while allowing for varying sized cardinalities among dimensions. Furthermore, we were able to utilize an open source Compact Hilbert SFC implementation called Uzaygezen [15]. Contributions to Uzaygezen were made by members of Google's Dublin Engineering Team and the word itself is Turkish for "space wanderer."

3   Contributions

3.1   Locality Preservation of Multi-dimensional Values

In the introduction, we briefly discussed that the GeoWave framework uses SFCs to decompose bounding hyperrectangles into ranges in a single dimension. Here we will go into further detail about how this is accomplished and how it benefits us in terms of locality preservation. For simplicity, we will use a two-dimensional example with a uniform cardinality to example to explain this process; however, GeoWave is capable of performing this process over any number of dimensions with mixed cardinalities.

GeoWave uses recursive decomposition to build what are effectively quadtrees when applied in two-dimensions. A quadtree is a data index structure in which two-dimensional data is recursively broken down into four children per region [12]. These trees are built over a gridded index in which each elbow (discrete point) in the Hilbert SFC maps to a cell in the grid. These cells provide a numbering system that gives us the ability to refer to each individual grid location, or bucket, with a single value. The curves are composable and nested, giving us the ability to store curves and grids at various levels of precision where the number of bits of the Hilbert SFC determines the grid. Although our "quadtrees" lack the traditional direct parent-child relationships between progressive levels, this tiered method in which we store each level has a similar effect (Fig. 1).

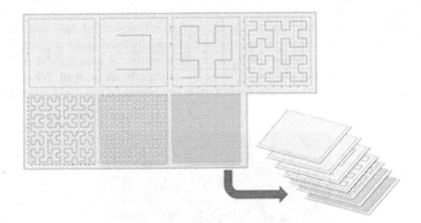

Fig. 1. Visualization of a tiered, hierarchical SFC storage system

Now that we have a tiered, hierarchical, multi-resolution grid instance, we can use it to decompose a single query into a highly optimized set of query ranges. As a simple example, we step though the query process with a two dimensional fourth order Hilbert SFC that corresponds to a 16 by 16 cell grid ($2^4 = 16$). Laid out over a map of the world that grid would have a resolution of 11.25° latitude and 22.5° longitude per cell. Once a query bounding box has been selected, the first step is to find the minimally continuous quadrant bounding the selection window. In this example, a user has selected an area bounded by the range of (3,13) -> (6,14) on the 16 by 16 grid. The Hilbert value at each of the four corners would be:

- (3,13) = 92
- (3,14) = 91
- (6,13) = 109
- (6,14) = 104

We then take the minimum and maximum Hilbert values, as they are the most likely to be in separate quadtree regions, and perform an exclusive or (XOR) operation to find the first differing region. To be precise, we take the Hilbert values as base 2 and use bitwise operations to define x as the number of bits set from the right (front padding a 0 if the result is odd). We find the starting and ending intervals of the minimally continuous quadrant bounding the selection window by applying a 1, followed by x number of 0 s as the lower bound and a 1 followed by x number of 1 s as the upper bound. In the example above, we have 91 and 109 as our minimum and maximum Hilbert values. It follows that (01011011 XOR 01101101) = 00110110, which has six bits, so therefore, the bounds for our minimally continuous quadrant are: 01000000 = 64 and 01111111 = 127.

The previous step takes us to the deepest node of the quadtree that fully contains the queried range. Now, we recursively decompose each region into four sub-regions and check to see if the sub-region fully describes its overlapping portion of the original query. In the example shown in Figs. 2 and 3, the darkest shaded region represents the query range. We first broke down the minimally continuous quadrant into four sub-regions, then broke down each of those into four sub-regions of their own. At this point, we only continue decomposing regions that partially overlap the original search area. Sections that fully overlap the bounding box are left at their current tier, and sections that do not overlap any portion of the bounding box are skipped. We continue to decompose the area until we have broken the query bounding box down into a series

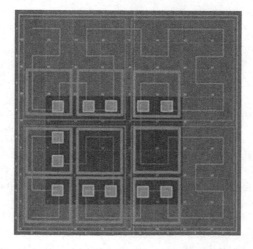

Fig. 2. A query is first decomposed into sub-regions

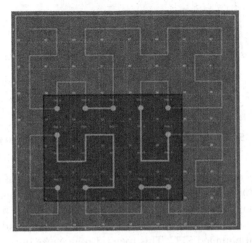

Fig. 3. A query broken decomposed into a series of ranges

of ranges along our Hilbert curve. This knowledge allows us to search only the keys within the original bounding box despite discontinuities along the curve.

This example shows how GeoWave uses SFCs in bounded dimensions like latitude and longitude; however, unbounded dimensions like time, elevation, or perhaps velocity may add another wrinkle to the process. In order to normalize real world values to fit a space filling curve the values must be bounded. We solve this issue by binning using a defined periodicity. A bin simply represents a bounded period for each dimension. The default periodicity for time in the GeoWave framework is a year, but other periods can easily be set for use cases were it would prove more efficient. The efficiencies gained around choosing a periodicity are outside the scope of this paper, but the short conclusion one can draw is that it is beneficial to choose a periodicity based on the precision of your queries. Consider it significantly suboptimal at a course-grained level to have use cases in which a single query intersects several individual periods, and suboptimal at a fine-grained level to have use cases with a query that will sub-select significant data that is within the precision of a single cell (Fig. 4).

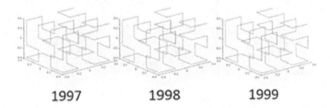

1997   1998   1999

Fig. 4. Bins for a one year time periodicity

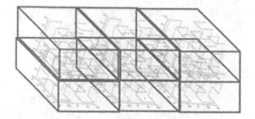

Fig. 5. Composition of multiple unbounded dimensions

Each bin is a hyperrectangle representing ranges of data (based on the set periodicity) and labeled by points on a Hilbert SFC. Bounded dimensions such as latitude or longitude are assumed to have only a single bin. However, if multiple unbounded dimensions are used, combinations of the bins of all of the unbounded dimensions will define each individual bounded SFC. This method may cause duplicates to be returned in scans of entries that contain extents that cross periods necessitating their removal during the fine grained filtering process – preferably performed on the data node for implementations that allow custom server-side filtering (Fig. 5).

3.2   Spatial Subsampling for Map Rendering

Earlier in this paper, we discussed the challenge of rendering massive datasets and gave a brief description of how we tackle the issue. In this section, we will go into further detail about the concepts we use in the GeoWave framework to solve these challenges. In order to visualize a dataset without having to process all of it, we must first break down tile requests into a request or range per pixel. Figure 6 shows this process for a subset of the pixels in a 256 by 256 pixel tile. Here you can see a representation of the correspondence of our index grid to our pixel grid.

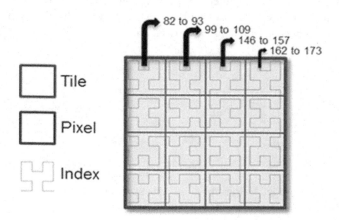

Fig. 6. Visualization of conversion from tile request to pixel request

We could define a query per pixel at this point, but we would still need to load all of the data. This bottleneck has not been improved yet – we are just returning the same amount of data in pixel-sized chunks instead of tiles. The final piece of the puzzle is to short circuit, or skip, the read of other values within the precision of a pixel when any value is returned matching all filter criteria. The logic of this server-side row skipping is visualized in Fig. 7, and is implemented differently based on the datastore being used. As an example, we can extend the SkippingIterator class in Accumulo to scan the start of a range until a value is found, allow this value to stream back to the client, and seek to the start of the next pixel (key range) instead of reading the next key-value pair in the current range. This same functionality is carried out in HBase by applying a custom filter that behaves similarly, acting as a skipping filter that runs and seeks within the scope of a server-side scan.

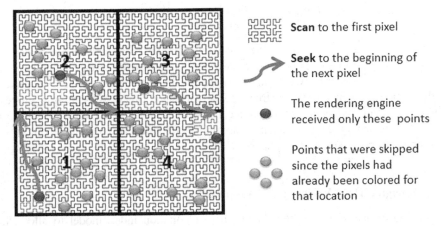

Fig. 7. Visualization of server-side row skipping in the GeoWave framework

While seeking is less performant than sequential reading of data (this relationship is borne out in the testing results in Sects. 4.2 and 4.3), there comes a point where the sheer amount of data that has to be read far outweighs any performance penalty for random IO. In our testing this point occurs at the order of magnitude of millions of features but is heavily dependent on factors such as disk subsystem, IO caching, etc. The example above, while perfectly valid, does make certain assumptions that greatly simplify the situation and aren't always valid in real world cases. In the provided example, the pixel grid very conveniently aligned with the index grid. This greatly simplified the logic involved in "skipping to the next pixel." In the figure below, we demonstrate a situation where the index grid and the pixel grid do not align. In this case, we are not able to skip directly to the next pixel but instead must iterate through indexed "units."

Again, considering our Hilbert SFC in two dimensions is much like a quadtree, we exploit the nesting properties of it in order to try and match pixel size to grid size. Though it is not guaranteed to match exactly (as demonstrated in Fig. 8), we do have a guarantee that worse case performance will be no worse than a factor of four over ideal.

Fig. 8. Our index grid is not always perfectly aligned with the pixel grid

If it's more than a factor of four, we can simply move one level up our quadtree, treating the index grid size as units of 4. This does assume that the cardinality of your index is at least a power of two higher than the cardinality of your zoom level (pixels per dimension). When this is not true, the performance of the subsampling process reverts to standard sequential IO. Using a cardinality of 31 as in the software's default spatial dimensionality definitions, this maximum effective subsampling resolution is less than 2 cm per pixel at the equator.

3.3   Managing Data Variety and Complexity

We previously discussed how GeoWave uses a common index model to allow the comingling of disparate data types in the same table. Here we will expand on this concept and discuss how it is implemented in the GeoWave framework.

Figure 9 describes the default structure of entries in the datastore. The index ID comes directly from the tiered SFC implementation. Each element contains its corresponding SFC tier, bin, and compact Hilbert value. We do not impose a requirement that data IDs are globally unique but they should be unique for the individual adapter.

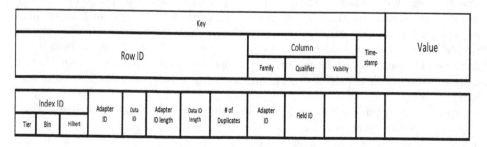

Fig. 9. Accumulo implementation of the GeoWave common model

In this way, an environment can be massively multi-tenant without imposing restrictions across individual usage. This pairing of "Adapter ID" and "Data ID" define a unique identifier for a data element that also creates a unique row ID. This is necessary as the indexed dimensions could be the same or similar, mapping to the same SFC value despite being different entities. Moreover, it is a fundamental requirement in a multi-tenant distributed key-value store to form unique row IDs for unique entries. It also allows us to rapidly filter out potential duplicates and false positives within server-side filtering and ensure complete accuracy of queries. Adapter and Data IDs can vary in length so we store their lengths as four byte integers to ensure proper readability. The number of duplicates is stored within the row ID as well to inform the de-duplication filter if this element needs to be temporarily stored to ensure no duplicates are sent to the caller. The Adapter ID is duplicated and used by the column family as the mechanism for adapter-specific queries to fetch only the appropriate column families. Accumulo has a concept of locality groups that triggers how column families are stored together and by default we use a locality group for each individual adapter (data type), but this behavior can be easily disabled if optimizing performance across many data types is preferred.

Regardless of the data type, we adapt this same persistence model to index it. This ensures that all GeoWave features will work the same way regardless of the data type. This is what allows the system to seamlessly store and retrieve many disparate data types within the same table.

We have already discussed that one of the benefits of this indexing method is the ability to perform fine grained filtering of the data. In this example, we will explain why this is necessary. In most of our previous cases, the bounding box for the query perfectly overlapped a Hilbert value or "bucket." In Fig. 10, that is not the case.

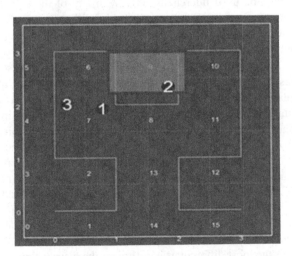

Fig. 10. Example of over-inclusion that will need to be filtered out in the fine-grained tuning

Since the bounding box overlaps a bucket, it will still "match" for that bucket since the potential to hit data is there. In this case, it would match for the range of 6 through 9. But we now need to make a second pass and check each bucket value to see if it overlaps the bounding box. In this case, we use the buckets as a rough filter to quickly reduce the amount of items we have to check. We then do a filter evaluation on the actual values at their native precision. In the case of complex geometric relationships, we use the popular Java Topology Suite (JTS) library to evaluate the filter. In this case, Point 1 and Point 2 both fall in bucket ranges of 6 through 9 and are evaluated using the fine-grained filtering process (but notice that Point 3 is ignored as it falls outside the range). BBOX overlaps Point 1 = false, so it is rejected. BBOX overlaps Point 2 = true, so it is included. It is important to note that all of this filtering is performed server side within the implementation constraints of the datastore. This allows the software to take advantage of a distributed cluster and parallelize the filtering process, as well as reduce unnecessary network traffic.

As a result, this reduces the time necessary to filter out false positives. Of note, this fine-grained filtering is only able to be performed across many disparate data types within a single scan on the server-side by using a definition of a common index model to store indexed values at native precision.

3.4 Key-Value Store Parity

GeoWave has established the goal of parity in function and performance among a variety of key-value stores. GeoWave's user-facing tools and developer-facing APIs are agnostic to the underlying datastore. The connection configuration is the only difference. A named store is configured once, via the command line tools, and that named entity can then be used interchangeably with any of the supported key-value stores by subsequent commands. We accomplish this by drawing on the commonalities of distributed key-value stores and microbenchmarking the differences.

One such commonality is that each of the supported stores can perform range scans more efficiently on a well-sorted range of keys than on a discontinuous set. However, the way in which these keys are sorted may effectively match typical query use cases resulting in a behavior referred to as hotspotting. Hotspotting is a situation where processing and data are isolated to specific resources. This is obvious in a single tenant use case where the only resources with activity are the ones within a single filter constraint, but also occurs in highly multi-tenant environments due to current events and common trends in usage patterns. Each datastore has slightly different ways to more randomly distribute data. Apache Cassandra [3] and Amazon's DynamoDB [1] for example force this random distribution, at least under default and recommended configuration, by requiring a "hash" key that does not have any implication on sort order. Additionally, they allow a "range" key that will benefit from sort order. With Apache Accumulo [2], Apache HBase [4], and Google's Cloud BigTable [5], there is a single row key. BigTable is the most different as it optimizes the distribution of data implicitly. Apache Accumulo and Apache HBase, in practice, often use a concept called pre-splitting. Pre-splitting a table enables the user to define key ranges that will immediately fall within partition boundaries. Using this methodology a hash or random prefix can be associated with each entry that defines its partition and then a subsequent sort ordered suffix can be appended to this. In this way, randomization is achieved to the extent necessary to ensure proper resource utilization and range scans can effectively be performed within each partition. A user or developer within the GeoWave framework is able to define a general purpose indexing approach that includes a partitioning strategy and a multi-dimensional numeric data strategy. GeoWave keeps the sort order portion contributed by an index strategy separate from the portion that doesn't imply sort order. Within Accumulo and HBase GeoWave can append the key appropriately. Within Cassandra and DynamoDB it can use the appropriate column type.

It cannot be expected that all behaviors within these datastores are similar. For example, DynamoDB and BigTable are externally hosted and the others are open source projects. We have already indicated how BigTable handles optimizing data distribution for the user, but DynamoDB also has interesting characteristics. For example, the payment model is very different – a user pays for provisioned throughput rather than defining a size for a cluster. This has direct implications on query optimizations within GeoWave. For example, if a user is throttled by read throughput the consideration to decompose a hyperrectangle into the most fine-grained set of ranges seems like an appropriate tradeoff, whereas if the user is not limited by read throughput it may be more efficient to minimize ranges (with the understanding that each range implies a separate REST invocation to AWS). Moreover, it is a reasonable expectation for each datastore to exhibit different behaviors for using many ranges. We will go into more detail on the implications of too many or too few ranges within the context of each datastore in Sect. 4.3.

4 Experimental Evaluation

4.1 Locality Preservation Performance

While the benefits of effective mapping between multi-dimensional objects and the sort order of keys within a distributed key-value store may be apparent to some, these benefits are easily demonstrable within the GeoWave framework. The cluster utilized in this benchmark is comprised of one master node and 20 data nodes; each node is an m3.xlarge instance type running in Amazon Web Services (AWS). In the following benchmark, we ingest data that is inherently four-dimensional with values for a three-dimensional coordinate space in X, Y, and Z, as well as a temporal component T. Each data entry also contains 500 bytes of extended metadata that is retrieved. One billion randomly generated data elements are ingested into GeoWave using HBase as the datastore and four different indexing definitions that exercise the underlying compact Hilbert curve to varying degrees of dimensionality. In the following graph, randomly generated queries of intentionally varied sizes are executed against each of the dimensional definitions. The time from defining query constraints to returning the last data element is plotted on the Y-axis on a logarithmic scale.

In Fig. 11, we can easily see the benefit of locality preservation of multi-dimensional objects. It is also apparent that this benefit can often be most impactful on more selective queries. On the query returning zero results, for example, it would be

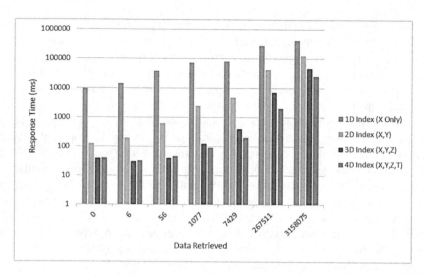

Fig. 11. Graph of response times for queries of four-dimensional data on varying index types (The source code to produce these results is publicly available (https://github.com/rfecher/dimensionality-experiments))

safe to assume that the single dimensional index filtered quite a bit out in the X dimension. But of what is left, there would be a GeoWave custom filter on the server side needed to evaluate and filter the false positives. While more effective than returning false positives to the client, it is considerably less effective than simply utilizing effective multi-dimensional representations for keys and ranges.

It is worth noting that for relatively tightly constrained queries the 3D index behaved nearly identical to the 4D index. This is understood to imply that with very tightly constrained queries there are very few, if any, entries that are additionally constrained by the fourth dimension. However, as the queries become less restrictive there is a significant benefit in applying the additional constraints along the fourth dimension to effectively and completely filter the key space.

4.2   Map Pixel-Based Spatial Subsampling Performance

To demonstrate the power and time savings of GeoWave spatial subsampling, we will record the amount of time (in seconds) that it takes to render various numbers of GPS locations contributed to the OpenStreetMap (OSM) project with and without subsampling. This OSM GPX dataset is a crowd sourced set of GPS derived trajectories primarily used to determine connectivity of road networks. This data is open source and can be downloaded from: http://planet.openstreetmap.org/. In this example, we set up a 10 node distributed computing cluster (one Master and nine Workers) using AWS Elastic Map Reduce (EMR). We chose to use Accumulo for our datastore, which was backed by non-provisioned IOPS EBS volume. Each node used an m3.2xlarge EC2 instance and the global set of GPX data used for this test was 21 GB compressed,

Table 2. Effects of spatial subsampling on time taken to render various numbers of features

Number of features	Completion time with subsampling (seconds)	Completion time without subsampling (seconds)
2.9 Billion	15.39 (SD = 1.21)	Out of memory
1.0 Billion	12.97 (SD = 1.83)	Out of memory
374 Million	6.76 (SD = 1.44)	Out of memory
76 Million	2.92 (SD = 1.16)	398.15 (SD = 32.12)
20 Million	1.64 (SD = 0.87)	64.15 (SD = 8.54)
3 Million	0.78 (SD = 0.49)	1.64 (SD = 0.25)
312,116	0.67 (SD = 0.35)	0.141 (SD = 0.07)
65,429	0.61 (SD = 0.10)	0.113 (SD = 0.09)
9,843	0.56 (SD = 0.14)	0.121 (SD = 0.09)

256 GB uncompressed, and 153 GB in HDFS as an Accumulo database file after ingest. The rendering process was handled by the popular open source software GeoServer. Table 2 shows the results of the comparison.

Each value shown in Table 2 is the average of 20 runs with the standard deviation reported afterwards. The "Out of memory" errors were on the GeoServer instance. The WMS process memory allocation was increased to 4 GB from the GeoServer default of 64 MB, but that was still not sufficient to prevent the error.

The results shown in Table 2 are as expected and the errors encountered further demonstrate the challenges of rendering this amount of data without techniques such as smart subsampling. When the dataset was quite small (e.g., hundreds of thousands of features), the spatial subsampling method was slower due to the impact of random seeks. As the number of features being rendered grew, however, the spatial subsampling method returned results in incredibly faster times. The traditional render method quickly became untenable. Even at the highest feature number that we were still successfully able to render using the traditional method, it took over six and a half minutes to finish.

4.3 Differences in Multi-range Scans Among Key-Value Stores

It is beyond the scope of this paper to comprehensively detail what a range scan implies on each datastore. But as a generalization, it is fair to say that an individual range scan within each datastore has a cost and it is unfair to assume that they behave the same in this respect. This is motivation for the following microbenchmark. As an example of the challenges faced in optimizing query range decomposition, let's suppose that a sparsely populated table is queried using a particular set of 5000 ranges that well-represent a query hyperrectangle. On this table it may intersect 1–5 entries per range, but on a different, more densely populated, table this same set of ranges intersects thousands of entries per range. Considering the first table is sparsely populated in this hyperrectangle, it would additionally be fair to assume that combining similar ranges to end up with 10 ranges will introduce no, or very few, false positives. It is a considerable advantage given the cost of each range in the sparsely populated table to combine the ranges. However, the effect of combining the ranges down to only 10 ranges to represent the hyperrectangle in the densely populated table is of considerable impact in the loss in specificity of the keys. The false positives produced far outweigh any cost of additional ranges. Some of this can be informed by the histograms of data that GeoWave stores as a statistic on each write, but this is an example of the complex tradeoffs involved that necessitate the following microbenchmarks.

The impacts of multi-range scans are benchmarked on a cluster with one m3.xlarge master node and five m3.xlarge data nodes for Accumulo, HBase, and Cassandra. The client accessing DynamoDB is an m3.xlarge instance with sufficiently high read throughput that the scan is not being throttled by AWS. BigTable is not included in this particular benchmark in part because BigTable optimizes data placement over time to distribute reads equally across all nodes. This is a challenge to control in a benchmark. Furthermore, BigTable's client API appears to be evolving with multi-range scan support in the near future (currently a separate scan must be executed for each range).

This microbenchmark is performed independently of GeoWave. It writes 100 million lexicoded numbers as keys to the underlying datastore where the values are a random set of 500 bytes. In DynamoDB and Cassandra, the default hash key load balancing is used with a single constant hash key and the lexicoded numbers are used for the range key. It executes scans with varying numbers of ranges. To be precise, the scans are repeated using powers of 10 between 1 and 100 million for the total number of ranges and either 10 million or 100 million total results are expected. The cluster size and volume of data in this microbenchmark were chosen to demonstrate the theoretical trends that should hold as compute resources are proportionally scaled with respect to data volume. The trend was consistent, and the median response times are plotted in Fig. 12 with logarithmically scaled axes.

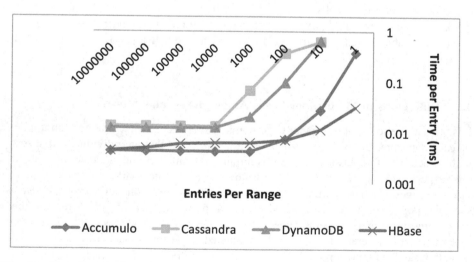

Fig. 12. Plot of results from the multi-range scan benchmark on multiple datastores (The source code to produce these results is publicly available (https://github.com/rfecher/range-sensitivity-experiments))

The relative difference between local minima and maxima across data stores is unimportant for this purpose and can be in part explained by varying degrees of configuration to appropriately control the experiment. As one example of this, index and block caching are disabled to avoid the effects of higher variance in repeated sampling. The important conclusion that we have drawn is that as the ratio of results per range decreases on varying sizes of multi-range scans, there is clearly an inflection point in the amount of time per entry. Moreover, the observed inflection point is different for each datastore. This is generalized to be considered a result of implicit costs for each individual range and implies that some datastores operate more efficiently on bulk multi-range scans if you have at least 10, 100, or 1000 entries per range. This is much less efficient in the extreme case if each range intersects a single entry. This is typically not as problematic in range scans on a single numeric dimension as you would likely be able to well-represent a query with a single, or at most, a few ranges. As the number of dimensions increases, the number of ranges must also increase to represent bounding hyperrectangles and it becomes critical to understand the performance characteristics of many ranges within each datastore.

5 Conclusions

In this work, we discussed how we leverage the GeoWave framework to provide efficient solutions to many of the implicit challenges in utilizing distributed computing frameworks with geospatial and temporal data. By using a tiered and binned SFC indexing system to index multidimensional data into single dimensional key-value stores, we are able to preserve locality in every dimension. We implement techniques like spatial subsampling to minimize the common bottleneck created at the map-rendering engine. By using an agnostic common indexing model, GeoWave allows for heterogeneous data types to be stored and analyzed in the same table. GeoWave has proven to be an incredibly powerful tool to bridge the gap between distributed computing and GIS. It has helped to lower the learning curve for these technologies and provides a fast, resource-efficient framework that is compatible with many of the most popular and best researched distributed datastores. Research is currently being done to further increase the number of compatible datastores as well as to add new analytic capabilities to the framework.

References

1. Amazon DynamoDB: Amazon DynamoDB (2017). https://aws.amazon.com/dynamodb/
2. Apache Accumulo: Apache Accumulo (2017). https://accumulo.apache.org/
3. Apache Cassandra: Apache Cassandra (2017). http://cassandra.apache.org/
4. Apache HBase: Apache HBase (2017). https://hbase.apache.org/
5. Cloud BigTable: Cloud BigTable (2017). https://cloud.google.com/bigtable/
6. Dean, J., Ghemawat, S.: MapReduce: simplified data processing on large clusters. Google Inc. (2004)

7. Eldawy, A., Mohamed, M.: The ecosystem of SpatialHadoop. SIGSPATIAL Spec. **6**(3), 3–10 (2015)
8. GeoServer: GeoServer (2017). http://geoserver.org/
9. Hamilton, C.H., Rau-Chaplin, A.: Compact Hilbert indices: space-filling curves for domains with unequal side lengths. Inf. Process. Lett. **105**, 155–163 (2008)
10. Haverkort, H., Walderveen, F.: Locality and bounding-box quality of two-dimensional space-filling curves. Comput. Geom. **43**, 131–147 (2010)
11. Indyk, P., Motwani, R., Raghavan, P., Vempala, S.: Locality-preserving hashing in multidimensional spaces, p. 618. ACM (1997)
12. Kim, H., Kang, S., Lee, S., Min, J.: The efficient algorithms for constructing enhanced quadtrees using MapReduce. IEICE Trans. Inf. Syst. **99**(4), 918–926 (2016)
13. Nishimura, S., Das, S., Agrawal, D.: MD-HBase: a scalable multi-dimensional data infrastructure for location aware. In: IEEE MDM 2011, vol. 1 (2011)
14. Paiva, J., Ruivo,, P., Romano, P., Rodrigues, L.: AUTOPLACER: scalable self-tuning data placement in distributed key-value stores. ACM Trans. Auton. Adapt. Syst. **9**(4) (2014). Article No. 19
15. Uzaygezen: Uzaygezen (2017). https://github.com/aioaneid/uzaygezen

Indexing and Aggregation

Sweeping-Based Temporal Aggregation

Danila Piatov$^{(\boxtimes)}$ and Sven Helmer$^{(\boxtimes)}$

Free University of Bozen-Bolzano, Bolzano, Italy
{danila.piatov,sven.helmer}@unibz.it

Abstract. The Timeline Index, which supports temporal joins, time travel, and temporal aggregation on constant intervals, has the potential of becoming a universal index for temporal database systems. Here we present a family of plane-sweeping algorithms that extend the set of operators supported by Timeline-Index-based databases to temporal aggregation on fixed intervals, such as a sliding windows or GROUP BY ROLLUP aggregation, and improve the existing algorithm for computing MIN/MAX temporal aggregates on constant intervals. Our method for selective aggregation relies on a new and very space-efficient data structure called MAX Skyline. In an empirical evaluation we show that our approach is superior to existing techniques, in some cases we improve the performance by orders of magnitude.

1 Introduction

There is a great need for efficient temporal operator implementations in modern-day database management systems, as temporal data is found in many financial and business applications running on top of these systems. For example, Kaufmann states that there are several temporal queries in the hundred most expensive queries executed on SAP ERP [9] and he continues by saying that this does not even include complex query types, such as temporal aggregation, which have to be implemented in the application layer. According to [9], customers of SAP desperately need (advanced) temporal operators for efficiently running queries pertaining to legal, compliance, and auditing processes. Facebook Timeline is also identified as a less critical, data-intensive temporal application domain.

Although temporal operators have been introduced into the SQL:2011 standard, the provided implementation is far from complete. While approaches for grouping and aggregating temporal data have already been widely discussed in academia (see Sect. 2 on related work), many systems are just starting to adopt and integrate these types of operations into their frameworks [10,15]. One of the reasons for this is that many of the proposed techniques are rather complex and highly specialized.

Looking at the example of a temporal relation **r** in Fig. 1, we can see why aggregation (and grouping) is much harder on it than on its non-temporal counterpart. Here line segments denote tuple validity intervals and the y-axis represents tuple values; we will describe this in more detail in Sect. 3.1. Temporally aggregating tuples on constant intervals means that we aggregate values for

M. Gertz et al. (Eds.): SSTD 2017, LNCS 10411, pp. 125–144, 2017.
DOI: 10.1007/978-3-319-64367-0_7

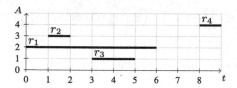

Fig. 1. Temporal relation **r**

time-spans in which the set of valid (active) tuples stays the same. For instance, in the interval $[0,1)$ tuple r_1 is valid (resulting in a sum of 2), then in $[1,2)$ tuples r_1 and r_2 are active (leading to a sum of 5), in the interval $[2,3)$ we are back to just r_1, from time point 3 to time point 5 we have the tuples r_1 and r_3, yielding a sum of 3, and so on. The situation is complicated further by defining custom aggregation intervals (i.e. fixed intervals) and using various aggregation functions. (We give a more formal definition in Sect. 3).

The Timeline Index is currently being used as the temporal index in a prototype of a commercial temporal DBMS. Only a single Timeline Index per relation is needed to support time travel, temporal aggregation on constant intervals and temporal and interval overlap joins [10,14]. We propose a generally applicable framework for temporal aggregation based on a plane-sweeping technique and the Timeline index to efficiently compute different types of temporal aggregation. We do so more efficiently and support more grouping scenarios than the current Time-Index-based algorithms. In Sect. 4 we show how to compute aggregate functions, such as SUM, COUNT, MAX, and MIN using just a single scan over the data. In particular, our contributions are the following:

- We develop a family of plane-sweeping algorithms for temporal aggregation that employ the Timeline Index. This extends the existing Timeline approach with aggregation on fixed intervals, supporting such operations as sliding windows and GROUP BY ROLLUP aggregation, and improves the computation of MAX/MIN aggregates on constant intervals.
- We introduce MAX Skyline, a very efficient data structure for maintaining MAX/MIN aggregates in temporal aggregation, which provides better functionality, performance, robustness, and space efficiency than the existing solutions for plane-sweeping algorithms.
- In an experimental evaluation, which is described in Sect. 5, we compare our approach to several other methods, illustrating that we are faster than many state-of-the-art techniques and on-par with highly specialized approaches such as SB-trees [16].

2   Related Work

Early work on temporal aggregation for constant intervals includes the *time index* [6], the *PA-tree* [11,13], and the *aggregation tree* [12]. Later on, Yang and Widom [16] proposed the *SB-tree* for aggregation on constant intervals and extensions for aggregation on fixed intervals in the form of dual SB-trees, JSB-trees, and MSB-trees. Böhlen et al. [3] generalized temporal aggregation operators and defined the *TMDA* operator that supports temporal aggregation on both constant and fixed intervals.

Table 1. State-of-the-art algorithms and operators they support

Family	Constant intervals		Fixed intervals	
	SUM/COUNT/AVG	MIN/MAX	SUM/COUNT/AVG	MIN/MAX
SB-tree [16]	SB-tree	SB-tree	JSB-tree & dual SB-trees	MSB-tree
TMDA [3]	TMDA-CI^c	TMDA-CI	TMDA-FI	TMDA-FI
Timeline index, existing [10]	Optimal algorithm	Top-K list based	*none*	*none*
Timeline index, our solution	*not needed*	Sect. 4.2	Sect. 4.3	Sect. 4.3

Kaufmann et al. [10] developed an algorithm for temporal aggregation of constant attributes on constant intervals using the *Timeline Index*—a structure very similar to the Endpoint Index used in our approach. The index is basically a chronological sequence of events of tuples starting and ending. The algorithm scans the index once, and for each event it adjusts the running aggregates and outputs a constant interval with the current aggregate values. We want to stress the generality of the Timeline framework, which supports a wide range of temporal operators. A summary of state-of-the-art algorithms and the operators they support is presented in Table 1.

3 Problem Formalization

3.1 Temporal Relations

A *temporal relation* **r** is a relation in which each tuple contains two additional attributes T_s and T_e that denote the inclusive left and the exclusive right (SQL:2011) endpoints of the tuple validity interval $T = [T_s, T_e)$.

Consider the temporal relation **r** in Fig. 1. It consists of four tuples that contain one explicit attribute A (the value of which is shown on the y-axis). In our notation, the tuples look like the following: $r_1 = \langle 2, [0, 6) \rangle$, $r_2 = \langle 3, [1, 2) \rangle$, $r_3 = \langle 1, [3, 5) \rangle$, and $r_4 = \langle 4, [8, 9) \rangle$. For the further examples, we assume that the tuples denote file history in a directory, and that their explicit attribute values are file sizes in kilobytes. The intervals indicate the period of time the file exists in the directory and is unchanged.

3.2 Temporal Aggregation on Constant Intervals

The simplest aggregation query is to apply an aggregation function to all tuples valid at a particular point in time (*active tuples*). This is called *instant (temporal) aggregation*. For example, for $t = 1$ we sum up the values of r_1 and r_2, resulting in the answer 5 kilobytes (directory size at time 1). If we want to have a complete picture, i.e., an overview of the aggregate value for every point in time t, we combine identical aggregation results for contiguous time instants into a single interval, thus creating a temporal relation. We generate one temporal tuple per time interval in which the set of argument tuples is constant. Therefore, this is called *temporal aggregation on constant intervals* [3].

The result of a SUM aggregation function on constant intervals applied to the relation **r** from Fig. 1 is shown in Fig. 2 (green line). It represents the history of the size of the directory. We note that by definition [3] constant intervals are not defined for time spans with no argument tuples. We therefore generate no output for those time spans (such as $[(6,8))$.

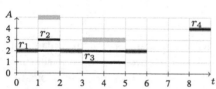

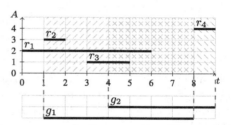

Fig. 2. The result of SUM aggregation on constant intervals (in green). (Color figure online)

Fig. 3. Temporal aggregation on fixed intervals

3.3 Temporal Aggregation on Fixed Intervals

Instead of aggregating tuples valid at an instant, we might want to aggregate tuples that were valid some time during a specific time-span. An example of such a query would be "SELECT MAX(A) FROM r WHERE T OVERLAPS $[0,2)$". In our example this gives us the size (3 kilobytes) of the biggest file that was in the directory during the timespan ranging from 0 (inclusive) to 2 (exclusive). Again, we are interested in a set of answers rather than a single one. Researchers have generalized the temporal aggregation on time spans and defined the *temporal aggregation on fixed intervals* [3]. When performing such an aggregation we assume that an additional temporal *group* relation is explicitly given, which defines a set of custom intervals on which to perform aggregation (these intervals may overlap). An output tuple is produced for each group tuple. A special case of aggregation on fixed intervals is, for instance, aggregation using a sliding window or a GROUP BY ROLLUP-type query.

Figure 3 shows relation **r** with an example group relation **g**. Here, g_1 overlaps with r_1, r_2, and r_3. Thus, the aggregate for g_1 is computed using those three argument tuples. The aggregate for g_2 considers the tuples r_1, r_3, and r_4, as it overlaps with them.

4   Sweeping-Based Temporal Aggregation

Temporal aggregation implemented with the help of a sweep-line algorithm and the Timeline Index has been partially explored by Kaufmann et al. [10], who show how to process temporal aggregates on constant intervals. We go further by extending the Timeline framework with aggregation on fixed intervals and also show how to improve the performance of aggregation on constant intervals significantly. For ease of exposition (and without loss of generality), we present a slightly simplified version of a Timeline Index, an Endpoint Index, here. After its introduction, we continue with a description of the algorithms running on top of it.

4.1   Endpoint Index

The idea of the Endpoint Index is that intervals, which can be seen as points in a two-dimensional space, are mapped onto one-dimensional *endpoints* or *events*. Let **r** be an interval relation with tuples $r_i, 1 \leq i \leq n$. A tuple r_i in an Endpoint Index is represented by two events of the form $e = \langle timestamp, type, tuple\_id \rangle$, where *timestamp* is T_s or T_e of the tuple, *type* is either start or end, and *tuple_id* is the tuple identifier, i.e., the two events for a tuple r_i are $\langle r_i.T_s, start, i \rangle$ and $\langle r_i.T_e, end, i \rangle$. For instance, for $r_3.T = [3, 5)$, the two events are $\langle 3, start, 3 \rangle$ and $\langle 5, end, 3 \rangle$, which can be seen as "at time 3 tuple 3 started" and "at time 5 tuple 3 ended". Since events represent timestamps, we can impose a total order among events, where the order is according to *timestamp* and ties are broken by *type* (end < start). Endpoints with equal timestamps and types but different tuple identifiers are considered equal. An Endpoint Index for interval relation **r** is built by first extracting the interval endpoints from the relation and then creating the ordered list of events $[e_1, e_2, \ldots, e_{2n}]$ sorted in ascending order.

Consider interval relation **r** from Fig. 1. The Endpoint Index for it is [$\langle 0, start, 1 \rangle$, $\langle 1, start, 2 \rangle$, $\langle 2, end, 2 \rangle$, $\langle 3, start, 3 \rangle$, $\langle 5, end, 3 \rangle$, $\langle 6, end, 1 \rangle$, $\langle 8, start, 4 \rangle$, $\langle 9, end, 4 \rangle$].

4.2   Temporal Aggregation on Constant Intervals

The basic idea of using an Endpoint Index for aggregation on constant intervals is to replay the events in the order they happened, while maintaining a running aggregate value. After handling all events for a specific time point, we report the current value of the running aggregate as the output. (See Algorithm 1 for pseudocode.)

We model the running aggregate used in the algorithm as an abstract data type supporting three operations:

Algorithm 1. Sweeping-based temporal aggregation on constant intervals

---

    input   : Argument relation **r** and its Endpoint Index **e**, consisting of endpoints
                  $\langle timestamp,\ type,\ tuple\_id \rangle$

```
 1  var e ← first(e)                              // current endpoint of r
 2  var t ← e.timestamp
 3  var aggregate ← new Running aggregate of some type
 4  var activeCount ← 0
 5  while exists(e) do
 6   │  while exists(e) and t = e.timestamp do
 7   │   │  var r ← r[e.tuple_id]                  // load the tuple by id
 8   │   │  if e.type = start then
 9   │   │   │  activeCount ← activeCount + 1
10   │   │   │  aggregate.tupleStarted(r)
11   │   │  else
12   │   │   │  activeCount ← activeCount − 1
13   │   │   │  aggregate.tupleEnded(r)
14   │   └  e ← advance(e)
15   │  if exists(e) then
16   │   │  var next ← e.timestamp
17   │   │  if activeCount > 0 then
18   │   │   │  var value ← aggregate.getRunningValue
19   │   │   └  output(⟨value, [t, next)⟩)
20   └   └  t ← next
```

---

1. `tupleStarted(r)`: handles the start event of tuple r. The running aggregate needs to adjust its value by taking into consideration the newly started tuple;
2. `tupleEnded(r)`: handles the end event of tuple r. The running aggregate needs to undo the effect of tuple r on the aggregate value.
3. `getRunningValue()`: returns the current value of the aggregate, i.e., the aggregation result of all tuples that have started but have not yet ended.

While there are still unprocessed events in the Endpoint Index, we fetch the next batch of events that have the same timestamp and load their corresponding tuples. Then, depending on their type, either the `tupleStarted` or `tupleEnded` method is called to adjust the running aggregate. Finally, we generate a new output tuple to reflect the changing aggregate value.

The Timeline Index additionally supports *checkpoints*, which are bitmap masks of active tuples [10]. They allow for faster handling of queries that cover only a subrange of the temporal domain.

Cumulative (SUM, COUNT, AVG) Aggregation on Constant Intervals. Let us consider a SUM aggregate. We allocate one variable for holding the value of the running aggregate, which is initialized with 0. The method `tupleStarted(r)` increments this variable by the value of the attribute $r.A_i$

to be aggregated. For `tupleEnded(r)`, we decrement the variable by $r.A_i$. The method `getRunningValue()` simply returns the current value of the variable. The `COUNT` and `AVG` aggregates are handled analogously. This method is employed by many algorithms [3,10,11] including the Timeline-index-based ones. The run time is proportional to the number of start and end events and it is hard to do better, as every event triggers a change in the output.

Selective (`MIN`, `MAX`) Aggregation on Constant Intervals. Implementing the selective `MAX` (and `MIN`) aggregate is more complicated, in particular the implementation of the method `tupleEnded(r)`: when we remove the tuple with the current `MAX` value, we have to reconstruct the currently valid `MAX` value, which cannot be done arithmetically.

Kaufmann et al. [10] keep a sorted list of the Top-K values to do this, adding and removing values as tuples with values in the current Top-K list start and finish. There is a problem when a lot of tuples with large values finish and the Top-K list becomes empty. For that reason, all values which are not added to or removed from the Top-K list are maintained in two separate unordered lists. These two lists are then used to repopulate a Top-K list when it runs dry. In an extreme case, when $K = 100\%$, this solution becomes a tree of active values and the inserted and deleted lists can be omitted. Note that we cannot use a max-heap instead, as it does not support efficiently finding and removing an arbitrary element. While this approach works correctly, there are a few issues with it. First of all, for optimal performance the parameter k has to be determined (Kaufmann et al. use 0.01% of the number of distinct values in the data set). Independently of this, there is a space overhead for maintaining the two unordered lists (each of them might contain almost all the values from the data set).

Instead of using the Top-K list we propose a special data structure called *MAX Skyline*. Compared to a Top-K list, a MAX Skyline has a better performance, a smaller memory footprint, and does not require the tuning of any parameters. Consider the following set-up. We have a set of temporal tuples with values A and intervals that are unbounded from the left, i.e. T has the form $(-\infty, T_e)$. We want to have an incremental index structure that supports two operations:

1. `add(`T_e`, `A`)`: add tuple $\langle A, (-\infty, T_e) \rangle$ to the index;
2. `getMaxAt(t)`: query the index for the maximal value of tuples valid at time t, considering all tuples added so far.

Let us take the first three tuples r_1, r_2, and r_3 from Fig. 1 and pretend for a moment that they are unbounded from the left. If we plot `getMaxAt(t)` as a function of t for those tuples, we get the thick orange line in Fig. 4. Basically, this results in a skyline as defined in [4].

Because MAX Skyline is a non-increasing step function, we only need to store the right endpoints of each step to maintain it. For instance, for the skyline in Fig. 4, we only need to memorize the points $(2, 3)$ and $(6, 2)$ (see Fig. 5). We store the points in a tree ordered and indexed by the timestamp. The function `getMaxAt(t)` can be implemented in the following way. First, we search for

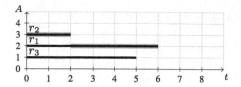

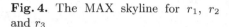

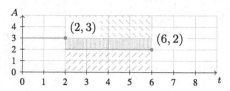

Fig. 4. The MAX skyline for r_1, r_2 and r_3

Fig. 5. The implementation of MAX skyline

the point that immediately follows the timestamp t (the chronologically first point that has a timestamp strictly bigger than t). Then we return the value of the point as the result. For example, in Fig. 5, for any $t \in [2, 6)$ we find point $(6, 2)$, which gives us the correct result of 2. The complexity of this algorithm is $O(\log(n))$.

We use C++ `std::map` as the implementation of the tree. It maps point timestamps to their values. The standard function `std::map::upper_bound`(t) does exactly what we need—it returns the first element with timestamp strictly bigger than t. If no such element exists, it returns a special value `std::map::end`.

The implementation of `add`(T_e, A) is more complex. Potentially, we are inserting a new point (T_e, A) with expiration timestamp T_e and value A into the skyline. To decide if we should insert the point we first find the immediate neighbors to the left and to the right of T_e in the MAX Skyline. Without loss of generality, we assume for the moment that both neighbors exist. If there is already a point with timestamp T_e, it is considered the left neighbor and the next point is considered the right neighbor. After having found the neighbors, one of three things happens. If the value of the new point is less than or equal to the value of the right-hand neighbor (lower area shaded in blue in Fig. 5), we discard it, as it has no effect on the maximum at any point in time. If the value of the new point is between the values of the left and right neighbor (middle area shaded in green), we include it in the skyline. Finally, if the value is greater or equal to the value of the left neighbor, we insert the new point into the skyline and also remove all the left neighbors from the skyline that have values smaller or equal. See Algorithm 2 for a more detailed description of `add`(T_e, A), considering all the corner cases of missing neighbors. Function `begin` returns the first point in the tree (or `end` if the tree is empty). Function `erase`(p) erases the point p from the tree. Function `emplace_hint`(p, `key`, `value`) inserts a new key-value pair (point) into the tree, using p as a hint about where to create the new point (and therefore has amortized constant complexity). The non-standard function `previousNode` returns a point's left neighbor (and the last point in the tree for `end`, undefined if the tree is empty, implemented by decrementing the STL iterator).

The amortized complexity of `add`(T_e, A) is $O(\log n)$, even though at first glance we seem to be looping through all the elements. This is because, cumulatively, the loop is executed fewer times than the number of calls of `add`(T_e, A), i.e., less than one iteration on average.

Algorithm 2. MAX Skyline :: add(T_e, A)

---

input : Tree *map* keeping the points of MAX Skyline, tuple value A, the exclusive
timestamp of expire T_e

1 **var** *right* ← *map*.upper_bound(T_e)
2 **if** *right* ≠ *map*.end **and** $A \leqslant right.value$ **then**
3 | **return**   // value is below or equal to the right neighbor's value
4 **while** *right* ≠ *map*.begin **do**   // while exists left neighbor
5 | **var** *left* ← *right*.previousNode
6 | **if** *left.value* > A **then**   // stop if it has higher value
7 | | **break**
8 | *map*.erase(*left*)   // otherwise erase it
9 *map*.emplace_hint(*right*, T_e, A)   // insert point (T_e, A) just before node *right*

For implementing a running aggregate for MAX aggregation on constant intervals for Algorithm 1 using the MAX Skyline we need one instance of it, called skyline in the following, and we map the three operations of a running aggregate to this skyline as follows:

1. tupleStarted(r): skyline.add($r.T_e$, $r.A$);
2. tupleEnded(r): nothing to do;
3. getRunningValue(now): skyline.getMaxAt(now).

The value now is available in Algorithm 1 in the form of t. Alternatively, it can be derived by obtaining T_s from the most recently started tuple or T_e from the most recently finished tuple, whichever event happened more recently.

Why does this work, even though MAX Skyline assumes that tuples are unbounded from the left? We observe that we only add to the skyline values of tuples that have actually started and only query for current (or future) moments in time, we never look backward beyond the starting timestamp of a tuple that was just inserted. Moreover, the MAX Skyline will only be queried for future moments up to (but not including) the point at which the next tuple starts. All future tuples are added to the skyline before querying any of their relevant values.

Consider the situation in Fig. 6, applying our algorithm to relation **r**. The current time t is equal to 1, at which point r_1 and r_2 have been inserted into the skyline (which is shown in orange). The operation getMaxAt(1) reports the correct value of 3. When r_2 finishes at $t = 2$, getMaxAt(2) returns 2 and then we never query the skyline for any timestamps less than 2 again. Also, the value of tuple r_4 is never queried before we actually insert it at $t = 9$.

The worst-case complexity of Algorithm 1 implemented with the help of a MAX Skyline is $O(n \log n)$. However, in the general case it is closer to a linear run time, as many of the tuples added to the skyline can be discarded right away. We only reach the worst case if every single tuple in a relation eventually becomes the MAX value at some point in time. In our empirical evaluation in Sect. 5.4 we show that the size of the MAX Skyline is actually very small.

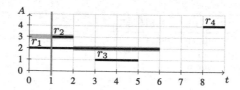

Fig. 6. MAX skyline after the start of r_2

4.3 Temporal Aggregation on Fixed Intervals

In contrast to the aggregation on constant intervals, here we have to group tuples by custom time spans that they overlap. Moreover, multiple grouping intervals can be active simultaneously. Because of these differences we cannot apply Algorithm 1.

Cumulative Aggregation on Fixed Intervals. For visualizing our method, the starting point is the situation depicted in Fig. 7. Basically, for a COUNT aggregate function we are interested in the number of overlapping argument tuples for a group tuple g_i. For instance, for g_2 in Fig. 7 this would be 3. We can express this as the number of argument tuples that are active at the moment g_i starts (green line for g_2 in Fig. 7) plus the number of argument tuples that start while g_i is active (i.e., after g_i started and before it ends, cf. area shaded blue in Fig. 7). For g_2, these numbers are 2 and 1.

Let us define the total number of started and ended argument tuples at time t as $S(t)$ and $E(t)$, respectively. Then the number of argument tuples active at the start of g_i is equal to $S(g_i.T_s) - E(g_i.T_s)$ and the number of argument tuples starting during g_i is equal to $S(g_i.T_e) - S(g_i.T_s)$. Putting this together, we get $S(g_i.T_e) - E(g_i.T_s)$.

We take Endpoint Indexes for both relations **r** and **g** and then perform an interleaved scan on both of them (similar to a sort-merge join). During the scan we maintain two counters: C_s for the number of started argument tuples and C_e for the number of finished argument tuples, incrementing C_s whenever we encounter a starting event of an argument tuple, C_e when encountering an end event. When a group tuple g_i starts, we save it in a set of active group tuples together with the current value of C_e, represented as $g_i.C_e$. In our example, $g_1.C_e = 0$ and $g_2.C_e = 1$. When a group tuple g_i ends, we remove it from the active tuple set and output it with the value $C_s - g_i.C_e$, which is the correct COUNT aggregate result. For instance, in Fig. 7, when g_1 finishes, C_s is equal to 3, giving us the answer $3 - 0 = 3$. When g_2 finishes, $C_s = 4$, giving us $4 - 1 = 3$. Algorithm 3 provides a more detailed description. The worst-case time complexity of the algorithm is $O(n_r + n_g)$, where n_r and n_g are the cardinalities of argument and group relations, respectively.

For the SUM aggregate function, instead of maintaining the counters C_s and C_e for the number of started and ended argument tuples, we accumulate the values of the explicit attribute that is to be aggregated in two sums, S_s and

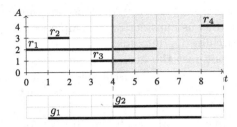 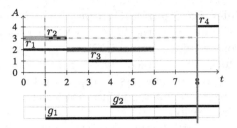

Fig. 7. COUNT aggregation on fixed intervals

Fig. 8. MAX aggregation for g_1

S_e. (In order to be able to access attribute values, we have to actually load the tuples.) For calculating an AVG aggregate function, we need to keep four values updated: C_s, C_e, S_s, and S_e.

Algorithm 3. COUNT aggregation on fixed intervals

---

input   : Argument relation **r**, group relation **g**, and their Endpoint Indexes $\mathbf{e}^r$
             and $\mathbf{e}^g$, consisting of endpoints $\langle timestamp,\ type,\ tuple\_id \rangle$

1   **var** $active^g \leftarrow$ **new** hash map of tuple identifiers to tuples
2   **var** $e^r \leftarrow$ first$(\mathbf{e}^r)$ // current endpoint of r
3   **var** $e^g \leftarrow$ first$(\mathbf{e}^g)$ // current endpoint of g
4   **var** $C_s \leftarrow 0$ // number of argument tuples that started
5   **var** $C_e \leftarrow 0$ // number of argument tuples that ended
6   **while** exists(e_r) **or** exists(e_g) **do**
7   **if not** exists(e^r) **or** (exists(e_g) **and** $e_g < e_r$) **then**
8   **var** $i \leftarrow e^g.tuple\_id$
9   **if** $e^g.type = start$ **then**
10   **var** $g_i \leftarrow \mathbf{g}[i]$ // load the tuple by id
11         $g_i.C_e \leftarrow C_e$ // memorize current C_e
12         $active^g[i] \leftarrow g_i$ // add to active tuples
13   **else**
14   **var** $g_i \leftarrow active^g[i]$ // get the tuple
15         $active^g$.remove(i) // save the memory
16   g_i.COUNT $= C_s - g_i.Ce$ // aggregate value
17   output(g_i)
18       $e^g \leftarrow$ advance(e^g)
19   **else**
20   **if** $e^r.type = start$ **then**
21         $C_s \leftarrow C_s + 1$
22   **else**
23         $C_e \leftarrow C_e + 1$
24       $e^r \leftarrow$ advance(e^r)

Selective Aggregation on Fixed Intervals. As before, we demonstrate our technique with the help of a `MAX` aggregate. Let us consider our example relations **r** and **g** from Fig. 8. We are now searching for maximum values during multiple simultaneously active time spans, not just the current value. Consequently, we cannot use a Top-K list here. Nevertheless, we are still able to utilize a single MAX Skyline instance (referred to as `skyline` in the following) for this computation.

Similarly to the cumulative aggregation case, we take Endpoint Indexes for both relations **r** and **g** and then perform an interleaved scan on the two indexes. As we go along, we add starting argument tuples to the MAX Skyline. When encountering the right endpoint of a group tuple g_i, we output its aggregate value as `skyline.getMaxSince(`$g_i.T_s$`)` as the output of the algorithm.

As an example, let us consider the moment in time when we reach the end of g_1 as shown in Fig. 8 by a red vertical line. The first three argument tuples were already added to the skyline, which at this point looks as indicated by the orange lines. We retrieve the `MAX` value for g_1 by querying the skyline for the maximal value after g_1 started ($t = 1$), yielding the correct result of 3.

To understand why this works in general let us consider a group tuple g_i at moment $g_i.T_e$. To determine the result we have to find the maximum value among all argument tuples r_j that overlap with g_i, which are those that have ended after g_i started and have started before g_i ended. More formally, a tuple r_j is relevant if $g_i.T_s < r_j.T_e$ and $r_j.T_s < g_i.T_e$. Clearly, any argument tuple r_j completely before $g_i.T_s$ (i.e., $r_j.T_e \leqslant g_i.T_s$) is not relevant. Such tuples only

---

Algorithm 4. `MAX` aggregation on fixed intervals

    input   : Group interval relation **g**, argument interval relation **r**, and their Endpoint Indexes e^g and e^r, consisting of endpoints $\langle timestamp, type, tuple\_id \rangle$

    output : Result tuples are passed to function `output`

1   **var** $skyline \leftarrow$ **new** MAX Skyline
2   **var** $e^g \leftarrow$ `first(`e^g`)`
3   **var** $e^r \leftarrow$ `first(`e^r`)`
4   **while** `exists(`e^g`)` **or** `exists(`e^r`)` **do**
5   **if not** `exists(`e^r`)` **or** `exists(`e^g`)` **and** $e^g < e^r$ **then**
6   **if** $e^g.type = $ end **then**
7   **var** $g \leftarrow$ **g**$[e^g.tuple\_id]$
8            $g.\max \leftarrow skyline.$`getMaxAt(`$g.T_s$`)`
9   `output(`r`)`
10        $e^g \leftarrow$ `advance(`e^g`)`
11   **else**
12   **if** $e^r.type = $ start **then**
13   **var** $r \leftarrow$ **r**$[e^r.tuple\_id]$
14            $skyline.$`add(`$r.T_e,\ r.A$`)`
15         $e^r \leftarrow$ `advance(`e^r`)`

influence the MAX Skyline before time $g_i.T_s$. Tuples which start after $g_i.T_e$ have not even been added to the skyline yet, so they are not considered. Therefore, those and only those argument tuples that are relevant to us form the MAX Skyline on interval $[g_i.T_s, g_i.T_e)$. Due to the fact that all tuples in the skyline are unbounded from the left, all relevant argument tuples are also valid for the time $g_i.T_s$. Therefore we query the skyline at that point. Algorithm 4 shows the whole process in more detail.

Again, the performance depends on the average size of the skyline. So, while theoretically the worst-case complexity of the algorithm is $O((n_r + n_g) \log n_r)$, where n_r and n_g are the cardinalities of the argument and grouping relations, respectively, in the average case it is much better, since the size of a skyline is usually much smaller than n_r.

5   Empirical Evaluation

5.1   Environment

All algorithms were implemented in-memory in C++ by the same author and were compiled with GCC 4.9.2 using -O3 optimization flag to 64-bit binaries. The execution was performed on a machine with two Intel Xeon E5-2667 v3 processors under Linux. All experiments used 16-byte tuples containing two 32-bit timestamp attributes (T_s and T_e) and two 32-bit attributes on one of which the aggregation took place. The experiments were also repeated on a seven-year-old Intel Xeon X5550 processor and on a notebook processor i5-4258U, showing a similar behavior.

5.2   Competitors

In order to make our results comparable to the most recent work on temporal aggregation we compare our approach with the following state-of-the-art methods: SB-tree-based approaches [16], TMDA-CI/-FI [3], and the Top-K list [10] (see also Table 1).

For cumulative aggregation on fixed intervals Yang and Widom propose two different SB-tree variants: dual SB-trees and JSB-trees. As there is no empirical evaluation comparing the two, we implemented both and found that for our set-up JSB-trees outperform dual SB-trees by 30–40%, therefore we only use JSB-trees for comparison with the other algorithms. All SB-trees were implemented in-memory with each block having a capacity of 32 children (this size gave the best performance in our case). We implemented TMDA-CI and TMDA-FI as in-memory algorithms and stripped out some complex grouping criteria not needed for our evaluation to make them more efficient.

5.3   Test Workloads

Synthetic Datasets. Synthetic datasets give us more control over investigating the impact of certain parameters (e.g. cardinality) on the performance of the

Table 2. Real-world dataset statistics

Dataset	n	$\lvert r.T \rvert$			$r.T$	$r.A$
		Min	Avg	Max	Domain	#Distinct
flight	58 k	61	8 k	86 k	812 k	878
inc	84 k	2	184	574	9 k	801
web	1.2 M	1	60 M	352 M	352 M	301 k
feed	3.7 M	1	432	8.5 k	8.6 k	5 k
basf	5.3 M	1	127 k	16 M	16 M	14
weather	83 M	300	300	300	252 M	58 k

Table 3. MAX skyline size depending on the size of r

n	Avg size	Max size
10^3	6–7	14–19
10^4	7–9	20–22
10^5	10–11	28–29
10^6	11–14	30–35
10^7	13–16	35–39
10^8	16–18	45–46

algorithms. The time intervals of the tuples are uniformly distributed within the range $[1, 10^9]$, while their duration follows three different uniform distributions (depending on the dataset): short intervals ranging from 1 to 10^3, medium from 1 to 10^6, and wide from 1 to 10^9. The values to be aggregated are uniformly distributed within the domain $[0, 2^{31} - 1]$. For aggregation on fixed intervals both argument and group relations follow the same distribution, but are generated independently.

Real-World Datasets. We use six real-world datasets that differ in size and data distribution. The main properties of them are summarized in Table 2. Here n is the number of tuples, $\lvert r.T \rvert$ is the tuple interval length, "$r.T$ domain" is the domain width of the dataset and "$r.A$ #distinct" is the number of distinct values in the dataset.

The dataset *weather* consists of air temperature measurements over the course of eight years by 120 weather stations located in South Tyrol. The value of each tuple is the temperature measured at one station for a five-minute window. Another dataset, *basf*, contains NMR spectroscopy data describing the resonating frequencies of different atomic nuclei [8]. As these frequencies can shift, depending on the bonds an atom forms, they are defined as intervals. The value of each tuple is the number of lines found in a specific part of the spectrum (denoted by the tuple interval). Rather than using time as a domain, this dataset uses frequencies. The Incumbent (*inc*) dataset [7] records the history of employees assigned to projects over a sixteen year period at a granularity of days. The *feed* dataset records the history of measured nutritive values of animal feeds over a 24 year period at a granularity of days; a measurement remains valid until a new measurement for the same nutritive value and feed becomes available [5]. The *web* dataset [1] records the history of files in the SVN repository of the Webkit project over an eleven year period at a granularity of seconds. The valid times indicate the periods in which a file did not change. The *flight* dataset [2] is a collection of international flights for November 2014, start and end of the intervals represent plane departure and arrival times with minute precision.

For the aggregation on fixed intervals, the same relation was used as argument and as group relation. When varying the size of the group relation, we subsample it (taking every n-th tuple).

5.4 Results

Before doing an overall comparison of the different approaches, we discuss the storage requirements of MAX Skyline in more detail and also show how to tune the parameter k of a Top-K list.

Size of MAX Skyline. The worst-case complexity for inserting a tuple into the MAX Skyline is $O(\log n)$. However, very often we do not have to insert a new value, so the average complexity is much smaller. We measured the size of the MAX Skyline (in data points) aggregating our synthetic datasets and observed that even when the size of the relation went up to 100 million tuples, the MAX Skyline only reached a size of 45 to 46 points and the average size (averaged over all skyline operations) is only around 16 to 18 points; see full results in Table 3. We repeated these experiments with our real world datasets and obtained similar results.

Top-K List Tuning. For tuning k Kaufmann et al. used 0.01% of the number of distinct values found in the relation [10]. We checked this and found that it only works for very short intervals, very often we need a bigger value for k to make the Top-K list efficient. In Fig. 9 we show the performance of a MAX aggregation on constant intervals using 0.01%, 0.1%, 1%, 10% and 100% for k. As can be clearly seen in Fig. 9, a Top-K list with $k = 0.01\%$ is always found at the upper end of the run time. The spike in the medium interval case is caused by a Top-K list running empty and having to be repopulated. Real world datasets showed similar results. Overall, we achieved the best results for $k = 10\%$, which we therefore used for the further comparisons.

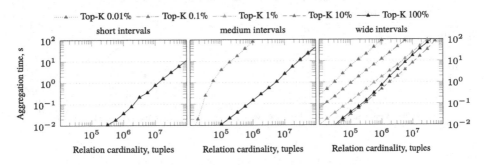

Fig. 9. Constant intervals, MAX, tuning Top-K list

Selective Aggregation on Constant Intervals. Next, we compare the performance of MAX Skyline to the Top-K list, SB-tree, and TMDA-CI algorithms. The run time results for the synthetic datasets are shown in Fig. 10, those for the real-world datasets in Fig. 11. These figures include the run time of index construction plus querying $(i+q)$ and just querying (q).

For the synthetic datasets (Fig. 10), we make the following observations. The MAX Skyline clearly outperforms the Top-K list, especially for medium and wide intervals, in some cases by up to an order of magnitude (note the logarithmic scale). TMDA-CI is clearly lagging behind, especially for medium and wide intervals, for which it is slower by orders of magnitude. The SB-tree without index construction is fastest (for medium and wide intervals it is so fast that it dropped out of the bottom of the figures). While at first glance it seems to be the clear winner, there are some drawbacks to it. Basically, an SB-tree index for MAX aggregation on constant intervals materializes the precomputed answer for one attribute, which means we have to build a different index for every attribute. On top of this, such an index is append-only. It is slower to create and query an SB-tree than it is to just query an Endpoint Index utilized by MAX Skyline. Even if we have to build an Endpoint Index, we are still competitive. For the Endpoint Index we only need to build one index per relation, it can be

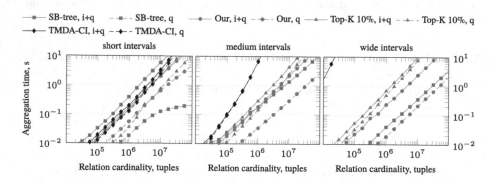

Fig. 10. Constant intervals, MAX, varying cardinality or **r**, synthetic dataset

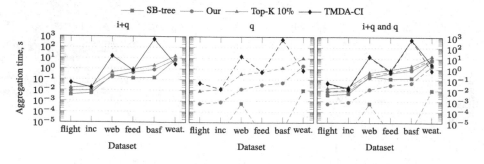

Fig. 11. Constant intervals, MAX, real world

updated without restrictions, and also be used much more universally, e.g. for other temporal operators, such as overlap joins.

In Fig. 11 we depict the results for the real-world datasets in three plots: one plot with index construction, one without, and one combining the two for better direct comparison. It shows a picture similar to the one for synthetic datasets.

Selective Aggregation on Fixed Intervals. Figure 12 compares our approach to an MSB-tree and TMDA-FI, showing that the SB-tree variant is not the uncontested winner anymore for this scenario. MAX Skyline is able to outperform the MSB-tree, with TMDA-FI lagging behind significantly (Top-K is missing, as it cannot support fixed intervals because by construction it only handles a single value). While in Fig. 12 the cardinality of the argument relation r and the grouping relation g is the same, in Fig. 13 we keep the size of r fixed at 10^8 tuples and vary the size of g, as we expect g to be smaller than r in practice. The performance of an MSB-tree improves a lot when we decrease the size of the grouping relation, since every group tuple triggers a query. We see a similar picture for the real-world datasets, the results of which are shown in Fig. 14. MSB-trees without index construction are fastest, followed by MAX Skyline and then TMDA-FI. Nevertheless, the arguments against SB-trees given in Sect. 5.4 still hold.

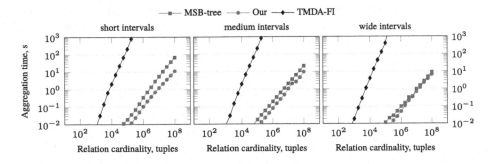

Fig. 12. Fixed intervals, MAX, varying r and g cardinalities, querying only

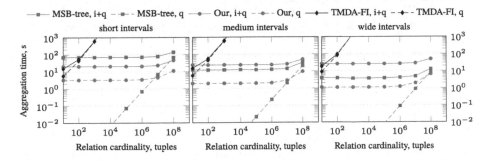

Fig. 13. Fixed intervals, MAX, $|r| = 10^8$, varying g cardinality

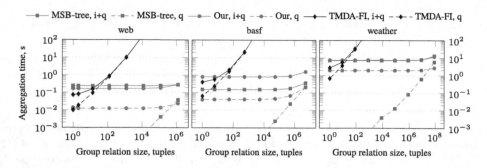

Fig. 14. Fixed intervals, MAX, real world data

Cumulative Aggregation on Fixed Intervals. Finally, we compare our approach with JSB-trees and TMDA-FI for cumulative aggregation (in this case
AVG) on fixed intervals. We structure the presentation of the results in the same
way as those for selective aggregation: Fig. 15 shows the run times for using the
same cardinality for **r** and **g** without constructing indexes, in Fig. 16 we vary
the size of **g**, and Fig. 17 depicts the results for real-world datasets. Generally,
the results look similar to those for selective aggregation, although it should be
noted that our approach generally performs better than an SB-tree (a JSB-tree
in this case). For ad-hoc queries (i.e., both, JSB-trees and our approach have
to build indexes on the fly) we are an order of magnitude faster. In the likely
scenario that an Endpoint Index already exists, the difference increases to two
orders of magnitude.

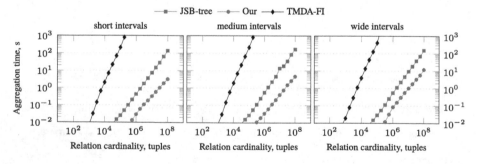

Fig. 15. Fixed intervals, AVG, varying **r** and **g** cardinalities, querying only

6   Conclusion

We developed a family of algorithms for temporal aggregation that uses Endpoint
Indexes, which have made their way into commercial temporal database proto-
types in the form of Timeline Indexes [10]. There, Timeline Indexes have already

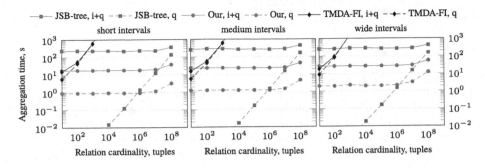

Fig. 16. Fixed intervals, AVG, $|r| = 10^8$, varying **g** cardinality

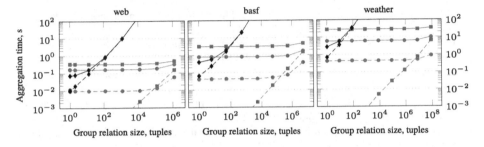

Fig. 17. Fixed intervals, AVG, real world data

proven their usefulness as a general index structure for temporal databases supporting a range of temporal operators, such as overlap joins, time travel, and aggregation on constant intervals. We show how they can also support cumulative and selective aggregation on fixed intervals, for selective aggregation we introduce a new main-memory data structure used during the processing, called MAX Skyline. MAX Skylines also improve the performance for evaluating selective aggregates on constant intervals significantly: they outperform the previously employed Top-K approach by an order of magnitude.

We also compared our approach experimentally with several state-of-the-art techniques, such as TMDA-FI, TMDA-CI, and SB-tree-based approaches. We are able to outperform TMDA-FI and TMDA-CI by several orders of magnitude. For SB-trees the situation is more complex. When precomputed SB-trees are available, they are often the fastest solution. When both, SB-trees and Endpoint/Timeline Indexes, have to be built on-the-fly, our approach is comparable with the SB-trees for selective aggregation and an order of magnitude faster for cumulative aggregation. Finally, when SB-trees are not available and Timeline Indexes are, our approach is up to one or two orders of magnitude faster. However, our approach has a decisive advantage: the algorithms will run on any system using a Timeline Index, while the SB-tree-based approaches require the construction of indexes for every attribute and every type of aggregation function. On top of that, SB-tree-based indexes for MIN and MAX are append-only.

References

1. The webkit open source project (2012). webkit.org
2. Behrend, A., Schüller, G.: A case study in optimizing continuous queries using the magic update technique. In: SSDBM (2014)
3. Böhlen, M., Gamper, J., Jensen, C.S.: Multi-dimensional aggregation for temporal data. In: Ioannidis, Y., Scholl, M.H., Schmidt, J.W., Matthes, F., Hatzopoulos, M., Boehm, K., Kemper, A., Grust, T., Boehm, C. (eds.) EDBT 2006. LNCS, vol. 3896, pp. 257–275. Springer, Heidelberg (2006). doi:10.1007/11687238_18
4. Borzsony, S., Kossmann, D., Stocker, K.: The skyline operator. In: ICDE, pp. 421–430 (2001)
5. Dignös, A., Böhlen, M.H., Gamper, J.: Overlap interval partition join. In: SIGMOD, pp. 1459–1470 (2014)
6. Elmasri, R., Wuu, G.T.J., Kim, Y.-J.: The time index: an access structure for temporal data. In: VLDB, pp. 1–12 (1990)
7. Gendrano, J.A.G., Shah, R., Snodgrass, R.T., Yang, J.: University information system (UIS) dataset. In: TimeCenter CD-1 (1998)
8. Helmer, S.: An interval-based index structure for structure elucidation in chemical databases. In: Melin, P., Castillo, O., Aguilar, L.T., Kacprzyk, J., Pedrycz, W. (eds.) IFSA 2007. LNCS (LNAI), vol. 4529, pp. 625–634. Springer, Heidelberg (2007). doi:10.1007/978-3-540-72950-1_62
9. Kaufmann, M.: Storing and Processing Temporal Data in Main Memory Column Stores. Ph.D. thesis, ETH Zurich (2014)
10. Kaufmann, M., Manjili, A.A., Vagenas, P., Fischer, P.M., Kossmann, D., Färber, F., May, N.: Timeline index: a unified data structure for processing queries on temporal data in SAP HANA. In: SIGMOD, pp. 1173–1184 (2013)
11. Kim, J.S., Kang, S.T., Kim, M.H.: On temporal aggregate processing based on time points. In: IPL, pp. 213–220, September 1999
12. Kline, N., Snodgrass, R.: Computing temporal aggregates. In: ICDE, pp. 222–231 (1995)
13. Moon, B., Fernando, I., López, V., Immanuel, V.: Scalable algorithms for large temporal aggregation. In: ICDE, pp. 145–154 (2000)
14. Piatov, D., Helmer, S., Dignös, A.: An interval join optimized for modern hardware. In: ICDE, pp. 1098–1109, May 2016
15. Saracco, C.M., Nicola, M., Gandhi, L.: A matter of time: temporal data management in DB2 10. Technical report, IBM (2012)
16. Yang, J., Widom, J.: Incremental computation and maintenance of temporal aggregates. VLDB J. **12**(3), 262–283 (2003)

Indexing the Pickup and Drop-Off Locations of NYC Taxi Trips in PostgreSQL – Lessons from the Road

Jia Yu$^{(\boxtimes)}$ and Mohamed Sarwat

School of Computing, Informatics, and Decision Systems Engineering,
Arizona State University, Tempe, AZ 85281, USA
{jiayu2,msarwat}@asu.edu

Abstract. In this paper, we present our experience in indexing the drop-off and pick-up locations of taxi trips in New York City. The paper presents a comprehensive experimental analysis of classic and state-of-the-art spatial database indexing schemes. The paper evaluates a popular spatial tree indexing scheme (i.e., GIST-Spatial), a Block Range Index (BRIN-Spatial) provided by PostgreSQL as well as a new indexing scheme, namely Hippo-Spatial. In the experiments, the paper considers five evaluation metrics to compare and contrast the performance of the three indexing schemes: storage overhead, index initialization time, query response time, maintenance overhead, and throughput. Furthermore, the benchmark takes into account parameters that affect the index performance, which include but is not limited to: data size, spatial query selectivity, and spatial area density, The paper finally analyzes the experimental evaluation results and highlights the key insights and lessons learned. The results emphasize the fact that there is no one size that fits all when it comes to indexing massive-scale spatial data. The results also prove that modern database systems can maintain a lightweight index (in terms of storage and maintenance overhead) that is also fast enough for spatial data analytics applications. The source code for the experiments presented in the paper is available here: https://github.com/DataSystemsLab/hippo-postgresql.

1 Introduction

The volume of available geospatial data increased tremendously and such data keeps evolving in unprecedented rates. For instance, New York City Taxi and Limousine Commission has recently released a taxi dataset (abbr. NYC Taxi) [1]. The dataset contains close to 200 Gigabytes of New York City Yellow Cab and Green Taxi trips. The dataset contains detailed records of over 1.1 billion individual taxi trips in the city from January 2009 through December 2016. Each record includes pick-up and drop-off dates/times, pick-up and drop-off precise location coordinates, trip distances, itemized fares, and payment method.

This work is supported by the National Science Foundation Grant 1654861.

© Springer International Publishing AG 2017
M. Gertz et al. (Eds.): SSTD 2017, LNCS 10411, pp. 145–162, 2017.
DOI: 10.1007/978-3-319-64367-0_8

(a) NYC Taxi Trips Heat Map   (b) Taxi Trips in the Laguardia Airport Regions

Fig. 1. NYC taxi trips

Figure 1(a) depicts a heat map of the NYC taxi trips. To make sense of the NYC taxi data, the first step is to digest the dataset in a database system. The user can then issue spatial queries using SQL, e.g., find all Taxi trips to Laguardia airport (see Fig. 1(b)).

To speed up such queries, a user may build a spatial index, e.g., R-tree, on the location or geometry attribute. Even though classic database indexes [8, 13] improve the query response time, they usually yield close to 15% additional storage overhead. Although the overhead may not seem too high for small databases, it results in non-ignorable cost in massive-scale spatial database scenarios, e.g., Taxi trips locations. Moreover, existing database systems take a lot of time in initializing and bulk loading the spatial index (e.g., R-Tree or Quad-Tree [10, 20, 24]) especially when the size of indexed spatial data reaches hundreds of Gigabytes or more. Furthermore, spatial indexes supported by state-of-the-art spatial database systems, e.g., PostGIS [4], are designed with the implicit assumption that the underlying spatial data does not change much. However, many modern applications constantly insert new spatial data into the database, e.g., inserting a new taxi trip record. Maintaining a database index incurs high latency since the DBMS has to locate and update those index entries affected by the underlying table changes. For instance, maintaining an R-Tree searches the tree structure and perhaps performs a set of tree nodes splitting or merging operations. That requires plenty of disk I/O operations and hence encumbers the time performance of the entire DBMS in update intensive application scenarios.

In this paper, we present our experience in indexing the drop-off and pick-up locations of NYC taxi trips. The paper presents a comprehensive experimental analysis of classic and state-of-the-art spatial database indexing schemes supported in PostgreSQL (a popular open source database system) [5, 23]. The paper evaluates a popular spatial tree indexing scheme (i.e., GiST-Spatial [2, 14, 15, 18]), a Block Range Index [3] (denoted as BRIN-Spatial) provided by PostgreSQL as well as a new indexing scheme, namely Hippo-Spatial [27, 28]. The experiments consider five main evaluation metrics, briefly described as follows: (1) Storage overhead: the extra storage space occupied by the spatial index structure, (2) Index initialization time: the time the system takes to create and bulk load the index. (3) Query response time: the time the database system takes to search the index and retrieve the corresponding spatial data (i.e., Taxi trips), (4) Maintenance overhead: the time the system takes to maintain the spatial index in response to data insertion or deletion. (5) Throughput: the number of data access operations, given a hybrid query/update workload, the database system can process on the indexed table in a given time period. Furthermore, the benchmark takes into account parameters that affect the index performance, which include but is not limited to: data size, spatial query selectivity, and spatial area density, The paper finally analyzes the experimental evaluation results and highlights the key insights and lessons learned. The results emphasize the fact that there is no one size that fits all when it comes to indexing massive-scale spatial data. The results also prove that modern database systems can maintain a lightweight index (in terms of storage and maintenance overhead) that is also fast enough for spatial data analytics applications.

The rest of the paper is organized as follows. Section 2 describes the spatial database indexing approaches considered in the benchmark. Section 3 describes the experimental environment and setup. Sections 4, 5, 6 and 7 explains the experimental evaluation results and their analysis. Finally, Sect. 8 highlights the key lessons learned from the benchmark.

2   Studied Spatial Database Indexing Schemes

This section gives an overview of the spatial database indexing schemes considered in the analysis. Section 2.1 summarizes the Generalized Search Tree (GiST-Spatial) indexing scheme while Sect. 2.2 gives an overview of Block Range Indexes (BRIN-Spatial). Section 2.3 highlights the details of the Hippo-Spatial indexing scheme.

2.1   Generalized Search Tree (GiST-Spatial)

Index structure. A Generalized Search Tree is a balanced search tree that accepts arbitrary data types including spatial data [15]. It holds the similar <key, pointer> tree structure like B-Tree or R-Tree [9,13,19,26] but the key varies according to the data type. To index spatial data [11–13,16,29], the key is 2 dimensional rectangle which is the Minimum Bounding Rectangle (MBR) of its child nodes. In a non-leaf node, the pointer points to its child node while in a leaf node, the pointer points to the parent table tuple. In other words, GiST-Spatial's basic idea is to group nearby spatial objects together and use a upper tree node stores their Minimum Bounding Rectangle (MBR) as well as pointers. Let m be the minimum allowed child nodes, N be the number of records, $Levels = \lceil log_m N \rceil$. A GiST-Spatial can contain up to $\sum_{i=1}^{Levels} \lceil N/m^i \rceil$ nodes. The GiST-Spatial bulk-loading [17] algorithm runs in a bottom-up fashion, which is indeed faster than inserting tuple one by one. The bulk loading algorithm generates plenty of tree nodes besides tuples pointers and, in practice, it writes many temporary files onto disk for scalability.

Index search. The index search algorithm takes as input a spatial rectangular range predicate. The algorithm starts at the root node and traverses the child nodes that satisfy the spatial predicate. The algorithm then prunes subtrees in GiST-Spatial, which possess MBRs that do not intersect with the spatial query predicate. The algorithm performs this step recursively until it reaches the tree leaf level and finally returns all spatial objects that lie within the spatial query range. The tree structure of GiST-Spatial offers fast index search on highly selective queries at the cost of excessive indexing and maintenance overhead.

Index maintenance. Insertion/Deletion in GiST-Spatial is similar to the R-Tree in the sense that it maintains a balanced tree structure. The update algorithm first traverses the tree and finds the node where the new index entry <key, pointer> should be inserted in/deleted from. For insertion, in case there is no space available, the target tree node will be split and GiST-Spatial will adjust other tree nodes to ensure the tree balance is preserved. For deletion, GiST-Spatial will merge tree nodes with extra space caused by data deletion and also adjust the tree nodes.

2.2 Block Range Index (BRIN-Spatial)

Index structure. As opposed to GIST, BRIN-Spatial is a sparse index [6,21, 22,25] that only stores pointers to disk pages in the indexed table. BRIN-Spatial groups pages into a fixed disk page range unit (128 pages per range by default). Each index entry in BRIN-Spatial contains two components: a static disk page range (e.g., page 1–10) and a Minimum Bounding Rectangle (MBR) that encloses all spatial data tuples that are recorded in the page range. The index initialization algorithm scans the indexed table only once to generate BRIN-Spatial. For each page range, BRIN-Spatial reads all tuples to construct the MBR for each index entry. An example of BRIN-Spatial is given in Fig. 2.

Index search. Given a spatial range query, the query processor only searches the index entries for which the MBRs intersect with the spatial query predicate. It is highly recommended to ensure that the indexed spatial objects physically maintain their spatial locality in a certain way, e.g., sorting by longitude/latitude coordinate or Hilbert curve. In that case, the index entries keep the minimal MBR overlap between each other. Under this premise, BRIN-Spatial is able to prune lots of disk pages without scanning them.

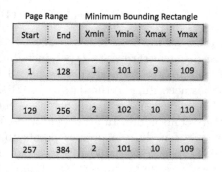

Page Range		Minimum Bounding Rectangle			
Start	End	Xmin	Ymin	Xmax	Ymax
1	128	1	101	9	109
129	256	2	102	10	110
257	384	2	101	10	109

Fig. 2. BRIN-Spatial

Index maintenance. For a newly inserted tuple, BRIN-Spatial first finds the page range it belongs to and then checks the tuple against the MBR of this page range. If the tuple is outside the MBR, BRIN-Spatial updates the MBR to cover the tuple otherwise BRIN-Spatial does nothing. For deletions, BRIN-Spatial does not update any index entries after delete tuples to improve performance. The underlying database will make a note on deleted tuples and make sure these tuples disappear from the returned tuples even if BRIN-Spatial returns them by mistake.

2.3   Hippo-Spatial

Index structure. Hippo[1] is a data-aware sparse index [28]. In context of spatial data, each Hippo (denoted as Hippo-Spatial) index entry is composed of two components: a dynamic disk page range and a histogram-based page range summary (depicted in Fig. 3). In the summary, specifically, the simplified histogram (called partial histogram), each bit shows whether the corresponding two dimensional bucket presents (1) in this page range or not (0). The histogram-based summary is extracted from the two complete load balanced 1D histograms on X and Y axises, respectively (visualized histograms given in Fig. 3). Such histograms are widely supported and naturally maintained by most existing DBMSs and execute with no much extra cost. Two 1D histogram buckets, one from X axis and one from Y, represent a 2D bucket. We number a 2D histogram bucket by its 1D buckets on X and Y. For example, bucket (1,1) represents the bucket on the lower-left corner of Fig. 3 histogram. Hippo-Spatial iterates each parent table tuple and groups as ranges contiguous similar pages (in terms of data distribution). In the partial histogram of each page range, distinct histogram buckets hit by tuples are marked as 1 in corresponding bits. Hippo-Spatial ensures that the partial histogram in each index entry has the same density:

$$Partial\ histogram\ density\ (D)\ =\ \frac{\#\ Buckets_{value=1}}{\#\ Buckets_{complete\ histogram}}$$

Index search. When a spatial range query is issued, the system first locates the histogram buckets cover/intersect/covered by the query predicate and outputs a partial histogram similar to the histogram-based summary maintained for each index entry. Then, the search algorithm reads each index entry and filters out the index entries for which the histogram-based summary has no common buckets with the query predicate. For all index entries that match the query predicate, the search algorithm inspects the corresponding disk pages and the qualified data tuples are returned.

Index maintenance. When a new tuple is inserted, Hippo-Spatial updates the index entries in an eager manner. It first finds the 2D histogram bucket where the tuple falls in and then runs a binary search on index entry sorted list to locate the page range which the tuple belongs to. The sorted list maintains a list of index entry pointers that are sorted in the ascending order of their start page ID. If this tuple hits a distinct histogram bucket, the partial histogram in Hippo-Spatial index entry will set the corresponding bit to 1; if no distinct buckets hit, Hippo-Spatial does nothing instead. On the other hand, Hippo-Spatial deletion

---

[1] Source code: https://github.com/DataSystemsLab/hippo-postgresql

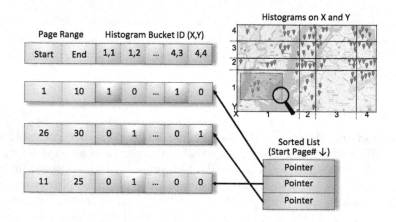

Fig. 3. Hippo-Spatial - index structure

runs in a lazy manner. This means Hippo-Spatial updates the index entries only for a batch of deletion operations. During the update, Hippo-Spatial scans the index entries and in case some tuples in a certain page range are deleted, Hippo-Spatial will re-summarize all pages in this page range and update the index entry.

3   Experimental Environment

We conduct the experiments on PostgreSQL 9.6 and PostGIS 2.3 with 128 MB default buffer pool. After fully loading the NYC city taxi trip dataset into PostgreSQL, the corresponding NYC Taxi trips table occupies 25 million PostgreSQL disk pages on the test machine. The size of PostgreSQL default buffer pool is rather small while the operating system memory is too large to be ignored. To avoid the impact of pre-cached data, we clear OS cache before each single transaction. We leverage the `EXPLAIN ANALYZE`, a PostgreSQL built-in performance analysis tool, to capture the execution time of all transactions and count the disk I/O operations. We use the default PostgreSQL 9.6 settings in all experiments. We use the `(CREATE INDEX)` (given below) to build the specified index on top of the NYC taxi trip table in PostgreSQL:

```
CREATE INDEX hippo_idx ON NYCTaxi USING Hippo (PickUpLocation);
```

All indexes are built on the NYC Taxi dataset pick-up location (i.e., latitude and longitude coordinate) attribute. For the sake of GiST-Spatial and BRIN-Spatial, the latitude and longitude coordinates are represented by a single coordinate attribute in PostgreSQL compatible geometry format. Two Hippo-Spatial indexes with the same configuration are built on latitude and longitude, respectively. BRIN-Spatial allows a parameter called Pages Per Range (P) which specifies the number of parent table pages summarized by each index entry. We use 32, 128 (default) and 512 to tune BRIN-Spatial. Hippo-Spatial accepts a parameter named Density (D) to control the partial histogram density inside each index entry. Its performance is also impacted by the number of buckets in the complete histogram (H). We choose three parameter combinations to tune Hippo-Spatial: (1) $D = 20\%$ $H = 400$ (default setting) (2) $D = 40\%$ $H = 400$ (3) $D = 20\%$ $H = 800$. All indexes use their default settings unless otherwise stated.

We issue a spatial range query on NYC Taxi table with a particular query window and qualified tuples are returned to the psql front-end. The format used in the experiments is:

```
EXPLAIN ANALYZE SELECT count(*) FROM NYCTaxi WHERE <predicate>;
```

The predicate represents a spatial range query window targeted at the pick-up attribute written in an index-dependent format. All insertions work in an eager manner to ensure the query correctness. In the experiments, we use the (INSERT INTO NYCTaxi VALUES (aTrip)) SQL command to inserts a new Taxi trip tuple in the NYC taxi trips table. We also use (COPY NYCTaxi FROM aFile) command to insert a batch of tuples in a single operation in order to avoid unnecessary I/O. Nonetheless, it still performs the insertion/index update tuple by tuple. In PostgreSQL, a DELETE operation just makes a note on the deleted tuples and hides them from the output instead of immediately removing them physically. That is due to the fact that clearing and recycling deleted tuples' physical space is a time-consuming process. All physical deletions and corresponding index updates only happen when the VACUUM command is invoked. The VACUUM command runs periodically but also accepts manual invocation from the user.

4 Studying the Indexing Overhead

This section studies the indexing overhead incurred by the three compared indexing schemes. We build three indexes on different sizes of the New York Taxi Trip data and record the corresponding overhead (Fig. 4) including index size and index initialization time. Results of using different index parameters are described in Figs. 5 and 6.

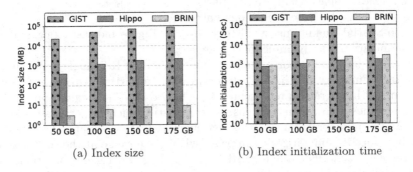

(a) Index size   (b) Index initialization time

Fig. 4. Indexing overhead on different data scales (logarithmic scale)

4.1 Index Size

As depicted in Fig. 4a, Hippo-Spatial occupies close to two orders of magnitude less storage space than GiST-Spatial. That happens due to the fact that GiST-Spatial stores the pointers of hundreds of millions of Taxi trips in the table and maintains a Minimum Bounding Rectangle in each tree node. On the other hand, Hippo-Spatial only stores disk page ranges and MBR summaries. A tuple pointer is a physical address that consists of a disk page ID and slot ID. Once the index search is completed, GiST-Spatial collects the pointers and passes them to the DBMS. Given a tuple pointer, the DBMS directly jumps to the specified address and retrieves the embedded tuple without any rechecks. Retrieving a small amount of pointers during queries is fast, yet storing 1.1 billion tuple pointers in an index is very space-consuming. In addition, each MBR is represented by four double values, minimum X and Y, maximum X and Y, also occupies non-negligible storage space. On the contrary, each Hippo-Spatial index entry only contains a disk page range and a concise summary. Generally speaking, a disk page may store 50–100 tuples, and that is why Hippo-Spatial incurs much less storage overhead.

As given in Fig. 4a, Hippo-Spatial leads to more storage overhead than BRIN-Spatial. That happens because Hippo-Spatial, as opposed to BRIN-Spatial, is data-aware and hence speeds up the search process. Each Hippo-Spatial index entry stores a histogram-based page summary instead of a simple MBR. Nonetheless, the extra storage space occupied by Hippo-Spatial is relatively small since its size is less than 1% of the indexed table.

Figure 5 studies the storage overhead of both BRIN-Spatial and Hippo-Spatial using different parameter settings. For instance, Hippo-D20%-H400 denotes a hippo index with density set to 20% and the number of histogram buckets set to 400 and BRIN-P128 denotes a BRIN-Spatial index with 128 pages per range. Hippo-Spatial occupies 100 times larger disk space than BRIN-Spatial. That makes sense because each index entry in Hippo-Spatial maintains a histogram-based summary of a dynamic page range

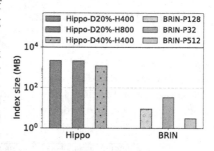

Fig. 5. Index size (log. scale)

while BRIN-Spatial only stores the Minimum Bounding Rectangle per each page range. Each summary in Hippo-Spatial represents a partial histogram and each bucket in this histogram is represented by a single bit. Although Hippo-Spatial compresses these partial histograms, they are still much larger than a simple MBR. As the number of pages per range increases, BRIN-Spatial occupies less disk space since it summarizes more pages within one range at the cost of slower query response time. For different Hippo-Spatial parameter combinations, The higher the histogram density, the more pages each Hippo-Spatial index entry summarizes. That will also lead to more tuples being summarized by each index entry. Maintaining the same density but increasing the total number of histogram buckets leads to an increase in the storage space occupied by Hippo-Spatial. That happens because more complete histogram buckets also leads to more tuples hitting more distinct buckets in each partial histogram.

4.2   Index Initialization Time

Figure 4b depicts the index initialization time incurred by creating each of the three indexes in PostgreSQL. The system takes the same time to bulk load Hippo-Spatial and BRIN-Spatial because each of them scans the indexed table tuple by tuple and summarizes each encountered tuple using an in-memory validation operation. The only difference is that, given a tuple, Hippo-Spatial finds the histogram bucket to which the tuple belongs using binary search while BRIN-Spatial checks whether the retreived

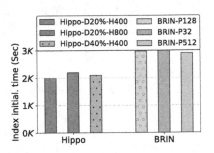

Fig. 6. Initialization time

tuple is covered by the temporary MBR and updates the MBR if needed. Moreover, PostgreSQL spends two orders of magnitude more time to bulk load GiST-Spatial compared to BRIN-Spatial and Hippo-Spatial. This happens because the initialization algorithm in GiST-Spatial is rather complex and requires a large number of temporary disk files to decide the boundries of the minimum bounding rectangles. Hence, the intensive disk I/O cost encumbers the initialization performance of GiST-Spatial.

Figure 6 depicts how a variety of parameters settings impact the initialization time of both BRIN-Spatial and Hippo-Spatial. Hippo-Spatial takes 30% less initialization time than BRIN-Spatial. That happens due to the fact that the index initialization algorithm makes use of a temporary in-memory data structure (denoted TmpEntry) to store the to-be-persisted index entry. For BRIN-Spatial and Hippo-Spatial, TmpEntry keeps summarizing new incoming tuples and updates MBR for BRIN-Spatial (partial histogram for Hippo-Spatial) if needed. This process continues until BRIN-Spatial reaches pages per range limit or Hippo-Spatial reaches the density limit. Then, TmpEntry will be serialized and persisted to disk. However, in most cases of Hippo-Spatial, the TmpEntry data structure is rarely updated because TmpEntry only notes distinct histogram buckets hit by the scanned tuples. Unlike Hippo-Spatial, BRIN-Spatial initialization algorithm keeps updating the MBR as long as the newly summarized tuple not fully covered by the MBR. Such frequent TmpEntry updates lead to the gap in the initialization time.

5   Evaluating the Query Response Time

This section studies the query execution performance time using each of the three considered indexing indexes. To identify the proper scenarios for different

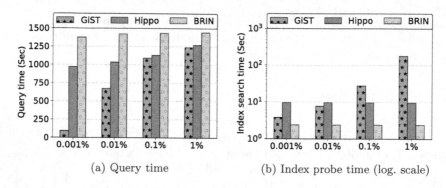

(a) Query time   (b) Index probe time (log. scale)

Fig. 7. Varying the spatial range query selectivity factor

indexes, we define two metrics of spatial range query: spatial range query selectivity and query range area size. The categorized results are given in Figs. 7 and 9.

5.1 Varying the Spatial Range Query Selectivity Factor

This section studies the impact of varying the spatial range query selectivity factor on the query response time. The selectivity factor of a given spatial range query is calculated as the ratio of the total NYC taxi trips returned by running the spatial range query over the total number trips stored in the database. We vary the average spatial range query selectivity from 0.001%, 0.01%, 0.1% to 1%. To generate the query workload with average selectivity, we first create GiST-Spatial index on the pick-up/drop-off location and randomly select a set of query points from the table. Then, we use each query point to issue a K Nearest Neighbors (KNN) searches on the NYC taxi table. The number K refers to the number of tuples returned by 0.001%–1% selectivity queries. For each KNN query, the returned K^{th} nearest neighbor and its mirror point against the query point represent a query range window that has the specified range selectivity. The generated spatial range queries are then used to run the experiments and the reported query execution time in Fig. 7 represents the average time PostgreSQL took to run the query workload.

As shown in Fig. 7, GiST-Spatial exhibits two orders of magnitude faster query execution performance than Hippo-Spatial and BRIN-Spatial on highly selective queries (0.001% selectivity factor). As the spatial range query selectivity factor becomes higher (lower selectivity), the query execution time gap between GiST-Spatial and Hippo-Spatial diminishes. For 0.1% and 1% selectivity factors, Hippo-Spatial is able to achieve similar query execution performance to that of GiST-Spatial. That happens due to the fact that, for highly selective queries (e.g., 0.001% selectivity), GiST-Spatial's balanced tree structure is able to prune disjoint subtrees and retrieve only a small amount of qualified NYC taxi tuples to recheck. On the other hand, Hippo-Spatial still has much more possible qualified page to inspect. For less selective queries (selectivity factor 0.1% and 1%), GiST-Spatial also has to retrieve more tuples for further inspection and that is why it has similar performance to that of Hippo-Spatial. However, BRIN-Spatial exhibits the slowest query execution performance as compared to GiST-Spatial and Hippo-Spatial. The main reason is that all Minimum Bounding Rectangles store with each index entry in BRIN-Spatial span the entire New York City metropolitan area and BRIN-Spatial actually inspects almost all disk pages occupied by the NYC taxi table to process queries with different selectivities.

Figure 7b describes the index probe time on different selectivity factors. The index probe time refers in particular to the time these indexes spend on searching index entries when a query is issued. That excludes the time the database system takes to read the data pages. For GiST-Spatial, the index probe time stands for the time GiST-Spatial used to find all qualified tuple pointers. The upcoming GiST-Spatial refine and data page retrieval phase is taken care of by PostgreSQL. For BRIN-Spatial and Hippo-Spatial, the index probe time represents the time these indexes spend on traversing all index entries. It is obvious that the index probe time for BRIN-Spatial and Hippo-Spatial is constant for all spatial range selectivity factors. That happens due to the fact that BRIN-Spatial and Hippo-Spatial always scan all index entries. On the other hand, for higher selectivity factors, GiST-Spatial have to expand its probe range and go to lower tree levels. Figure 7b shows that the index probe time of GiST-Spatial, in fact, increases exponentially.

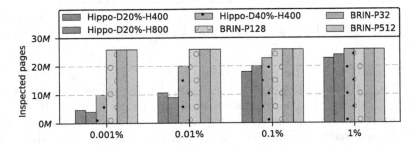

Fig. 8. Inspected data pages on different query selectivities

Figure 8 depicts the total number of inspected data pages using different index parameters. Both BRIN-Spatial and Hippo-Spatial need to inspect possible qualified pages for retrieving the truly qualified tuples. As given in Fig. 8, Hippo-Spatial inspects less pages than BRIN-Spatial. To be precise, Hippo-Spatial with 20% density inspects up to 6 times less NYC taxi data pages on 0.001% and 0.01% selectivity factors and BRIN-Spatial inspects up to 40% more disk pages for queries with 0.1% and 1% selectivity factors. That happens because Hippo-Spatial is able to prune more data pages since it only inspects page ranges which have joint histogram buckets with the spatial query predicate. Hippo-Spatial with 40% density and Hippo-Spatial with 800 histogram buckets (i.e., Hippo-D40%-H800) experience slower query execution time as compare to

Hippo-Spatial. The partial histograms of Hippo-D40%-H800 are too full and too many bits set to 1. That increases the probability that each index entry in Hippo-Spatial has joint buckets with the spatial query predicate. It is also worth noting that BRIN-Spatial in general (with various parameters setting) inspects the same number of data pages since it always inspects the entire table due to its data-agnostic nature.

5.2 Varying the Spatial Range Area Size

This section studies the impact of varying the size of the spatial range area. The range area represents the area covered by the issued spatial range query. In Sect. 5.1, we discussed the query response time for different query selectivity factors. However, users rarely issue spatial queries in strict accordance to the selectivity factor. Assume that a user observe the NYC taxi dataset on a web browser. The user usually searches dense areas. In fact, spatial data is alway highly skewed and sparse areas such as deserts are less interesting for analysts. We define two types of queries:

- random area spatial query (studied in Fig. 9b): To generate such queries, we issue spatial range queries in random locations that lie within the New York City region.
- dense area spatial queries (studied in Fig. 9a): To generate this workload, we limit the spatial queries to dense locations (e.g., Manhattan). A dense location contains a large number of Taxi trips. For instance, the hottest/densest data areas in New York Taxi dataset are Times Square, JFK airport and Laguardia airport.

Furthermore, we vary the range area size from $10^{-5}\%$ to 0.01%. Larger range area such as 0.001% or 0.01% exposes the region of a city while smaller range area such $10^{-5}\%$ exhibits the nearby businesses of our current location. Results are given in Fig. 9. As it turns out in Fig. 9b, GiST-Spatial achieves the best query execution performance for queries generated in random locations within NYC.

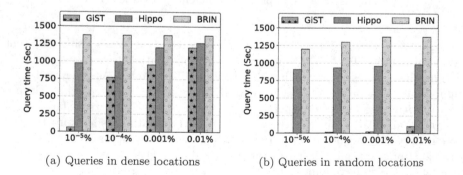

(a) Queries in dense locations   (b) Queries in random locations

Fig. 9. Query time issued in different spatial areas

That happens because spatial data is always skewed and most spatial range queries only return few tuples. On the contrary, in Fig. 9a, GiST-Spatial takes much more time for queries issued in dense areas of NYC. That is due to the fact that the number of taxi trip records in the Manhattan (i.e., dense) area are far more than other areas in New York City. Moreover, Hippo-Spatial exhibits just a bit slower query execution performance than GiST-Spatial. BRIN-Spatial, on the other hand, exhibits the slowest query execution performance since it has to inspect a large fraction of data pages.

6 Studying the Index Maintenance Overhead

This section studies the index maintenance overhead of all considered indexing schemes. We study the overhead incurred by two main index maintenance operations, i.e., insertion (see Fig. 10a) and deletion time (see Fig. 10b).

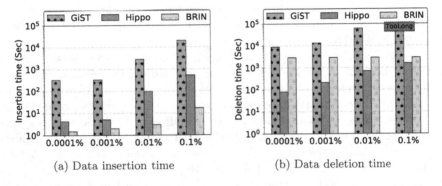

(a) Data insertion time   (b) Data deletion time

Fig. 10. Index maintenance performance on different data update percentage

6.1 Insertion Time

This section studies the time the database system takes to update the index when new taxi trip inserted in the NYC taxi table. Note that updating the index due to tuple insertion is deemed necessary to ensure the correctness of future queries. This section compares the three indexing schemes after inserting a certain amount of tuples in the NYC taxi table. We vary the number of inserted tuples as ratio of the original data size, i.e., 0.0001%, 0.001%, 0.01% and 0.1% tuples of the index NYC taxi table and insert them using the COPY FROM SQL clause.

As depicted in Fig. 10a, GiST-Spatial exhibits the highest index maintenance overhead when new tuples are inserted. That happens because GiST-Spatial spends too much time on locating the proper tree node. Furthermore, GiST-Spatial spends a non-ignorable amount of time on splitting the tree nodes to accommodate the newly inserted key. Frequent tree structure traverse and adjustments result in tremendous disk I/Os. Hippo-Spatial and BRIN-Spatial exhibit more than two orders of magnitude less maintenance overhead for insertion. That is due to the fact that both Hippo-Spatial and BRIN-Spatial possess a flat index structure which is relatively less complex than GiST-Spatialand hence easier to maintain. A newly inserted tuple leads to updating at most a single index entry. On the other hand, Hippo-Spatial takes more time to insert a new tuple in contrast to BRIN-Spatial. That happens because Hippo-Spatial checks each new tuple against the complete histogram and updates the corresponding on-disk partial histogram if this new tuple hits a new distinct histogram bucket. On-disk updates happens more frequently in Hippo-Spatial since BRIN-Spatial only does physical entry updates when the new tuple is outside the corresponding MBR.

6.2 Deletion Time

In this section, we evaluate the time PostgreSQL takes to maintain each of the three tested index structures in response to deleting a tuple(s) from the NYC taxi trip table. Similar to Sect. 6.1, we vary the percentage of deleted tuple to take 0.0001%, 0.001%, 0.01% and 0.1% values.

As shown in Fig. 10b, Hippo-Spatial achieves close to two orders of magnitude better performance than GiST-Spatial) in handling the DELETE operation. For the sake of batch deletion, Hippo-Spatial re-summarizes an index entry that contain many deleted tuples in one go meanwhile GiST-Spatial searches for the affected tree nodes and sometimes merges the affected tree nodes in response to tuple deletion. On the other hand, BRIN-Spatial follows a naive lazy update strategy that rebuilds the entire index after a fixed number of tuples is deleted from the indexed table. That explains why Hippo-Spatial achieves close to an order of magnitude better performance BRIN-Spatial on low deletion percentages. The performance gap slightly decreases when a large percentage of the table is deleted because Hippo-Spatial has to re-summarize most index entries in that case, which is equivalent to re-building the whole index.

6.3 Hybrid Workload Performance

Figure 11 compares the performance of three indexes in hybrid query/update workloads. We generated five query/update workloads that vary the percentage of issued search operations as compared to the update operations, named after the percentage of search operations in the entire workload: 10%, 30%, 50%, 70% and 90%. Each workload consists of a thousand operations, which represent either index search or data update operations. In the experiments, we measure the system throughput achieved for each workload. The throughput is measured in terms of the number of operations per second. In each workload, the average spatial query selectivity factor is set to 0.01% while the average number of updated tuples is set to 0.01%

As it turns out in Fig. 11, GiST-Spatial yields the lowest system throughput. That happens because GiST-Spatial spends too much time on index maintenance. BRIN-Spatial works faster than GiST-Spatial due to fast index maintenance although it incurs high latency when performing search operations. Hippo-Spatial consistently achieves the highest system throughput, as compared to BRIN-Spatial and GiST-

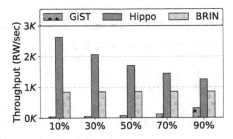

Fig. 11. Throughput

Spatial. That is explained by the fact that Hippo-Spatial exhibits better index maintenance performance than GiST-Spatial and also exhibits a competitive query response time. Although Hippo-Spatial is outperformed by BRIN-Spatial when performing insertion operations, Hippo-Spatial still achieves higher throughput than BRIN-Spatial given its relatively superior query execution performance and fast data deletion operations. In summary, we can conclude that Hippo-Spatial and BRIN-Spatial are more suitable for update-involved workloads while Hippo-Spatial outperforms BRIN-Spatial due to better query response time and faster data deletion.

7   Summary of Results

We summarize the results of the experimental evaluation as follows (see Table 1):

- *Indexing overhead:* Indexing overhead includes two factors: index storage overhead and initialization time. Hippo-Spatial and BRIN-Spatial occupy orders

Table 1. Summery of results

Metric	GiST-Spatial	Hippo-Spatial	BRIN-Spatial
Storage overhead	84 GB	2 GB	10 MB
Initialization time	28 h	30 min	45 min
Selectivity query	✓ 0.001% selectivity	✓ Selectivity between 0.01% and 1%	✗
Dense area query	✓ 10^{-5}% range query area	✓ range query area ≥ 10^{-4}%	✗
Index insertion	6 min for inserting 10^{-4}% data	4 s for inserting 10^{-4}% data	1 s for inserting 10^{-4}% data
Index deletion	2 h for deleting 10^{-4}% data	2 min for deleting 10^{-4}% data	Index rebuilt
Hybrid workload	✓ Query-intensive	✓ Balanced workload and update-intensive	✓ Update-intensive

of magnitude smaller index size as compared to GiST-Spatial. In addition, the index initialization time taken by the system to create GiST-Spatial is two orders of magnitude higher than the others.

- *Query response time:* Hippo-Spatial is two orders of magnitude slower than GiST-Spatial on very highly selective queries (selectivity factor $\leq 0.001\%$) but still holds competitive query response time on queries with selectivity factor between 0.01% and 0.1%. Another observation is that GiST-Spatial executes order of magnitude faster performance when executing spatial range queries over very small area such as $10^{-5}\%$. Also, Hippo-Spatial achieves competitive query time on larger range query area such as $10^{-4}\%$ and 0.001%. BRIN-Spatial always exhibits a slow query execution performance for all query area sizes.

- *Index maintenance overhead:* The data insertion time taken by Hippo-Spatial is ten times more than the time take by BRIN-Spatial, yet still 10 times faster than GiST-Spatial on all update percentages (0.0001% to 0.1%). Hippo-Spatial deletion speed is more than an order of magnitude faster than GiST-Spatial and 2–10 times faster than BRIN-Spatial. In hybrid workloads, Hippo-Spatial achieves two orders of magnitude higher throughput than GiST-Spatial on update-intensive workloads (10%, 30% queries) while GiST-Spatial has higher throughput on query-intensive workloads.

8   Key Insights and Learned Lessons

Through extensive experiments, we presented a comprehensive analysis of classic and state-of-the-art spatial database indexing schemes supported in PostgreSQL, GiST-Spatial, Hippo-Spatial and BRIN-Spatial. Below, we share our key insights through the following learned lessons:

- *Do not create GiST-Spatial (i.e., spatial tree index) when the database system is deployed on a storage device with high $ per GB.* The storage overhead introduced by GiST-Spatial created over the NYC taxi dataset is 84 GB (see Table 1), which is close to 50% of the original data size. Note that the dollar cost increases dramatically when the DBMS is deployed on modern storage devices (e.g., SSD and Non-Volatile-Ram) since they are still more than an order of magnitude expensive than classic Hard Disk Drives (HDDs). As per Amazon.com and NewEgg.com, the dollar cost per storage unit for HDD and SSD are 0.04 and 1.4 $/GB, respectively. Instead, the user may consider Hippo-Spatial and BRIN-Spatial to reduce the overall storage cost since these indexes only occupy between 0.1 and 1 % as compared to the original dataset.

- *Do not use BRIN-Spatial or Hippo-Spatial for Yelp-like applications.* Applications like Yelp usually issue very highly selective spatial range queries that retrieve point-of-interests (e.g., 0.001% range query selectivity) and present them to the end-user. As per the experiments, GiST-Spatial is deemed a perfect indexing scheme for Yelp-like applications given its superior performance in executive highly selective spatial range queries (see Table 1).

Furthermore, spatial data (i.e., Point-of-Interests) in Yelp are not dense. That is due to the fact that every longitude and latitude location on the surface of the earth contains a few (usually one) Point-of-Interests (or buildings).

– *Use Hippo-Spatial for spatial analytics applications over dynamic and dense spatial data.* NASA constantly collects Earth science data (e.g., weather, pollution, socioeconomic data) [7]. Earth science data is quite dense and new data is inserted into the system on a daily basis. Furthermore, since geospatial data in such applications is typically consumed as aggregate visualizations (e.g., Heatmap, Cartogram), spatial range queries on such data are not quite selective (selectivity factor between 0.1% and 1%) as in Yelp-like applications. Having said that, Hippo-Spatial is deemed the perfect for such data given: (1) its small storage footprint and low maintenance overhead compared to GiST-Spatial and (2) its superior query execution performance over selective queries and higher throughput compared to BRIN-Spatial.

References

1. New york city taxi and limousine commission. http://www.nyc.gov/html/tlc/html/about/trip_record_data.html
2. Aoki, P.M.: Generalizing "search" in generalized search trees. In: Proceedings of the 14th International Conference on Data Engineering, pp. 380–389. IEEE (1998)
3. Block range index. https://www.postgresql.org/docs/9.6/static/brin.html
4. Postgis - spatial and geographic objects for postgresql. http://postgis.net
5. Postgresql: a powerful, open source object-relational database system. https://www.postgresql.org/
6. Bontempo, C., Zagelow, G.: The IBM data warehouse architecture. Commun. ACM **41**(9), 38–48 (1998)
7. Earth science data. https://earthdata.nasa.gov
8. Comer, D.: Ubiquitous b-tree. ACM Comput. Surv. CSUR **11**(2), 121–137 (1979)
9. Corral, A., Vassilakopoulos, M., Manolopoulos, Y.: Algorithms for joining R-trees and linear region quadtrees. In: Güting, R.H., Papadias, D., Lochovsky, F. (eds.) SSD 1999. LNCS, vol. 1651, pp. 251–269. Springer, Heidelberg (1999). doi:10.1007/3-540-48482-5_16
10. Finkel, R.A., Bentley, J.L.: Quad trees: a data structure for retrieval of composite keys. Acta Inf. **4**(1), 1–9 (1974)
11. Fusco, F., Stoecklin, M.P., Vlachos, M.: Net-fli: on-the-fly compression, archiving and indexing of streaming network traffic. VLDB J. **3**(1–2), 1382–1393 (2010)
12. Goldstein, J., Ramakrishnan, R., Shaft, U.: Compressing relations and indexes. In: Proceedings of the International Conference on Data Engineering, ICDE, pp. 370–379. IEEE (1998)
13. Guttman, A.: R-trees: a dynamic index structure for spatial searching. In: Proceedings of the ACM International Conference on Management of Data, SIGMOD, pp. 47–57. ACM (1984)
14. Hellerstein, J.M.: Generalized search tree. In: Liu, L., Tamer Özsu, M. (eds.) Encyclopedia of Database Systems, pp. 1222–1224. Springer, US (2009)
15. Hellerstein, J.M., Naughton, J.F., Pfeffer, A.: Generalized search trees for database systems, September 1995

16. Kamel, I., Faloutsos, C.: Hilbert R-tree: an improved R-tree using fractals. In: Proceedings of the International Conference on Very Large Data Bases, VLDB, September 1994
17. Kamel, I., Khalil, M., Kouramajian, V.: Bulk insertion in dynamic R-trees. In: Proceedings of the International Symposium on Spatial Data Handling, SDH, pp. 31–42 (1996)
18. Kornacker, M., Mohan, C., Hellerstein, J.M.: Concurrency and recovery in generalized search trees. ACM SIGMOD Rec. **26**, 62–72 (1997). ACM
19. Lee, M.-L., Hsu, W., Jensen, C.S., Cui, B., Teo, K.L.: Supporting frequent updates in R-trees: a bottom-up approach. In: Proceedings of the International Conference on Very Large Data Bases, VLDB, pp. 608–619, September 2003
20. Samet, H., Webber, R.E.: Storing a collection of polygons using quadtrees. ACM Trans. Graph. TOG **4**(3), 182–222 (1985)
21. Sidirourgos, L., Kersten, M.L.: Column imprints: a secondary index structure. In: Proceedings of the ACM International Conference on Management of Data, SIGMOD, pp. 893–904. ACM (2013)
22. Ślezak, D., Eastwood, V.: Data warehouse technology by infobright. In: Proceedings of the ACM International Conference on Management of Data, SIGMOD, pp. 841–846. ACM (2009)
23. Stonebraker, M., Rowe, L.A.: The design of postgres. In: Proceedings of the ACM International Conference on Management of Data, SIGMOD, pp. 340–355. ACM (1986)
24. Tayeb, J., Ulusoy, Ö., Wolfson, O.: A quadtree-based dynamic attribute indexing method. Comput. J. **41**(3), 185–200 (1998)
25. Weiss, R.: A technical overview of the oracle exadata database machine and exadata storage server. Oracle White Paper. Oracle Corporation, Redwood Shores (2012)
26. Xu, X., Han, J., Lu, W.: RT-tree: an improved R-tree indexing structure for temporal spatial databases. In: Proceeding of the International Symposium on Spatial Data Handling, SSDH, pp. 1040–1049, July 1990
27. Yu, J., Moraffah, R., Sarwat, M.: Hippo in action: scalable indexing of a billion New York city taxi trips and beyond. In: Proceedings of the International Conference on Data Engineering, ICDE. IEEE (2017)
28. Jia, Y., Sarwat, M.: Two birds, one stone: a fast, yet lightweight, indexing scheme for modern database systems. Proc. VLDB Endowment **10**(4), 385–396 (2016)
29. Zukowski, M., Heman, S., Nes, N., Boncz, P.: Super-scalar RAM-CPU cache compression. In: Proceedings of the International Conference on Data Engineering, ICDE, pp. 59–59. IEEE (2006)

Towards Spatially- and Category-Wise k-Diverse Nearest Neighbors Queries

Camila F. Costa$^{(\boxtimes)}$ and Mario A. Nascimento

Department of Computing Science, University of Alberta, Edmonton, Canada
{camila.costa,mario.nascimento}@ualberta.ca

Abstract. k-nearest neighbor (k-NN) queries are well-known and widely used in a plethora of applications. In the original definition of k-NN queries there is no concern regarding diversity of the answer set, even though in some scenarios it may be interesting. For instance, travelers may be looking for touristic sites that are not too far from where they are but that would help them seeing different parts of the city. Likewise, if one is looking for restaurants close by, it may be more interesting to return restaurants of different categories or ethnicities which are nonetheless relatively close. The interesting novel aspect of this type of query is that there are competing criteria to be optimized. We propose two approaches that leverage the notion of linear skyline queries in order to find spatially- and category-wise diverse k-NNs w.r.t. a given query point and which return all optimal solutions for any linear combination of the weights a user could give to the two competing criteria. Our experiments, varying a number of parameters, show that our approaches are several orders of magnitude faster than a straightforward approach.

1 Introduction

k-Nearest neighbor (k-NN) queries [14] have been extensively studied by researchers in several areas, such as databases and data mining, information retrieval, pattern recognition and statistics. Namely, given a query point q, a set of points P and a distance metric, a k-NN query finds the set of k points $P' \in P$ such that no other point in $P \setminus P'$ is closer to q than those in P' according to that metric.

One potential drawback of k-NN queries is the fact that the k-NNs are determined based *only* on their distance to the query and no assumption is made on how they relate to each other. Providing homogeneous result sets, i.e., ones where the elements are very similar to each other, may not add much information to the query result as whole. One such scenario is document ranking [3], where returning k documents that are close to the query, but diverse among themselves is likely to increase the amount of overall information one gathers.

This concern has motivated some research to incorporate diversity into similarity searches, i.e., to find elements that are as close as possible to the query,

This research has been partially supported by NSERC, Canada and CNPq's Science Without Borders program, Brazil.

M. Gertz et al. (Eds.): SSTD 2017, LNCS 10411, pp. 163–181, 2017.
DOI: 10.1007/978-3-319-64367-0_9

while, at the same time, being as diverse as possible. The relation between diversity and closeness has been mainly investigated within the domain of information retrieval, e.g., [3,5], recommendation systems, e.g., [19], web retrieval, e.g., [13] and spatial query processing [11,20].

In this paper we investigate k-Diverse NNs (kDNNs) queries by considering two different notions of diversity, namely spatial and categorical diversities, in typical spatial k-NN queries. In *spatial diversity* the diversity between two data points is given by the distance between them. For instance, a tourist visiting a city may want to explore points-of-interest that are not too far from his/her location, but that at the same time cover different parts of the city. Regarding *categorical diversity*, the diversity is modeled by the difference between categories (or labels) of data points. As an example, consider a user looking for restaurants close by, in which case it may be interesting to return different types of restaurant so that the user could make a decision based on diversified options.

As we shall discuss shortly, previous works that dealt with similar problems, i.e., balancing closeness and diversity, suffer from two main shortcomings: (1) they rely on an user-provided linear combination of the relative importance of closeness over diversity, and (2) they do not provide optimal solutions, given the problem is NP-hard [4]. The approaches we propose in this paper overcome both such shortcomings. In order to find *all* results that are *optimal* under *any* given arbitrary combination of two competing criteria (e.g., closeness and diversity) we rely on the notion of skyline queries [2], more specifically linear skylines [15]. The result set of a skyline query contains elements which are not dominated by any other element. In the context of kDNN queries, a solution is not dominated if there is no one with higher closeness *and* diversity than its own. Linear skyline queries have been proposed as a way to reduce the amount of results to a more intuitive set. Two important properties of linear skylines is that they typically return (1) a much smaller subset of the conventional skyline which (2) are optimal under any linear combination of the competing criteria. The linear skyline obtained from the conventional skyline shown in Fig. 1a can be visualized in Fig. 1b.

The main contributions of this paper are two algorithms to find optimal answers for spatially- and category-wise kDNN queries using linear skylines.

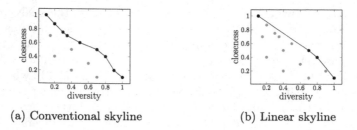

(a) Conventional skyline   (b) Linear skyline

Fig. 1. Illustration of conventional vs linear skylines, where the linked dark dots denote the frontier of non-dominated solutions returned by each approach.

In our proposed algorithms the candidate subsets are generated in decreasing order of closeness to the query point, i.e., increasing order of distance. This facilitates checking whether a candidate is part of the skyline or not, since all posterior subsets must have higher diversity than the previous ones in order to be non-dominated. Our extensive experiments, investigating several different parameters, show that our proposals find all optimal solutions for any given linear combination of the optimization criteria several orders of magnitude faster than a straightforward approach.

The remainder of this paper is structured as follows. In Sect. 2 we present a brief discussion of related work. Next we formally define kDNN queries. Our proposed solutions are presented in Sect. 4. The experimental evaluation and results are shown and discussed in Sect. 5. Finally, Sect. 6 presents a summary of our findings and suggestions for further work.

2 Related Work

The concept of incorporating diversity into similarity searches has its origins in information retrieval. The Maximal Marginal Relevance (MMR) model [3] is one of the earliest proposals to consider diversity to re-rank documents in the answer set. At each step, the element with higher marginal relevance is selected. A document has high marginal relevance if it is both relevant to the query and has minimal similarity to previously selected documents.

In [18] Vieira *et al.* survey several other approaches, but since they claim that their proposed methods, Greedy Marginal Contribution (GMC) and Greedy Randomized with Neighborhood Expansion (GNE), are superior (and also due to restricted space) we will review those two only. Similarly to MMR, GMC constructs the result set incrementally and the elements are ranked based on their maximum marginal contribution (a concept similar to maximal marginal relevance in [3]). In the GNE approach, a predefined number of result sets is found and the one that maximizes the optimization function is returned. Differently from GMC, in each iteration the element chosen is a random element among the top ranked ones. For each result set R constructed, the algorithm then performs swaps between elements in R and elements that offer a great chance to be more diverse. If in any of these swaps lead to a better result, according to the optimization function, then it is set as the best solution. We choose to compare our proposed approaches to GNE because, although it is an approximate solution, it is the one that presented the highest precision w.r.t. the optimal solution (obtained through an exhaustive brute-force approach). Moreover, it can support both types of diversity considered in this paper.

In the context of spatial diversity, several other approaches have been proposed. Jain *et al.* [10] consider the problem of providing diversity in the results of k-NN queries. The result set is constructed incrementally and the elements are analyzed in decreasing order of closeness. Lee *et al.* [12] presented the Nearest Surrounder query, which aims at finding the nearest objects from a query point from different angles. The work proposed by Abbar *et al.* [1] strives to find the most diverse set within a predefined radius in a Hamming space. Kucuktunc *et al.* [11] investigated the diversified k-NN problem based on angular similarity. Finally, Zhang *et al.* [20] studied the problem of diversified spatial keyword search on road networks. These works focused on spatial diversity only, i.e. do not support the categorical diversity dimension considered in this paper. Moreover, all the approaches presented above propose approximate solutions to the problem of optimizing a given linear combination of similarity and diversity. We, on the other hand, find the set of all optimal solutions for any linear combination of these criteria, i.e., the linear skyline, introduced next.

The skyline operator was first introduced in [2]. Given a d-dimensional data set, the skyline query returns the points that are not dominated by any other point. In the context of kDNN queries, a solution S is not dominated if there is no solution S' with higher closeness and diversity than S. One interesting aspect of skyline queries is that the user does not need to determine beforehand weights for closeness and diversity. The skyline query provides the user with a set of multiple and equally interesting solutions in the sense they are all non-dominated, for arbitrary weights. The users then make their decision based on varied options. A drawback of skyline queries is that it may return a large number of points, which may make it difficult for the user to interpret the results and choose the solution that better suits their preference. To overcome this problem [9,16,17] have proposed to reduce the amount of results by focusing on finding the k skyline points that best diversify the skyline result. On the other hand, [15] proposes the more pragmatic concept of linear skyline queries. A linear skyline consists of a relatively small subset of the conventional skyline which is optimal under *all* linear combination functions of the criteria to be optimized. Thus, in this paper we rely on the notion of linear skylines.

3 Problem Definition

The k-Diverse Nearest Neighbor (kDNN) problem incorporates diversity into the traditional k-NN problem. The goal is to find a set of solutions with k elements that are close to the query q while being as diverse among them as possible.

Let $P = \{p_1, p_2, \ldots p_n\}$ be a set of data points. Throughout this paper we refer to the closeness of a point $p_i \in P$ w.r.t the query q as the opposite of their Euclidean distance $d(p_i, q)$. Thus maximizing the closeness of p_i w.r.t q is the same as minimizing $d(p_i, q)$. Using this definition, the distance from q to a subset $S \subseteq P$ is defined as the maximum distance between q and any point in S, i.e., $d(q, S) = \max_{s_i \in S}\{d(q, s_i)\}$.

We consider two different notions of diversity in this paper: spatial and categorical. In spatial diversity, also adopted in [7,20], the diversity between two points $p_i, p_j \in P$ is given by the distance between them, i.e., $div(p_i, p_j) = d(p_i, p_j)$. The idea is to find points that are well spatially distributed and consequently cover different areas.

Regarding categorical diversity, we assume that there is a set of categories $G = \{g_1, g_2, ..., g_{|G|}\}$ and that each point $p_i \in P$ is labeled with a category denoted by $g(p_i) \in G$. Note that categories are non-exclusive, i.e., more than one point may belong to the same category. We model the diversity between categories as a matrix $M_{|G| \times |G|} = (m_{p,q})$, and for simplicity we assume that $0 \leq m_{p,q} \leq 1$ and in particular $m_{p,q} = 0$ if $p = q$. We can then represent the diversity between two categories g_p and g_q by the element $m_{p,q}$ in M. The diversity between a pair of points $p_i, p_j \in P$ is then given by $m_{g(p_i),g(p_j)}$, i.e., the diversity between their categories. Finally, the diversity within a set $S \subseteq P$ is defined as the minimum pairwise diversity between the points of S, i.e., $div(S) = \min\limits_{(p_i,p_j) \in S \wedge (p_i \neq p_j)} \{div(p_i, p_j)\}$.

Let us now define the notions of conventional and linear skyline domination.

Definition 1. *Let R and T be subsets $R, T \subseteq P$ of size k. Then, in the context of kDNN queries, R dominates T, denoted as $R \prec T$, if*

$$d(q, R) < d(q, T) \quad and \quad div(R) \geq div(T) \quad or$$
$$d(q, R) \leq d(q, T) \quad and \quad div(R) > div(T)$$

That is, R is better in one criteria and at least as good as T in the other one. Thus the conventional skyline, i.e., the set of non-dominated subsets, is given by

$$\{S \subseteq P \ s.t. \ \nexists R \subseteq P : R \prec S, |S| = |R| = k\}.$$

In our proposed approaches we first find the conventional skyline and then, in a posterior step, we extract the linear skyline from the conventional one.

Definition 2. *Let SK be a conventional skyline. A subset $SK' \subseteq SK$ linearly dominates a solution $T \in SK$, denoted as $SK' \prec_{lin} T$, iff*

$$\forall w \in \mathbb{R}^2 \ \exists R \in SK' : w^T r > w^T t$$

where w is a 2-dimensional weight vector and r (t) is a vector representing the closeness and diversity of R (T). The maximal set of linearly non-dominated solutions is referred to as linear skyline [15].

Note that in contrast to conventional domination, testing for linear domination requires comparing a solution to more than one other solution.

Finally, the problem addressed in this paper can now be defined as follows.

Problem Definition. *Given a query q, a positive integer k and a set of data points P, the k-Diverse NN (kDNN) query aims at finding the set of all k-sized linearly non-dominated subsets $S \subseteq P$.*

4   Proposed Solutions

As our main contribution in this paper we present two algorithms to the optimal kDNN problem using linear skylines. Both algorithms find the conventional skyline with the linear skyline being extracted in a posterior step. They analyze the candidate subsets in increasing order of $d(q, S)$, which means that the following subsets necessarily need to have a higher diversity than the previous ones in order to be non-dominated. This allows us to establish a lower bound to the diversity of the following non-dominated result sets. Throughout this paper this lower bound is denoted $minDiv$. Both proposed solutions, referred to as *Recursive Range Filtering (RRF)* and *Pair Graph (PG)*, use such lower bound to filter out dominated subsets. *RRF* works recursively by combining a partial solution S' with points in a candidate set C, one at time, that have a higher diversity to S' than $minDiv$. *PG* strives to reduce the number of generated combinations by pruning subsets that contain pairs of elements $(p_i, p_j) \subseteq P$ such that $div(p_i, p_j) \leq minDiv$ and therefore could not be in a non-dominated solution.

Before introducing *RRF* and *PG* we establish as a comparison baseline a straightforward, brute-force algorithm *(BF)*. *BF* generates all possible subsets $S \subseteq P$ of size k in an arbitrary order. If S is not dominated by any other solution in the *skyline* set, then S is added to *skyline*. When a new solution S is added to *skyline*, we check if no previous solutions in *skyline* are dominated by S. If such solutions exist, they are removed from *skyline*.

Our proposed solutions, *RRF* and *PG*, are general in the sense that they work for both categorical and spatial diversity. However, note that for categorical diversity only the closest points from each category are needed in order to find the whole skyline set. Farthest points would lead to solutions with lower closeness and that would not have higher diversity, since the diversity between two points is determined by their categories. Therefore, such solutions would be dominated. This means that the algorithms can stop once the closest points from each category have been found. *RRF* and *PG* are both based on Algorithm 1, presented next. They differ in how they find the set with maximum diversity for a fixed closeness, which is done by the method *findSetMaxDiv* used within Algorithm 1. Since each of our proposed solutions has its own implementation of *findSetMaxDiv*, they are presented in turn next.

Algorithm 1 takes as input a query point q, the size of the answer sets k and the set of data points P. The first subset S generated is the closest one, i.e., the one with minimum $d(q, S)$ (line 1). The elements $s_i \in S$ are removed from P (line 2) and added to the set of previously examined points P' (line 3). S is then added to *skyline*, since it is guaranteed to be a non-dominated solution, and $minDiv$ is set to $div(S)$. The variable $minDiv$ represents the minimum diversity that the following solutions must have in order to be non-dominated. All subsets S' with $div(S') \leq minDiv$ are discarded since they are dominated solutions and do not belong to the *skyline*. Next, the remaining elements $p_d \in P$ are examined in increasing order of $d(q, p_d)$ (lines 6 and 7). For each p_d, the method *findSetMaxDiv* looks for the set S with maximum $div(S) > minDiv$ such that $p_d \in S$ and $S \setminus p_d \subseteq C$. Where C is a subset of P' containing all $p_i \in P'$ such that $div(p_d, p_i) > minDiv$ (line 9). This step eliminates the elements that can not be combined with p_d in order to obtain a non-dominated solution.

Note that $d(q, S) = d(q, p_d)$ since p_d is the farthest point from q in S. Also, for a particular p_d, all subsets examined by *findSetMaxDiv* have the same distance, since they necessarily contain p_d. This implies that for each p_d there is up to one non-dominated solution, which is the one with highest diversity. At each step, the current p_d is removed from P and added to P'. Moreover, if a non-dominated solution S is found by *findSetMaxDiv*, S is added to *skyline* and *minDiv* is updated to $div(S)$ (lines 11 to 14).

Algorithm 1. MinDistMaxDiv

Data: Query q, set of data points P and result set size k
Result: All non-dominated solutions $S \subseteq P$, $|S| = $ k
1   $S \leftarrow k\text{-}NN(q, P)$;
2   $P \leftarrow P \setminus S$;
3   $P' \leftarrow S$;
4   $skyline \leftarrow S$;
5   $minDiv \leftarrow div(S)$;
6   **while** $|P| \; \textit{¿} \; 0$ **do**
7   $\quad$ $p_d \leftarrow \underset{p_i \in P}{\mathrm{argmin}}(dist(q, p_i))$;
8   $\quad$ $P \leftarrow P \setminus p_d$;
9   $\quad$ $C \leftarrow$ all $p_i \in P'$ such that $div(p_d, p_i) > minDiv$;
10 $\quad$ $S \leftarrow findSetMaxDiv(args)$;
11 $\quad$ **if** S *is not null* **then**
12 $\quad\quad$ $skyline \leftarrow skyline \cup S$;
13 $\quad\quad$ $minDiv \leftarrow div(S)$;
14 $\quad$ **end**
15 $\quad$ $P' \leftarrow P' \cup p_d$;
16 **end**
17 **return** skyline;

Next, we prove that the set of solutions generated by Algorithm 1 is correct and complete. For this we assume that the *findSetMaxDiv* method is correct and finds the corresponding non-dominated solution for a given $p_d \in P$, if such solution exists. We prove such assumption, i.e., the correctness of the *findSetMaxDiv* method used in each of our proposed solutions in the following subsections, when *RRF* and *PG* are presented.

Theorem 1. *The skyline set does not contain dominated solutions.*

Proof. The first solution S found by Algorithm 1 is the closest one and thus is not dominated. All consecutive solutions can not dominate previous solutions in terms of closeness, since they are found in increasing order of distance. Therefore, those solutions must have a higher diversity than minDiv, the highest diversity found so far, in order to be non-dominated. For each point $p_d \in P$, a new solution S is returned by the $findSetMaxDiv$ method only if $div(S) > minDiv$.

Moreover, findSetMaxDiv finds the corresponding non-dominated solution for p_d, i.e., the one with highest diversity, if such solution exists. Therefore, no dominated solutions can be part of skyline. ☐

Theorem 2. *All non-dominated solutions are in the skyline set.*

Proof. Let us suppose by contradiction that there is a non-dominated solution S that is not part of the skyline. Assuming that $d(q, S) = d(q, p_d)$, $p_d \in P$, implies that S was not found by the findSetMaxDiv method when p_d was the current element with minimum distance in P. This is a contradiction since findSetMaxDiv is guaranteed to find the corresponding non-dominated solution for any $p_d \in P$ if such solution exists. (This last statement is proven in the following, in the context of each proposed method.) ☐

4.1 Recursive Range Filtering (RRF)

The *findSetMaxDiv* method used in the *RRF* solution, which in this context we denote as *RRF-findSetMaxDiv* to remove any ambiguity, assumes that p_d, the point in P with minimum distance, is part of the non-dominated result set S, if such set exists. Also, as explained above, p_d is the farthest element from q in S. A candidate set C, containing points that can be combined with the partial set $S' = \{p_r\}$ in order to obtain a non-dominated solution, is given as input to *RRF-findSetMaxDiv*. One element c_i from C is chosen at a time to also be part of S'. Elements that can not be combined with c_i are removed from C and the algorithm is recursively called for the new partial set S' and candidate set C.

Algorithm 2 shows how the *RRF-findSetMaxDiv* method works. It takes as input a partial set S', which initially only contains the point p_d, a set C of candidate points that can be combined with S', the number n of elements that need to be added to S' to form a set of size k and the current set with maximum diversity *setMaxDiv*, which is initially empty. If $n > 1$ (line 9), the candidates c_i in C are examined one at a time. More specifically, we check whether the candidate set C_i obtained after adding c_i to S' contains at least $n - 1$ points (line 14). If so, the function is recursively called for the new partial set S'_i and candidate set C_i (line 16). Otherwise, it is not possible to obtain a non-dominated solution S such that $S'_i \subseteq S$ and therefore S'_i can be discarded. Note that we avoid creating the same subset more than once by not adding previously examined candidate points $c_j \in C$ such that $1 \leq j < i$ to C_i (lines 11 and 12). That is because all the possible sets containing $S' \cup c_j$ have already been created, including the ones that contain c_i, and adding c_j to the candidate set C_i would just create duplicates and make the algorithm less efficient.

In the base case (line 1), only one point still need to be added to S'. We then simply create a solution S for each $c_i \in C$ and set the one with highest diversity to *setMaxDiv*. When no other set of size k can be built, the algorithm returns *setMaxDiv* (line 21).

Figure 2 illustrates one iteration of Algorithm 2 for $k = 4$, $p_d = p_6$ and considering spatial diversity. The points $P' = \{p_1, p_2, p_3, p_4, p_5\}$ have already been

---

Algorithm 2. *RRF-findSetMaxDiv*

Data: Partial set S', candidate set C, $n = k - |S'|$, $setMaxDiv$ and $minDiv$

1 **if** *n=1* **then**
2   **for** $c_i \in C$ **do**
3          $S \leftarrow S'$;
4          $S \leftarrow S \cup c_i$;
5   **if** $div(S) > div(setMaxDiv)$ **then**
6             $setMaxDiv \leftarrow S$;
7   **end**
8   **end**
9 **else**
10   **for** $c_i \in C$ **do**
11   **if** $|C| - i - 1 \geq n - 1$ **then**
12             $C_i \leftarrow C[i+1, |C|]$;
13             $C_i \leftarrow$ all $c_j \in C_i$ such that $div(c_i, c_j) > minDiv$;
14   **if** $|C_i| \geq n - 1$ **then**
15                 $S'_i \leftarrow S' \cup c_i$;
16                 $setMaxDiv \leftarrow$
                   $findSetMaxDiv(S'_i, C_i, n - 1, setMaxDiv, minDiv)$;
17   **end**
18   **end**
19   **end**
20 **end**
21 **return** $setMaxDiv$;

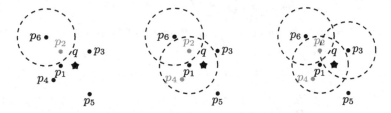

Fig. 2. *RRF-findSetMaxDiv* execution for $p_d = p_6$, $k = 4$ and spatial diversity.

examined and the following skyline solutions have been generated $skyline = \{(\{p_1, p_2, p_3, p_4\}, d = 0.9, div = 0.5), (\{p_2, p_3, p_4, p_5\}, d = 1.12, div = 1)\}$. Therefore at this point $minDiv = 1$.

The method takes as input $S' = \{p_6\}$, $C = \{p_1, p_3, p_4, p_5\}$, $n = 3$ $(k - 1)$ and $setMaxDiv = \emptyset$. Note that, as shown in Fig. 2, p_2 can not be combined with p_6 and thus is not part of C. The first candidate point p_1 is added to S', which discards p_4 from C. The algorithm is then recursively called for $S' = \{p_6, p_1\}$, $C = \{p_3, p_5\}$. The candidate point p_3 is added to S' and the only remaining candidate is $C = \{p_5\}$. The base case, where $n = 1$, is reached and a solution is created for each point in C. The solution $S_1 = \{p_6, p_1, p_3, p_5\}$ with $div(S_1) = 1.12$ is built and becomes the current $setMaxDiv$. When the second candidate point of $C = \{p_3, p_5\}$, i.e., p_5, is considered to be added to $S' = \{p_6, p_1\}$ (omitted in the example), the corresponding candidate set will be empty. As explained above, this avoids creating the same set S_1 twice. Next, the second candidate from $C = \{p_1, p_3, p_4, p_5\}$ is added to $S' = \{p_6\}$, creating the new candidate set $C = \{p_4, p_5\}$. Note that p_1 is not added to C because all the possible sets containing p_1 have already been generated. The solution $S_2 = \{c_3, c_4, c_5, c_6\}$ with $div(S_2) = 1.35$ is found and becomes the current $setMaxDiv$. When p_4 and p_5 are considered to be added to $S' = \{p_6\}$ (omitted in the example), they are pruned because there are not enough points to be combined with them in order to create a non-dominated solution that has not yet been generated.

Theorem 3. *For any p_d, the element in P with minimum distance, the method RRF-findSetMaxDiv finds the corresponding non-dominated solution S, where $dist(q, S) = dist(q, p_d)$, if such a solution exists.*

Proof. Let us assume that S exists and the method does not find it. This means that at least one $s_i \in S$, $s_i \neq p_d$ has been discarded by RRF-findSetMaxDiv. Let us analyze the two possible cases:

1. $div(s_i, s_j') \leq minDiv$ holds for at least one $s_j' \in S'$, $S' \subseteq S \setminus s_i$.
   This is a contradiction to the assumption that S is a non-dominated solution and thus $div(s_i, s_j) > minDiv$ for all $s_i, s_j \in S$.
2. There is no subset $S' \subseteq S \setminus s_i$ such that $S' \cup s_i$ can be combined with at least $n = k - |S'| - 1$ points.
   This is also a contradiction to the fact that S is a solution with k points and thus $S' \cup s_i$ can be combined with $S \setminus \{S' \cup s_i\}$.

Thus, such set S must be found by the method RRF-findSetMaxDiv.   □

4.2   Pair Graph Solution (PG)

The idea behind the PG solution is to combine pairs of elements (p_i, p_j) such that $div(p_i, p_j) > minDiv$ in order to construct sets of size k that can potentially be a non-dominated solution. Two pairs can be combined if they have an element in common. For instance, (p_i, p_j) can be combined with (p_j, p_l) if $div(p_i, p_j) > minDiv$ and $div(p_j, p_l) > minDiv$, obtaining the set $S_3 = \{p_i, p_j, p_l\}$ of size three. Similarly, S_3 can be combined with a pair (p_l, p_m) to obtain the set $S_4 = \{p_i, p_j, p_l, p_m\}$ of size four if $div(p_l, p_m) > minDiv$. However, note that there is no guarantee that $div(S_4) > minDiv$, unless $div(p_i, p_l) > minDiv$, $div(p_i, p_m) > minDiv$ and $div(p_l, p_m) > minDiv$.

A non-dominated solution $S \subseteq P$ for which the candidate point p_d is the farthest one from q, can only contain previously examined elements $p_i \in P'$ such that $div(p_d, p_i) > minDiv$, i.e., the points in the candidate set C. Therefore, the set of pairs $\mathcal{P}$ needed for following the strategy aforementioned are all (p_d, p_i) for $p_i \in C$ and all (p_i, p_j) for $p_i, p_j \in C$.

In order to facilitate the construction of sets of size greater than two by combining the pairs in $\mathcal{P}$, we build a graph $G(V, E)$, where the set of vertices is $V = C \cup p_d$. Also, for each pair $(p_i, p_j) \in \mathcal{P}$, we create an edge (p_i, p_j) and add it to the set of edges E. The cost of an edge (p_i, p_j) is given by $div(p_i, p_j)$. Finding sets of size k is then equivalent to finding paths with k vertices in G.

Figure 3 shows the graph G built for the example shown in Fig. 2, i.e., for $p_d = p_6$, $k = 4$ and spatial diversity. Since all the sets of size $k = 4$ must contain p_6, in order to find paths in G that may represent a non-dominated solution, we assume that the starting point is p_6. Note that the possible paths of size 4 starting from p_6 are $PT_1 = (p_6, p_1, p_3, p_5)$, $PT_2 = (p_6, p_3, p_4, p_5)$ and $PT_3 = (p_6, p_1, p_3, p_4)$. The diversity of PT_1 is 1.14 because this is the minimum cost of an edge between any two vertices of PT_1. Since $1.14 > 1 = minDiv$, PT_1 is a candidate to be added to skyline. Similarly, the diversity of PT_2 is 1.35, which makes PT_2 a better candidate than PT_1. However, PT_3 does not have diversity higher than $minDiv$, since there is no edge (c_1, c_4) in E and thus $div(c_1, c_4) \leq minDiv$. Therefore, PT_2 represents the set with highest diversity.

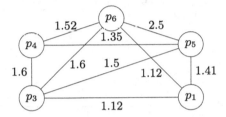

Fig. 3. Graph G built for the example shown in Fig. 2.

Algorithm 3 shows all steps of the $PG\text{-}findSetMaxDiv$ method. In order to find paths of a given size k, we use a solution based on Dijkstra's classical algorithm. First, all edges $(u, v) \in E$, representing paths of size 2, are added to a priority $queue$ (lines 5 to 7). The paths PT in $queue$ are ordered by decreasing order of an upper bound $ub(PT)$ that represents an optimistic value to the diversity of a set of size k that contains the vertices of PT. Let us assume that the size of PT is $p \leq k$ and that $l = k - p$. As exemplified above, in order to obtain a non-dominated solution of size k that contains the vertices of PT, PT can only be extended by adding l vertices that are neighbors of all vertices of PT. Let MN be this set of mutual neighbors. If $|MN| < l$, then it is not possible to find a non-dominated solution that contains the vertices of PT, therefore PT can be discarded. Otherwise, for each $pt_i \in PT$ let $div^l(pt_i, mn_j)$ be its l-th largest diversity to an element $mn_j \in MN$. We refer to the minimum of these values as $div(PT, MN)$. Since $div(PT, MN)$ is an upper bound to the diversity of PT to l other elements, if $div(PT, MN) \leq minDiv$, PT can also be discarded. The same occurs if $div(PT) \leq minDiv$. Therefore, the upper bound $ub(PT)$ to the diversity of a set S of size k containing the vertices of PT is given by $\min\{div(PT), div(PT, MN)\}$.

At each step the path PT with highest $ub(PT)$ is removed from $queue$ and expanded if it contains less than k vertices (lines 9 and 12). Let v be the last

---

Algorithm 3. *PG-findSetMaxDiv*

Data: Candidate set C, result set size k, point p_d and $minDiv$

1 $\mathcal{P} \leftarrow$ all pairs (p_d, p_i), $p_i \in C$;
2 $\mathcal{P} \leftarrow \mathcal{P} \cup$ all pairs $(p_i, p_j) \subset C$, $div(p_i, p_j) > minDiv$;
3 $G(V, E) \leftarrow buildGraph(\mathcal{P})$;
4 $queue \leftarrow \emptyset$;
5 **for** $(v, u) \in E$ **do**
6 $\quad$ | $\quad$ $queue$.add$((v, u))$;
7 **end**
8 **while** $|queue|$ ¿ 0 **do**
9 $\quad$ | $\quad$ $PT \leftarrow queue.dequeue()$;
10 $\quad$ | $\quad$ **if** $|PT| = k$ **then**
11 $\quad$ | $\quad$ | $\quad$ **return** PT;
12 $\quad$ | $\quad$ **else if** $|PT|$ ¡ k **then**
13 $\quad$ | $\quad$ | $\quad$ $v \leftarrow PT[|PT|]$;
14 $\quad$ | $\quad$ | $\quad$ **for** $(v, u) \in E$ **do**
15 $\quad$ | $\quad$ | $\quad$ | $\quad$ **if** $ub(PT \cup u) > minDiv$ **then**
16 $\quad$ | $\quad$ | $\quad$ | $\quad$ | $\quad$ $queue$.add$(PT \cup u)$;
17 $\quad$ | $\quad$ | $\quad$ | $\quad$ **end**
18 $\quad$ | $\quad$ | $\quad$ **end**
19 $\quad$ | $\quad$ **end**
20 **end**
21 **return** *null*;

---

vertex in the path PT. For each neighbor u of the vertex v, PT is extended by adding u. If $ub(PT \cup u) > minDiv$ then the path $(PT \cup u)$ is added to the queue (lines 15 and 16). If PT has size k (line 10), PT is returned. Note that no other posterior path PT' of size k could have higher diversity than PT. Since PT' is not the first set of size k removed from $queue$, then $ub(PT') \leq ub(PT)$, which means that in an optimistic scenario the diversity of PT' is at most equal to $div(PT)$. If no path of size k is found, the algorithm returns *null*.

Theorem 4. *For any p_d, PG-findSetMaxDiv finds the non-dominated solution S, where $d(q, S) = d(q, p_d)$, if such a solution exists.*

Proof. We divide this proof into two parts. First we prove that the algorithm does not remove any pair of points necessary for finding S, if S exists. Then we show that given all the necessary pairs, PG-findSetMaxDiv finds S.

Let us suppose that S exists and the algorithm discards a pair $(s_i, s_j) \subset S$. There are two possible cases:

1. *$s_i = p_d$ or $s_j = p_d$*
   This means that PG-findSetMaxDiv discarded a pair containing p_d. Since S is a non-dominated solution, $div(p_d, s_p) > minDiv$ for any $s_p \in S, s_p \neq p_r$. However, all pairs (p_d, s_p) such that $div(p_d, s_p) > minDiv$ are maintained by our algorithm, a contradiction to the assumption that (s_i, s_j) was discarded.

2. $s_i \neq p_d$ and $s_j \neq p_d$

   The pair (s_i, s_j) is only discarded by the algorithm if $div(p_d, s_i) \leq minDiv$ and/or $div(p_d, s_j) \leq minDiv$. Since $\{p_d, s_i, s_j\} \subseteq S$ and S is a non-dominated solution, $div(p_d, s_i) > minDiv$ and $div(p_d, s_j) > minDiv$, a contradiction.

   Therefore no necessary pair is discarded.

   Next We need to prove that PG-findSetMaxDiv is able to find S. We do so, again by contradiction, assuming that $S = (s_1, s_2, ..., s_i, s_{i+1}, ..., s_k)$ exists and it is not found. This means that S is not removed from the queue in the method. This happens if, during the expansion of a s_i, $2 \leq i < k$, the partial solution $S' = (s_1, s_2, ..., s_i, s_{i+1})$ is not inserted in the queue, which in turn occurs in the following situations:

1. *There is no edge $(s_i, s_{i+1}) \in E$.*
   This is a contradiction since as proven above, no pair $(s_i, s_j) \subseteq S$ is discarded by the method and consequently there is an edge $(s_i, s_{i+1}) \in E$.
2. *$ub(S') \leq minDiv$.*
   Therefore $div(S') \leq minDiv$ and/or $div(S', MN) \leq minDiv$. Since $S' \subseteq S$, $div(S') \geq div(S)$. As S is a non-dominated solution, $div(S) > minDiv$ and consequently $div(S') > minDiv$, a contradiction. If $div(S', MN) \leq minDiv$ then there is no non-dominated solution of size k that contains S'. A contradiction to the fact that S is non-dominated and $S' \subseteq S$. $\square$

4.3   From Conventional Skylines to Linear Skylines

In order to extract the linear skyline from the conventional one we follow a strategy based on the one proposed in [15]. In that paper the authors find linearly non-dominated paths in a bi-criteria road network. However, their goal was to minimize two competing costs associated with each path, while we aim at maximizing both closeness and diversity of solutions. Moreover, in that paper the linear skyline is constructed during the network expansion, while we first find the conventional skyline and then filter the linear skyline from it.

Due to lack of space we omit the details of the extraction process but summarize it as follows. Let SK be the conventional skyline, a solution $D \in SK$ is linearly dominated by two solutions $R, T \in SK$ if the 2-dimensional cost vector d representing D lies below the line between r and t, or more formally: let n be the normal vector of the line between R and T, such that $n^T r = n^T t$, then $\{r, t\} \prec_{lin} d$ iff $n^T d < n^T r = n^T t$ [15].

The procedure for extracting the linear skyline from SK then consists of eliminating all the points of SK that lie below a line between any two other skyline points [15]. More specifically, the solutions from SK are inserted into a list LS, that represents the linear skyline, in increasing order of diversity. After, let us assume that LS contains i elements and a new solution S^{i+1} is inserted at position $i + 1$. We verify whether this insertion leads to the removal of other solutions already in LS by performing a left traversal in LS. We first check if the left neighbor of S^{i+1}, i.e., S^i, is linearly dominated by verifying whether $\{s^{i-1}, s^{i+1}\} \prec_{lin} s^i$ holds. If so, S^i is removed from LS and S^{i-1} becomes the left neighbor of S^{i+1}. We then check whether $\{s^{i-2}, s^{i+1}\} \prec_{lin} s^{i-1}$. This process will be terminated when the first element of LS has been examined or the current left neighbor of S^{i+1} is a linearly non-dominated solution.

5   Experiments

We performed experiments using synthetically generated data sets with different distributions and sizes. (Even though real datasets could be used, synthetic datasets allowed us to investigate our proposal with respect to a broad range of parameters.) We first investigate how our proposed solutions perform in comparison to the brute force solution (*BF*) mentioned in Sect. 4. Then we show how our *PG* and *RRF* compare to the previously proposed approximate approach *GNE* in terms of processing time and quality of the results. The algorithms were developed in Java and the experiments were conducted on a Linux-based virtual machine with 4 vCPU and 8 GB main memory.

Given the kDNN problem's NP-hardness [4], and similarly to what was done in previous works, e.g., [1,6,18], we assume that the data set given as input to our algorithms is a candidate set $CS \subseteq P$ with elements that are relevant to the user[1]. This filtering step discards elements that are far from the query point and would not add any value to the results.

In the experiments that follow we varied the following parameters: the number of candidate points $|CS|$ between 100 and 1000 (with 500 being the default), the number k of points in each subset between 3 and 10 (with the default being 5) and the data distribution as Uniform (default), Gaussian and Clustered. The clustered data set was created using the generator described in [8] based on multivariate normal distribution. The data set generated contains three clusters with roughly the same number of points. As usual, when varying one parameter the others were kept at their default values.

Finally, it is important to stress that (1) all the graphs shown next present the average results in logarithmic scale, and (2) if an algorithm is not able to find the skyline in less than 10 min the computation is aborted and (simplistically) assumed to take 10 min when computing the average.

5.1   Spatial Diversity

We first evaluated how our proposed solutions behave when the spatial diversity is considered, i.e. when the dissimilarity between two points is given by their Euclidean distance. Figure 4 shows the results of this experiment.

Effect of k. As expected and shown in Fig. 4a, the running time of all solutions increases with k. However, the performance of *BF* degrades much faster since it generates all the possible subsets of size k in order to find the ones that belong to

---

[1] For simplicity we assume that CS is the set of $|CS|$-NN wrt the query point.

the skyline. For $k > 3$ BF is not able to find the skyline in less than 10 min. RRF showed to be around 30% faster than PG for $k = 3$ and slower for larger values. Most of the RRF's execution time is spent in the filtering process. For each point added to a partial set, a range query centered at that point is performed in order to discard points that can not be combined with the current partial set in order to obtain a non-dominated solution of size k. Therefore, the larger the value of k, the greater the number of range filtering performed. Moreover, RRF finds all the possible sets of size k with diversity higher than $minDiv$. On the other hand, in the PG approach, partial subsets that offer a greater chance to be part of the solution with maximum diversity are expanded first. Once the first solution of size k is found, the algorithm stops. This pruning is more effective for larger values of k. As shown in Fig. 4a, for $k = 3$ the gains from such pruning are not outweighed by the extra cost of computing the upper bound to the diversity of partial subsets and the costs involved in the queue operations.

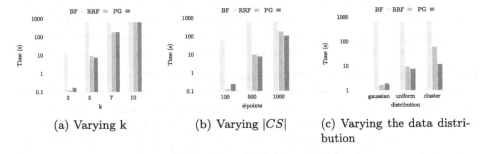

(a) Varying k (b) Varying $|CS|$ (c) Varying the data distribution

Fig. 4. Running time for spatial diversification.

Effect of the number of candidate points. Figure 4b shows that the running time of all solutions also increases with the number of candidate points. This is simply because there is a larger number of subsets that could be part of the skyline, requiring the algorithms to examine more combinations. As can be seen, BF performs poorly even for a small number of points. Similar to the previous experiment, RRF performs better for a lower number of points, while PG is more efficient when more points are considered for the same reason explained above.

Effect of the data distribution. As shown in Fig. 4c, BF is not affected by the data distribution since it generates all combinations of size k regardless of how the data is distributed. On the other hand, RRF and PG perform better for a Gaussian distribution. Since there is a higher concentration of points around the mean of the distribution, a range around a point in a dense area will include more points, which in turn will be eliminated from the result set and consequently less combinations will need to be tested. In an uniform distribution the points are more scattered in space, which implies that a range will likely include less points, increasing the number of possible combinations that need to be checked. Finally, in the clustered distribution there is more than one concentration of points, which means that, although a range around a point will probably eliminate points in the same cluster, it will most likely not include points from other clusters. Therefore, since the diversity between points in different clusters will tend to be high, the number of subsets that can potentially be a non-dominated solution will also increase, requiring more combinations of points to be generated and checked.

5.2 Categorical Diversity

Next, we evaluated how our proposed approaches behave for categorical diversity varying the number of categories and the distribution of the diversities between two categories. We have considered data sets with 5, 20 and 100 categories, where the default value is 20. Regarding the diversity distribution, we have generated diversity matrices where the diversity between two categories is either the same for any pair of categories, follows an uniform distribution or an exponential distribution. The uniform distribution is used as default. In the following experiments we do not report the results for BF since it is not affected by these parameters and was shown above to be a non-competitive approach.

Effect of the number of categories. As shown in Fig. 5a, the running time of both approaches increases with the number of categories. Since for categorical diversity only the closest points from each category are needed, the higher the number of categories, the higher the number of points that need to be examined in order to find the skyline set. Also note that for the same reason, the execution time is much shorter than when a spatial diversity is considered. Moreover, RRF outperforms PG for all examined values. This can be explained by the fact that RRF is more efficient for a smaller number of points, as shown in a previous experiment, and only up to 100 points, i.e., the number of categories, are needed in this experiment in order to find the skyline set.

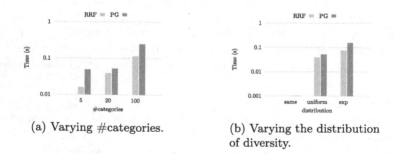

(a) Varying #categories.   (b) Varying the distribution
                                     of diversity.

Fig. 5. Running time for categorical diversification.

Effect of the distribution of diversities. Figure 5b shows that both solutions present better performance when the diversity between any pair of categories is the same. This is simply because once a solution where no two points belong to the same category is found, it is not possible to find a posterior solution with higher diversity. Thus the range filtering performed in both algorithms will eliminate all remaining points. Compared to an uniform distribution, in an exponential distribution there are more pairs of categories with small diversity and fewer with high diversity. This means that it may take longer to find solutions with higher diversity, which in turn reduces the pruning power of both solutions since subsets are discarded based on the highest diversity found so far.

5.3 Effectiveness and Efficiency

As mentioned in Sect. 2, *GNE* provides an approximated result whereas we obtain the optimal one. Nonetheless, for the sake of completeness we evaluated how sub-optimal were *GNE* results as well as how much faster they were obtained. We fixed λ, the trade-off between similarity and diversity given as input to the *GNE* algorithm, at 0.5. We assumed that the similarity of a point $p_i \in P$ w.r.t. q is given by $sim(p_i, q) = 1 - \frac{d(p_i, q)}{d_m}$, where d_m is the maximum distance used for normalization. Since *GNE* requires both criteria to be on the same scale, we also normalized the diversity value. Due to limited space we only present a summary of our findings.

Regarding spatial diversity, the quality of the results produced by *GNE* decreases when k increases. For instance, for $k = 3$, *GNE* is around 20% worse than our optimal approaches, while for $k = 7$ it is about 54% worse. The same behavior was observed when we varied the number of candidate points, where *GNE's* results were up to 40% worse (for 1000 candidate points). Lastly, *GNE* presented better results for a Gaussian distribution, being around 21% worse than our optimal approaches, followed by the uniform distribution, where it was about 33% worse. For a clustered distribution, *GNE's* results were around 46% worse than the optimal ones our proposed approaches provide.

With relation to categorical diversity, the quality of the results produced by *GNE* increases with the number of categories. For instance, for 5 categories, *GNE* is around 54% worse than our optimal approaches, while for 100 categories it is only 7% worse. Regarding the diversity of the categories, *GNE* presented better results for the uniform distribution, being about 20% worse than our optimal approaches, but it was around 90% worse for an exponential distribution.

GNE's sub-optimality is compensated by its efficiency. It is able to find an approximate solution in less than 1 s in all cases we evaluated. However, it is important to stress that while it is faster, it finds an *approximate* solution for *one* single linear combination of the optimized criteria whereas our solutions find all *optimal* solutions for *any* such linear combination.

6 Conclusion

We have addressed the kDNN query, i.e., a variation of k-NN queries. This problem consists of finding elements that are as close as possible to the query point, while, at the same time, being as diverse as possible. We have considered two different notions of diversity, namely spatial and categorical. Previously proposed solutions are approximate and require the user to determine a specific weight for each type of diversity. Our proposed approaches based on linear skylines find *all* results that are *optimal* under *any* given arbitrary linear combination of the two competing diversity criteria.

We proposed two solutions to kDNN queries using the notion of skylines. *Recursive Range Filtering (RRF)* strives to reduce the number of combinations generated by recursively combining a partial set with candidate points that are diverse enough to be part of a non-dominated solution. *Pair Graph (PG)* works by pruning subsets that contain pairs of points that have low diversity and could not be together in a non-dominated solution. Experiments varying a number of parameters showed that (1) both *RRF* and *PG* are orders of magnitude faster than a straightforward solution, and (2) *PG* outperforms *RRF* for higher values of k and a greater number of points and *RRF* is more efficient otherwise.

As future work we would like to investigate kDNN queries within road networks, the use of real datasets, as well as to explore further the notion of diversity in other domains, e.g., text retrieval.

References

1. Abbar, S., et al.: Diverse near neighbor problem. In: SOCG, pp. 207–214 (2013)
2. Börzsönyi, S., et al.: The skyline operator. In: ICDE, pp. 421–430 (2001)
3. Carbonell, J., Goldstein, J.: The use of MMR, diversity-based reranking for reordering documents and producing summaries. In: SIGIR, pp. 335–336 (1998)
4. Carterette, B.: An analysis of NP-completeness in novelty and diversity ranking. Inf. Retrieval **14**, 89–106 (2011)
5. Clarke, C.L., et al.: Novelty and diversity in information retrieval evaluation. In: SIGIR, pp. 659–666 (2008)
6. Gollapudi, S., Sharma, A.: An axiomatic approach for result diversification. In: WWW, pp. 381–390 (2009)
7. Gu, Y., et al.: The moving k diversified nearest neighbor query. In: TKDE, pp. 2778–2792 (2016)
8. Handl, J., Knowles, J.: Cluster generators for large high-dimensional data sets with large numbers of clusters (2005). http://dbkgroup.org/handl/generators
9. Huang, Z., et al.: A clustering based approach for skyline diversity. Expert Syst. Appl. **38**(7), 7984–7993 (2011)
10. Jain, A., Sarda, P., Haritsa, J.R.: Providing diversity in K-nearest neighbor query results. In: Dai, H., Srikant, R., Zhang, C. (eds.) PAKDD 2004. LNCS (LNAI), vol. 3056, pp. 404–413. Springer, Heidelberg (2004). doi:10.1007/978-3-540-24775-3_49
11. Kucuktunc, O., Ferhatosmanoglu, H.: λ-diverse nearest neighbors browsing for multidimensional data. In: TKDE, pp. 481–493 (2013)
12. Lee, K.C.K., Lee, W.-C., Leong, H.V.: Nearest surrounder queries. In: ICDE, p. 85 (2006)
13. Rafiei, D., Bharat, K., Shukla, A.: Diversifying web search results. In: WWW, pp. 781–790 (2010)
14. Roussopoulos, N., Kelley, S., Vincent, F.: Nearest neighbor queries. ACM SIGMOD Rec. **24**, 71–79 (1995)

15. Shekelyan, M., Jossé, G., Schubert, M., Kriegel, H.-P.: Linear path skyline computation in bicriteria networks. In: Bhowmick, S.S., Dyreson, C.E., Jensen, C.S., Lee, M.L., Muliantara, A., Thalheim, B. (eds.) DASFAA 2014. LNCS, vol. 8421, pp. 173–187. Springer, Cham (2014). doi:10.1007/978-3-319-05810-8_12
16. Tao, Y.: Diversity in skylines. IEEE Data Eng. Bull. **32**(4), 65–72 (2009)
17. Valkanas, G., Papadopoulos, A.N., Gunopulos, D.: Skydiver: a framework for skyline diversification. In: EDBT, pp. 406–417 (2013)
18. Vieira, M.R., et al.: On query result diversification. In: ICDE, pp. 1163–1174 (2011)
19. Yu, C., Lakshmanan, L., and Amer-Yahia, S.: It takes variety to make a world: diversification in recommender systems. In: EDBT 2009, pp. 368–378 (2009)
20. Zhang, C., et al.: Diversified spatial keyword search on road networks. In: EDBT, pp. 367–378 (2014)

Spatio-Temporal Functional Dependencies for Sensor Data Streams

Manel Charfi$^{(\boxtimes)}$, Yann Gripay, and Jean-Marc Petit

Université de Lyon, CNRS, INSA-LYON, LIRIS,
UMR5205, 69621 Villeurbanne, France
{manel.charfi,yann.gripay,jean-marc.petit}@insa-lyon.fr

Abstract. Nowadays, sensors are cheap, easy to deploy and immediate to integrate into applications. Since huge amounts of sensor data can be generated, selecting only relevant data to be saved for further usage, e.g. long-term query facilities, is still an issue. In this paper, we adapt the declarative approach developed in the seventies for database design and we apply it to sensor data streams. Given *sensor data streams*, the key idea is to consider both spatio-temporal dimensions and Spatio-Temporal Functional Dependencies as first class-citizens for designing sensor databases on top of any relational database management system. We propose an axiomatisation of these dependencies and the associated attribute closure algorithm, leading to a new normalization algorithm.

1 Introduction

Thousands and even millions of sensors can be deployed easily, generating data streams that produce cumulatively huge volumes of data. Whenever a sensor produces a data (temperature, humidity...), two dimensions are of particular interest: the *temporal* dimension to stamp the value produced at a particular time and the *spatial* dimension to identify the location of the sensor. Both dimensions have different granularities organized into hierarchies that may vary according to special needs of applications (see an example in Fig. 1).

Example 1. We consider a running example of sensor data streams from intelligent buildings. In each building different sensors (temperature, luminosity, humidity...) are deployed. At the scale of several buildings, a huge number of sensors exist, each one sending values at its own rate. Let us consider a temperature sensor data stream as given in Table 1, where *location* and *time* follow the granularity hierarchies presented in Fig. 1. Without loss of generality, we shall assume throughout the paper that the `location` attribute contains a string value containing the concatenation of all its "granules".

The stream given in Table 1 is representative of a wide variety of sensor data streams. We consider static sensors in this paper, i.e. sensors set to a particular place to sense the environment. We focus on applications requiring long term storage of data streams, in opposition to monitoring applications which have to react

© Springer International Publishing AG 2017
M. Gertz et al. (Eds.): SSTD 2017, LNCS 10411, pp. 182–199, 2017.
DOI: 10.1007/978-3-319-64367-0_10

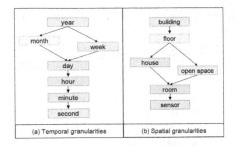

Fig. 1. Temporal and spatial dimensions

Table 1. An instance of *temperatureSensors*

Temperature	Location	Time
21	oxygen:f1:h1:livingRoom:s11	2016/03/02 11:59:00
20	oxygen:f1:h1:kitchen:s21	2016/03/02 11:59:30
20	oxygen:f1:h1:livingRoom:s12	2016/03/02 12:01:00
23	oxygen:f1:h1:kitchen:s22	2016/03/02 12:01:00
20	oxygen:f1:h1:livingRoom:s11	2016/03/02 12:01:30
24	oxygen:f1:h1:bathroom:s31	2016/03/02 12:02:00
20	oxygen:f1:h1:livingRoom:s12	2016/03/02 12:02:30
23	oxygen:f1:h1:kitchen:s21	2016/03/02 12:15:00

in quasi real-time using for example continuous query [1, 2]. Many types of systems can be used for long term storage, from classical file systems (in json, flat text. . .) to Relational Database Management Systems (RDBMS). In the former, a file is created for a given period of time (e.g. week, month). In the latter, data streams are stored as classical tables with specific attributes to cope with spatio-temporal aspects of the stream. In both cases, the burden is let to application designers who have to pose complex queries (e.g. full-text search or SQL-like queries) to deal with temporal and spatial dimensions. Whenever the data volume is high, the query processing time can be also prohibitive. Selecting only relevant data from sensor data streams to be saved for further usage, e.g. long-term query facilities, is still an issue since huge amounts of sensors can be deployed for a specific application. The challenges of applying traditional database management systems to extract business value from such data are still there. In this setting, the problem we are interested in is the following: *Given a set of sensor data streams, how to build a relevant sensor database for long-term reporting applications?*

In our work, we focus on the implementation of a declarative approach that aims to guarantee the storage of the "relevant" sensor data at application-specific granularities. Instead of relaying on query workload or data storage budget to define an optimized database [20], we borrow the declarative approach developed in the seventies for database design using functional dependencies as constraints.

Our aim is to apply this approach in order to transform real-time sensor data streams into a sensor database. We argue that such constraints augmented with the spatial and the temporal dimensions are required to keep only the "relevant data" from incoming sensor streams. The sensor database satisfies the specified constraints while approximating data stream values. Thus, the data approximation and data reduction are "controlled" by the set of constraints. As the required results concern data over long periods of times (e.g. couple of months to several years), approximating sensor data streams should decrease storage space while allowing to express "relevant queries" more easily. We introduce Spatio-Temporal Functional Dependencies (STFDs) which extend Functional Dependencies (FDs) with the temporal and the spatial dimensions.

Example 2. Let us take back our running example. The building manager considers that the "temperature of each room remains the same over each hour", leading to the following STFD: $location^{room}, time^{hour} \rightarrow temperature$ whose syntax captures the intended meaning, i.e. it does not exist two different temperature values on the same room during a given hour.

Such constraints are straightforward to understand and convey the semantics allowing to decide what are the relevant data to keep in the database. Classical database normalization techniques can be revisited in this setting to obtain so-called *granularity-aware sensor databases*.

At given spatio-temporal granules (e.g. for a given hour and a given house), we have different alternatives to choose data, each one could be seen as an *aggregation* of values, from simple ones like *first, minimum, average within each spatio-temporal granule* to more elaborated ones. Thus, depending on the application context, we can imagine more complex aggregations that allow to define complex functions (e.g. average of the 3 first values corresponding to a given valid domain) and even to avoid noisy and incomplete data issues. We call these annotations Semantic Value Assumptions (SVA) which are based on *semantic assumptions* of temporal databases [3]. So, if we consider temperature values of each room at the scale of an hour, we have to precise which value will be associated to each couple (room,hour): it may be the average of all values from all sensors in this room at all the minutes of this hour. Once a sensor database has been built, we have a data exchange problem [7]: how to load data from the stream to the database? To do that, the main problem is to decide which values to pick up from the stream. This decision is easily made thanks to SVAs.

Contribution
We aim to establish a sensor data storage system that eases the storage of spatio-temporal data streams. To the best of our knowledge, this is the first contribution talking advantage of spatio-temporal constraints defined at database design time to automatically produce an approximated database saving "relevant" data with respect to users' constraints. Our main objective is at the end to have a reduced database that contains a summary of sensor data streams in accordance to application-predefined requirements. Thus, given a set of sensor

data streams, we propose a declarative approach to build a representative sensor database on top of any RDBMS with the following key features:

- Both spatio-temporal granularity hierarchies and STFD are considered as first class-citizens.
- A specific axiomatisation of STFD and an associated attribute closure algorithm, leading to an efficient normalization algorithm are introduced.
- A middleware to load on-the-fly relevant data from sensor streams into the granularity-aware sensor database is proposed.

We have implemented a prototype to deal with both database design and data loading. We have conducted experiments with synthetic and real-life sensor data streams coming from Intelligent Building.

Paper organization
Section 2 gathers preliminaries. In Sect. 3, we define the formalism of STFDs leading to the design of the granularity-aware sensor database. In Sect. 4, we sketch the architecture of our declarative system and the experiments conducted on intelligent building data streams. The related work is presented in Sect. 5. Finally, in Sect. 6 we conclude and expose some perspectives.

2   Preliminaries

To define spatio-temporal granularity hierarchies, we first define a general notion of granularity, borrowed from time granularity definition [3], then a partial order on a set of granularities for which we require a lattice structure.

Let T be a countably infinite set and $\leq$ a total order on T. A granularity is a mapping G from $\mathbb{N}$ to $\mathcal{P}(T)$ where $\mathcal{P}(T)$ is the powerset of T. A non-empty subset $G(i)$, $i \in \mathbb{N}$, of a granularity G is called granule. The granules in a granularity do not overlap. A granularity G is *finer than* a granularity H (H is coarser than G), denoted $G \preceq H$, if for each integer i, there exists an integer j such that $G(i) \subseteq H(j)$. Intuitively this means that each granule of H holds a set of granules of G. Let $\mathcal{G}$ be a set of granularities and $\preceq$ a partial order on $\mathcal{G}$. We assume the set $(\mathcal{G}, \preceq)$ is a lattice, meaning that each two-element subset $\{G_1, G_2\} \subseteq \mathcal{G}$ has a join (i.e. least upper bound) and a meet (i.e. greatest lower bound). For $X \subseteq \mathcal{G}$, $glb(X)$ denotes the greatest lower bound of X in $\mathcal{G}$. The greatest and least elements of $\mathcal{G}$, or the top and bottom elements, are denoted by Top and $Bottom$, respectively. G is *collectively finer* than $\{G_1, \ldots, G_m\}$, denoted by $G \preceq_c \{G_1, \ldots, G_m\}$, if for each positive integer i, there exist k, j such that $G(i) \subseteq G_k(j), 1 \leq k \leq m$ and j a positive integer. Intuitively this means that there exists in the set of granularities $\{G_1, \ldots, G_m\}$ at least one granularity that, for each granule $G(i)$ taken independently, is coarser than G. As a particular case, $G \preceq G'$ implies $G \preceq_c \{G'\}$.

Application to spatio-temporal dimensions
For each considered dimension, a lattice of granularities is defined accordingly. For the sake of clearness, we assume without loss of generality that a total

order exists for time instants and sensors, meaning that two time instants (resp. two sensors) can always be compared. In this setting, we define two lattices, $(\mathcal{T}, \preceq_t)$ and $(\mathcal{S}, \preceq_s)$ leading to two granularity hierarchies. $\mathcal{T}$ is a set of temporal granularities and $\mathcal{S}$ a set of sensor location granularities. An example is given in Fig. 1 where the arrows connect the coarser to the finer granularities.

3   Database Modeling for Sensor Data

3.1   Granularity Aware Sensor Database

We assume the reader is familiar with database notations, see for example [11] for details. Hereinafter we use G as a spatial granularity and H as a temporal granularity. We extend the temporal database definitions given in [3] to take into account the spatial dimension. Let $\mathcal{U}$ be a universe (set of attributes) and $\mathcal{D}$ be a countably infinite set of constant values. A *spatio-temporal module schema* over $\mathcal{U}$ is a triplet $M = (R, G, H)$, where $R \subseteq \mathcal{U}$ is a relation schema, $G \in (\mathcal{S}, \preceq_s)$ is a spatial granularity and $H \in (\mathcal{T}, \preceq_t)$ is a temporal granularity. For a relation schema R, $Tup(R)$ is the set of all possible tuples defined over $\mathcal{D}$. A *spatio-temporal module* is a quadruple $\mathcal{M} = (R, G, H, \varphi)$, where (R, G, H) is a *spatio-temporal module schema* and φ is a mapping from $\mathbb{N} \times \mathbb{N}$ to $\mathcal{P}(Tup(R))$. Actually, the mapping function $\varphi(i, j)$ gives the tuples over attributes of R that hold at each couple of granules $(G(i), H(j))$. When clear from context, we shall use the time instants and sensor locations instead of integers to describe a particular mapping function, e.g. $\varphi(building_x{:}house_y{:}room_z{:}sensor_i, 2016/03/02\ 11{:}59{:}00)$ instead of $\varphi(i, j)$ for some integer values i and j.

Example 3. Consider the "raw" data stream given in Table 1. It can be represented at different granularities: e.g. with the module $\mathcal{M}_1=(R, G_1, H_1, \varphi_1)$ where $R=\{temperature\}$, $G_1=room$, $H_1=hour$ and the windowing function φ_1 is:

φ_1(oxygen:f1:h1:livingRoom,2016/03/02 11) = {<21>}
φ_1(oxygen:f1:h1:kitchen,2016/03/02 11) = {<20>}
φ_1(oxygen:f1:h1:livingRoom,2016/03/02 12) = {<20>}
φ_1(oxygen:f1:h1:kitchen,2016/03/02 12) = {<23>}
φ_1(oxygen:f1:h1:bathroom,2016/03/02 12) = {<24>}

A granularity-aware *sensor database schema* **R** over $\mathcal{U}$ is a fixed set of *spatio-temporal module schemas* over $\mathcal{U}$. A granularity-aware *sensor database* **d** is a finite set of *spatio-temporal modules* defined over **R**.

3.2   Spatio-Temporal Functional Dependency (STFD)

Dedicated FDs for sensor data streams have to take into account the temporal and spatial dimensions. Many extensions of FDs to temporal DB have been proposed but none of them extends FDs to both temporal and spatial dimensions. In the sequel, we extend temporal functional dependencies introduced in [3].

Intuitively, a STFD means that the X-values determine the Y-values within each granule of the spatio-temporal granularities.

Let $X, Y \subseteq \mathcal{U}$ and $(\mathcal{T}, \preceq_t)$, $(\mathcal{S}, \preceq_s)$ two granularity hierarchies. To express STFDs, we need to consider two special attributes, disjoint from $\mathcal{U}$, to take into consideration granularities. Let *location* and *time* be the special spatial and temporal attributes respectively. When clear from context, *location* and *time* will be abbreviated by L and T respectively.

Definition 1. *A spatio-temporal FD over $\mathcal{U}$ is an expression of the form:* $X, location^G, time^H \rightarrow Y$ *where* $G \in (\mathcal{S}, \preceq_s)$ *is a spatial granularity and* $H \in (\mathcal{T}, \preceq_t)$ *is a temporal granularity.*

We shall see that the case $X = \emptyset$ is meaningful for STFDs whereas the classical FD counterpart is almost useless (i.e. $\emptyset \rightarrow A$ means that in every possible relation, only one value for A is allowed).

Example 4. Regarding the temperature approximations, one may consider that the temperature of the same room does not change all along the same hour. This approximation can be represented as: $\emptyset, location^{room}, time^{hour} \rightarrow temperature$ (or simply: $location^{room}, time^{hour} \rightarrow temperature$).

The satisfaction of a STFD with respect to a module is defined as follows:

Definition 2. *Let $\mathcal{M} = (R, G, H, \varphi)$ be a spatio-temporal module, $X, Y \subseteq R$ and $f\colon X, location^{G'}, time^{H'} \rightarrow Y$ an STFD. f is satisfied by $\mathcal{M}$, denoted by $\mathcal{M} \models f$, if for all tuples t_1 and t_2 and positive integers i_1, i_2, j_1 and j_2, the following three conditions imply $t_1[Y] = t_2[Y]$: (1) $t_1[X] = t_2[X]$, (2) $t_1 \in \varphi(i_1, j_1)$ and $t_2 \in \varphi(i_2, j_2)$, and (3) $\exists\, i'$ such that $G(i_1) \cup G(i_2) \subseteq G'(i')$ and $\exists\, j'$ such that $H(j_1) \cup H(j_2) \subseteq H'(j')$.*

This definition extends classical FDs ($r \models X \rightarrow Y$) as follows:

(1) is the classical condition for FDs, i.e. the left-hand sides have to be equal on X for the two considered tuples.
(2) bounds tuples t_1, t_2 to be part of two spatio-temporal granules of $\mathcal{M}$ (equivalent to $t_1, t_2 \in r$).
(3) restricts eligible tuples t_1 and t_2 in such a way that the union of their spatial (resp. temporal) granules with respect to G (resp. H) has to be included in some granule of G' (resp. H').

Example 5. Let us consider the module $\mathcal{M}_2 = (R, G_2, H_2, \varphi_2)$ where $R = \{tempe\text{-}rature\}$, $G_2 = sensor$, $H_2 = second$. We have: $\mathcal{M}_2 \models location^{room}, time^{hour} \rightarrow temperature$ and $\mathcal{M}_2 \not\models location^{house}, time^{hour} \rightarrow temperature$. As for FDs, the non-satisfaction is easier to explain since we just need to exhibit a counter example. The two first tuples given in Table 1 form a counter-example since both sensors belong to the house $h1$ and have been produced at the hour 11 whereas $20 \neq 21$.

From these examples, we argue that STFDs are quite natural to express *declarative constraints over sensor data streams* and provide a powerful abstraction mechanism towards granularity-aware sensor database design.

3.3   Reasoning on STFDs

Inference Axioms for STFDs. In order to derive all the possible STFDs logically implied by a set of STFDs, we need to define the inference axioms corresponding to STFDs. We propose the three following finite axioms:

(A1) Restricted reflexivity:
  if $Y \subseteq X$ then $F \vdash X, L^{Top}, T^{Top} \to Y$

(A2) Augmentation:
  if $F \vdash X, L^G, T^H \to Y$ then $F \vdash X, Z, L^G, T^H \to Y, Z$

(A3) Extended transitivity:

$$\text{if } \begin{cases} F \vdash X, L^{G_1}, T^{H_1} \to Y \\ F \vdash Y, L^{G_2}, T^{H_2} \to Z \end{cases} \text{ then } F \vdash X, L^{G_3}, T^{H_3} \to Z$$

  where $G_3 = glb(\{G_1, G_2\})$ and $H_3 = glb(\{H_1, H_2\})$.

These three inference axioms for STFDs are a generalization of axioms of temporal FDs, shown to be sound and complete in [3]. The proof for STFDs is similar and is omitted in this paper.

Closure of Attributes. The closure of attributes plays a crucial role for reasoning on classical FDs and are generalized to STFDs as follows. Let **R** be a sensor database schema over $\mathcal{U}$, F a set of STFDs over $\mathcal{U}$ and $X \subseteq \mathcal{U}$. The closure of X with respect to F is denoted by X_F^+. X_F^+ contains elements of the form (B, G, H) over $\mathcal{U} \times \mathcal{S} \times \mathcal{T}$. X_F^+ is defined as follows:
  $X_F^+ = \{(B, G, H) \mid F \vdash X, L^G, T^H \to B$ such that there is no $F \vdash X, L^{G'}, T^{H'} \to B$ with $G' \preceq_s G', H \preceq_t H'$ and $(G \neq G'$ or $H \neq H')\}$.

Algorithm 1 computes the finite closure of a set of attributes X with respect to F. This algorithm is a generalization of the classical closure algorithm for FDs [11] taking into account the granularities. Its basic idea is to compute progressively the set X_F^+. Line 1 encodes the first axiom (A1). The following procedure is repeated over all STFDs until a fix point is reached (line 13). For each STFD of line 4, if $A_1, .., A_k$ appears in the current closure X_{prev}, then $B_1, .., B_m$ are added to X_F^+ with the corresponding spatial and temporal granularities. Line 14 ensures that the closure is composed of elements with incomparable granularities for the same attribute.

Example 6. Let us consider $\mathcal{U} = \{temperature, humidity, luminosity, CO2\}$, four sensor types sending the temperature, humidity, luminosity and CO_2 values, the granularity hierarchies given in Fig. 1 and the set F of STFDs:

$F = \{location^{room}, time^{hour} \to temperature; location^{house}, time^{day} \to humidity;$
$location^{room}, time^{day} \to luminosity; location^{room}, time^{minute} \to CO2;$
$location^{room}, time^{hour} \to humidity; location^{room}, time^{minute} \to temperature;$
$location^{house}, time^{hour} \to humidity; location^{sensor}, time^{hour} \to luminosity;$
$location^{room}, time^{hour} \to CO2 \}$

The closure of *temperature* w.r.t. F is: $temperature_F^+ = \{(temperature, Top, Top), (humidity, house, day), (luminosity, room, day), (CO2, room, hour)\}$.

Algorithm 1. ClosureAttribute

Require:
    F: a set of STFDs over $\mathcal{U}$
    $X \subseteq \mathcal{U}$
Ensure:
    X_F^+: the finite closure of X with respect to F

  1: $X_F^+ := \{(A, Top, Top) \mid a \in X\} \cup \{(\emptyset, Top, Top)\}$
  2: **repeat**
  3:    $X_{prev} := X_F^+$
  4:    **for each** $A_1, .., A_k, L^G, T^H \to B_1, .., B_m \in F$ **do**
  5:      **for each** $\{(A_1, G_1, H_1), ..., (A_k, G_k, H_k)\} \subseteq X_{prev}$ **do**
  6:        $G' := glb(G_1, .., G_k, G)$
  7:        $H' := glb(H_1, .., H_k, H)$
  8:        **for each** $B \in \{B_1, .., B_m\}$ **do**
  9:          $X_F^+ := X_F^+ \cup \{(B, G', H')\}$
10:        **end for**
11:      **end for**
12:    **end for**
13: **until** $X_F^+ = X_{prev}$
14: Minimize X_F^+ such that there is no two elements (A, G, H) and (A, G', H') with $G \preceq_s G', H \preceq_t H'$ and $(G \neq G'$ or $H \neq H')$

The closure of attributes with respect to a set F of STFDs is polynomial and allows to decide whether or not a given STFD is implied by F, as shown in the following property.

Property 1. $F \vdash X, L^G, T^H \to B$ **iff** $\exists Y \subseteq X_F^+$ such that $Y = \{(B, G_{i_k}, H_{j_l}) \mid i_k \in i_1..i_n, j_l \in j_1..j_m\}$, $G \preceq_c \{G_{i_1}, \ldots, G_{i_n}\}$ and $H \preceq_c \{H_{j_1}, \ldots, H_{j_m}\}$.

Example 7. Let us consider the set F of STFDs given in Example 6. We consider the following STFDs:
f_1 : $temperature, location^{room}, time^{hour} \to luminosity$; and
f_2 : $temperature, location^{openspace}, time^{day} \to humidity$;
As $(luminosity, room, day) \in temperature_F^+$ and $hour \preceq_t day$ we have $F \vdash f_1$. But $F \nvdash f_2$ since $temperature_F^+$ only contains $(humidity, house, day)$ and $openspace \npreceq_s house$.

Attribute closure is one of the technical contributions of the paper and is, to the best of our knowledge, a new result never addressed in related works.

3.4 Normalization

Our aim is to extend the well known *synthesis algorithm* for database design [11] from a set of classical FDs in our setting. With STFDs, we propose a normalization technique based on two main steps: first, computing a minimal cover of a set of STFDs and then producing spatio-temporal modules. Let F^+ be the closure of F. F^+ is defined by: $F^+ = \{X, L^G, T^H \to Y \mid F \vdash X, L^G, T^H \to Y\}$.

Definition 3. *A set F' of STFDs is a* cover *of a set F of STFDs if $F^+ = F'^+$. A cover F' of F is minimal if $\nexists$ a cover G of F such that $|G| < |F'|$.*

Algorithm 2. MinimalCover

Require:
    F: a set of STFDs over $\mathcal{U}$
Ensure:
    F': a minimal cover of F

1: $F' := \emptyset$
2: **for each** $X, L^G, T^H \rightarrow Y \in F$ **do**
3:     **for each** $(A, G', H') \in X_F^+$ **do**
4:         $F' := F' \cup \{X, L^{G'}, T^{H'} \rightarrow A\}$
5:     **end for**
6: **end for**
7: **for each** $X, L^{G'}, T^{H'} \rightarrow A \in F'$ **do**
8:     $F'' := F' \setminus \{X, L^{G'}, T^{H'} \rightarrow A\}$
9:     **if** $F'' \vdash \{X, L^{G'}, T^{H'} \rightarrow A\}$ **then**
10:         $F' := F' \setminus \{X, L^{G'}, T^{H'} \rightarrow A\}$
11:     **end if**
12: **end for**
13: **while** there exists $f, f' \in F'$ with the same left-hand-side **do**
14:     Merge f and f'
15: **end while**

Algorithm 2 generalizes the classical procedure to get a minimal cover of FD. We sketch the main steps to compute a minimal cover. First, we saturate the initial set of STFDs with STFDs induced by the closure of each set of attributes (line 2–6). Then, we apply classical minimization procedure (line 7–12) and finally, we merge STFDs whose left-hand sides are the same by taking the union of their right-hand sides (line 13–15) (not detailed in the Algorithm).

Example 8. We consider the set F of STFDs in example 6. Using Algorithm 2 we get the following minimal cover F' (details are omitted):

$$F' = \{location^{room}, time^{hour} \rightarrow temperature, CO2;$$
$$location^{house}, time^{day} \rightarrow humidity; location^{room}, time^{day} \rightarrow luminosity\}$$

The sensor database schema can be deduced from the obtained minimal cover: for each STFD, a module is generated. Algorithm 3 presents the main steps allowing to get a granularity-aware sensor database schema from a set of STFDs. Studying the properties of this decomposition is left for future work.

Algorithm 3. Normalization

Require:
    F: a set of STFDs over $\mathcal{U}$
Ensure:
    R: a granularity-aware sensor database schema

1: **R** $:= \emptyset$
2: $F' := MinimalCover(F)$
3: **for each** $A_1, .., A_k, L^G, T^H \rightarrow B_1, .., B_m \in F'$ **do**
4:     $R := \{A_1, .., A_k, B_1, .., B_m\}$
5:     $M := (R, G, H)$
6:     **R** $:= $ **R** $\cup M$
7: **end for**

Example 9. Continuing the previous example, we obtain a granularity-aware sensor database schema $\mathbf{R} = \{M_1, M_2, M3\}$ with:
$M_1 = (< temperature, CO2 >, room, hour)$, $M_2 = (< humidity >, house, day)$, and $M_3 = (< luminosity >, room, day)$.

It is worth noting that every module of such a database schema is easily implementable on top of any RDBMS. In the sequel, we denote the database schema obtained through the normalization process by the *abstract schema* because in our context we need to add more semantic information to this schema.

3.5   Semantic Value Assumption

Given a set of sensor data streams, constraints for long-term storage can be defined as a set of STFDs. We have seen that a granularity-aware sensor database schema can be obtained from them thanks to the proposed normalization algorithms which allow producing a database schema, i.e. abstract schema, from the user-defined inputs (dimensions, sensor stream schemas and STFDs). After that, the user annotates the abstract schema with the semantic information allowing to specify the "relevant" data at the right granularities. In fact, at given spatio-temporal granules, we have different alternatives to choose data, each one could be seen as an aggregation of values with some aggregate functions. These functions can be simple aggregations (e.g. first, max . . .) or more elaborated ones (e.g. average of the 3 first values corresponding to a given valid domain) depending on the application context. To do so, we introduce the so-called *"Semantic Value Assumptions"* (SVA) allowing to declaratively define the values to be selected. Annotating an abstract schema, obtained by normalization techniques based on STFDs, with user-defined SVAs leads to a *concrete schema*, which can be implemented on top of classical RDBMS. The definition of SVA is:

Definition 4. *Let* $M = (R, G, H)$ *be a spatio-temporal module schema and* $A \in R$. *A SVA is a triplet* (A, agg_fct, M) *where* agg_fct *is an aggregation function* (*first,avg. . .*) *over the spatio-temporal granularities* G *and* H.

Example 10. Let us consider M_2 in Example 9. The SVA $(humidity, first, M_2)$ means that the *first* humidity value per house per day has to be kept in M_2 and the SVA $(humidity, avg, M_2)$ means that the average of the humidity values per house per day has to be kept in M_2.

Whenever multiple SVAs exist for a given couple $(attribute, module\ schema)$, new attributes could be created in the target schema of the underlying RDBMS. These specific annotations allow, in a declarative manner, to annotate the database schema with the semantic information required to indicate which value is representative in each granule. SVAs can be very complex aggregations that allow to define complex functions and even to avoid noisy and incomplete data issues. This semantic information is important in database design level as well as in the data stream loading procedure. In fact, as the relevant tuples definition is ensured thanks to SVAs, for each SVA the system instantiates a specific data wrapper. It is possible to implement SVA in different manners namely using triggers or a dedicated middleware. As triggers do not scale to important data stream loads [1,8], we chose to implement SVA data wrappers in a middleware.

4   Prototype for Sensor Database

We propose a declarative system for both database design and data stream loading containing the following two levels:

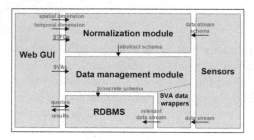

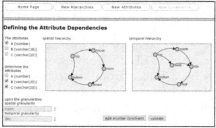

Fig. 2. An overview of our architecture   **Fig. 3.** STFD definition in our prototype

1. *Sensor database* **design:** Given a spatial and a temporal dimensions, the aim of this module is to determine a *granularity-aware sensor database* schema from a set of data streams, a set of STFDs and a set of SVAs.
   Once the granularity-aware sensor database schema is defined, this module allows to create the corresponding database relations in SQL, data description language implemented on top of any RDBMS.
2. **On-the-fly data loading:** Once the database is created, this level ensures the selection of the relevant data from the received sensor data streams. Thanks to STFDs and SVAs this middleware observes sensor data, chooses data to be stored and prunes the rest.

4.1   Implementation

We implemented in Java and Prolog a prototype containing the two levels presented previously. An overview of the proposed architecture is presented in Fig. 2 where we found the following main modules:

1. **Normalization module:** This module takes the user inputs (i.e. spatio-temporal dimensions, stream schemas and STFDs) and generates database schema using the algorithms presented in Sect. 3. This process leads to the database *abstract schema*. The reasoning about the spatio-temporal dimensions is proceeded through a Prolog environment. A *.pl* file, containing the *finer than* relationships between the different granularities, is generated from the input spatio-temporal dimensions. This file is checked whenever an algorithm needs to compare two granularities.

2. **Data management module:** At this stage, the user defines the SVAs corresponding to the proposed abstract database schema. This module updates the abstract schema with the corresponding semantic annotations which leads to the *concrete schema*. This module also ensures the selection of the "relevant" data (i.e. corresponding to the specification of the set of SVAs) from sensor data streams. Thus, for each SVA the system instantiates a specific *SVA data wrapper* which observes the sensor data stream concerning its attribute and identifies the accurate values. Once the user validates the obtained sensor database schema, the database relations are created. We used *Oracle 11G* for the implementation and the experiments.

3. **Sensor module:** This module is the interface between the sensors and the system. It gathers sensor data and links it to the user-defined dimensions.

4. **Web GUI:** This module allows the application manager to: **(a)** design the temporal and spatial dimensions, **(b)** from data streams at hand, define the stream schemas, **(c)** declare relevant STFDs from (a) and (b), **(d)** once an abstract schema exists, define a set of SVAs.

We believe that STFDs are natural and easy to express. Indeed, Fig. 3, contains a screen-shot from our prototype containing the user interface for STFD definition. As we can see the user just has to check the concerned attributes and select each granularity without caring about any syntax.

4.2   Ongoing Experiments

Data accuracy w.r.t. data reduction

In this section, we are mainly interested in studying the trade-off between data reduction and data accuracy with respect to some STFDs. To do so, we consider real-life sensor data. We have conducted real-life experiments in two buildings in our university. A total of around 400 heterogeneous physical sensors are deployed to measure temperature, humidity, $CO2/VOC$, presence, contact (for doors/windows), electricity consumption, weather conditions...

In these experiments, we consider 19 temperature sensors belonging to 7 different rooms. Each sensor sends a new value per minute. We are interested on data coming from these sensors all along one day (June, 1 2016). The idea is to compare the data reduction with respect to the considered set of STFDs. Then, to compare the considered data ("relevant data"), we observe the impact of this reduction upon the accuracy of the results. In our case, the initial sensor data stream spatio-temporal granularities are *sensor* and *minute*. All received sensor data is stored in a table called d_0 (i.e. raw data). We consider the following three sensor database tables and their corresponding STFDs:

1. sensor DB table d_1: contains the first temperature value per sensor per hour with respect to the STFD $location^{sensor}, time^{hour} \rightarrow temperature$,

2. sensor DB table d_2: contains the first temperature value per room per hour with respect to the STFD $location^{room}, time^{hour} \rightarrow temperature$, and

Table 2. Data reduction w.r.t. STFDs

Database table	Number of tuples	ratio: $\mid d' \mid$ / $\mid d \mid$
d_0	$27379 = \mid d \mid$	100%
d_1	456	$1,67\%$
d_2	162	$0,59\%$
d_3	7	$0,03\%$

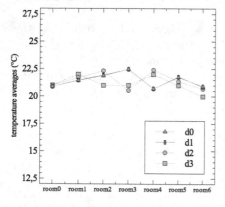

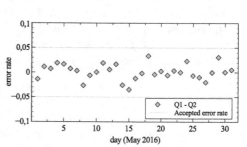

Fig. 4. Data accuracy w.r.t. data reduction

Fig. 5. Difference between the results of Q_1 and Q_2

3. sensor DB table d_3: contains the first temperature value per room per day with respect to the STFD $location^{room}, time^{day} \rightarrow temperature$.

Table 2 shows the important data reduction that may be done thanks to STFDs. As we can see, considering coarser granularities increases the data reduction. Next we are interested in data accuracy. Thus we aim to check how STFDs data approximation impacts data. To do so, we ask the different obtained database tables for the average of temperature per each room during the day. These averages are given in Fig. 4. According to this figure, approximating data with coarser granularities increases the error rate and decreases data accuracy. In order to evaluate the error rate and to check if the approximation deteriorates the evaluation of the temperature averages we are now interested in the difference between the results over *raw data* and a *sensor database*. Therefore, we executed the following queries:

1. query Q_1: computes on a *raw database* table the average of the temperature during each day of a given month for a given sensor, and
2. query Q_2: computes on a *sensor database* table (i.e. this table contains only one value per sensor per hour) the same average value.

The obtained difference values are given in Fig. 5. We can see that the difference is maintained under an error rate of 5% (most values being randomly disseminated on the $\pm 2\%$ rate stripe) which could be considered acceptable in a real life settings. It also evidences the fact that even though query Q_2 considers only one value per hour for the assessment of the average temperature per day, the output of the calculation sticks to the value obtained with a finer granularity. Actually, considering coarser granularities may not have a major effect on the final results, due to the fact that sensor values may be unchangeable over some spatio-temporal granules.

Data storage

We simulate now the sensors of an intelligent building containing 10 houses, each house contains 5 rooms. In each room we consider at least 2 sensors of each type (e.g. temperature, humidity...). Thus we simulate the operation of several hundreds of sensors, each sending data at a frequency of one value per minute. We took a sensor data stream, e.g. temperature, and we stored it in different relations with different temporal and spatial granularities. For instance, if we consider the temporal granularity *hour* the concerned relation contains one temperature value per hour per sensor. And, if we consider the spatial granularity *room* the concerned relation contains one temperature value per minute per room. The experiments of this section count the number of tuples in each relation. As expected, it is clear from Fig. 6 that with finer granularities we have more tuples. However, we can note that the temporal granularities are more discriminating than the spatial granularities. So as we have seen the use of STFDs with coarser granularities may reduce the number of tuples. This leads to more efficient queries as they have less tuples to scan. We also mention that, storing relevant data according to application requirements at specific spatio-temporal granularities enables the use of simpler queries (simple SELECT queries with simple WHERE clauses instead of queries with nested SELECT and JOINS). The stored relevant data can be considered as prior-computed query results.

Sensor data loading efficiency

In this section we are interested in the total execution time over predefined duration of our prototype while storing data in the appropriate database relations. Thus, we compare some implemented SVAs with the baseline solution, i.e. storing all received sensor data. Each sensor sends one value per minute.

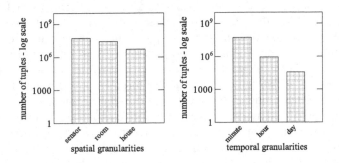

Fig. 6. Number of tuples when varying temporal and spatial granularities

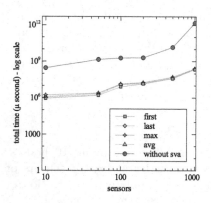
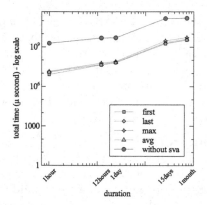

Fig. 7. Total execution time w.r.t. the number of sensors

Fig. 8. Total execution time w.r.t. the duration

The stored data in the following sets of experiments is approximated upto the spatio-temporal granularities *room* and *hour*. We compare the total execution time of the baseline solution with four SVAs (*first, last, maximum* and *average*). In this section we focus on the temperature stream. First, we focus on the variation of the total execution time with respect to the sensor number. We limit the scope of these experiments on a duration of 1 h. The obtained results are given in Fig. 7. Then, we vary the duration of our experiments for a fixed number of sensors (100 sensors). Using SVAs in order to select on-the-fly the relevant data is more efficient than the baseline solution: the total time in the case of SVAs is lower by a 10^3 order of magnitude in Fig. 7 and by a 10^2 order of magnitude in Fig. 8 (in a *log* scale).

5   Related Work

As far as we know there is no many contributions aiming at using the spatial and the temporal dimension in order to retrieve the relevant data from sensor data streams. Nowadays sensor data management can be seen from two points of view: real-time, i.e. continuous queries [1,2] versus historical [6,12,13] data management. In this paper, we were interested in long-term storage of sensor data. Our aim is to decrease the storage space and increase query efficiency of long-term reporting applications. Some approaches were interested in resolving storage problems that can result from the important amount of data generated by sensors. As far as we know, there is no formal approach for dealing with spatio-temporal stream querying considering different granularities.

TFDs have been mainly introduced in order to constraint the temporal data in temporal databases. There have been an important number of articles aiming at defining and characterizing TFDs namely [9,15–18]. The three first approaches [9,15,17] handle TFD without time granularity. In [15], the author defined a temporal relation as a temporal sequence of database states and extended each tuple with its updated version. The data model in [17] was extended with a valid time which represents a set of time points. The author presented the suitable definition of FD in the presence of valid time and defines two classes of temporal dependencies: Temporal Functional Dependencies (TFDs) and Dynamic Functional Dependencies (DFDs). In [9], data contains two time dimensions: valid time and transaction time. The authors handle the problem of expressing the functional dependencies with such data. These works do not consider granularity as a central notion as we do in this paper. Both [16,18] handled multiple time granularities. The authors in [16] and in [3] defined the time granularities and the different relationships between them. They defined the *temporal module schema* and the *temporal schema* as well as TFD. In [18], the author extended the dependencies presented in [17] using time granularity and object identity which is a time-invariant identity that relates the different versions of the same object.

Roll-up Dependencies (RUDs) [19] define dependencies with a higher abstraction level for OLAP DB. They extend TFDs to non temporal dimensions allowing each attribute to roll up through the different levels of its associated hierarchy. Algorithmic aspects such as attribute closure is not studied for RUD, as we do in this paper. We just need two dimensions, i.e. temporal and spatial. In fact, we distinct two particular attributes, for spatial and temporal dimensions, and we combine them with classical attributes (i.e. attributes without associated hierarchies). Moreover, unlike [19] which deals with schemas, we deal with attributes and we propose a normalization algorithm and a new closure algorithm of a set of attributes from a set of STFDs. Our approach is not comparable to OLAP since our main goal is to retrieve and organize sensor data and not to analyze multidimensional data from multiple perspectives.

We also defined a declarative structure, i.e. SVA, which annotates the generated database design in order to enrich it with semantic information about relevant data selection. SVAs are also useful to select on-the-fly relevant data. In fact, SVAs represent a sort of data exchange [7] mechanism inspired from interval-based semantic assumptions designed for temporal databases [3]. The point-based and interval-based semantic assumptions in [3] can be used to derive or compress temporal data.

The use of SVAs in order to choose the relevant tuples reminds us load shedding techniques. The load shedding process [14] intends to reduce the workload of the data stream management system by dropping tuples from the system. Several approaches proposed different tuples dropping strategies, e.g. random [10], with a priority order [4] or a semantic strategy [5]. The load shedding usually interferes in the physical plan of the query while our approach aims to interfere from the database design and to take into account predefined approximations. Our contribution can be thought as a "declarative load shedding process" since we allow to prune data stream from declarative constraints, instead of sampling techniques.

6   Conclusion

In this paper, we have considered the long-term storage problem of sensor data streams. We have presented a declarative approach to build a granularity-aware sensor database from sensor data streams on top of any RDBMS. Our core idea is to take into account the spatial and temporal aspects of sensor data streams thanks to Spatio-Temporal Functional Dependencies (STFDs) and to adapt the classical normalization techniques developed for relational databases to sensor databases. We have defined a dedicated normalization algorithm based on a novel closure algorithm for STFDs. The closure of attributes plays a crucial role in the generation of a minimal cover of a set of STFDs and thus in the production of normalized sensor database schemas. We have also defined Semantic Value Assumption (SVA), a declarative database schema annotation, allowing to specify the mechanism to load, on-the-fly and automatically, the relevant data into the sensor database. A prototype has been implemented in order to test both sensor database design from STFDs and data loading techniques. We have conducted experiments on real and synthetic data streams from intelligent buildings. We discussed the trade-off between the data accuracy and the data reduction.

We have highlighted our proposition in the context of intelligent buildings for domestic sensors. Nevertheless, our proposition relies upon clear theoretical foundations that enable to take both spatial and temporal dimensions into account for sensor data streams. The approach is quite versatile and could be adopted in a wide range of application contexts. Many extensions could be done, for instance to consider *mobile sensors* or to study specific properties of the decomposition algorithm for STFDs.

Acknowledgments. This work is supported by the ARC6 program of the Rhône-Alpes region, France.

References

1. Abadi, D.J., Carney, D., Çetintemel, U., Cherniack, M., Convey, C., Lee, S., Stonebraker, M., Tatbul, N., Zdonik, S.: Aurora: a new model and architecture for data stream management. VLDB J. Int. J. Very Large Data Bases **12**(2), 120–139 (2003)
2. Arasu, A., Babu, S., Widom, J.: The CQL continuous query language: semantic foundations and query execution. VLDB J. Int. J. Very Large Data Bases **15**(2), 121–142 (2006)
3. Bettini, C., Jajodia, S., Wang, S.: Time Granularities in Databases, Data Mining, and Temporal Reasoning. Springer, Heidelberg (2000)
4. Carney, D., Çetintemel, U., Cherniack, M., Convey, C., Lee, S., Seidman, G., Stonebraker, M., Tatbul, N., Zdonik, S.: Monitoring streams: a new class of data management applications. In: Proceedings of the 28th International Conference on Very Large Data Bases, VLDB 2002, pp. 215–226. VLDB Endowment (2002)
5. Das, A., Gehrke, J., Riedewald, M.: Approximate join processing over data streams. In: Proceedings of the 2003 ACM SIGMOD International Conference on Management of Data, SIGMOD 2003, pp. 40–51. ACM, New York (2003)

6. Diao, Y., Ganesan, D., Mathur, G., Shenoy, P.J.: Rethinking data management for storage-centric sensor networks. In: CIDR, vol. 7, pp. 22–31 (2007)
7. Fagin, R., Kolaitis, P.G., Popa, L.: Data exchange: getting to the core. ACM Trans. Database Syst. (TODS) **30**(1), 174–210 (2005)
8. Golab, L., Özsu, M.T.: Data stream management. Synth. Lect. Data Manage. **2**(1), 1–73 (2010)
9. Jensen, C.S., Snodgrass, R.T., Soo, M.D.: Extending existing dependency theory to temporal databases. IEEE Trans. Knowl. Data Eng. **8**(4), 563–582 (1996)
10. Kang, J., Naughton, J.F., Viglas, S.D.: Evaluating window joins over unbounded streams. In: Proceedings of 19th International Conference on Data Engineering, pp. 341–352. IEEE (2003)
11. Levene, M., Loizou, G.: A Guided Tour of Relational Databases and Beyond. Springer, London (2012)
12. Lewis, M., Cameron, D., Xie, S., Arpinar, B.: ES3N: a semantic approach to data management in sensor networks. In: Semantic Sensor Networks Workshop (2006)
13. Petit, L., Nafaa, A., Jurdak, R.: Historical data storage for large scale sensor networks. In: Proceedings of the 5th French-Speaking Conference on Mobility and Ubiquity Computing, pp. 45–52. ACM (2009)
14. Tatbul, N., Çetintemel, U., Zdonik, S., Cherniack, M., Stonebraker, M.: Load shedding in a data stream manager. In: Proceedings of the 29th International Conference on Very Large Data Bases, vol. 29, pp. 309–320. VLDB Endowment (2003)
15. Vianu, V.: Dynamic functional dependencies and database aging. J. ACM (JACM) **34**(1), 28–59 (1987)
16. Wang, X.S., Bettini, C., Brodsky, A., Jajodia, S.: Logical design for temporal databases with multiple granularities. ACM Trans. Database Syst. (TODS) **22**(2), 115–170 (1997)
17. Wijsen, J.: Design of temporal relational databases based on dynamic and temporal functional dependencies. In: Clifford, J., Tuzhilin, A. (eds.) Recent Advances in Temporal Databases. Workshops in Computing, pp. 61–76. Springer, London (1995). doi:10.1007/978-1-4471-3033-8_4
18. Wijsen, J.: Temporal FDS on complex objects. ACM Trans. Database Syst. (TODS) **24**(1), 127–176 (1999)
19. Wijsen, J., Ng, R.T.: Temporal dependencies generalized for spatial and other dimensions. In: Böhlen, M.H., Jensen, C.S., Scholl, M.O. (eds.) STDBM 1999. LNCS, vol. 1678, pp. 189–203. Springer, Heidelberg (1999). doi:10.1007/3-540-48344-6_11
20. Zilio, D.C., Rao, J., Lightstone, S., Lohman, G., Storm, A., Garcia-Arellano, C., Fadden, S.: DB2 design advisor: integrated automatic physical database design. In: Proceedings of the Thirtieth International Conference on Very Large Data Bases, vol. 30, pp. 1087–1097. VLDB Endowment (2004)

Recommendation

Location-Aware Query Recommendation
for Search Engines at Scale

Zhipeng Huang[1]([✉]) and Nikos Mamoulis[2]

[1] The University of Hong Kong, Hong Kong, China
zphuang@cs.hku.hk
[2] University of Ioannina, Ioannina, Greece
nikos@cs.uoi.gr

Abstract. Query recommendation is a popular add-on feature of search engines, which provides related and helpful reformulations of a keyword query. Due to the dropping prices of smartphones and the increasing coverage and bandwidth of mobile networks, a large percentage of search engine queries are issued from mobile devices. This makes it possible to provide better query recommendations by considering the physical locations of the query issuers. However, limited research has been done on location-aware query recommendation for search engines. In this paper, we propose an effective spatial proximity measure between a query issuer and a query with a location distribution obtained from its clicked URLs in the query history. Based on this, we extend two popular query recommendation approaches to our location-aware setting, which provides recommendations that are semantically relevant to the original query and their results are spatially close to the query issuer. In addition, we extend the bookmark coloring algorithm for graph proximity search to support our proposed approaches online, with a spatial partitioning based approximation that accelerates the computation of our proposed spatial proximity. We conduct experiments using a real query log, which show that our query recommendation approaches significantly outperform previous work in terms of quality, and they can be efficiently applied online.

1 Introduction

Keyword search, which allows a user to express her query with a few keywords, has become a fundamental tool in Web search engines. Recently, lots of interest from both research and industry has been drawn to the topic of *query recommendation*, which is closely related to keyword search. Besides ranking the Web pages according to their relevance to the query provided by the user, a search engine may provide several alternative formulations of the query, which can be more focused and interesting to the user. Query recommendation, as an add-on function to a search engine, has provided significant benefit to search engine users [11] in terms of providing better user experience.

Most of the existing work on query recommendation focuses on the analysis of *query logs*, which contain large amounts of historical information of search

© Springer International Publishing AG 2017
M. Gertz et al. (Eds.): SSTD 2017, LNCS 10411, pp. 203–220, 2017.
DOI: 10.1007/978-3-319-64367-0_11

engine users, including what query was issued by whom at what time and which URLs were subsequently clicked [5,6,8,14]. The query logs are often represented as *graphs* of queries and other related components, allowing graph analysis techniques to perform relevance search on query logs. For example, [5] built a graph of *queries* based on query logs. The weight of a graph edge that connects two queries is proportional to the times that the two queries are issued by a same user within a short period (i.e., the queries are in the same *search session*). Query recommendation is performed by applying Personalized PageRank proximity search on this graph starting from the original query.

Nowadays, as mobile devices are ubiquitous, many keyword search queries are expressed by mobile users and have *spatial intent*, i.e., the users require results that are physically close to their locations. However, very few studies have considered location information when performing query recommendation. The most recent work [17] proposes a solution based on a bipartite graph that connects queries to their clicked documents (or URLs). The edges of the graph are adjusted based on the location of the query issuer and then Personalized PageRank proximity search is applied to obtain the recommendations. The work of [17] only considers the locations of documents to derive the proximity between queries. In this work, we propose a more sophisticated approach that generates a spatial distribution for each query (based on its clicked URLs) and uses it to directly measure the proximity between each query and the query issuer. We extend two popular query recommendation approaches, i.e., query-flow graph [5] and term-query-flow graph [6], to our location-aware setting. The basic idea is to give higher preference to queries that have larger spatial proximity to the user during the random walk-based recommendation process, which finally leads to recommendations that are more spatially relevant compared to those suggested by the method of [17], as shown in our experimental results.

The main technical challenge is efficiency; search engines should provide instant responses to users; hence, query recommendation should also be conducted in sub-seconds. However, differently from the traditional query recommendation setting, which ignores user locations, we need to consider the spatial proximity between queries and the user, which can only be obtained online after the user issues her query. We first adopt Bookmark Coloring Algorithm (BCA) [4], a classic method for online Personalized PageRank based proximity search, to support our recommendation method. Then, we design an approximate version of the algorithm, which is based on a spatial partitioning and greatly reduces the cost without sacrificing the quality of the results.

We perform extensive experiments on a real query log from to verify the performance of our methods. We first conduct a user study, which shows that the users prefer recommendations generated by our proposed spatial proximity measure compared to four alternatives. Then, we compare our proposed location-aware query recommendation approaches with previous work, and show that our approaches achieve significantly better recommendations in terms of both semantic relevance and spatial proximity.

The contributions of our paper are summarized as follows:

- We consider the spatial proximity between a query and a search engine user in location-aware query recommendation. We propose an effective spatial proximity measure, shown to outperform alternatives in our user study.
- We extend two popular query recommendation approaches to support location-aware recommendation.
- We evaluate our proposed location-aware query recommendation methods, demonstrating that our methods outperform previous work significantly and that they can be applied online (within 300 ms).

The rest of the paper is organized as follows. Section 2 introduces some preliminaries and definitions. Section 3 presents our location-aware query recommendation model. Section 4 includes our algorithm for efficient query recommendation. Section 5 presents our experimental results. Section 6 reviews related work and Sect. 7 concludes the paper.

2   Preliminaries and Definitions

In this section, we introduce some necessary preliminaries and definitions, including *query logs* (Sect. 2.1), how we model and obtain the relevance of queries to locations (Sect. 2.2), and two popular location-agnostic query recommendation methods (Sects. 2.3 and 2.4).

2.1   Query Log

The query log of a keyword search engine is typically modeled as a set of records (q_i, u_i, t_i, C_i), where q_i is a query submitted by user u_i at time t_i, and C_i is the set of clicked URLs by u_i after q_i and before the user issues another query.

Following common practice [5,6,14], we can partition a query log into task-oriented *sessions*, where each session is a contiguous sequence of query records from the same user. Two contiguous queries are put in the same session if their time difference is at most t_θ (typically, t_θ is 30 min). Within the same session, we can assume that the user's search intent remains unchanged.

2.2   Obtaining Locations from a Query Log

Location Distribution of URLs. A webpage, corresponding to a URL, may contain information about one or more spatial locations. The *location distribution* of a URL d is a probability distribution p_d over a set of locations $L = \{(lat, lon) \mid lat, lon \in R\}$, where $\sum_{l \in L} p_d(l) = 1$. For the purposes of this paper, for each URL in the query log, we fetch the document and parse the content using GeoDict[1], a simple library/tool for pulling location information from unstructured text. This provides us with the location distribution for

---

[1] https://github.com/petewarden/geodict

each URL. Alternatively, other methods for extracting locations from text can be applied [9].

Location Distribution of Queries. We also model the location distribution of a query q_i as a probability distribution p_{q_i}, and we can obtain it from a linear combination of the distributions of the clicked URLs for q_i. Formally,

$$p_{q_i}(l) = \frac{\sum_{d_j \in C_{q_i}} p_{d_j}(l)}{\sum_{l' \in L} \sum_{d_j \in C_{q_i}} p_{d_j}(l')},$$

where C_{q_i} is the set of clicked URLs for query q_i. In this paper, we use the location distributions of the queries to facilitate the problem of recommending queries to a search engine user u, that are not only semantically relevant to the query issued by u, but also are spatially close to the physical location of u.

2.3   Query-Flow Graph

One of the most promising directions for performing query recommendation relies on the extraction of behavioral patterns in query reformulation from query logs. The query-flow graph (QFG in short) [5] is a graph representation of query logs, capturing the "flow" between query units. Intuitively, a QFG is a directed graph of queries, in which an edge (q_i, q_j) with weight w indicates that query q_j follows query q_i in the same session of the query log with probability w.

More formally, QFG is defined as a directed graph $G_{qf} = (Q, E, W)$, where Q is the set of nodes, with each node representing a unique query in the log, $E \subseteq Q \times Q$ is the set of edges, and W is a weighting function assigning a weight $w(q_i, q_j)$ to each edge $(q_i, q_j) \in E$. In G_{qf}, two queries q_i and q_j are connected if and only if there exists a session in the query log where q_j follows q_i. Figure 1 illustrates a QFG with three query nodes.

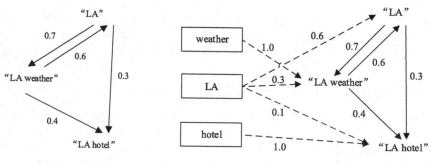

Fig. 1. QFG   **Fig. 2.** TQGraph

Recommendation via QFG. Given a query $q \in Q$, the top-k recommendations for q can be obtained by a random walk with restart (RWR) [13] process starting from q, as suggested in [5]. At each step of the RWR, the random walker moves to an adjacent node with probability $1 - \alpha$ via the transition matrix W, or *teleports* to the original node q with probability α. In this way, a RWR process defines a Personalized PageRank score $PPR(q, q', W)$ for each node q' as the probability that the RWR starting from q reaches node q'. In this way, the top-k recommendations can be the set of k nodes q' in Q, which have the maximum PPR scores w.r.t. q in QFG. In other words, the recommendation score $rec_q(q')$ for each $q' \in Q$ is defined as:

$$rec_q(q') = PPR(q, q', W), \tag{1}$$

where W is the transition matrix for the PPR, and queries having the top-k scores are recommended.

However, QFG has an obvious disadvantage for query recommendation; that is, it cannot make any recommendation to an input query q, if $q \notin Q$. In other words, if a query has not appeared in the query log before, QFG fails to generate any recommendation.

2.4 Term-Query-Flow Graph

Another popular method for query recommendation is the term-query-flow graph (TQGraph) [6], which basically extends the QFG method by considering a center-piece subgraph induced by terms contained into queries.

Formally, a TQGraph is a directed graph $G_{tqg} = G_{qf} \cup G_{tq}$, where G_{qf} is the QFG as described in Sect. 2.3, and G_{tq} is a bipartite graph of *term* nodes and *query* nodes. Specifically, the set of nodes in the TQGraph is $V_{tq} = Q \cup T$, where Q is the set of queries and T is the set of terms. Edge (t, q) exists in E_{tq}, if the term t is contained in query q. Figure 2 illustrates a TQGraph with three query nodes and three term nodes.

Recommendation via TQGraph. Given a query $q \in Q$, the top-k TQGraph recommendations for q are obtained by ranking all $q' \in Q$ based on their aggregate PPR scores w.r.t. each term $t \in q$. In other words, the recommendation score $rec_q(q')$ for each $q' \in Q$ is defined as follows:

$$rec_q(q') = \prod_{t \in q} PPR(t, q', W \cup E_{tq}) \tag{2}$$

We can see that TQGraph can generate recommendations for query q, as long as all the terms within q appear in the query log. Empirically, TQGraph has a much better *query coverage* compared to QFG, because it can also be used for queries that are asked for the first time.

3   Location-Aware Query Recommendation

In this section, we introduce our location-aware query recommendation models. Our goal is to provide recommendations that are spatially close to the query issuer. For this, we should first define spatial proximity between a user located at l_u and a query with location distribution p_q. Some of the alternatives are:

- (i) Expected Distance (ED). Formally, $ED(p_q, l_u) = \sum_{l \in L(q)} p_q(l) \times dist(l, l_u)$.
- (ii) Min Distance (MinD). Formally, $MinD(p_q, l_u) = min_{l \in L(q)} dist(l, l_u)$.
- (iii) Max Distance (MaxD). Formally, $MinD(p_q, l_u) = max_{l \in L(q)} dist(l, l_u)$.
- (iv) Mean Distance (MeanD). Formally, $MeanD(p_q, l_u) = dist(mean(p_q), l_u)$.

However, these four distance-based measures are not fully consistent with our goal of finding spatially close queries to the user. For example, suppose we have two queries q_1 and q_2 with the following location distribution:

q_1 Hong Kong: 0.6, New York: 0.3, Los Angeles: 0.1
q_2 Beijing: 0.8, Los Angeles: 0.2

One would argue that q_1 is more spatially relevant to a user u_1 located in Hong Kong, compared to q_2, as a great portion of q_1's distribution is very close to u_1, which means that the majority of URLs related to query q_1 contain information that is spatially relevant to u_1. However, using ED, MaxD or MeanD might not select q_1, because after considering all locations of q_1, its overall distance to Hong Kong is quite far. At the same time, MinD does not distinguish q_1 and q_2 for a user u_2 located in Los Angeles, because MinD neglects the support probability of each location. Hence, we should define a more appropriate spatial proximity measure that better captures the distance between a query and a search engine user. In this direction, we propose the following measure:

Definition 1 Spatial Proximity sim_s. *Given the location of the user $l_u \in L$, a range threshold r, and a location distribution p_q of a query q, the spatial proximity between l_u and q is the portion of p_q within distance r from l_u, i.e.,*

$$sim_s(q, l_u) = \sum_{dist(l_u, l') < r} p_q(l')$$

The range threshold r models the distance that the user is willing to travel in order to use a service offered by the query results. In the example above, assuming that we select r as a within-city travel distance, we will have: $sim_s(q_1, l_{u_1}) = 0.6$, $sim_s(q_2, l_{u_1}) = 0$, $sim_s(q_1, l_{u_2}) = 0.1$, $sim_s(q_2, l_{u_2}) = 0.2$. This is consistent with the intuition that u_1 is more related to query q_1, and u_2 is more related to query q_2. In our experiments, we use $r = 100\,km$ by default and compare the performances of our methods for different values of r.

After obtaining the spatial proximity between the user and the queries, we can adjust the weights on the edges of QFG to give higher preference to queries that have larger sim_s to the user.

Definition 2 Spatially Adjusted Weights. *Given a query $q \in Q$ issued at location l_u, the spatially adjusted weight for an edge of a QFG $(q_i, q_j) \in E$ is defined as:*

$$\tilde{w}(q_i, q_j) = \beta \times w(q_i, q_j) + (1 - \beta) \times sim_s(q_j, l_u)),$$

where β is a parameter that controls the relative importance of spatial proximity and $w(q_i, q_j)$ is the original weight of the edge (q_i, q_j) in the QFG.

With the linear function in Definition 2, we obtain a location-aware transition matrix $\tilde{W}$ from the original matrix W. Then, we can perform location-aware query recommendation based on $\tilde{W}$. In other words, the recommendation processes for spatial QFG (SQFG) and spatial TQGraph (STQGraph) are the same as their location-agnostic counterparts, except that we use $\tilde{W}$ instead of W.

Recommendation via SQFG. Given a query $q \in Q$ issued at location l_u, the top-k SQFG recommendations for q can be obtained by a location-aware Personalized PageRank w.r.t. node q, i.e.,

$$rec_q(q') = PPR(q, q', \tilde{W}), \tag{3}$$

where $rec_q(q')$ is the recommendation score for query q' and $\tilde{W}$ is the location-aware transition matrix for the PPR.

Recommendation via STQGraph. Given a query $q \in Q$ issued at location l_u, the top-k STQGraph recommendations for q can be obtained by ranking all $q' \in Q$ based on their aggregate PPR scores w.r.t. each term $t \in q$. In other words, the recommendation score $rec_q(q')$ for each $q' \in Q$ is defined as follows:

$$rec_q(q') = \prod_{t \in q} PPR(t, q', \tilde{W} \cup E_{tq}) \tag{4}$$

4   Location-Aware PPR

Both our SQFG and STQGraph models require computation of location-aware PPR over the spatially adjusted transition matrix $\tilde{W}$. However, this can be expensive, as the transition matrix $\tilde{W}$ depends on the location l_u of the search engine user, which can only be known at query time. Thus, traditional indexing techniques for efficiently computing PPR cannot be used in our setting. In this section, we first introduce a basic solution, by extending the Bookmark Coloring Algorithm (BCA) [4]. Then, we propose a more efficient, approximate version of the algorithm, which is based on a spatial partitioning approach.

4.1   BCA with Online Transition Matrix $\tilde{W}$

We extend the famous Bookmark Coloring Algorithm (BCA) [4] to compute the top-k PPR results based on the location-aware transition matrix $\tilde{W}$ on query time as our basic method. The basic idea of BCA is to model the RWR process as a bookmark coloring process, in which some portion of the ink in a processed node is sent to its neighbors, while the remaining ink is retained at the node.

Specifically, starting from the query node q with 1.0 units of ink, BCA keeps α portion in q and distributes the remaining $1 - \alpha$ portion to q's neighbors in the graph using the weights of the outgoing edges to determine the percentage of ink sent to each neighbor. The process is repeated for each node that receives ink, until the residue ink to be redistributed becomes a very small percentage of

---

Algorithm 1. BCA

Input: Transition matrix W, starting node q, user location l_u
Output: Apprximated PPR vector rec_q

1 PriorityQueue $que \leftarrow \emptyset$
2 Add q to que with $q.ink \leftarrow 1.0$
3 $R \leftarrow \emptyset$;
4 $Cache \leftarrow \emptyset$;
5 **while** $que \neq \emptyset$ *and* $que.top.ink \geq \epsilon$ **do**
6 | Deheap the first entry top from que;
7 | $R[top] \leftarrow R[top] + top.ink \times \alpha$;
8 | **for** $q' \in top.neighbors$ **do**
9 | | **if** $q' \in Cache$ **then**
10 | | | $sim_s(q', l_u) \leftarrow Cache[q']$;
11 | | **else**
12 | | | Compute $sim_s(q', l_u)$ using Definition 1;
13 | | | $Cache[q'] \leftarrow sim_s(q', l_u)$;
14 | | Compute $\tilde{w}(top, q')$ using Definition 2;
15 | | $q'.buf \leftarrow top.ink \times (1 - \alpha) \times \tilde{w}(top, q')$;
16 | | **if** $q'.buf \geq \epsilon$ **then**
17 | | | Add q' to que with ink $q'.buf$;
18 | | | $q'.buf \leftarrow 0$;

19 **return** R

---

the original 1.0 units. Different from traditional PPR computation using BCA, our transition matrix $\tilde{W}$ can only be obtained online by Definition 2, after we know the location of the query issuer l_u.

In our implementation, the spatially adjusted weights of each edge (q_i, q_j) are also computed online based on l_u, at the time when the query node q_i is distributing ink. This means that the computation of $\tilde{W}$ is done during BCA simulation. A node distributes ink only if the quantity of the ink exceeds a threshold ϵ (typically, $\epsilon = 10^{-5}$). BCA terminates when there are no more nodes to distribute ink.

We adopt the following two optimizations in our BCA implementation.

– **Lazy Updating Mechanism.** In the original BCA, a node distributes its ink aggressively, i.e., each neighbor q' of node top will be pushed into the priority queue que after receiving some portion of $top.ink$. On the other hand, we only care about the nodes with ink greater than ϵ. Based on these two observations, a lazy updating mechanism can reduce the number of non-necessary pushing without changing the final results; the pushing a node q' into the priority queue que is delayed until the ink it receives is greater than ϵ. If the amount of received ink is less than ϵ, q' only accumulates it in a buffer; as soon as the buffer's ink exceeds ϵ, q' is pushed into que.

- **Spatial Proximity Caching**. Every time when we need to distribute ink to a query node q', we need to compute the spatial proximity between q' and the location of the user l_u. However, the same query node may be processed multiple times in a single BCA call. In view of this, we cache the spatial proximities between the location of the user l_u and the query nodes that have been computed so far. By doing this, we only need to compute the spatial proximity for a query q' once during a BCA call.

Algorithm 1 details our implementation of BCA, including the two optimizations mentioned above. Priority query que maintains the nodes to be processed in descending order of their ink (Line 1). que initially contains only one node with ink amount 1.0 (Line 2), i.e., the starting node of the PPR. The nodes that have some retained ink are kept in a dictionary R, which is initially empty (Line 3). Termination conditions are checked at each iteration (Line 5). Within each iteration, we first dequeue from que the node with the most ink to distribute (Line 6). Then we leave α portion to its result (Line 7), and distribute the rest to its neighbors with weights $\tilde{w}$ (Lines 8–18). We first check the spatial cache whether q' has been computed before (Line 9). If so, the spatial proximity between q' and l_u can be directly got from the cache (Line 10). Otherwise we need to compute $sim_s(q', l_u)$ (Line 12) and save to the cache (Line 13). Finally, for each of the neighbors, if the received ink is greater than a threshold ϵ (Line 16), the corresponding query node will be pushed into que (Line 17) and the corresponding buffer is cleared (Line 18). Finally, the dictionary R is returned. In SQFG, where a single RWR search is applied, the k query nodes in R with the most retained ink are recommended. In STQGraph, the Rs of the RWR searches from all terms are summed up at each query node before computing and returning the top-k query nodes.

Partitioning Based Approximation. The previous two optimizations guarantee the same results as the original BCA and improve its performance. However, the cost of computing the spatial proximity $sim_s(q', l_u)$ between a query q' and the location of the user l_u at each iteration is still the bottleneck of the algorithm. Recall from Definition 1 that we need to enumerate all locations of the query q' in order to accumulate the distribution. To reduce this high cost, we propose to compute a partitioning based approximation of $sim_s(q, l)$ as follows:

$$\hat{sim}_s(q, l) = \sum_{cir(l_u) \text{ intersects } c} p_q(c), \tag{5}$$

where c is a spatial partition of locations, $cir(l_u)$ is the circle with l_u as center and r as radius, and $p_q(c) = \sum_{l' \in c} p_q(l')$ is the location distribution of q that falls into partition c.

We use a grid to partition the space. Hence, locations that fall into the same cell belong to the same partition. If the length of each grid cell is a, to compute $\hat{sim}_s$, we only need to accumulate $p_q(c)$ for at most $\lceil \frac{2r}{a} + 1 \rceil^2$ partitions.

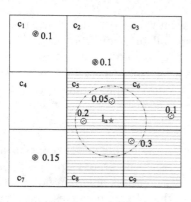

Table 1. Location distribution approximation

Cell	$p_q(c)$	Cell	$p_q(c)$	Cell	$p_q(c)$
c_1	0.1	c_2	0.1	c_3	0
c_4	0	c_5	0.25	c_6	0.1
c_7	0.15	c_8	0	c_9	0.3

Fig. 3. Example of partitioning

In our experiments, we use $a = r$, so the computational cost is much lower than computing the exact sim_s, which requires enumeration of all locations.

Figure 3 illustrates an example of our partitioning based approximation. The dots with number next to them represent the location distribution of a query q. Suppose a user is located at the starred location and the circle is defined by that location and the range threshold r. The shaded cells are those which intersect the circle and according to Definition 1, $sim_s(q, l_u) = 0.2 + 0.05 + 0.3 = 0.55$. After using our spatial partition, we can obtain an approximation of the location distribution as in Table 1. Then, an approximation of the spatial proximity is computed as $\hat{sim}_s(q, l_u) = 0.25 + 0.1 + 0 + 0.3 = 0.65$.[2]

5   Experimental Evaluation

5.1   Dataset

We use AOL in all our experiments. AOL is a well-known public query log from a major commercial search engine, which consists of Web search queries collected from 657 k users over a two months period in year 2006. This dataset is sorted by anonymous user ID and sequentially arranged, containing 20 M query instances corresponding to around 9 M distinct queries. After we sessionize the query log with $\theta_t = 30$ min, we obtain a total of 12 M sessions.

5.2   Methodology

We adopt the automatic evaluation process described in [14], to assess the performance of the tested methods. In a nutshell, we use part of the query log as training data to generate recommendations for a kept-apart query log

---

[2] We do not further refine to get an exact result by looking into the locations within the cells, because we believe that those locations near the range r from the user are still spatially relevant (see the location in cell c_6 of Fig. 3).

fragment (the test data). In the test query log, we denote by $q_{i,j}$ the j_{th} query in session s_i. We assume that all $\{q_{i,j} \mid j > 1\}$ are good recommendations for query $q_{i,1}$ which, in accordance to previous work [14].

Specifically, we use 90% of the query log for training, which contains 11 M sessions and 8.4 M distinct queries. We use the remaining 10% of the query log to generate testing queries. We first extract sessions with at least two queries, and randomly sample 10 K queries as our testbed. We take the first query of each session as input and the queries that follow as the ground truth recommendations. Formally, the ground truth for input query $q_{i,1}$ is $\{q_{i,j} \mid j > 1\}$, where $q_{i,j}$ is the j_{th} query appearing in the i_{th} session. While the objective of this evaluation approach may not necessarily be aligned with what a good recommendation could be on particular instances, by being entirely unsupervised and applied on a large number of sessions, it is a strong indicator of the techniques' performance. Note that we randomly assign the location of the query issuer l_u, as the dataset does not contain the location information of about the users.

We use the following three metrics to evaluate the performance of each method:

- *coverage*. This is the percentage of input queries that can be served with at least one recommendation.
- *precision@k*. This is the percentage of recommended queries in the top-k lists that are in the ground truth as described previously. Formally, $precision@k = \frac{\#HIT}{k \cdot \#query}$, where $\#HIT$ is the total number of recommended queries that are part of the ground truth, and $\#query$ is the number of input queries.
- $sim_s@k$. This is the average spatial proximity (see Definition 1) between the recommended queries in the top-k lists to the location of the query issuer l_u.

Competitors

- **LKS** [17]. LKS is the most recently proposed location-aware keyword suggestion approach. It first builds a bipartite graph of queries and URLs using the query log, and then performs location-aware random walk over the graph during online recommendation. We use the default settings of LKS, i.e., the restart probability α_{LKS} and the edge weight adjustment parameter β_{LKS} are both set to 0.5.
- **SQFG.** SQFG is our spatial QFG method as described in Sect. 3. By default, we set the spatial radius threshold $r = 100$ km, the restart probability $\alpha = 0.5$, and the spatial adjustment factor $\beta = 0.5$.
- **STQGraph.** STQGraph is our proposed spatial TQGraph approach as described in Sect. 3. We use the same default settings as in SQFG.
- **STQGraph*.** STQGraph* is our spatial partitioning based approximation approach. By default, we use 100 Km as the length of each grid cell, and all the other parameters are same as STQGraph.

5.3   User Study

We first conducted a user study to compare our proposed spatial proximity sim_s in Definition 1 with the four alternatives mentioned in Sect. 3. We first used our STQGraph to generate top-1 recommendations for 100 random queries with different spatial proximity measures. Then, for each of the recommended queries, we showed its location distribution as well as the location of the query issuer l_u to the participants. They were asked to rate the spatial relevance between the recommended query and the query issuer, using one of the following rating levels: 0 for not related at all, 1 for somehow related and 2 for very related. The recommended queries were shuffled before given to the participants, so that they could not know which spatial proximity measure was used to generate the recommended query. We asked 15 participants (HKU students) to rate the recommended queries, and each of the queries were given at least 3 ratings.

The results are shown in Fig. 4. We can see that our sim_s has the largest percentage of 2 s, which means that sim_s is acknowledged to be the best measure of spatial proximity. Out of the four alternatives, ED and MeanD received relatively better user feedback. This is because a smaller ED or MeanD implies smaller overall distance to the query issuer. MinD got the worst user feedback because MinD only considers the location l which is the closest to l_u, but not the probability $p_q(l)$, which could be too small.

From this user study, we can conclude that users prefer sim_s over the other four proximity measures. In the rest of our experiments, we use sim_s to evaluate the spatial quality of recommended queries.

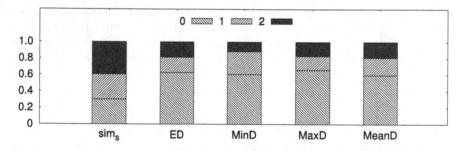

Fig. 4. User study on different spatial proximity measures

5.4　Effectiveness

We first compare the four tested methods. Then we test the parameter sensitivity of our STQGraph method to different parameter values.

Comparison Results. Table 2 compares all methods in terms of their *coverage*. We can see that SQFG has relatively low *coverage* compared the other three approaches. This is expected, because SQFG has the same disadvantage as QFG, i.e., it cannot provide recommendations to any previously unseen queries. STQGraph and LKS have similar *coverage*, much higher than that of SQFG. Note that STQGraph and STQGraph* have the same *coverage*, since our spatial partition based approximation only affects the ranking of the recommended queries.

Table 2. *coverage* results

Method	LKS	SQFG	STQGraph	STQGraph_P
coverage	36.8%	27.9%	37.1%	37.1%

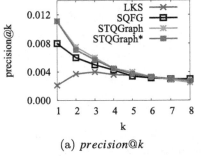

(a) *precision@k*

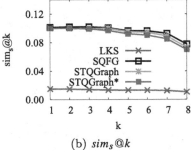

(b) *sim_s@k*

Fig. 5. Comparison.

Figure 5(a) shows the *precision@k* of all methods. Since typical search engines (e.g. Google and Yahoo!) show eight recommendations, we test values of k from 1 to 8. Observe that our STQGraph and STQGraph* methods have significantly larger *precision@k* compared to LKS. As k becomes larger, *precision@k* becomes smaller. This means our recommendation methods rank the recommended queries reasonably, as those with smaller ranks are more precise. STQGraph* has almost the same *precision@k* as STQGraph, which means that our spatial partition based approximation does not harm the recommendation quality in terms of semantic quality. The precision of SQFG is lower than that of STQGraph for small values of k.

Figure 5(b) shows the results of $sim_s@k$. Similar to the case of *precision@k*, $sim_s@k$ drops as k increases. LKS has very poor $sim_s@k$ result compared with our approaches. This is because LKS tends to recommend queries that share the same clicked URLs with the input query, without directly considering the location distribution of the recommended queries. SQFG, STQGraph* and STQGraph achieve almost the same $sim_s@k$.

Parameter Sensitivity. We now test the sensitivity of our STQGraph approach to the values of its parameters. As *coverage* result is only related to the connectivity of STQGraph and is not sensitive to the parameters, we only show the *precision@k* and $sim_s@k$ results.

- **Varying** α. Figure 6 shows the quality of STQGraph for various values of α. Observe that α does not influence *precision@k* very much. In addition, larger α leads to smaller $sim_s@k$. This is because a larger α gives higher weight to the adjacent queries to the input query in the graph, whereas potentially better queries exist at a larger distance.

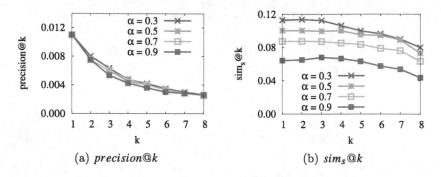

(a) *precision@k*   (b) *sim_s@k*

Fig. 6. Varying α

- **Varying β.** Figure 7 shows the quality of STQGraph for various values of β. We can see that larger β values lead to slightly higher *precision@k*. However, larger β values lead to smaller *sim_s@k*. This is because a larger β gives higher weight to the semantic relevance between queries and lower weight to the spatial proximity. Overall, a value of β close to 0.5 strikes a balance between the two factors giving good *precision@k* and *sim_s@k* at the same time.
- **Varying r.** Figure 8 plots the quality of STQGraph for various values of r. A larger r leads to a smaller *precision@k*. This is because a larger r will result in larger spatial proximity in general, which eventually puts less emphasis to the original weights on the edges of QFG. From Fig. 8(b), we observe that too large and too small r values lead to worse spatial proximity results. When we use a very small r, we get very small spatial proximity in general, leading to a worse *sim_s@k* result. When we use a very large r, we cannot distinguish queries that are actually close to the user, also leading to a worse *sim_s@k*. This result shows that we should choose an appropriate r for our method. Empirically, $r = 10^6$ (100 km) gives good results.

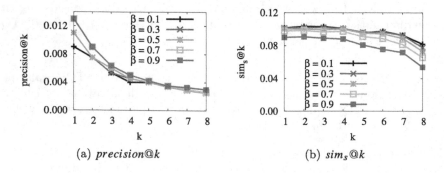

(a) *precision@k*   (b) *sim_s@k*

Fig. 7. Varying β

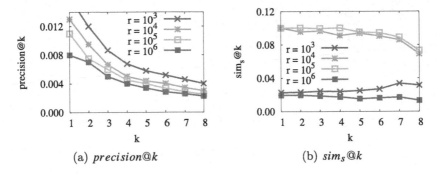

(a) *precision@k*   (b) *sim$_s$@k*

Fig. 8. Varying r

5.5   Efficiency

Now we test the efficiency of our optimizations and the approximation technique. We compare the overall running time of our STQGraph recommendation method, implemented with four versions of BCA for this purpose:

- **BCA.** The basic BCA algorithm without any optimizations.
- **BCA_L.** The BCA algorithm with the lazy update mechanism.
- **BCA_LC.** The BCA algorithm with the lazy update mechanism and spatial caching.
- **BCA_LCP.** The BCA algorithm with the lazy update mechanism, spatial caching and spatial partitioning based approximation. Note that this method corresponds to our STQGraph* method, which returns slightly different recommendations to STQGraph.

- **Varying α.** Figure 9(a) shows the average running time of STQGraph using the different versions of BCA for different values of α. We can see that all four versions terminate faster for larger values of α, which is consistent with our intuition. BCA_LCP is significantly faster than all other versions. When $\alpha = 0.5$, it takes only 0.3 s for a query, which indicates that our STQGraph* can provide instant query recommendations.
- **Varying β.** Figure 9(b) shows the running times for different values of β. A first observation is that the cost of the different versions of BCA is not much sensitive to β, as β only determines how much importance we put to spatial proximity. For the default values of α and r, BCA_LC takes around 1.0 s for each query, while BCA_LCP needs only 0.3s. Considering that STQGraph* achieves similar effectiveness to STQGraph, as shown in our previous experiments, STQGraph* (which uses BCA_LCP) is more suitable for real-time applications.
- **Varying r.** Figure 9(c) shows the running times for different values of r. Observe that the runtimes for all methods are not sensitive to the change of r. This is because r only influences the values of spatial proximity sim_s.

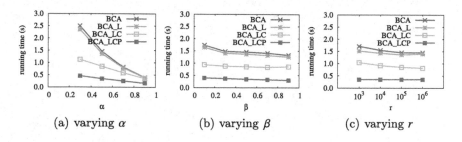

Fig. 9. Running time.

6   Related Work

Query autocompletion and query recommendation both aim at providing accurate query reformulation suggestions on-the-fly. In query autocompletion, only the most relevant *expansions* of the input query are shown, typically while the user is typing the query. Most existing works apply prefix-based recommendations and use trie-like index structures [3,7,18,19]. In this paper, we focus on query recommendation, where the suggestions are not necessarily expansions of the input query. There have been many works on query recommendation, and most of them rely on analyzing query logs to extract useful patterns that model user behavior. All these works boil down to modeling the similarity between queries, often using random walk based proximity measures on graphs that may include users, terms, queries and URLs.

Early approaches rely on clustering similar queries [1,20], where the similarity is defined using the query-URL graph or the term-vector representations of queries obtained from the clicked URLs. Later, [22] proposed the extraction and analysis of search sessions from the query log that capture the causalities between queries, and combined this with content-based similarity. In [2], the authors introduced the concept of *cover-graph*, a bipartite graph between queries and Web pages, where links indicate the corresponding clicks. [12] proposes recommending queries in a structured way for better satisfying exploratory interests of users, and [21] proposes a context-aware query recommendation model considering the relationship between queries and their clicks.

References [5,6] are two of the most influential works in query recommendation. Both of them exploit flow patterns in query logs, and use graph-based methods to perform query recommendation. [5] builds a graph of queries, termed the query-flow graph (QFG), in which the links model the transition probabilities between queries. [6] further extends the QFG to a term-query-flow graph (TQGraph), which also include nodes representing terms within queries. In this way, TQGraph can provide recommendations even for queries that never appeared in the query log. In both works, the top-k recommendations are obtained by performing random walk with restart (RWR) in the graphs.

Although many keyword search queries are sent from mobile users who have spatial search intent, there is limited research on location-aware query recommendation. In [16], the similarity between two queries is considered high if there are groups of similar users issuing these queries from nearby places at similar times. Google [15] keeps track of the locations where past queries are issued and determines the similarity between queries by also considering the proximity between the locations of the corresponding query issuers. [23] apply a learning model on the tensor representation of the user-location-query relations to predict the user's search intent. The most recent related work [17] proposes a location-aware keyword suggestion (LKS) method, which extends the idea of [10]. However, LKS only considers the location information for the documents (URLs), without considering that of queries. As we argue in this paper, it is more important to consider the spatial proximity between the user and the queries than the documents, because it is the queries we recommend to the user in the task of query recommendation. We experimentally compare our proposed methods with LKS and show that our methods provide better recommendations.

7 Conclusion

We study the problem of location-aware query recommendation for search engines. We first propose a spatial proximity measure between a keyword search query and a search engine user. Then, based on this proximity measure, we extend two popular query recommendation approaches (i.g., QFG and TQGraph) to apply for the location-aware setting. In this way, we can generate recommendations that are not only semantically relevant, but also spatially close to the query issued by a user at a specific location. In addition, we extend the Bookmark Coloring Algorithm to support efficient online query recommendation. We also propose an approximate version of the algorithm that uses spatial partitioning to accelerate the computation of our proposed spatial proximity measure. Experiments on a real query log show that our proposed methods significantly outperform previous work in terms of both semantic relevance and spatial proximity, and that our method can be applied to providing recommendations within only a few hundreds of milliseconds.

Acknowledgements. We thank the reviewers for their valuable comments. This work is partially supported by GRF Grant 17205015 from Hong Kong Research Grant Council. It has also received funding from the European Union's Horizon 2020 research and innovation programme under grant agreement No. 657347.

References

1. Baeza-Yates, R.A., Hurtado, C.A., Mendoza, M.: Query recommendation using query logs in search engines. In: EDBT Workshops on Current Trends in Database Technology (2004)
2. Baeza-Yates, R.A., Tiberi, A.: Extracting semantic relations from query logs. In: KDD (2007)
3. Bar-Yossef, Z., Kraus, N.: Context-sensitive query auto-completion. In: WWW (2011)

4. Berkhin, P.: Bookmark-coloring algorithm for personalized pagerank computing. Internet Math. **3**, 41–62 (2006)
5. Boldi, P., Bonchi, F., Castillo, C., Donato, D., Gionis, A., Vigna, S.: The query-flow graph: model and applications. In: CIKM, pp. 609–618. ACM (2008)
6. Bonchi, F., Perego, R., Silvestri, F., Vahabi, H., Venturini, R.: Efficient query recommendations in the long tail via center-piece subgraphs. In: SIGIR, pp. 345–354. ACM (2012)
7. Cai, F., Liang, S., de Rijke, M.: Time-sensitive personalized query auto-completion. In: CIKM (2014)
8. Cao, H., Jiang, D., Pei, J., He, Q., Liao, Z., Chen, E., Li, H.: Context-aware query suggestion by mining click-through and session data. In: KDD, pp. 875–883 (2008)
9. Chen, Y.-Y., Suel, T., Markowetz, A.: Efficient query processing in geographic web search engines. In: SIGMOD, pp. 277–288 (2006)
10. Craswell, N., Szummer, M.: Random walks on the click graph. In: SIGIR, pp. 239–246. ACM (2007)
11. Downey, D., Dumais, S.T., Horvitz, E.: Heads and tails: studies of web search with common and rare queries. In: SIGIR (2007)
12. Guo, J., Cheng, X., Xu, G., Shen, H.: A structured approach to query recommendation with social annotation data. In: CIKM, pp. 619–628. ACM (2010)
13. Haveliwala, T.H.: Topic-sensitive pagerank. In: WWW, pp. 517–526. ACM (2002)
14. Huang, Z., Cautis, B., Cheng, R., Zheng, Y.: KB-enabled query recommendation for long-tail queries. In: CIKM, pp. 2107–2112 (2016)
15. Myllymaki, J., Singleton, D., Cutter, A., Lewis, M., Eblen, S.: Location based query suggestion. US Patent 8,301,639, 30 October 2012
16. Ni, X., Sun, J., Chen, Z.: Mobile query suggestions with time-location awareness. US Patent Ap. 12/955,758, 31 May 2012
17. Qi, S., Wu, D., Mamoulis, N.: Location aware keyword query suggestion based on document proximity. TKDE **28**(1), 82–97 (2016)
18. Shokouhi, M.: Learning to personalize query auto-completion. In: SIGIR (2013)
19. Shokouhi, M., Radinsky, K.: Time-sensitive query auto-completion. In: SIGIR (2012)
20. Wen, J.-R., Nie, J.-Y., Zhang, H.-J.: Clustering user queries of a search engine. In: WWW (2001)
21. Yan, X., Guo, J., Cheng, X.: Context-aware query recommendation by learning high-order relation in query logs. In: CIKM, pp. 2073–2076. ACM (2011)
22. Zhang, Z., Nasraoui, O.: Mining search engine query logs for query recommendation. In: WWW, pp. 1039–1040 (2006)
23. Zhao, Z., Song, R., Xie, X., He, X., Zhuang, Y.: Mobile query recommendation via tensor function learning. In: IJCAI, pp. 4084–4090 (2015)

Top-k Taxi Recommendation in Realtime Social-Aware Ridesharing Services

Xiaoyi Fu[1(✉)], Jinbin Huang[1], Hua Lu[2], Jianliang Xu[1], and Yafei Li[3]

[1] Department of Computer Science, Hong Kong Baptist University,
Kowloon Tong, Hong Kong
{xiaoyifu,jbhuang,xujl}@comp.hkbu.edu.hk
[2] Department of Computer Science, Aalborg University, Aalborg, Denmark
luhua@cs.aau.dk
[3] School of Information Engineering, Zhengzhou University, Zhengzhou, China
ieyfli@zzu.edu.cn

Abstract. Ridesharing has been becoming increasingly popular in urban areas worldwide for its low cost and environment friendliness. In this paper, we introduce social-awareness into realtime ridesharing services. In particular, upon receiving a user's trip request, the service ranks feasible taxis in a way that integrates detour in time and passengers' cohesion in social distance. We propose a new system framework to support such a social-aware taxi-sharing service. It provides two methods for selecting candidate taxis for a given trip request. The grid-based method quickly goes through available taxis and returns a relatively larger candidate set, whereas the edge-based method takes more time to obtain a smaller candidate set. Furthermore, we design techniques to speed up taxi route scheduling for a given trip request. We propose travel-time based bounds to rule out unqualified cases quickly, as well as algorithms to find feasible cases efficiently. We evaluate our proposals using a real taxi dataset from New York City. Experimental results demonstrate the efficiency and scalability of the proposed taxi recommendation solution in real-time social-aware ridesharing services.

1 Introduction

Taxis play an important role as a transportation alternative between public and private transportations all over the world, delivering millions of passengers to their destinations everyday. Unfortunately, the number of taxis usually cannot meet the needs of people at peak time especially in urban areas, so that many people have to wait for a long time to grab a taxi. As a result, taxi ridesharing, which can reduce energy consumption and air pollutant emission, is considered as a promising solution to tackle this problem [1–4].

Many challenges exist in accomplishing a realtime taxi ridesharing service. From the riders' perspective, a rider seeks to share a ride with others with small detour and social comfort (e.g., having common friends or hobbies with other riders). Hence, if the service provider would like to encourage more users to

© Springer International Publishing AG 2017
M. Gertz et al. (Eds.): SSTD 2017, LNCS 10411, pp. 221–241, 2017.
DOI: 10.1007/978-3-319-64367-0_12

participate in ride-sharing, it has to improve the service quality by striking a balance between riders' social connections and their travel distances. In real life, the safety concern of sharing a ride with strangers may hinder the user from accepting taxi ridesharing. Moreover, it may be uncomfortable and awkward for passengers to travel with someone they do not know at all, especially when the trip is long. It is therefore of high interest to provide taxi rideshare services that consider both social factor and trust issue as well as offer several good options for users to choose the taxi they would like to take.

The other challenge of a realtime ridesharing service is to process ride requests in real time. Recently, the popularization of GPS-enabled mobile devices has enabled people to call a taxi anywhere and at any time. This involves two tasks: (i) finding the taxis that are able to accommodate the new request without violating the constraints of their current schedules and (ii) selecting several good taxis (e.g., top-k taxis) for the user to choose. Given a large number of taxis and a new trip request, it is time consuming to find the feasible schedules and calculate riders' social connections at the same time. Therefore, designing an efficient and scalable ridesharing algorithm with a proper metric to measure whether a taxi is suitable for a given ride request is very challenging.

The majority of previous studies [3–5] focus on designing efficient assignment algorithms with the objective of minimizing the travel/detour cost and maximizing the rate of matched requests. Companies like *Uber*, *Didi* and *Kuaidi* allow riders to submit ride request ahead of time. However, these previous works have not fully met the new requirements, e.g., improving the ridesharing experience by considering the social aspect. Intrinsically, people would like to share a ride with someone who makes them feel comfortable and safe. This paper is concerned with *social-aware* ridesharing which aims to improve the ridesharing experience.

To address the aforementioned challenges, we propose a novel type of dynamic ridesharing service, named *top-k Social-aware Taxi Ridesharing (TkSaTR)* service, which processes user requests not only based on route schedule, but also considering riders' social relationships. Generally, given a set of taxis moving in a road network, each taxi has zero or several riders. Upon receiving a new trip request, the system provides the top-k qualified taxis for the rider to select by considering both the spatial and social aspects. If no appropriate taxi is retrieved, the system asks the user to modify the trip request or resubmit it at a later time.

The rest of this paper is organized as follows. Section 2 surveys the related work. The *TkSaTR* problem is formulated in Sect. 3. We present two candidate taxis searching methods in Sect. 4. In Sect. 5, we present the taxi scheduling and top-k taxi selection approach. Experimental results are reported in Sect. 6, followed by the conclusion and future work in Sect. 7.

2   Related Work

Existing studies on ridesharing fall into three categories: static ridesharing, dynamic ridesharing and trust-conscious ridesharing.

2.1 Static Ridesharing

Most relevant early works studied static ridesharing where the information of drivers and ride requests are known in advance. Static ridesharing covers three typical application scenarios: slugging, carpooling, and dial-a-ride.

In slugging [6], a rider walks to the driver's origin location and departs at the driver's departure time. At the driver's destination, the rider alights and walks to her/his own destination. Ma et al. [1] studied the slugging problem and its generalization from a computational perspective.

Carpooling is a typical ridesharing form for daily commutes where drivers need to adapt their routes to riders' routes. The carpooling problem is proved to be NP-hard, and the exact methods can only work efficiently on small instances, e.g., it can be solved by using linear programming techniques [7,8]c. For large instances of the carpooling problem, heuristics are proposed [9–11]. For a many-to-many carpooling system in which there are multiple vehicle and rider types, Yan and Chen [12] proposed a time-space network flow technique to develop a solution based on Lagrangian relaxation.

In the dial-a-ride problem (DARP), no private car is involved and the transportation is carried out by non-private vehicles (such as taxis) that provide a shared service. In order to receive the service, each rider submits a request with the origin and destination locations. In turn, the service provider returns a set of routes with the minimum cost that satisfy all ride requests under some spatio-temporal constraints. A survey [13] summarized the early studies on DARP. In general, DARP is NP-hard and only instances with a small number of vehicles and requests can be solved optimally, and the approaches are often based on integer programming techniques [14]. These approaches are usually implemented in two phases: the first phase is to partition the requests and obtain an initial schedule, and the second phase is to improve the solution by local search.

2.2 Dynamic Ridesharing

Motivated by the recent mobile technologies, dynamic ridesharing services have been drawing increasing attention from both industry and academia [3–5,15]. In dynamic ridesharing systems, riders and drivers continuously enter and leave the system. Dynamic ridesharing algorithms match up them in real time. Generally, the existing works can be divided into to *centralized* ridesharing and *distributed* ridesharing.

In a centralized dynamic ridesharing system, a central service provider performs all the necessary operations to match riders to drivers. Agatz et al. [16] surveyed optimization techniques for centralized dynamic ridesharing services in which different optimization objectives (e.g., minimizing system-wide overall travel distance or travel time) and spatio-temporal constraints (with desired departure/arrival time or spatial proximity constraints) are considered. Rigby et al. [17] proposed an opportunistic user interface to support centralized ridesharing planning as well as preserving location privacy. Huang et al. [4]

formulated a centralized real-time ridesharing problem with service guarantee, and proposed a novel kinetic tree based solution.

The drawback of the centralized ridesharing is the lack of scalability. To address this issue, distributed ridesharing solutions have been developed. d'Orey et al. [18] modeled a dynamic and distributed taxi-sharing system and proposed an algorithm based on peer-to-peer communication and distributed coordination. Zhao et al. [15] proposed a distributed ridesharing service based on a geometry matching algorithm that shortens waiting time for riders and avoids traffic jams.

Note that existing proposals for dynamic ridesharing do not consider social connections among riders and drivers.

2.3   Trust-Conscious Ridesharing

A few recent studies addressed the trust issue in ridesharing [2, 9,19,20]. Alternative approaches include adopting reputation-based systems and profile check by associating with social networks such as Facebook [9]. Cici et al. [2] proposed to assign participants who are friends or friends of friends into one group in order to gain potential benefits of ridesharing. Li et al. [19] employed k-core as the primary social model and proposed algorithms to solve social-aware ridesharing group queries.

Table 1. Existing studies on ridesharing

Existing work	Characteristics		
	Social factor	Optimization objective	Real-time
T-share [5]	no	Travel distance	yes
Kinetic tree [4]	no	Trip cost	yes
SaRG [19]	yes	Travel cost of the group	no
TkSaTR [this paper]	yes	Ranking function	yes

Unlike our study presented in this paper, the ridesharing model used in [19] is slugging and cannot adapt to the dynamic environment.

Table 1 summarizes the characteristics of the most typical existing proposals for ridesharing. In this paper, we are interested in real-time taxi ridesharing services (TkSaTR) that consider the social factor and optimize the mix of spatial and social factors.

3   Problem Formulation

3.1   Definitions

In our setting, a road network is viewed as a time-dependent graph $G(N, E, W)$. Specifically, N is the set of nodes each representing a road intersection or a terminal point, and $E \in N \times N$ is the set of network edges each connecting two nodes. Representing a road segment from node n_i to n_j, edge $e_{i,j} \in E$ is associated with a time-dependent weight function $W_e(t)$ that indicates the travel cost along the edge $e_{i,j}$. In particular, $W_e(t)$ specifies how much time it takes to travel from n_i to n_j if a vehicle departs from n_i at time t.

To ease the explanation, we represent the road network as an undirected graph in the examples in this paper. Nevertheless, the algorithms we propose can handle the road network as a directed graph.

On the other hand, we model a social network as an undirected graph $G_S = (V_s, E_s)$, where a node $v_i \in V_s$ represents a user and an edge $(v_i, v_j) \in E_s$ indicates the social connection between two users v_i and $v_j \in V_s$.

Definition 1 (Trip Request). *A trip request is denoted by $tr = (t, o, d, pw, dt)$ where t indicates when tr is submitted, o and d represent the origin and destination, respectively, pw represents the time window when the rider wants to be picked up, and dt represents the latest drop off time.*

For a time window pw, we use $pw.e$ and $pw.l$ to denote its earliest and latest time, respectively. In practice, a rider may only need to specify $tr.d$ and $tr.dt$. The ridesharing service can automatically record $tr.t$ and $tr.o$ (if the rider is GPS-enabled). We may also set $tr.pw.e$ to $tr.t$ and $tr.pw.l$ to a time later than $tr.pw.e$ by a default time period, e.g., 5 min.

Definition 2 (Schedule). *A schedule s of n trip requests $tr_1, tr_2, ..., tr_n$ is a sequence of origins and destinations such that for each request tr_i, either (i) both $tr_i.o$ and $tr_i.d$ exist in the sequence and $tr_i.o$ precedes $tr_i.d$, or (ii) only $tr_i.d$ exists in the sequence.*

Each taxi in operation is associated with a schedule that changes dynamically over time. For example, if two trip requests tr_1 and tr_2 are assigned to a taxi, the initial schedule can be $(tr_1.o, tr_2.o, tr_2.d, tr_1.d)$. After the taxi picks the rider at $tr_1.o$, its associated schedule is updated to $(tr_2.o, tr_2.d, tr_1.d)$ accordingly. Each taxi continuously maintains its geographic location $T.l$, the number of on-board riders $T.n$, and its schedule $T.s$.

Definition 3 (Satisfaction). *Given a taxi T and a trip request tr, T satisfies tr if and only if:*

(1) $T.n$ is less than the capacity of T;
(2) T is able to pick up the rider of tr at $tr.o$ on or before $pw.l$ and drop off the rider at $tr.d$ no later than $tr.dt$;
(3) T can serve all riders currently in $T.s$ without violating any time constraints imposed by their trip requests.

It is noteworthy that a taxi T does not satisfy a request tr if T cannot arrive at $tr.o$ during time window $tr.pw$.

When a trip request tr comes into the ridesharing service, there can be many taxis that satisfy tr. To this end, we employ a ranking function $f(tr, T_i)$ that scores a taxi T_i with respect to request tr by considering (i) the travel distance to transport the new rider from $tr.o$ to $tr.d$, and (ii) the social connectivity between the new rider and existing riders in $T_i.s$. The former considers the *spatial* aspect, whereas the latter concerns the *social* aspect. The ranking function $f(tr, T_i)$ can be any monotonic aggregate function that combines social score S_i and spatial score D_i for taxi T_i. We detail scores S_i and D_i in Sect. 3.2. We give two types of ranking functions:

$$f(tr, T_i) = \omega \cdot S_i + (1 - \omega) \cdot D_i \tag{1}$$

$$f(tr, T_i) = S_i \times D_i \tag{2}$$

The top-k taxis selection algorithms proposed in this paper are independent of the concrete definition of $f(tr, T_i)$. In the rest of this paper, we use Eq. 1 as the ranking function. In Eq. 1, parameter $\omega \in [0, 1]$ specifies the relative importance of the social and spatial factors. When $\omega > 0.5$, the social cohesion of the ridesharing group is more important than the travel distance.

Based on the definitions above, we formally define our research problem:

Problem 1 (**Top-k Social-aware Taxi Ridesharing Query**). Given a set of taxis each having its current schedule of riders, a road network $G(N, E, W)$, a social network $G_S = (V_s, E_s)$, and a trip request tr, a top-k social-aware taxi ridesharing query (TkSaTR query for short) returns the top-k taxis satisfying tr with the highest score calculated by the ranking function $f(tr, T_i)$.

In this paper, we study the online social-aware ridesharing problem, i.e., the system does not know the information of upcoming trip requests and a submitted trip request needs to get response from the system in real time. The performance of an online algorithm is usually analyzed by *competitive ratio* [21]. An algorithm $\mathcal{A}$ is called c-competitive for a constant $c > 0$, if and only if, for any input $\mathcal{I}$ the result of $\mathcal{A}(\mathcal{I})$ is at most c times worst than the globally optimal solution. We show that no online algorithm can achieve a good competitive ratio for our social-aware rideahring problem.

Theorem 1. *There is no deterministic online algorithm for the social-aware ridesharing problem that is c-competitive ($c > 0$).*

Proof. Suppose an algorithm $\mathcal{A}$ is c-competitive, then the output of $\mathcal{A}$ must be at most c times worst than the optimal solution for every input. Consequently, to show $\mathcal{A}$ is not c-competitive, we only need to find one input that $\mathcal{A}$ cannot provide the solution no worse than c times to the optimal solution.

We assume that an adversary knows every decision $\mathcal{A}$ makes and can submit trip requests as $\mathcal{A}$'s input. For simplicity, we assume there is only one taxi locating at point $(0, 0)$. We also assume that the earliest departure time is equal to the trip request submission time and the latest departure time is w later than the earliest departure time (i.e., the maximum waiting time is w). Initially, there are two trip requests tr_1 and tr_2 with origins $(w, 0)$ and $(-w, 0)$ respectively. $\mathcal{A}$ can make three possible choices for the taxi: (1) move towards tr_1, (2) move towards tr_2 and (3) stay still. If choice 1 is selected, then the adversary would submit n more trip requests at $(-w - 1, 0)$ at time $t = 1$ and their destinations are similar to tr_2's. These n riders and tr_2 are friends with each other. Under this circumstance, the global optimal solution can serve $n + 1$ trip requests and riders' social cohesion is the highest (every rider knows each other), while $\mathcal{A}$ can at most complete one trip request. Similar arguments can be made if choice 2 or choice 3 is selected. By submitting more trip requests in a similar manner, the adversary can make $\mathcal{A}$'s solution unboundedly worse than the global optimal solution. Therefore, the assumption of $\mathcal{A}$ is c-competitive is invalid and the proof completes. □

3.2 Spatial and Social Scores

For the spatial score, suppose $T_i.s'$ is the schedule after the new request tr is assigned to taxi T_i. We use the following equation to capture the average detour of the riders who are assigned to T_i.

$$
D_i = \frac{\sum_{tr_i \in T_i.s'} t_f(tr_i.o, tr_i.d)}{\sum_{tr_i \in T_i.s'} t_{share}(tr_i.o, tr_i.d)} \tag{3}
$$

For a trip request tr_i already in T_i's schedule, $t_f(tr_i.o, tr_i.d)$ is the travel time along the fastest path between tr_i's origin and destination (i.e., the travel time without using ridesharing), and t_{share} is the travel time from tr_i's origin to its destination through the ridesharing enabled by T_i.

Let R_i be the union of the user issuing the new trip request tr and all those involved in taxi T_i's current schedule. Let $sd(u_j, u_k)$ represent the social distance, i.e., the number of hops, between users u_j and u_k in a social network $G_S = (V_s, E_s)$. As an extreme case, if there is no path between nodes u_j and u_k, $sd(u_j, u_k)$ will be $dia(G_S) + 1$ where $dia(G_S)$ is the diameter of G_S. Accordingly, we define the social score as $S_i = \frac{|R_i| \cdot (|R_i| - 1)}{\sum_{j=1}^{|R_i|} \sum_{k=1}^{|R_i|} sd(u_j, u_k)}$. It is normalized to the range of $(0, 1]$.

3.3 Solution Overview

A brute-force approach for a TkSaTR query enumerates all the taxis to find the top-k taxis. Given a trip request tr, we check the taxis one by one and retrieve those as the candidates that can reach $tr.o$ before time $tr.pw.l$. Subsequently, for each candidate taxi T_i's schedule $T_i.s$, we enumerate all the permutations of inserting $tr.o$ and $tr.d$ into $T_i.s$, and find those taxis that satisfy tr. These taxis are called *feasible taxis*. Finally, we calculate the ranking score for each feasible taxi and return the top-k taxis with the highest ranking scores. If there is no feasible taxis with respect to a request tr, the system may suggest the user to modify the trip request or resubmit it later. Apparently, this brute-force approach is inefficient as it does not identify and rule out unpromising taxis aggressively.

We propose an efficient framework to solve the problem, as illustrated in Fig. 1. It consists of two major modules: Candidates Searching to find candidate taxis for a new trip request and Optimal Scheduling to add the request to candidate taxis' schedules and return the top-k taxis. We design two methods for selecting candidate taxis. The grid-based method rules out unpromising taxis

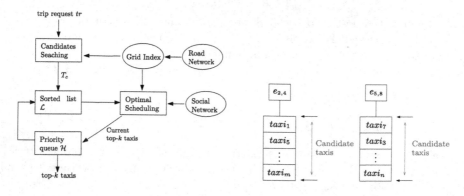

Fig. 1. System framework   **Fig. 2.** Selected candidate taxis

using a grid index and returns a relatively larger candidate set, whereas the edge-based method expands the selection through the road network and obtains a smaller candidate set. For the scheduling module, we derive travel time based bounds to rule out unqualified cases (taxis and schedules) quickly. We also design algorithms to insert $tr.o$ and $tr.d$ into $T_i.s$ efficiently without violating the relevant time constraints. We only calculate the concrete ranking score for feasible and most promising taxis.

4   Candidate Taxis Searching

The candidate taxis searching process is intended to find a candidate set of taxis that are likely to satisfy a given trip request tr. In this section, we propose two candidate searching methods. The *Edge-based Candidates Selection* searches for candidate taxis by expanding from tr's origin location to reachable edges in the road network. The *Grid-based Candidates Selection* utilizes a grid index to prune unpromising taxis that are too far away from tr's origin location.

4.1   Edge-Based Candidates Selection

In this method, we maintain a list $e.L_t$ for each road network edge e. The list contains the IDs of taxis that are currently located on edge e. Taxis are appended to the list when it enters e, and a taxi is removed from the list when it leaves e. As a result, all IDs in the list $e.L_t$ are sorted in ascending order of the taxis' entering time.

An example is shown in Fig. 4. Suppose that a trip request tr arrives at time t_{cur} with its origin location at n_4. Any taxi that is currently on edges from which n_4 is reachable before $tr.pw.l$ is retrieved as a candidate taxi. This time constraint is captured in Eq. 4, where n_x indicates a taxi's current location.

$$t_{cur} + t(n_x, n_4) \le tr.pw.l \tag{4}$$

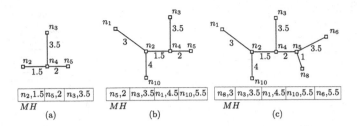

Fig. 3. Searching candidate taxis

Suppose that $tr.pw.l = tr_{cur} + 2$, which means $t(n_x, n_4)$ must less be than or equal to 2. The edge-based selection expands the search from n_4 to other nodes in the ascending order of their travel time to n_4. The expansion stops once it encounters a node n_x that fails to satisfy Eq. 4. To enhance the performance, a min-heap MH is used to store the nodes visited so far, using the minimum travel time to n_4 as the priority. Initially, n_4's adjacent nodes n_2, n_3, n_5 are en-heaped into MH. This selection stops when the de-heaped node fails to satisfy Eq. 4. As node n_2 has the minimum travel time to n_4, it is de-heaped and processed first. As it satisfies Eq. 4, the taxis in $e_{2,4}.L_t$ are retrieved as candidate taxis as shown in Fig. 2. Subsequently, nodes n_1 and n_{10} are en-heaped with their travel time to n_4, as shown in Fig. 3(b). Furthermore, n_5 is de-heaped and n_8 and n_6 are en-heaped as shown in Fig. 3(c). Next, n_8 is de-heaped as it has the minimum travel time to n_4 in MH. As $t(n_8, n_4)$ is 2, Eq. 4 is violated and the selection process stops.

4.2 Grid-Based Candidates Selection

In the grid-based candidates selection, we utilize a grid index to maintain the information and speed up the candidates searching process. The whole road network is divided by a grid. For each grid cell g_i, the information about the road network and taxis is stored in two tables: the *edge table* T_{edge} and the *node table* T_{node}.

Specifically, table T_{edge} stores the following information of each edge e within cell $g_{i,j}$: (1) the nodes n_i and n_j connecting e (if n_i is in $g_{i,j}$), (2) the grid cell in which n_j is located, (3) the travel time function $W_e(t)$, and (4) the ID list L_t of the taxis currently on e. All taxi IDs are sorted in ascending order according to their entering time to e. A taxi T_i is removed from $e.L_t$ when it leaves e. The travel time function $W_e(t)$ implies the travel time on edge e for the current timestamp t. On the other hand, table T_{node} maintains the coordinate (x, y) of each node n_i in $g_{i,j}$.

To illustrate the grid index, a road network with 17 edges and 17 nodes is shown in Fig. 4. The whole road network is divided into 4×4 grid cells, from $g_{0,0}$ to $g_{3,3}$. Take $g_{1,2}$, a gray rectangle, as an example. The contents of T_{edge} and T_{node} for $g_{1,2}$ are shown in Table 2 (the taxi lists ta of edges are omitted due to space limit). Note that only the cells enclosing the nodes of an edge need to

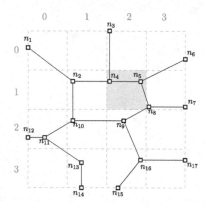

Table 2. Information of $g_{1,2}$

T_{edge}

e	n_i	n_j	$T(e)$	$g_{m,n}$
$e_{4,2}$	n_4	n_2	1.5	$g_{1,1}$
$e_{4,3}$	n_4	n_3	3	$g_{0,2}$
$e_{4,5}$	n_4	n_5	2	$g_{1,2}$
$e_{5,6}$	n_5	n_6	3.5	$g_{0,3}$
$e_{5,8}$	n_5	n_8	2	$g_{1,3}$

T_{node}

n	(x,y)
n_4	(30,52)
n_5	(30,73)

Fig. 4. A road network

maintain the relevant T_{edge} and T_{node}. For example, the information about edge $e_{1,2}$ is found in cell $g_{0,0}$ or $g_{1,1}$, but not in cell $g_{1,0}$.

With these tables, the grid-based candidates selection process is straightforward. After receiving a trip request tr whose origin is in grid cell $g_{i,j}$, the system first searches for the candidate taxis which are likely to satisfy tr. The taxis within $\left\lceil \frac{(tr.pw.l-tr.t) \cdot D}{cs} \right\rceil$ cells are the candidates as only these taxis may be able to satisfy the pickup time constraint. Here, D is the maximum velocity of a taxi and cs is the grid cell size. Any taxi beyond these cells is too far away to reach tr's origin location before tr's latest pickup time.

5   Taxi Scheduling and Top-k Taxi Selection

Given the set T_C of candidate taxis with respect of a trip request tr, the purpose of the taxi scheduling process is to find the top-k taxis satisfying tr with the highest score calculated by the ranking function $f(tr, T_i)$. We start with describing the overall procedure of top-k taxis selection.

5.1   Overall Procedure of Top-k Taxis Selection

After the candidate taxis are obtained in set T_C, we calculate the social score for each taxi in T_C and sort all candidate taxis as follows. For each candidate taxi $T_i \in T_C$, we calculate its *ranking score upper bound* as $RSUB_i = \omega \cdot S_i + (1 - \omega) \cdot D_i^+$ where S_i is T_i's social score and D_i^+ is the upper bound of T_i's spatial score D_i, both with respect to the new request tr. It is apparent that $RSUB_i \geq f(tr, T_i) = \omega \cdot S_i + (1 - \omega) \cdot D_i$ for $\omega \in (0, 1)$ (Eq. 1). We elaborate on how to derive D_i^+ in Sect. 5.2. All candidate taxis are put in a sorted list $\mathcal{L}$ in descending order of their $RSUB_i$ values.

Subsequently, we sequentially process each candidate taxi T_i in $\mathcal{L}$. In particular, we find for T_i the optimal schedule that satisfies the request tr and yields the optimal (largest) spatial score. This will be detailed in Sect. 5.4. Afterwards, we calculate the final ranking score for taxi T_i using $f(tr, T_i)$ and push T_i into a priority queue $\mathcal{H}$ that uses the ranking score $f(tr, T_i)$ as the priority. The overall procedure stops when $\mathcal{H}$ contains at least k taxis and the current candidate taxi T_i from $\mathcal{L}$ to process has an $RSUB_i$ value no greater than the k-th final ranking score in $\mathcal{H}$. In such a case, all remaining candidates in T_i cannot have higher ranking scores than the k-th taxi in $\mathcal{H}$. As a result, the top-k taxis are already found in $\mathcal{H}$.

5.2   Spatial Score Upper Bounds

In order to derive the upper bound for the spatial score defined in Eq. 3, we estimate the lower bound for $t_{share}(tr_j.o, tr_j.d)$ used in the denominator of Eq. 3. To ease the presentation, let s be $tr_j.o$ and e be $tr_j.d$, and $t(s, e)$ be $t_{share}(tr_j.o, tr_j.d)$. We derive two lower bounds for $t(s, e)$.

A straightforward way is to consider the Euclidean distance from s to e, denoted as $d_{Euc}(s, e)$ instead of the complex road network distance. Suppose that we know the maximum speed v_{max} a taxi can travel at in the road network. Then we have

Lemma 1 (Euclidean Travel Time Lower Bound). $TTLB_{Euc}(s, e) = \frac{d_{Euc}(s, e)}{v_{max}} \leq t(s, e)$.

An alternative is to use the grid index and derive a lower bound accordingly. For each grid cell, we identify its boundary node that is immediately connected to some node in a different cell. Any path linking a node in a cell g_i with some node in another cell g_j must go through at least two boundary nodes in g_i and g_j, respectively. We derive the *Grid Travel Time Lower Bound* based on this observation. For each pair of cells g_i and g_j, we pre-compute the fastest path distance from each boundary node in g_i to each boundary node in g_j, and store the smallest one as $sgd_{i,j}$, i.e., $sgd_{i,j} = \min_{b_i \in g_i, b_j \in g_j} t(b_i, b_j)$. In each cell g_i, we pre-compute and store the fastest path distance from each node to its nearest boundary node b_{ni}. Then we have

Lemma 2 (Grid Travel Time Lower Bound)
   $TTLB_{grid}(s, e) = t(s, b_{ni}) + sgd_{i,j} + t(b_{nj}, e) \leq t(s, e)$.

Proof. For any path from node s in grid cell g_i to node e in grid cell g_j, it consists of three parts: (1) the path from s to some boundary node b'_{ni} in g_i, (2) the path from b'_{ni} to some boundary node b'_{nj} in g_j, (3) the path from b'_{nj} to e.

Because $t(s, b_{ni}) \leq t(s, b'_{ni})$, $sgd_{i,j} = \min_{b_i \in g_i, b_j \in g_j} t(b_i, b_j) \leq t(b'_{ni}, b'_{nj})$, and $t(b'_{nj}, e) \leq t(b_{nj}, e)$, hence $t(s, b_{ni}) + sgd_{i,j} + t(b_{nj}, e) \leq t(s, e)$, $TTLB_{grid}(s, e)$ is the lower bound of $t(s, e)$. □

Consider the example shown in Fig. 5, where s and e are located in grid cells g_i and g_j, respectively. The closest pair of boundary nodes between g_i and g_j is b_2 and b_3. For origin node s, its nearest boundary node in g_i is

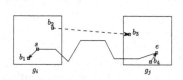

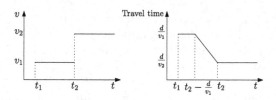

Fig. 5. Boundary node estimator   **Fig. 6.** Travel speed and travel time function

b_1, and destination node e's nearest boundary node in g_j is b_4. As a result, $TTLB_{grid}(s,e) = t(s,b_1) + t(b_2,b_3) + t(b_4,e) \leq t(s,e)$.

When one of the travel time lower bounds is used to replace $t_{share}(tr_j.o, tr_j.d)$ in Eq. 3, the equation gives D_i^+ that is an upper bound for the original D_i.

5.3   Time-Dependent Fastest Path Calculation

In our setting, the road network is modeled as a time-dependent graph in which the travel time may change for the same route in different periods of time. In this section, we present how to compute the fastest path from an origin s to a destination e starting at timestamp t.

Consider an edge $e_{i,j}$ with length d. Suppose that the travel speed allowed on $e_{i,j}$ is v_1 during $[t_1, t_2)$ and v_2 after t_2. Consequently, the travel time on $e_{i,j}$ (i.e., from one end to the other of $e_{i,j}$) is a continuous, piecewise-linear function of the departing ticme t from n_i [22]. Specifically, the time-dependent weight function for $t \in [t_1, t_2]$ is:

$$W_{e_{i,j}}(t) = \begin{cases} \frac{d}{v_1}, & t \in [t_1, t_2 - \frac{d}{v_1}] \\ (1 - \frac{v_1}{v_2})(t_2 - t) + \frac{d}{v_2}, & t \in [t_2 - \frac{d}{v_1}, t_2] \end{cases}$$

The relationship between speed and travel time on $e_{i,j}$ is illustrated in Fig. 6. In case that an edge contains more than two different speed patterns, $W_{e_{i,j}}(t)$ is still a continuous piecewise linear function of t with more than two linear segments.

Based on this, our procedure of fastest path calculation in a time-dependent graph is as follows. We keep a set N_E of expanded nodes and a priority queue F of frontier nodes. Initially, N_E is empty and F contains only the origin node s. Subsequently, the computation employs iterations to expand to reachable nodes from s. Each iteration chooses one node n from F, expands it by adding its unvisited neighbors to F, and then moves n to N_E. To choose the next node from F, instead of choosing n_i where the travel time from s to n_i is the smallest as in Dijkstra's algorithm, we choose the node n_j such that the travel time from s to n_j plus the estimated travel time from n_j to e is the smallest. Note that the estimate here must be a lower bound of the actual travel time in order to ensure correctness. Also, the closer the estimate is to the actual travel time, the more efficient the fastest path calculation is.

5.4 Optimal Schedule

This section introduces how to find the *optimal schedule* of a candidate taxi T given a trip request tr. The *optimal schedule* of a taxi T is the schedule with the highest spatial score after inserting tr among all the possible schedules.

To find the optimal schedule of a taxi T, intuitively we need to try all possible ways to find the way of insertion with the highest spatial score. For simplicity, here we assume the precedence relation in the current schedule remains unchanged during the scheduling process. Then inserting tr into $T.s$ can be done via two steps: (1) insert $tr.o$ into $T.s$; (2) insert $tr.d$ into $T.s$. The system checks the *feasibility* (i.e., whether T can satisfy tr or not) of all possible insertion ways for T and computes the spatial score D of those feasible ones. The spatial score of the optimal schedule of T (i.e., the highest spatial score) would be stored as the spatial score of T. For example, Fig. 7 shows one possible way to insert tr into a taxi schedule $tr_1.o \rightarrow tr_2.o \rightarrow tr_1.d \rightarrow tr_2.d$. In order to insert $tr.o$ between $tr_1.o$ and $tr_2.o$ optimally, we consider the fastest path according to current traffic condition rather than the shortest path in the road network.

For each insertion possibility, the system needs to check whether it is feasible or not. As shown in Fig. 7, to insert $tr.o$ after $tr_1.o$, the algorithm first evaluates if the taxi T is able to reach $tr.o$ before $tr.pw.l$. If not, the insertion fails. Otherwise, the system further checks whether some successor in the schedule would be delayed due to inserting $tr.o$. In $T.s$, if any other rider's time constraint is violated due to inserting $tr.o$ into $T.s$, then the insertion also fails.

If $tr.o$ is inserted successfully, the system then inserts $tr.d$ similarly. The system calculates spatial scores for all schedules which are feasible after inserting $tr.d$ and $tr.d$, the schedule with the highest spatial score will be stored as the optimal schedule of taxi T.

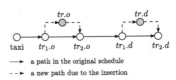

Fig. 7. One possible insertion of a ride request into a schedule

5.5 Hopping Algorithm

In order to insert a trip request tr into a schedule s and find the feasible insertion ways more efficiently, we introduce the *Hopping Algorithm* to find all the feasible insertion ways with some prune techniques. An early stop condition and two pruning conditions are derived, which are utilized in the hopping algorithm.

Early Stop Condition. Given a taxi T with schedule s and a new incoming trip request tr, if there exists some tr_x in s that satisfies both following conditions, then T definitely cannot satisfy tr: (1) $tr.dt < tr_x.pw.e$; (2) $tr.pw.e + t_f(tr.o, tr.d) > tr_x.pw.l$.

Condition (1) indicates that T must drop off tr at $tr.d$ before picking up tr_x at $tr_x.o$. In condition (2), the inequation suggests that T cannot reach $tr.d$ on or before tr_x's latest departure time even if it departs at $tr.pw.e$ and takes the fastest path from $tr.o$ to $tr.d$. Hence, tr cannot be inserted into s because inserting tr will break the time constraint of tr_x in s.

Algorithm 1. Hopping algorithm

Require: a trip request tr, a taxi T
Ensure: the set *result* of all feasible schedules after inserting tr into $T.s$
1: $posLeft \leftarrow 0, posRight \leftarrow s.size(), result \leftarrow \emptyset$
2: **while** $posRight \geq 0$ **do**
3: 　　**if** $tr.pw.e + t_f(tr.o, tr.d) > tr_{posRight+1}.dt$ **then**
4: 　　　　break 　　　　　　　　　　　　　　　　　▷ Pruning Condition 2
5: 　　**else**
6: 　　　　$posRight--$
7: **while** $posLeft \leq T.s.size()$ **do**
8: 　　**if** $tr_{posLeft}$ is a pick up point and $tr.dt < tr_{posLeft}.pw.e$ and $tr.pw.e + t_f(tr.o, tr.d) > tr_{posLeft}.pw.l$ **then**
9: 　　　　return $\emptyset$ 　　　　　　　　　　　　　　　▷ Early Stop Condition
10: 　　Schedule $s \leftarrow T.s$, Insert $tr.o$ at $posLeft$
11: 　　**if** $t_{arrival} > tr.pw.l$ **then**
12: 　　　　break 　　　　　　　　　　　　　　　　　▷ Pruning Condition 1
13: 　　**else if** insertion of $tr.o$ succeeds **then**
14: 　　　　$posValid \leftarrow posRight$
15: 　　　　**if** $posLeft \geq posRiget$ **then**
16: 　　　　　　$posValid \leftarrow posLeft$
17: 　　　　**for** each possible insertion position i such that $i > posValid$ in s **do**
18: 　　　　　　**if** insertion of $tr.d$ at i succeeds **then**
19: 　　　　　　　　$result \leftarrow result \cup s$, remove $tr.d$ from s 　　　　▷ Reset s
20: 　　$posLeft++$
21: return $result$

Taking the schedule in Fig. 7 as an example, if $tr.dt < tr_1.pw.e$ and $tr.pw.e + t_f(tr.o, tr.d) > tr_1.pw.l$, then tr cannot be inserted into this schedule no matter which positions $tr.o$ or $tr.d$ is inserted into.

Pruning Condition 1. Note that for a trip request tr, when inserting $tr.o$ between points i and j in a schedule s, if the projected arrival time along the fastest path at $tr.o$ is later than $tr.pw.l$ (i.e., $t_{arrival} > tr.pw.l$), then the insertion of all the points after i in the schedule cannot succeed either. Likewise, if the taxi cannot drop off the rider at $tr.d$ before $tr.dt$ when inserting $tr.d$ between i and j, then $tr.d$ can not be inserted after any successor of i. Based on this observation, the system is able to prune some insertion ways directly.

Pruning Condition 2. Given a trip request tr and a schedule s, when trying to insert $tr.d$ before some drop-off point $tr_x.d$ in s, if inequality $tr.pw.e + t_f(tr.o, tr.d) > tr_x.dt$ holds, then $tr.d$ cannot be inserted into any position before $tr_x.d$.

Given the Early Stop Condition and Pruning Conditions above, we propose an algorithm named Hopping Algorithm to find all the feasible schedules after inserting tr into a taxi T's schedule $T.s$. Its pseudo code is given in Algorithm 1. Basically, it utilizes two pointers, *posLeft* and *posRight*, to indicate possible insertion positions pruned by the two pruning conditions, respectively. Given a trip request tr and a taxi T, the Hopping Algorithm returns the set $result$ that contains all the feasible schedules after inserting tr into $T.s$. In $T.s$, positions before $posLeft$ are the possible positions where $tr.o$ could be inserted. Likewise, positions after $posRight$ are the possible positions where $tr.d$ could be inserted. Initially, $posLeft$ is at the left most position of $T.s$ and $posRight$ is at the right most position (lines 1). The algorithm first finds the position of $posRight$ according to Pruning Condition 2 (lines 2–6). Then $posLeft$ moves from left to right most position in $T.s$. If the Early Stop Condition is satisfied, then the algorithm returns $\emptyset$ which means tr cannot be inserted into $T.s$ (lines 8–9). When the Early Stop Condition is violated, the algorithm attempts to insert $tr.o$ at $posLeft$. If the projected arrival time at $tr.o$ is later than $tr.pw.l$, then $tr.o$ cannot be inserted into the current position and the positions after $posLeft$ in $T.s$ according to Pruning Condition 1 (lines 10–12). If $tr.o$ is inserted successfully, then the algorithm attempts to insert $tr.d$ from the right most position to $posRight$ and add the feasible schedules into $result$ (lines 13–20).

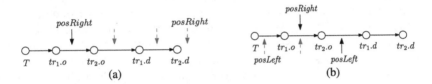

Fig. 8. An example of hopping algorithm

We take the schedule in Fig. 7 as an example to explain the procedure of the Hopping Algorithm. After initialization, the algorithm first determines the position of $posRight$. As shown in Fig. 8a, initially $posRight$ stays at the right most position in $T.s$ (after $tr_2.d$) and moves towards left until Prune Condition 2 is held (shown by gray dotted arrows). Suppose that $tr.pw.e + t_f(tr.o, tr.d) > tr_2.dt$, $posRight$ stops moving. This indicates that $tr.o$ cannot be inserted into any position before $tr_2.o$. Accordingly, the final position of $posRight$ is shown by the black arrow in Fig. 8a. As shown in Fig. 8b, after determining the position of $posRight$, $posLeft$ first stays before $tr_1.o$. As the Early Stop Condition is not satisfied and $tr.o$ can be inserted at $posLeft$ successfully, the algorithm checks the positions after $posValid$ (equals to $posRight$ for now) to see if $tr.d$ can be inserted. The schedules with successfully inserted $tr.d$ are added into $result$ set. Subsequently, $posLeft$ moves towards right and stays between $tr_1.o$ and $tr_2.o$, this procedure is repeated. It is noteworthy that, if $posLeft$ moves to the same position as $posRight$ or the position after $posRight$, $posValid$ points to the position of $posRight$ since $tr.d$ cannot be inserted before $tr.o$ in $T.s$. When $posLeft$ is between $tr_2.o$ and $tr_1.d$ (shown by the black arrow in Fig. 8b), the taxi cannot reach $tr.o$ before $tr.pw.l$ (Pruning Condition 1 holds) and $posLeft$ stops moving. The schedules in $result$ are the feasible schedules after inserting tr into $T.s$.

6   Experimental Evaluation

In this section, we experimentally evaluate the performance of our approach with the **NN** (i.e., Nearest Neighbors) approach and the **T-Share** approach [5]. We also evaluate the performance of four algorithms proposed in this paper: the brute-force approach (*BF* for short) presented in Sect. 3.3, the optimal schedule algorithm with edge-based candidate selection (*OE*), the Hopping Algorithm with edge-based candidate selection (*HE*) and the Hopping Algorithm with grid-based candidate selection (*HG*). The NN approach keeps finding the next nearest taxi. It stops either the current taxi satisfies the new trip and gets the trip, or no taxi can accept the trip and the trip is rejected by the system. All algorithms are implemented in Java, and run on a workstation with Xeon X5650 2.67 GHz CPU and 32 GB RAM.

6.1   Experimental Settings

We conduct the experiments using the real road network of New York, which contains 264,346 nodes and 366,923 edges. We evaluate the algorithms on a large scale taxi dataset containing 1,445,285 trips made by New York taxis over one month (January, 2016) [23]. We make use of the friendship network extracted from *Gowalla* [24], which consists of 196,591 nodes and 950,327 edges, to simulate users' social connections.

We classify the road segments into three categories: (1) highways, (2) roads inside downtown and (3) roads outside downtown. We assign driving velocities to the road segments as shown in Table 3.

We study the effect of the following parameters: the number of taxis, the max waiting time, the value of k and the max number of riders. For all experiments, we assume the rider always chooses the top-1 taxi. The default (bold) and other tested settings are shown in Table 4. We measure the following metrics: (1) *Taxi-sharing rate* as the percentage of trip requests participating in taxi-sharing among all trip requests, (2) *Average social score* of the trip requests which are satisfied by the system, and (3) *Query time* as the clock time for returning the top-k taxis for a given trip request.

6.2   Experimental Results

Taxi-Sharing Rate. In this set of experiments we compare the taxi-sharing rate of our approach with those of NN and T-Share. Since the four algorithms proposed in this paper perform the same in taxi-sharing rate and social score, we choose HG as the representative of our methods.

As shown in Fig. 9, our method enables more riders to share rides with others (i.e., higher taxi-sharing rate) compared to NN and T-Share under different settings, especially when the time constraint is strict or the number of taxis is large. As mentioned before, our objective is to find the best taxis that integrates both detour and riders' social cohesion. Therefore, when the number of taxis is large, our approach still works to assign a new rider to the taxi with a high

Table 3. The velocity model

Category of road	Velocity
Highway	7am–10am and 4pm–7pm: 60 km/h Otherwise 100 km/h
Roads in downtown	7am–10am and 4pm–7pm: 30 km/h Otherwise 60 km/h
Roads outside downtown	50 km/h

Table 4. Parameters settings

Parameter	Settings
# of taxis	500, 1500, **5000**, 10000, 20000
Max waiting time	1 min, 3 min, 5 min, **10 min**, 20 min
k	1, **3**, 5, 10, 20
Max riders	2, 3, **4**, 5, 6

ranking score (i.e., the average detour and the average social distance of riders in the taxi are both better) and the taxi-sharing rate remains stable. On the other hand, NN simply assigns a rider to the nearest taxi and T-Share seeks less detour for a new rider, as the number of taxis increases. Thus, both NN and T-Share approaches tend to assign a rider to an empty taxi. Hence the taxi-sharing rate of these two approaches drops dramatically.

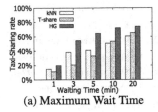
(a) **Maximum Wait Time**

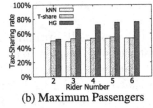
(b) **Maximum Passengers**

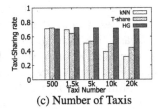
(c) **Number of Taxis**

Fig. 9. Comparing taxi-sharing rate of the algorithms

Social Score. As mentioned, the objective of TkSaTR query is to find the taxis that maximize the ranking score combining the spatial and social aspects. In this experiment, we compare the average social score of taxis which are chosen by the system for each trip request. Again, as our proposed algorithms yield the same social score, we take HG as the representative.

Figure 10 shows that our method achieves a higher social score (almost 40% higher) than the other approaches in almost all experimental settings. The main reason is that our method is designed to make a *social-aware* assignment, i.e., finding the top-k taxis with the highest scores considering both spatial and social factors, whereas the other approaches are not designed to achieve that goal. NN simply assigns a new trip request to the nearest feasible taxi and T-Share only considers spatial detour. Hence, the average social cohesion of riders in one taxi of these two approaches is not as good as that of our method.

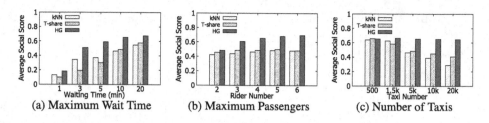

Fig. 10. Comparing average social score of the algorithms

Query Time. To evaluate the scalability of our system, we evaluate the average query processing time. Figure 11a shows that when more taxis are added, the scalability of T-Share suffers. The efficiency of our approach also decreases when the number of taxis increases, as more taxis in total tend to increase the number of candidate taxis needed to be checked in the optimal schedule model. In contrast, NN only considers the taxis near the origin of a new trip request and simply chooses the nearest feasible taxi, and therefore its performance remains stable when the number of taxis grows.

Figure 11b shows the saved distance (compared to no ridesharing) by three approaches for 1,000 queries in New York City. The saved distance increases as the number of taxis grows for all three approaches. Our method does not save as much as the other approaches since it focuses on *social-awareness* and sometimes may sacrifice spatial advantage to achieve more social cohesion. In contrast, T-Share aims to find the spatially optimal taxi (i.e., the taxi with the minimum additional travel distance with respect to a trip request). Nevertheless, our method still saves around 740 Km travel distance for 1000 queries when the taxi number is 10,000. Given that there are 13,237 taxis in New York City and there are 33,825 taxi requests per day on average (learned from the dataset), the saving achieved by our method is nevertheless over 13 million kilometers in distance per year, which equals 1 million liters of gas per year (supposing a taxi consumes 8 liter gas per 100 km) and 2.4 k ton of carbon dioxide emission per year (supposing each liter of gas consumption generates 2.3 kg carbon dioxide).

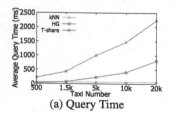

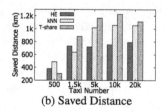

Fig. 11. Query time performance and saved distances

TkSaTR Algorithms Comparison. In this section, we experimentally evaluate the performance of the four algorithms proposed in this paper, namely BF, OE, HE, and HG.

As shown in Fig. 12a, HE and HG outperform the other algorithms with the increasing number of taxis. This is because the Hopping Algorithm used by HE and HG utilizes pruning techniques to reduce the number of possible insertion ways needed to be checked and thus mitigates the scheduling workload. Among the two candidate selection methods, the grid-based HG performs better than the edge-based HE as HG saves time in selecting candidates by utilizing the grid index. On the other hand, the scalability of BF suffers when the number of taxis is large. It enumerates all the taxis to find the top-k ones and therefore the computation load increases dramatically when encountering a large number of taxis.

To study the efficiency of the Hopping Algorithm, we evaluate the pruning times (the number of reduced feasibility checks) and the pruning rate (the percentage of reduced feasibility checks) of 1,000 queries in comparison to the approach without using Hopping Algorithm. As shown in Fig. 12b, the pruning rate is over 95% when the number of taxis is less than 1,500. Although the pruning rate drops when the number of taxis increases, the total number of checks reduced by the Hopping Algorithm is still considerably high. This suggests that the Hopping Algorithm effectively reduces the amount of computation during the optimal scheduling process.

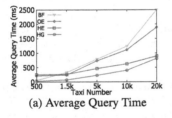

(a) Average Query Time

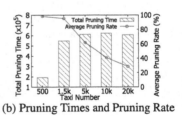

(b) Pruning Times and Pruning Rate

Fig. 12. Comparing performance of algorithms proposed in this paper

7   Conclusion

In this paper, we studied the problem of realtime top-k social-aware taxi ridesharing. Unlike existing studies, we introduce social-awareness into realtime ridesharing services. Our objective is to find the top-k taxis that take into account both spatial concern and social concern. With the proposed service, riders can select their preferred taxi among the top-k ones ranked in a manner that integrates spatial and social aspects. We validated our proposed solution with a large scale New York City taxi dataset. The experimental results demonstrate the effectiveness and efficiency of our proposal.

As for future work, we plan to extend the service by establishing riders' profiles that consider not only the friendships but also their interest. Another interesting direction is to process incoming trip requests within a short period of time (e.g., 5 s) in batch to improve the system throughput.

Acknowledgments. This work is supported by Hong Kong RGC grants 12200114, 12201615, 12244916 and NSFC grant 61602420.

References

1. Ma, S., Wolfson, O.: Analysis and evaluation of the slugging form of ridesharing. In: Proceedings of the 21st ACM SIGSPATIAL, pp. 64–73 (2013)
2. Cici, B., Markopoulou, A., Frias-Martinez, E., Laoutaris, N.: Assessing the potential of ride-sharing using mobile and social data: a tale of four cities. In: Proceedings of the 2014 ACM International Joint Conference on Pervasive and Ubiquitous Computing, pp. 201–211 (2014)
3. Ma, S., Zheng, Y., Wolfson, O.: Real-time city-scale taxi ridesharing. IEEE Trans. Knowl. Data Eng. **27**(7), 1782–1795 (2015)
4. Huang, Y., Bastani, F., Jin, R., Wang, X.S.: Large scale real-time ridesharing with service guarantee on road networks. Proc. VLDB Endowment **7**(14), 2017–2028 (2014)
5. Ma, S., Zheng, Y., Wolfson, O.: T-share: a large-scale dynamic taxi ridesharing service. In: IEEE 29th International Conference on Data Engineering (ICDE), pp. 410–421 (2013)
6. Badger, E.: Slugging-the people's transit (2011)
7. Baldacci, R., Maniezzo, V., Mingozzi, A.: An exact method for the car pooling problem based on lagrangean column generation. Oper. Res. **52**(3), 422–439 (2004)
8. Calvo, R.W., de Luigi, F., Haastrup, P., Maniezzo, V.: A distributed geographic information system for the daily car pooling problem. Comput. Oper. Res. **31**(13), 2263–2278 (2004)
9. Agatz, N., Erera, A., Savelsbergh, M., Wang, X.: Sustainable passenger transportation: Dynamic ride-sharing (2010)
10. Tsubouchi, K., Hiekata, K., Yamato, H.: Scheduling algorithm for on-demand bus system. In: Information Technology: New Generations, 2009, ITNG 2009, pp. 189–194. IEEE (2009)
11. Yuan, N.J., Zheng, Y., Zhang, L., Xie, X.: T-finder: a recommender system for finding passengers and vacant taxis. IEEE TKDE **25**(10), 2390–2403 (2013)
12. Yan, S., Chen, C.Y.: An optimization model and a solution algorithm for the many-to-many car pooling problem. Ann. Oper. Res. **191**(1), 37–71 (2011)
13. Cordeau, J.F., Laporte, G.: The dial-a-ride problem: models and algorithms. Ann. Oper. Res. **153**(1), 29–46 (2007)
14. Xiang, Z., Chu, C., Chen, H.: A fast heuristic for solving a large-scale static dial-a-ride problem under complex constraints. Eur. J. Oper. Res. **174**(2), 1117–1139 (2006)
15. Zhao, W., Qin, Y., Yang, D., Zhang, L., Zhu, W.: Social group architecture based distributed ride-sharing service in vanet. Int. J. Distrib. Sens. Netw. **10**(3), 650923 (2014)
16. Agatz, N., Erera, A., Savelsbergh, M., Wang, X.: Optimization for dynamic ridesharing: a review. Eur. J. Oper. Res. **223**(2), 295–303 (2012)

17. Rigby, M., Krüger, A., Winter, S.: An opportunistic client user interface to support centralized ride share planning. In: Proceedings of the 21st ACM SIGSPATIAL, pp. 34–43 (2013)
18. d'Orey, P.M., Fernandes, R., Ferreira, M.: Empirical evaluation of a dynamic and distributed taxi-sharing system. In: 15th International IEEE Conference on Intelligent Transportation Systems, pp. 140–146. IEEE (2012)
19. Li, Y., Chen, R., Chen, L., Xu, J.: Towards social-aware ridesharing group query services. IEEE Trans. Serv. Comput. **PP**(99), 1 (2015)
20. Bistaffa, F., Farinelli, A., Ramchurn, S.: Sharing rides with friends: a coalition formation algorithm for ridesharing (2015)
21. Sleator, D.D., Tarjan, R.E.: Amortized efficiency of list update and paging rules. Commun. ACM **28**(2), 202–208 (1985)
22. Kanoulas, E., Du, Y., Xia, T., Zhang, D.: Finding fastest paths on a road network with speed patterns. In: ICDE 2006, p. 10, April 2006
23. TLC: NYC TLC trip data. http://www.nyc.gov/html/tlc/html/about/trip_record_data.shtml
24. SNAP: Gowalla. https://snap.stanford.edu/data/loc-gowalla.html

P-LAG: Location-Aware Group Recommendation for Passive Users

Yuqiu Qian[1]([✉]), Ziyu Lu[1], Nikos Mamoulis[2], and David W. Cheung[1]

[1] The University of Hong Kong, Hong Kong, Hong Kong
{yqqian,zylu,dcheung}@cs.hku.hk
[2] University of Ioannina, Ioannina, Greece
nikos@cs.uoi.gr

Abstract. Consider a group of users who would like to meet to a place in order to participate in an activity together (e.g., meet at a restaurant to dine). Such meeting point queries have been studied in the context of spatial databases, where typically the suggested points are the ones that minimize an aggregate traveling distance. Recently, meeting point queries have been enriched to take as input, besides the locations of users, also some preference criteria (e.g., expressed by some keywords). However, in many applications, a group of users may require a meeting point recommendation without explicitly specifying any preferences. Motivated by this, we study this scenario of group recommendation for such passive users. We use topic modeling to infer the preferences of the group on the different points of interest and combine these preferences with the aggregate spatial distance of the group members to the candidate points for recommendation in a unified search model. Then, we propose an extension of the R-tree index, called TAR-tree, that indexes the topic vectors of the places together with their spatial locations, in order to facilitate efficient group recommendation. We propose and compare three variants of the TAR-tree and a compression technique for the index, that improves its performance. The proposed techniques are evaluated on real data; the results demonstrate the efficiency and effectiveness of our methods.

1 Introduction

With the proliferation of smart mobile devices, recommendation services are becoming location-aware; advertisements and suggestions to users are not generated only based on the (explicitly expressed or implicitly derived) user preferences, but also based on the user locations. For example, a mobile user who likes green tea would be recommended a nearby green tea shop (if she wants to take a break), instead of a more famous one, which is too far.

In this paper, we consider the scenario of providing recommendations to a *group* of users who would like to meet and enjoy an activity or event together. By considering only the locations of the individual users in the group, a meeting point query can be modeled and solved as an *group nearest neighbor* search [23],

© Springer International Publishing AG 2017
M. Gertz et al. (Eds.): SSTD 2017, LNCS 10411, pp. 242–259, 2017.
DOI: 10.1007/978-3-319-64367-0_13

where the suggested places are the ones that minimize the average or maximum distance to the users of the group. In addition, *spatial skyline* queries [27] can be used to suggest the places which are not dominated by other places in the space defined by their distances from the users in the group.

Given the fact that the recommended places by spatial-only criteria may not be consistent with the preferences of the group, recent work has considered additional, non-spatial search criteria that are explicitly expressed by the group. For example, the group may provide a set of keywords and require that the recommended places are textually relevant to the keywords in addition to being spatially near the group members. In this direction, the *textually relevant spatial skyline* was proposed in [28], which includes all objects that are not dominated by others with respect to their aggregate spatial distance to the group users and their similarity to a set of given keywords. A similar, *location-aware group preference* query was proposed in [17]; the objective is to find a nearby destination that belongs to a specific category (e.g., a bar), which is also close to places that satisfy the explicitly given preferences of each user.

In a real scenario of recommending a place to a group of users, the users could be *passive*, i.e., they might not express their preferences explicitly. For example, it might be hard for the group to reach a consensus for the set of keywords to use in the search, or some users in the group might find it difficult to express their preferences by keywords or by ranking functions on the features of the places. Hence, previous work on meeting point recommendation largely ignores the preferences of silent users. In this paper, we *learn* these preferences by analyzing the check-in/reviewing history of users and use them in meeting point recommendation. It is worth mentioning that, although geographic information has been considered in previous work on recommending venues to groups of passive users (e.g., see [20]), recommender systems do not consider the case where the users of the group are presently at different locations.

Hence, in this paper, we consider the problem of location-aware group recommendation for *passive users* (P-LAG for short), assuming that the members of the group are spatially dispersed and they have to meet in order to enjoy an activity or event at the recommended place. Although the locations of the users are known, the preferences of the group should be *inferred* based on any previously collected information related to the activities of the users on the candidate places (e.g., check-in history in a geo-social network such as Foursquare, place visits derived by analyzing GPS trajectories, posted reviews that indicate place visits and preferences). For this purpose, we use a *topic modeling* approach [5], which represents each user by a topic vector that models his/her preferences to each topic. A topic is a hidden property of the places; accordingly, the candidate places are also represented by topic vectors. Thus, for each user and each place we can model the preference of the user to the place by the similarity between the corresponding vectors. For the group of users, we compute an aggregate topic vector and use it to model the preferences of the group.

Our second contribution is an extension of the classic R-tree, called TAR-tree, to index the topic vectors of the places together with their spatial locations.

We propose three variants of the TAR-tree and a compression technique for the index, that greatly improves search performance. The effectiveness of the TAR-tree is largely due to the skew in the topic vectors and the spatial auto-correlation of the topics, which we analyze experimentally. Finally, we conduct a comparative study, using two real datasets, which demonstrates the efficiency and effectiveness of our proposed methods.

The remainder of this paper is organized as follows. Section 2 provides a formal definition of the top-k location-aware group recommendation problem in a passive user setting and discusses a naive solution as well as our framework. In Sect. 3, we present our approach for extracting the topic vectors of users and places, in a preprocessing phase. Online query evaluation is studied in Sect. 4. Section 5 evaluates the effectiveness and efficiency of the proposed framework. In Sect. 6, we discusses related work. Finally, Sect. 7 concludes the paper.

2   Problem Definition

The P-LAG problem we are focusing on in this paper is to find k venues (i.e., places) which are (i) spatially close to the current location of each group user and (ii) consistent to the preferences of the group. Without loss of generality, Euclidean distance $dist(p_i, q_j)$ is used as spatial distance in this work. A formal definition of the P-LAG query is given below:

Definition 1 (*P-LAG query*). *Consider a set of venues $P = \{p_1, p_2, ..., p_n\}$, each with a topic vector $p_i.\psi$. Given a group $Q = \{q_1, q_2, .., q_m\}$ of users having an aggregate topic vector $Q.\psi$, the P-LAG query finds the k venues $p_i \in P$ which minimize the distance function $Dist(p_i, Q) = f(\sum_{q_j \in Q} dist(p_i, q_j), \omega(p_i.\psi, Q.\psi))$, where $dist(p_i, q_j)$ is the spatial distance between p_i and the location of user q_j, $\omega(p_i.\psi, Q.\psi)$ is the similarity between topic vectors $p_i.\psi$ and $Q.\psi$.*

Notice that the ranking function $Dist(p_i, Q)$ can be any monotonic aggregate function $f(\cdot, \cdot)$ which considers both spatial information and user preferences, such as weighted sum $\alpha \times \sum_{q_j \in Q} dist(p_i, q_j) + (1 - \alpha) \times (1 - \omega(p_i.\psi, Q.\psi))$ [30] and weighted distance $\sum_{q_j \in Q} dist(p_i, q_j)/\omega(p_i.\psi, Q.\psi)$ [28]. In this paper, we use the latter definition, because it is parameter-free and it does not require normalization. Still, the approaches proposed in this paper are independent of how $f(\cdot, \cdot)$ is defined.

To support P-LAG queries, we follow the two-step framework illustrated in Fig. 1. First, the topic vectors of both users and venues (i.e., user preference vectors and venue topic vectors) are automatically extracted with the help of a model that uses previous trajectory/check-in information. This is an offline process executed once based on the available historical data. The process can be repeated in regular time intervals in order to keep the topic vectors up-to-date. The second component of our framework addresses online query evaluation. Once a group of users are formed, the top-k venues for the group are identified and recommended within milliseconds.

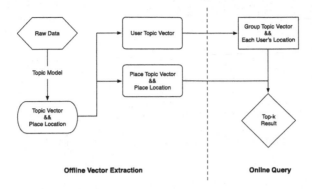

Fig. 1. Overall framework

3 Vector Extraction

3.1 Topic Vector Extraction

Our framework first extracts topic vectors for all users and places. For this purpose, we apply Latent Dirichlet Allocation (LDA) [5] on the history of user visits to places, which can be obtained by past check-in or trajectory records. Hence, the raw input data is a collection of $\langle u, l, t \rangle$ records, where u represents a user, l is a location (i.e., place) visited by u, and t is the time of the visit.

For each user $u \in U$, we synthesize a *document* D_u by aggregating all records of u, and regard each location $l \in L$ visited by u, as a word in D_u. Hence, for each user u, we have a document $D_u = \{u, L_u\}$, where L_u includes the identifiers of all places visited by u. Overall, we obtain a collection D of documents D_u, one for each $u \in U$.

Figure 2 shows the graphical representation of the Location-LDA model we have designed, which uses two distributions: a user-topic distribution θ and a topic-location distribution ϕ. The topic-location distribution represents the topic vector for each location, e.g., ϕ_l for location l. The user-topic distribution represents the topic vector for each user, e.g., θ_u for user u. The latent topics Z are extracted based on the check-in preferences of users (i.e., they are not related to the spatial features of the places). α, β are prior parameters for the two distributions; we use $\alpha = 50/K$ and $\beta = 0.01$, where K is the number of topics.

3.2 Topic Vector Analysis

In this section, we conduct two spatial analysis studies on the extracted topic vectors that lead us to important observations. The first study shows the correlation between the geographical distance and topic similarity of two locations. The second one shows that different topics have different spatial distributions. We use the topic vectors extracted from Yelp-USA, which is a subset of the Yelp

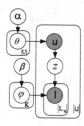

Fig. 2. Graphical representation of our Location-LDA model

Dataset Challenge dataset[1]. The vectors have a dimensionality $K = 20$, which is the default value in our experiments (see Sect. 5).

Correlation between Spatial Distance and Topic Proximity. Intuitively, locations that are geographically close to each other should have overlapping topics with high probability. To confirm this, we sampled 2000×2000 locations pairs from the check-in data of Yelp-USA. For each pair of locations (places) (p_i, p_j), we compute their distance $dist(p_i, p_j)$ in the Euclidean space and the similarity between their topic vectors $w(p_i.\psi, p_j.\psi)$. The topic similarity is computed as the dot product of the vectors.

We divided the location pairs into several groups based on their geographic distances and calculated the average topic similarity for each group. As shown in Fig. 3, topic similarity is positively correlated to spatial proximity. The size of each point in the plot represents the number of location pairs that fall in the corresponding distance range. Note that location pairs that are within 1 km to each other are significantly more related than others.

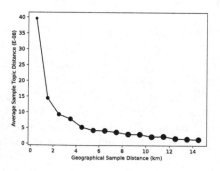

Fig. 3. Correlation between spatial distance and topic proximity

Geographical Topic Heatmap. By visualization analysis, we observed that the spatial distributions of different topics differ significantly. In other words, different topics form different spatial clusters even in a map of relatively small scale.

---

[1] http://www.yelp.com/dataset_challenge.

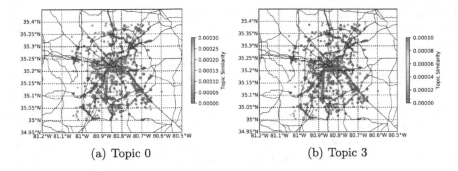

(a) Topic 0   (b) Topic 3

Fig. 4. Topic value heatmap in Charlotte

For example, in Fig. 4, we use heatmaps to visualize the spatial distribution of different topics in locations of Charlotte. Each place is colored based on its value of topic 0 (Fig. 4(a)) and topic 3 (Fig. 4(b)). Observe that different topics cover different spatial regions and different spatial regions have relevance to a given topic.

4 Indexing and Search for P-LAG

Deriving topic vectors for venues and users as described in Sect. 3 allows us to search for the best places to recommend to a given group of users, based on spatial distance and topic similarity (see Definition 1). Specifically, for a group $Q = \{q_1, q_2, .., q_m\}$ of spatially dispersed users, we should compute the k places $p_i \in P$ that minimize the objective function $Dist(p_i, Q) = \sum_{q_j \in Q} dist(p_i, q_j)/\omega(p_i.\psi, Q.\psi)$.

A naive algorithm (NA) traverses all venues $p_i \in P$ one by one, computes the objective function for each of them, and maintains the set of k venues with smallest $Dist(p_i, Q)$. This approach is generally expensive because it computes $Dist(p_i, Q)$ exhaustively for each place p_i. In addition, the topic vector similarity component of $Dist(p_i, Q)$ is more expensive to compute compared to the spatial distance component. Therefore, a natural approach for P-LAG queries is to access the candidate venues for recommendation in a spatial order, while deriving bounds for the k-th distance; this enables us to stop search early by avoiding examining the places whose spatial distances are too large.

4.1 Basic R-tree Approach

An intuitive approach for supporting P-LAG queries efficiently is to spatially index all venues $p_i \in P$, e.g., by an R-tree [4,11]. With the help of the R-tree, we can derive a lower bound for the group spatial distance $\sum_{q_j \in Q} dist(e, q_j)$ to each entry e in the R-tree (see [23]). However, we can access the topic vector of each place, only after accessing the place at leaf level of the tree, which gives us no information about the vector similarity between the places under an R-tree entry e and $Q.\psi$ during search. This makes it hard to prune sets of candidate places at the non-leaf levels of the R-tree. A basic approach for alleviating this problem is to compute and use the maximum value of $\omega(p_i.\psi, Q.\psi), \forall p_i \in P$, i.e. $\omega(p^*.\psi, Q.\psi)$, where $p^*.\psi$ is a vector having the maximum values of each dimension in the topic space. It is easy to prove $\omega(p^*.\psi, Q.\psi)$ as a upper bound of all $\omega(p_i.\psi, Q.\psi), p_i \in P$ since $\omega(p_i.\psi, Q.\psi) = \sum_j p_i.\psi_j \cdot Q.\psi_j \leq \sum_j p^*.\psi_j \cdot Q.\psi_j = \omega(p^*.\psi, Q.\psi)$. Therefore, $\omega(p^*.\psi, Q.\psi)$ is also a upper bound and $Dist(e, Q) = \sum_{q_j \in Q} dist(e, q_j)/\omega(p^*.\psi, Q.\psi)$ can be utilized as a lower bound to prune non-qualifying R-tree nodes.

Algorithm 1 illustrates our algorithmic framework for P-LAG queries. Line 17 computes and uses a lower distance bound of each R-tree entry. This bound for the basic R-tree approach (i.e., $\sum_{q_j \in Q} dist(e, q_j)/\omega(p^*.\psi, Q.\psi)$) is too loose, motivating us to study additional ways for tightening the bound and speeding up P-LAG search.

Algorithm 1. P-LAG Algorithmic Framework

1: **procedure** RTREE APPROACH(P, Q, k)
2:     MinHeap $H \leftarrow \varnothing$ ▷ nodes to visit
3:     $R \leftarrow \varnothing$ ▷ top-k result set
4:     $kDist \leftarrow Inf$ ▷ k-th $Dist(p_i, Q)$ in R
5:     Add $root$ of RTree to H
6:     **while** H is not empty **do**
7:         $e \leftarrow$ deHeap(H)
8:         **if** $e.dist > kDist$ **then**
9:             **return** R
10:         $N \leftarrow readNode(e)$
11:         **if** e is an leaf node **then**
12:             **for** each object p' in N **do**
13:                 $p'.dist \leftarrow \dfrac{\sum_{q_j \in Q} dist(p', q_j)}{\omega(p'.\psi, Q.\psi)}$
14:                 Update R and $kDist$ with p'
15:         **else**
16:             **for** each entry e' in N **do**
17:                 $e'.dist \leftarrow$ lower bound of $Dist(e', Q)$
18:                 **if** $e'.dist < kDist$ **then**
19:                     Add e' to H
20:     **return** R

4.2 TAR-tree Approach: Topic-Aware R-tree

Our approach for improving our framework is to embed vector information into the R-tree, making it a TAR-tree (i.e. Topic-aware R-tree), which helps deriving tighter bounds for pruning the search space effectively.

B-TAR-tree: Ball Bounding. One way of doing so is to borrow the idea behind the Ball-tree [22], and use ball bounding to group multi-dimensional topic vectors $p_i.\psi$, such that $p_i \in P$. For each entry e in the R-tree, we need to find the bounding ball with the minimum radius that contains all places in the subtree pointed by e. The minimum balls are computed in a bottom up fashion similar to the ball tree, but following the original structure of the R-tree.

Following the proof in [25], it is easy to get an upper bound for vector similarity for each entry e in the B-TAR-tree, as $\omega(e.\psi, Q.\psi) \leq \omega(e.\mu, Q.\psi) + e.\lambda \cdot \|q\|$, where $e.\mu$ is the center of minimum ball and $e.\lambda$ is the radius of the minimum ball of entry e. Therefore, we can derive a bound for $Dist(e, Q)$ as $Dist(e, Q) = \sum_{q_j \in Q} dist(e, q_j)/\omega(e.\psi, Q.\psi) \geq \sum_{q_j \in Q} dist(e, q_j)/(\omega(e.\mu, Q.\psi) + e.\lambda \cdot \|q\|)$ to be used in line 17 of Algorithm 1. However, the ball bounding approach has the side effect of storing extra information at the non-leaf levels of TAR-tree: a vector $e.\mu$ as well as a number $e.\lambda$. Hence, each non-leaf node has a much smaller fanout compared to the original R-tree and the tree's height becomes larger.

N-TAR-tree: Norm Bounding. As an alternative of the ball bounding technique, which requires the storage of one extra vector for each MBR, we suggest a second approach (norm bounding), which only needs to store one number per MBR. Specifically, for each non-leaf entry e in the R-tree pointing to a node S with entries $\{s_1, s_2, ..., s_f\}$, it is easy to show that $\omega(s_i.\psi, Q.\psi) = \|s_i.\psi\| \cdot \|Q.\psi\| \cdot \cos(s_i.\psi, Q.\psi) \leq \|s_i.\psi\| \cdot \|Q.\psi\|, \forall s_i \in S$. Let $e.\phi$ be the maximum $\|s_i.\psi\|, \forall s_i \in S$. We can get a tighter bound for $\omega(e.\psi, Q.\psi)$ compared to $\omega(p^*.\psi, Q.\psi)$ of the basic R-tree approach, since $\omega(e.\psi, Q.\psi) \leq e.\phi \cdot \|Q.\psi\|$, where $e.\phi$ is the maximum norm value of entries $s_i \in S$ (i.e., a single number).

R-TAR-tree: Rectangle Bounding. The third proposed TAR-tree variant uses rectangle bounding. For each entry e in a R-tree node, we store a vector $e.\psi$, which stores for each topic dimension the maximum value of that dimension in the subtree pointed by e. Then, we have $\omega(e.\psi, Q.\psi) \geq \omega(s_i.\psi, Q.\psi), \forall s_i \in S$, where $S = \{s_1, s_2, ..., s_f\}$ is the set of entries in the node pointed to by e.

As an illustration, in Fig. 5, M_1 and M_2 are two entries of a TAR-tree, each pointing to a leaf node containing a triple of venues; each venue v_1 to v_6 has a topic distribution vector (shown next to each v_i), whereas the corresponding vector of each M_i has the maximum of values for each topic of the entries in M_i. By using the minimum spatial distances of the user locations to each M_i and the vector of M_i, we can derive bounds for the spatial distances and the vector

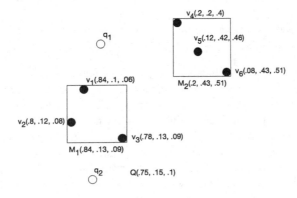

Fig. 5. Example of TAR-tree

similarities of each venue in M_i. These bounds can be used to prune M_i, if it is worse than the k-th venue found so far.

4.3   Vector Compression in the TAR-tree

All three TAR-tree variants we have discussed extend the R-tree to store extra topic information into its entries. However, the potential benefit of getting tighter bounds can be outweighed by the side effect of higher storage requirements, leading to a slower method. In view of this, we propose a compression technique that exploits the skew and the spatial autocorrelation of topic vectors.

To demonstrate the skew of topic vectors in practice, we first conduct an experiment on the Foursquare [10] dataset we used in our experiments. For 20-dimensional topic vectors, Fig. 6(a) shows the average distribution of their values after ranking them from larger to smaller. Note that the largest 2–3 scalars are significantly larger than the remaining ones. We repeat this experiment, this time for the non-leaf entries of the above-leaf level of the TAR-tree (Fig. 6(b)). The skew at the vectors of the non-leaf entries indicates that the vectors in the same leaf node are correlated. This is due to the spatial autocorrelation of the topics (see our analysis in Sect. 3.2).

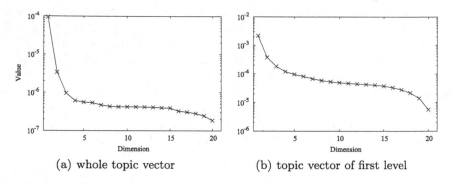

(a) whole topic vector   (b) topic vector of first level

Fig. 6. Skewness of topic vectors stored in the TAR-tree

Recall that, in the R-TAR-tree, we store a vector $e.\psi$ for each entry e in an R-tree node, whose value at each dimension is the maximum of the corresponding dimensions in all its children entries. Then, we use $\omega(e.\psi, Q.\psi)$ as the largest approximate vector topic value of all its children nodes. To reduce the storage cost for these vectors, which is linear to their dimensionality, we propose a compression technique. Take $e.\psi = (v_0, v_1, ..., v_{19})$ whose vector size is 20 as a example (shown in Fig. 7). After rearranging the scalars in decreasing value, we get a list $v_{t_0}, v_{t_1}, ..., v_{t_{19}}$, where v_{t_i} is the value of dimension t_i in the original $e.\psi$ vector and $v_{t_i} \geq v_{t_{i+1}}$ for each $i \in [0, 18]$. Instead of storing the entire 20-dimensional vector $e.\psi$, we can reduce the storage size by only storing the top-m (m is called compression factor) largest values in it together with their indexes t_i and the $(m+1)$-th largest value. This way, the storage cost of each topic vector can be reduced to $2m + 1$. With the compressed vector, we have a looser upper bound for vector similarity as $w(e.\psi, Q.\psi) = \sum_j v_j \cdot Q.\psi_j = \sum_{j=0}^{m-1} v_{t_j} \cdot Q.\psi_{t_j} + \sum_{j=m}^{|e.\psi|-1} v_{t_j} \cdot Q.\psi_{t_j} \leq \sum_{j=0}^{m-1} v_{t_j} \cdot Q.\psi_{t_j} + \sum_{j=m}^{|e.\psi|-1} v_{t_m} \cdot Q.\psi_{t_j}$, where v_{t_m} denotes the $(m+1)$-th largest value.

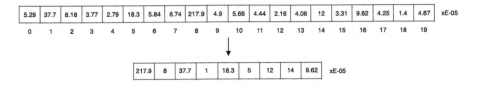

Fig. 7. Example of compression

This vector compression technique is used only at the non-leaf entries of the tree. We denote this approach by R-TAR-tree+. However, non-leaf nodes are significantly fewer than leaf-nodes, hence the space saved by the compression is limited. Therefore, we consider also applying this compression to place topic vectors at the leaves of the tree. Note that after this change the tree only indexes the vectors approximately. We denote this approach by R-TAR-tree-C. Since topic vector extraction is an approximation technique by nature, we do not expect much loss of accuracy in practice.

5 Experimental Evaluation

We evaluate the performance of all approaches proposed in this paper, namely the basic R-tree approach (Sect. 4.1), the three variants of TAR-tree approach (Sect. 4.2), and the two compression approaches (Sect. 4.3). All tested methods were implemented in Java and the experiments were conducted on an Intel(R) Core(TM) i5-3570 CPU @ 3.40 GHz machine, with 16 GB of RAM.

5.1 Datasets

We used two real datasets, Foursquare [10] and Yelp-USA[2], in our evaluation. Foursquare is a popular geo-social network with more than 50 million users and 10 billion check-ins[3]. Yelp-USA is a subset of the dataset used in the Yelp Dataset Challenge, containing all USA located records. The details of the datasets are shown in Table 1. In the topic vector extraction phase, for each dataset, we mark 80% of the data as the training set and use the remaining data to test the accuracy of our Location-LDA model. To generate the user groups for the queries, we sample users by bounding the average distance to each other; we use the real learnt topic vectors for the users.

---

[2] http://www.yelp.com/dataset_challenge.
[3] https://foursquare.com/about/.

Table 1. Statistics of the datasets

Dataset	Yelp-USA	Foursquare
# of records	1,525,924	2,290,997
# of users	432,536	11,326
# of venues	38,694	187,218

5.2 Efficiency Analysis

We evaluate the performance of different methods under various parameter settings, i.e. varying the result set size k, the topic vector dimensionality K, the group size, the average distance between the users in the group measured in degrees of latitude and longitude, and the compression factor m. The details of the parameters are listed in Table 2.

Table 2. Parameters (default values in bold)

Parameter	Value
k	5, 10, **20**, 50, 100
Dimensionality of topic vectors (K)	5, 10, **20**, 50, 100
Group size	3, 5, **10**, 15, 20
User maximum distance	0.05, 0.1 **0.2**, 0.5, 1
Compression parameter (m)	1, 2, **4**, 6, 8

Effect of k. To study the effect of the result set size k, we fix the other parameters to their default values, and vary k from 5 to 100. As shown in Fig. 8, the cost increases with k, which is consistent with our expectation. We observe that the basic R-tree method is much faster compared to the naive sequential scan approach and the TAR-tree methods further reduce the query time and I/O. Different TAR-tree methods do not have big differences and generally R-TAR-tree+ performs best among all exact methods. R-TAR-tree-C improves the efficiency of R-TAR-tree+ up to 2 times; as we will see in Sect. 5.3 its accuracy is very high.

Effect of vector dimensionality. The second parameter of which the effect we investigate is the vector dimensionality. As shown in Fig. 9, on both datasets, the average query time increases as the dimensionality increases. The I/O cost shows a similar trend (we omit the plots due to space constraints). At the same time, we observed (in experiments not shown here) that the topic extraction quality does not improve much when we increase the dimensionality to above 20. Therefore, 20 is a good enough value in terms of quality.

Effect of group size. The third parameter that we study is the group size $|Q|$; in general, the groups can be formed by any number of users. Observe from

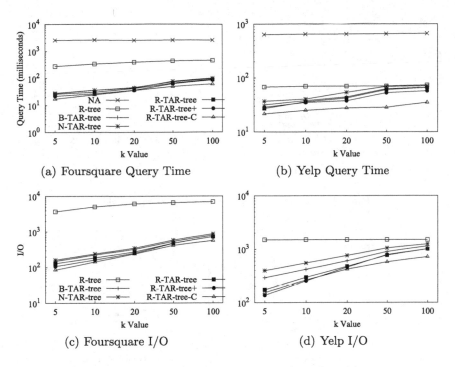

Fig. 8. Comparison when varying the value of k

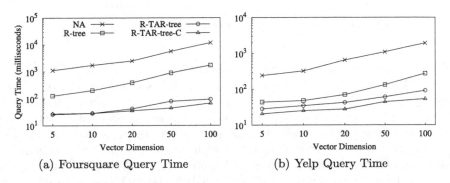

Fig. 9. Comparison when varying vector dimensionality

Fig. 10 that the query time decreases when the group size increases. The reason is that the increase in the group size increases the importance of the spatial distance component, which helps to prune the search space faster.

Effect of user maximum distance. The fourth parameter is the maximum distance between the users in the group, which we vary from 0.05 to 1° of latitude and longitude and sample instances randomly so that their distances satisfy this bound. Similar to the effect of the group size, the query time and I/O increases when distance increases as shown in Fig. 11.

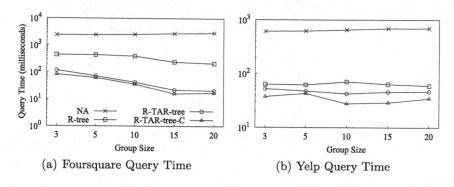

(a) Foursquare Query Time   (b) Yelp Query Time

Fig. 10. Comparison when varying user group size

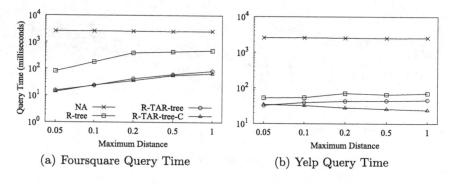

(a) Foursquare Query Time   (b) Yelp Query Time

Fig. 11. Comparison when varying group average distance

Effect of compression size. The compression techniques used in R-TAR-tree+ and R-TAR-tree-C are different in nature. For R-TAR-tree+, only vectors in non-leaf nodes of the tree are compressed, rendering it an exact method, while R-TAR-tree-C is an approximate method since it compresses all vectors. The compression factor m therefore plays different role in them as shown in Fig. 12: the query time of R-TAR-tree-C increases with compression size, while the query time of R-TAR-tree+ first decreases and then increases. When m is too small, the approximate vectors in non-leaf nodes are too loose to be useful, resulting in higher I/O cost. However, when m increases, the approximation becomes better, which brings the final kinked pattern.

5.3 Effectiveness

In this section, we study the performance of R-TAR-tree-C since it is an approximate method. We use Precision@k $= |E_k \cap R_k|/k$ for measuring the accuracy of the result R_k computed by R-TAR-tree-C compared to the exact top-k result E_k. Table 3 shows that when the compression factor m equals 4, the result set is nearly same as that of exact search for topic vector dimensionality of 20.

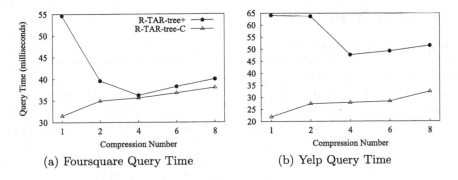

Fig. 12. Comparison when varying the compression parameter m

Table 3. Precision@k when varying the compression parameter m

Dataset	Foursquare					Yelp				
Precision@k	1	2	4	6	8	1	2	4	6	8
1	0.110	0.290	1.0	1.0	1.0	0.0	0.010	0.990	1.0	1.0
5	0.322	0.612	1.0	1.0	1.0	0.018	0.070	0.998	1.0	1.0
10	0.563	0.781	1.0	1.0	1.0	0.123	0.270	0.999	1.0	1.0
20	0.841	0.937	1.0	1.0	1.0	0.494	0.653	1.0	1.0	1.0

5.4 Storage Requirements

Although the TAR-tree method is more efficient than the basic R-tree, it needs more storage. Table 4 shows the space occupied by the TAR-tree versions and the R-tree for different vector dimensionality. R-TAR-tree has higher storage size compared to R-tree, but R-TAR-tree+ reduces the cost a little using compression. The differences are not big because the R-tree also stores the topic vectors of places in the leaf nodes to avoid random accesses for fetching them. Finally, R-TAR-tree-C compresses all vectors and therefore has the least storage requirement among all methods.

Table 4. Storage size when varying vector dimension

Dataset	Foursquare					Yelp				
Vector dimensionality	5	10	20	50	100	5	10	20	50	100
R-tree	47.0M	70.2M	118.6M	263.1M	492.5M	9.8M	14.6M	24.5M	54.3M	101.6M
R-TAR-tree	47.4M	71.4M	122.3M	284.0M	578.0M	10.0M	14.9M	25.5M	59.0M	119.5M
R-TAR-tree+	47.2M	70.2M	119.7M	269.1M	514.9M	9.8M	14.7M	24.6M	55.1M	106.3M
R-TAR-tree-C	35.4M	42.6M	56.5M	100.2M	182.1M	7.4M	8.9M	11.8M	20.8M	37.7M

6   Related Work

Group Recommendation. Recommender systems [1] have been successfully deployed in a wide range of applications, such as inferring the preferences of a given user on a set of items (e.g., products, services, events, venues) and recommending the items with the highest probability to be liked by the user. Group recommendation is also supported, with the input of a set of users, recommending the items that are most likely to be favored by the input group. Early approaches [15,21], for each item, combine the predicted ratings of all group members to derive a single representative rating for the group and then suggest to the group the items with the highest representative ratings. In [3] the agreement between group members is also considered. More recently [16], the social relationships between members are used to enhance the quality of group recommendations. Topic-based group recommendation models [18,35] estimate the impact (i.e., influence) that every user in the group has on the other group members for different topics (i.e., aspects) of the objects to be recommended. For the case where the recommended items are venues, the spatial preferences of the group have to be considered. In [35], the distances between candidate venues and those visited in the past by the users of the group are used as a factor in the group recommendation model. To our knowledge, no previous work considers the current locations of users in the target group.

Group Spatial Search. Nearest neighbor (NN) search [26] is one of the most common spatial queries, well studied in both Euclidean [12] and spatial network spaces [24]. Given a query location q and a collection P of spatial points of interest, NN query retrieves the nearest spatial object in P to q (or the k nearest ones in kNN search). NN queries, were extended to aggregate NN (ANN) queries in [23]. An ANN query takes as input a group $Q = \{q_1, q_2, ..., q_m\}$ of m query locations (e.g., representing different users who want to dine together). The objective is to find the object p in P that minimizes an aggregation of the distances from each $q_i \in Q$ to p (e.g., $\sum_{q_i \in Q} dist(q_i, p)$ or $\max_{q_i \in Q} dist(q_i, p)$). The distance function can be Euclidean distance [23] or spatial network distance [34]. Papadias et al. [23] propose an algorithm for ANN queries, which applies on the original Euclidean space and assumes that P is indexed by an R*-tree. Yiu et al. [34] present algorithms which adapt the top-k retrieval methods of Fagin et al. [9] to compute ANN queries in spatial networks.

Later, the spatial skyline query (SSQ) [27] has been proposed as an alternative of ANN search, which utilizes the concept of skyline query [6]. Similar to ANN search, the input is a set P of points of interest and a set Q of m query locations, representing the locations of a group of users who want to meet. A point $p_i \in P$ is said to spatially dominate another point $p_j \in P$ if for each $q \in Q$, $dist(q, p_i) \leq dist(q, p_j)$. Intuitively, in this case, p_i would be a better point to meet compared to p_j in the eyes of every user in the group. SSQ reports as the spatial skyline the largest subset of P which contains only points that are not spatially dominated by others in P. Voronoi Diagrams are precomputed and indexed to facilitate the efficient computation of spatial skylines in [27,29]. [8] extended the techniques of [27] to apply on a road network, where Euclidean distance is replaced by shortest path distance. Later, [36] proposed a novel index and a filter-and-refine approach for this problem. Dynamic skyline queries in general metric spaces have been investigated in [7]. However, both ANN or SSQ do not consider the explicit or implicit preferences of the users.

Preference-based Meeting Point Search. [28] extends SSQ to a Spatio-Textual Skyline (STS) query, which allows users to find places that are both close to the locations of the group users and relevant to a set of user-defined keywords. Three different models for integrating textual relevance into spatial skylines are proposed. Among them, model STD (Spatio-Textual Dominance), which replaces the spatial distance measure of the derived dimensions by a combined spatio-textual distance, is experimentally shown to be the best one. Another recent work [17] proposed a Location-aware Group Preference (LGP) query, which suggests a place labeled with a specified category feature (e.g., hotel) to a group of users. Each user in the group has a location and a set of additional preferences. The query result should belong to the specified category and should be near the current location of the users and close to places satisfying the additional preferences of users. In this project, we assume that the group members are spatially dispersed and study methods for venue recommendation that consider (i) the traveling cost of the group members to the suggested venues based on their locations and (ii) the implicit preferences of users for the venues.

Spatial Topic Modeling. Spatial topic models have been extensively studied in the past decade. Yin et al. [33] proposed a model based on Probabilistic Latent Semantic Indexing (PLSI), after observing that geographical distributions can help model topics while topics group different geographical regions. Hong et al. [13] proposed a sparse generative model to uncover geographical topic patterns on Twitter. Liu et al. [19] proposed the spatio-temporal topic model to capture microscopic and macroscopic patterns of check-ins. Ahmed et al. [2] presented a hierarchical topic model which captures regional variations of topics. Spatial topic models are typically used for location recommendation [14,31,32]. For example, Yin et al. [31] proposed a system (LCARS) that exploits both local preferences and item content information for spatial item recommendation. Since our main focus is to model the implicit user preferences with spatial topic modeling, we adapted the most commonly used Latent Dirichlet Allocation (LDA) model [5] for simplicity.

7 Conclusion

This paper is the first study on location-aware group recommendation queries for passive users (P-LAG problem), which builds on previous work on meeting point recommendation. We follow a two-step framework: offline topic vector extraction and online querying based on appropriate indexing. Three variants of a TAR-tree approach are proposed, as well as a vector compression technique that improves search performance and reduces the storage requirements. In the future, we plan to test alternative approaches for preference extraction or elicitation and apply our framework with alternative definitions of spatial distance (e.g., travel distance on spatial networks).

Acknowledgements. We thank the reviewers for their valuable comments. This work is partially supported by GRF Grants 17201414 and 17205015 from Hong Kong Research Grant Council. It has also received funding from the European Union's Horizon 2020 research and innovation programme under grant agreement No 657347.

References

1. Aggarwal, C.C.: Recommender Systems: The Textbook. Springer, Switzerland (2016)
2. Ahmed, A., Hong, L., Smola, A.J.: Hierarchical geographical modeling of user locations from social media posts. In: WWW, pp. 25–36. ACM (2013)
3. Amer-Yahia, S., Roy, S.B., Chawlat, A., Das, G., Yu, C.: Group recommendation: semantics and efficiency. PVLDB **2**(1), 754–765 (2009)
4. Beckmann, N., Kriegel, H.-P., Schneider, R., Seeger, B.: The r*-tree: an efficient and robust access method for points and rectangles. In: SIGMOD, vol. 19, pp. 322–331. ACM (1990)
5. Blei, D.M., Ng, A.Y., Jordan, M.I.: Latent dirichlet allocation. JMLR **3**, 993–1022 (2003)
6. Borzsony, S., Kossmann, D., Stocker, K.: The skyline operator. In: ICDE, pp. 421–430. IEEE (2001)
7. Chen, L., Lian, X.: Dynamic skyline queries in metric spaces. In: EDBT, pp. 333–343. ACM (2008)
8. Deng, K., Zhou, X., Tao, H.: Multi-source skyline query processing in road networks. In: ICDE, pp. 796–805. IEEE (2007)
9. Fagin, R., Lotem, A., Naor, M.: Optimal aggregation algorithms for middleware. J. Comput. Syst. Sci. **66**(4), 614–656 (2003)
10. Gao, H., Tang, J., Liu, H.: gscorr: modeling geo-social correlations for new check-ins on location-based social networks. In: CIKM, pp. 1582–1586. ACM (2012)
11. Guttman, A.: R-trees: a dynamic index structure for spatial searching, vol. 14. ACM (1984)
12. Hjaltason, G.R., Samet, H.: Distance browsing in spatial databases. TODS **24**(2), 265–318 (1999)
13. Hong, L., Ahmed, A., Gurumurthy, S., Smola, A.J., Tsioutsiouliklis, K.: Discovering geographical topics in the twitter stream. In: WWW, pp. 769–778. ACM (2012)
14. Hu, B., Ester, M.: Spatial topic modeling in online social media for location recommendation. In: ACM RecSys, pp. 25–32. ACM (2013)
15. Jameson, A., Smyth, B.: Recommendation to groups. In: Brusilovsky, P., Kobsa, A., Nejdl, W. (eds.) The Adaptive Web. LNCS, vol. 4321, pp. 596–627. Springer, Heidelberg (2007). doi:10.1007/978-3-540-72079-9_20
16. Li, K., Lu, W., Bhagat, S., Lakshmanan, L.V., Yu, C.: On social event organization. In: SIGKDD, pp. 1206–1215. ACM (2014)
17. Li, M., Chen, L., Cong, G., Gu, Y., Yu, G.: Efficient processing of location-aware group preference queries. In: CIKM, pp. 559–568. ACM (2016)

18. Liu, X., Tian, Y., Ye, M., Lee, W.-C.: Exploring personal impact for group recommendation. In: CIKM, pp. 674–683. ACM (2012)
19. Liu, Y., Ester, M., Qian, Y., Hu, B., Cheung, D.W.: Microscopic and macroscopic spatio-temporal topic models for check-in data. TKDE (2017)
20. Lu, Z., Li, H., Mamoulis, N., Cheung, D.W.: Hbgg: a hierarchical Bayesian geographical model for group recommendation. In: SDM (2017)
21. O'connor, M., Cosley, D., Konstan, J.A., Riedl, J.: Polylens: a recommender system for groups of users. In: Prinz, W., Jarke, M., Rogers, Y., Schmidt, K., Wulf, V. (eds.) ECSCW 2001. Springer, Dordrecht (2001). doi:10.1007/0-306-48019-0_11
22. Omohundro, S.M.: Five balltree construction algorithms. International Computer Science Institute Berkeley (1989)
23. Papadias, D., Shen, Q., Tao, Y., Mouratidis, K.: Group nearest neighbor queries. In: ICDE, pp. 301–312. IEEE (2004)
24. Papadias, D., Zhang, J., Mamoulis, N., Tao, Y.: Query processing in spatial network databases. In: VLDB, pp. 802–813. VLDB Endowment (2003)
25. Ram, P., Gray, A.G.: Maximum inner-product search using cone trees. In: ACM SIGKDD, pp. 931–939. ACM (2012)
26. Roussopoulos, N., Kelley, S., Vincent, F.: Nearest neighbor queries. In: SIGMOD, vol. 24, pp. 71–79. ACM (1995)
27. Sharifzadeh, M., Shahabi, C.: The spatial skyline queries. In: VLDB, pp. 751–762. VLDB Endowment (2006)
28. Shi, J., Wu, D., Mamoulis, N.: Textually relevant spatial skylines. TKDE **28**(1), 224–237 (2016)
29. Son, W., Lee, M.-W., Ahn, H.-K., Hwang, S.: Spatial skyline queries: an efficient geometric algorithm. In: Mamoulis, N., Seidl, T., Pedersen, T.B., Torp, K., Assent, I. (eds.) SSTD 2009. LNCS, vol. 5644, pp. 247–264. Springer, Heidelberg (2009). doi:10.1007/978-3-642-02982-0_17
30. Wu, D., Cong, G., Jensen, C.S.: A framework for efficient spatial web object retrieval. PVLDB **21**(6), 797–822 (2012)
31. Yin, H., Sun, Y., Cui, B., Hu, Z., Chen, L.: Lcars: a location-content-aware recommender system. In: SIGKDD, pp. 221–229. ACM (2013)
32. Yin, H., Zhou, X., Shao, Y., Wang, H., Sadiq, S.: Joint modeling of user check-in behaviors for point-of-interest recommendation. In: CIKM, pp. 1631–1640. ACM (2015)
33. Yin, Z., Cao, L., Han, J., Zhai, C., Huang, T.: Geographical topic discovery and comparison. In: WWW, pp. 247–256. ACM (2011)
34. Yiu, M.L., Mamoulis, N., Papadias, D.: Aggregate nearest neighbor queries in road networks. TKDE **17**(6), 820–833 (2005)
35. Yuan, Q., Cong, G., Lin, C.-Y.: Com: a generative model for group recommendation. In: SIGKDD, pp. 163–172. ACM (2014)
36. Zou, L., Chen, L., Özsu, M.T., Zhao, D.: Dynamic skyline queries in large graphs. In: Kitagawa, H., Ishikawa, Y., Li, Q., Watanabe, C. (eds.) DASFAA 2010, Part II. LNCS, vol. 5982, pp. 62–78. Springer, Heidelberg (2010). doi:10.1007/978-3-642-12098-5_5

Data Mining

Grid-Based Colocation Mining Algorithms on GPU for Big Spatial Event Data: A Summary of Results

Arpan Man Sainju and Zhe Jiang$^{(\boxtimes)}$

Department of Computer Science, The University of Alabama,
Tuscaloosa, AL 35487, USA
asainju@crimson.ua.edu, zjiang@cs.ua.edu

Abstract. This paper investigates the colocation pattern mining problem for big spatial event data. Colocation patterns refer to subsets of spatial features whose instances are frequently located together. The problem is important in many applications such as analyzing relationships of crimes or disease with various environmental factors, but is computationally challenging due to a large number of instances, the potentially exponential number of candidate patterns, and high computational cost in generating pattern instances. Existing colocation mining algorithms (e.g., Apriori algorithm, multi-resolution filter, partial join and joinless approaches) are mostly sequential, and thus can be insufficient for big spatial event data. Recently, parallel colocation mining algorithms have been developed based on the Map-reduce framework. However, these algorithms need a large number of nodes to scale up, which is economically expensive, and their reducer nodes have a bottleneck of aggregating all instances of the same colocation patterns. Another work proposes a parallel colocation mining algorithm on GPU based on the iCPI tree and the joinless approach, but assumes that the number of neighbors for each instance is within a small constant, and thus may be inefficient when instances are dense and unevenly distributed. To address these limitations, we propose a grid-based GPU colocation mining algorithm that includes a novel cell aggregate based upper bound filter, and two refinement algorithms. We prove the correctness and completeness of proposed GPU algorithms. Preliminary results on both real world data and synthetic data show that proposed GPU algorithms are promising with over 30 times speedup on up to millions of instances.

1 Introduction

Given a set of spatial features and their instances, co-location mining aims to find subsets of features whose instances are frequently located together. Examples of colocation patterns include symbiotic relationships between species such as Nile Crocodiles and Egyptian Plover, as well as environmental factors and disease events (e.g., air pollution and lung cancer).

Societal applications: Colocation mining is important in many applications that aim to find associations between different spatial events or factors.

© Springer International Publishing AG 2017
M. Gertz et al. (Eds.): SSTD 2017, LNCS 10411, pp. 263–280, 2017.
DOI: 10.1007/978-3-319-64367-0_14

For example, in public safety, law enforcement agencies are interested in finding relationships between different crime event types and potential crime generators. In ecology, scientists analyze common spatial footprints of various species to capture their interactions and spatial distributions. In public health, identifying relationships between human disease and potential environmental causes is an important problem. In climate science, colocation patterns help reveal relationships between the occurrence of different climate extreme events. In location based service, colocation patterns help identify travelers that share the same favourite locations to promote effective tour recommendation.

Challenges: Mining colocation patterns from big spatial event data poses several computational challenges. First, in order to evaluate if a candidate colocation pattern is prevalent, we need to generate its instances. This is computationally expensive due to checking spatial neighborhood relationships between different instances, particularly when the number of instances is large and instances are clumpy (e.g., many instances are within the same spatial neighborhoods). Second, the number of candidate colocation patterns are exponential to the number of spatial features. Evaluating a large number of candidate patterns can be computationally prohibitive. Finally, the distribution of event instances in the space may be uneven, making it hard to design parallel data structure and algorithms.

Related work: Colocation pattern mining has been studied extensively in the literature, including early work on spatial association rule mining [1,2] and colocation patterns based on event-centric model [3]. Various algorithms have been proposed to efficiently identify colocation patterns, including Apriori generator and multi-resolution upper bound filter [3], partial join [4] and joinless approach [5], iCPI tree based colocation mining algorithms [6]. There are also works on identifying regional [7–9] or zonal [10] colocation patterns, and statistically significant colocation patterns [11–13], top-K prevalent colocation patterns [14] or prevalent patterns without thresholding [15]. Existing algorithms are mostly sequential, and can be insufficient when the number of event instances is very large (e.g., several millions). Recently, parallel colocation mining algorithms have been proposed based on the Map-reduce framework [16] to handle a large data volume. However, these algorithms need a large number of nodes to scale up, which is economically expensive, and their reducer nodes have a bottleneck of aggregating all instances of the same colocation patterns. Another work proposes a GPU based parallel colocation mining algorithm [17] using iCPI tree [18–20] and the joinless approach, but this method assumes that the number of neighbors for each instance is within a small constant (e.g., 32), and thus can be inefficient when instances are dense and unevenly distributed.

To address limitations of related work, we propose GPU colocation mining algorithms based on a grid index, including a cell-aggregate-based upper bound filter and two refinement algorithms. Proposed cell-aggregate-based filter is easier to implement on GPU and is also insensitive to pattern *clumpiness* (the average number of overlaying colocation instances for a given colocation instance) compared with the existing multi-resolution filter. We use a GPU platform due to its better energy efficiency and pricing compared to Map-reduce based clouds.

Contributions: We make the following contributions in the paper: (1) We designed and implemented parallel colocation mining algorithms on GPU, including a novel cell-aggregate based upper bound filter that is easier to implement on GPU and also insensitive to pattern clumpiness (i.e., number of colocation instances within the same neighborhood), two parallel refinement algorithms based on prefix-based HashJoin and grid search; (2) We provided theoretical analysis of the correctness and completeness of proposed algorithms; (3) We conducted extensive experimental evaluations on both real world and synthetic data with various parameter settings. Preliminary results show that proposed GPU algorithms are promising with over 30 times speedup on up to millions of instances.

Scope and outline: We focus on spatial colocation patterns defined by the event-centric models [3]. Other colocation definitions such as Voronoi diagram based are beyond our scope. We also assume the underlying space is Euclidean space. Also, in this paper, we are only concerned with the comparison of computational performance of various colocation mining algorithms.

The outline of the paper is as follows. Section 2 reviews basic concepts and the definition of the colocation mining problem. Section 3 introduces our proposed GPU colocation pattern mining algorithms, and analyzes the theoretical properties of algorithm correctness and completeness. Section 4 summarizes our experimental evaluation of proposed algorithms on both synthetic datasets and a real world dataset. Section 5 discusses memory bottleneck issues in our approach. Section 6 concludes the paper with potential future research directions.

2   Problem Statement

2.1   Basic Concepts

This subsection reviews some basic concepts based on which the colocation mining problem can be defined. More details on the concepts are in [3].

Spatial feature and instances: A *spatial feature* is a categorical attribute such as a crime event type (e.g., assault, drunk driving). For each spatial feature, there can be multiple *feature instances* at the same or different point locations (e.g., multiple instances of the same crime type "assault"). In the example of Fig. 1(a), there are three spatial features (A, B and C). For spatial feature A, there are three instances (A_1, A_2, and A_3). Two feature instances are *spatial neighbors* if their spatial distance is smaller than a threshold. Two or more instances form a *clique* if every pair of instances are spatial neighbors.

Spatial colocation pattern: If the set of instances in a clique are from different feature types, then this set of instances is called a *colocation (pattern) instance*, and the corresponding set of features is a *colocation pattern*. The *cardinality* or *size* of a colocation pattern is the number of features involved. For example, in Fig. 1(a), (A_1, B_1, C_1) is an instance of colocation pattern (A, B, C) with a size or cardinality of 3. If we put all the instances of a colocation pattern as different rows of a table, the table is called an *instance table*. For example, in Fig. 1(b), the instance table of colocation pattern (A, B) has three row instances, as shown in the third table of the bottom panel. A spatial colocation pattern is *prevalent* (significant) if its feature instances are *frequently* located within the same neighborhood cliques. In order to quantify the prevalence or frequency, an interestingness measure called participation index has been proposed [3].

The *participation ratio* of a spatial feature within a candidate colocation pattern is the ratio of the number of unique feature instances that participate in colocation instances to the total number of feature instances. For example, in Fig. 1, the participation ratio of B in candidate colocation pattern $\{A, B\}$ is $\frac{2}{3}$ since only B_1 and B_2 participate into colocation instances ($\{A_1, B_1\}, \{A_3, B_2\}$). The *participation index* (PI) of a candidate colocation pattern is the minimum of participation ratios among all member features. For example, the participation index of the candidate colocation pattern $\{A, B\}$ in Fig. 1 is the minimum of $\frac{3}{3}$ and $\frac{2}{3}$, and is thus $\frac{2}{3}$. We use "candidate colocation patterns" to refer to those whose participation index values are undecided.

2.2 Problem Definition

We now introduce the formal definition of colocation mining problem [3].

Given:

- A set of spatial features and their instances
- Spatial neighborhood distance threshold
- Minimum threshold of participation index: θ

Find:

- All colocation patterns whose participation index are above or equal to θ

Objective:

- Minimize computational time cost

Constraint:

- Spatial neighborhood relationships are defined in Euclidean space

Figure 1 provides a problem example. The input data contains 12 instances of 3 spatial features A, B, and C. The neighborhood distance threshold is d. The prevalence threshold is 0.6. The output prevalent colocation patterns include $\{A, B\}$ (participation index 0.67) and $\{B, C\}$ (participation index 0.67). Colocation mining is similar to association rule mining in market basket analysis [21], but is different in that there are no given "transactions" in continuous space. Generation colocation instance tables ("transactions") is the most computationally intensive part.

3 Proposed Approach

This section introduces our proposed GPU colocation mining algorithm. We start with the overview of algorithm structure, and then describe the main part for parallel algorithms implemented in GPU. We prove the correctness and completeness of proposed algorithm.

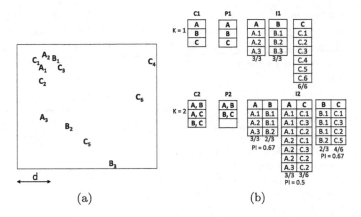

Fig. 1. A problem example with inputs and outputs. (a) Input spatial features and instances; (b) Candidate and prevalent colocation patterns, instance tables

3.1   Algorithm Overview

The overall structure of our algorithm is similar to the one proposed by Huang et al. in 2004 [3]. The novelty of our algorithm is that we design a novel upper-bound filter based on aggregated counts of feature instances in grid cells. Compared with the existing multi-resolution filter [3], our upper bound filter is easier to parallelize on GPU and does not rely on the assumption that colocation instances are clumpy into a small number of cells.

The overall structure of our algorithm is shown in Algorithm 1. The algorithm identifies all prevalent colocation patterns iteratively. Candidate colocation patterns and their instance tables of cardinality $k + 1$ are generated, based on prevalent patterns and their instance tables of cardinality k. Each candidate pattern of cardinality $k + 1$ is then evaluated based on the participation index computed from its instance table. For cardinality $k = 1$, prevalent colocation patterns simply consist of the set of input features, and their instance tables are the instance lists for each feature (step 1 in Algorithm 1). Step 2 generates candidate patterns C_{k+1} of size $k + 1$ based on prevalent patterns P_k using Apriori property [21] (i.e., a candidate pattern of size $k + 1$ cannot be prevalent and thus needs not to be generated if any subset pattern of size k is not prevalent). Step 4 builds a grid index on spatial point instances with the cell size equal to the distance threshold. Step 5 counts the number of instances for each feature in every cell. This will be used in our upper bound filter. Step 7 starts the iteration. As long as the set of candidate patterns C_{k+1} is not empty, the algorithm evaluate each candidate pattern $c \in C_{k+1}$. When evaluate a candidate pattern, the algorithm first computes an upper bound of its participation index in parallel using GPU kernels based on the grid index (step 9–10). If the upper bound is below the threshold, the candidate pattern is pruned out. Otherwise, the algorithm runs into a refinement phase, generating the pattern instance table $I_{k+1}.c$ and computing the participation index PI. We design two different parallel refinement algorithms to speed up instance table generation: one using the grid to rule out unnecessary joins, the other using prefix-based hash joins (steps 13 to 16). After all prevalent patterns of cardinality $k+1$ are identified, the algorithm go to the next iteration (steps 19–22). Figure 1(b) illustrates the execution trace for $k = 1$ and $k = 2$.

Algorithm 1. Parallel-Colocation-Miner

Input: A set of spatial features F
Input: Instances of each spatial features $I[F]$
Input: Neighborhood distance threshold d
Input: Minimum prevalence threshold θ
Output: All prevalent colocation patterns P

1: Initialize $P \leftarrow \emptyset$, $k \leftarrow 1$, $C_k \leftarrow F$, $P_k \leftarrow F$
2: Initialize $C_{k+1} \leftarrow$ APRIORIGEN(P_k, $k + 1$)
3: Initialize $P_{k+1} \leftarrow \emptyset$
4: Initialize instance tables I_k ($k = 1$) by feature instances
5: Overlay a regular grid with cell size $d \times d$ (total N cells)
6: Compute $CountMap[N \times |F|]$ in one round instance scanning
7: **while** $|C_{k+1}| > 0$ **do**
8:   **for each** $c \in C_{k+1}$ **do**
9:     Initialize $PCountMap[N \times |c|] \leftarrow 0$
10:     $Upperbound =$ PARALLELCELLAGGREGATEFILTER($CountMap,PCountMap,c$)
11:     **if** $Upperbound \geq \theta$ **then**
12:       $BitMap \leftarrow 0$ //initialize bitmap for instances of each feature
13:       **if** Hash Join Refinement **then**
14:         $(I_{k+1}.c, PI) \leftarrow$ PARALLELHASHJOINREFINE(I_k, c)
15:       **else if** Grid Search Refinement **then**
16:         $(I_{k+1}.c, PI) \leftarrow$ PARALLELGRIDSEARCHREFINE($I_k, CInstances, c, BitMap$)
17:       **if** $PI \geq \theta$ **then**
18:         $P_{k+1} = P_{k+1} \cup c$
19:   $P \leftarrow P \cup P_{k+1}$ //add prevalent patterns to results
20:   $k \leftarrow k + 1$; $C_k \leftarrow C_{k+1}$; $P_k \leftarrow P_{k+1}$, $I_k \leftarrow I_{k+1}$ //prepare next iteration
21:   $C_{k+1} \leftarrow$ APRIORIGEN(P_k, $k + 1$)
22:   $P_{k+1} \leftarrow \emptyset$
23: **return** P

3.2 Cell-Aggregate-Based Upper Bound Filter

The proposed cell aggregate based upper bound filter first overlays a regular grid with its cell size equal to the distance threshold (shown in Fig. 2), and then computes an upper bound of participation index based on aggregated counts of feature instances in cells. Proposed filter is different from the existing multi-resolution filter [3] in that the computation of upper bound is not based on generating coarse scale colocation instance tables. There are two main advantages of cell aggregate based filter on GPU: first, it is easily parallelizable and can leverage the large number of GPU cores; second, its performance does not rely on the assumption that pattern instances are clumpy into a small number of cells, which is required by the existing multi-resolution filter.

To introduce proposed cell aggregate based filter, we define a key concept of **quadruplet**. A quadruplet of a cell is a set of four cells, including the cell itself as well as its neighbors on the right, bottom, and right bottom. For a cell that is located on the right and bottom boundary of the grid, not all four cells exist and its quadruplet is defined empty (these cells will still be covered by other quadruplets). For example, in Fig. 2, the quadruplet of cell 0 includes cells $(0, 1, 4, 5)$, while the quadruplet of cell 15 is an empty set.

Based on the concept of quadruplet, we can check all potential colocation instances by examining all quadruplets. When examining a quadruplet, our filter computes the aggregated count of instances for every feature in the candidate pattern. If the aggregated count for any feature is zero, then there cannot exist colocation instances in the quadruplet. Otherwise, we pretend that all these feature instances participate into colocation pattern instances. This tends to over-estimate the participating instances of a colocation pattern (an "upper bound"), but avoids expensive spatial join operations.

Algorithm 2 shows details of proposed filter. The algorithms have three main variables, including *CountMap*, *PCountMap*, and *Quadruplet Aggregate*. *CountMap* records the true aggregated instance count for each feature in every cell. *PCountMap* records the instance count for each feature in every cell that potentially participates in the candidate colocation pattern. *QuadrupletAggregate* is a local array for each cell to record the aggregated count of instances within the quadruplet for each pattern feature. Specifically, steps 1 to 11 computes potential number of participating instances for each feature in each cell in parallel. A kernel thread is allocated to each cell. For a specific cell i, the kernel first gets the quadruplet (step 2). Step 3 initializes a local array *QuadrupletAggregate* with zero values. Steps 4 to 8 compute the aggregated count of instances for each pattern feature ($QuadrupletAggregate[f]$). If aggregated count of any pattern feature is zero, then there cannot be any candidate pattern instance in the quadruplet and thus the parallel kernel thread terminates (step 8). Otherwise, all feature instances in the quadruplet can potentially participate into colocation pattern instances. Steps 9 to 11 record the potential participating instances from each cell in the quadruplet. This is done by copying instance counts from *CountMap* to *PCountMap* for the 4 cells in the quadruplet. It is worth noting that duplicated-counting on the same cell is avoided since different GPU kernel threads may over-write the count for a cell with the same value. Finally, steps 12 to 14 compute the upper bound of participation index based on counts of potential participating instances in *PCountMap*. We use built in GPU library to compute the total counts of distinct participating instances in step 13.

Figure 2 provides an illustrative execution trace of Algorithm 2. Figure 2(a) shows the input spatial instances overlaid with a regular grid. The distance threshold is d. Assume that the candidate colocation pattern is (A, B). Figure 2(b) shows how the filter works. The *CountMap* array stores the

---

Algorithm 2. ParallelCellAggregateFilter

Input: $CountMap$, feature instance count in cells
Input: $PCountMap$, participating feature instance count in cells
Input: PR, participation ratio
Input: Candidate colocation pattern c
Output: Upper bound of participation index, $upperBound$

1: **for each** cell i **do in parallel**
2:     $QuadrupletCells =$ GETQUADRUPLET(cell i)
3:     $QuadrupletAggregate[\|c\|] \leftarrow 0$
4:     **for each** feature $f \in c$ **do**
5:         **for each** cell $j \in QuadrupletCells$ **do**
6:             $QuadrupletAggregate[f] \leftarrow QuadrupletAggregate[f] + CountMap[j][f]$
7:         **if** $QuadrupletAggregate[f] == 0$ **then**
8:             **finish** the parallel thread for cell i //no pattern instance in the quadruplet
9:     **for each** feature $f \in c$ **do**
10:        **for each** cell $j \in QuadrupletCells$ **do**
11:            $PCountMap[j][f] \leftarrow CountMap[j][f]$ //participating instance count
12: **for each** feature $f \in c$ **do**
13:     $PR[f] =$ PARALLELSUM($CountMap[\ ][f])/|I_1.f|$
14: $upperBound =$ MIN(PR)
15: **return** $upperBound$

---

number of instances for feature A and B in each cell. A GPU thread is assigned to each cell to compute the counts of feature instances within its quadruplet. For example, the leftmost GPU thread is assigned to cell 0. The aggregated instance count for this quadruplet $((0, 1, 4, 5))$ is shown by the leftmost $QuadrupletCount$ array, with 2 instances for A and 1 instance for B. Since instances from both features exist, the number potential participating instances in these four cells ($PCountMap$) are copied from corresponding cell values in $CountMap$, as shown by the fork branches close to the bottom. In contrast, the quadruplet of cell 1 $((1, 2, 5, 6))$ does not contain instances of A, and thus cannot contain colocation pattern instances.

Lemma 1. *The participation index of a colocation pattern in the cell-aggregate-based filter is an upper bound of the true participation index value.*

Proof. The proof is based on the following fact. We create an upper bound to the true number of neighboring points in neighboring cells (quadruplet) by assuming that all pairs of points of neighboring cells are within the distance threshold, which coincides with the cell size. Of course, some of them will not, but it is impossible for points not within neighboring cells to be neighboring with respect to the distance threshold.   □

Theorem 1. *The cell aggregate based upper bound filter is correct and complete.*

Proof. The proof is based on Lemma 1. The algorithm is complete (it does not mistakenly prune out any prevalent pattern) due to the upper bound property.

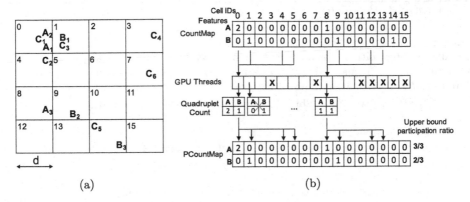

Fig. 2. Grid-aggregate based Upper Bound Filter: (a) A regular grid (b) An execution trace of upper bound filter

The algorithm is correct since it computes the exact participation index of a candidate pattern if it passes the upper bound filter.   □

3.3   Refinement Algorithms

The goal of the refinement phase is to generate the instance table of a candidate colocation pattern, and to compute participation index. Generating colocation instance tables is the main computational bottleneck, and thus is done in GPU. As shown in Algorithm 1, we have two options for refinement algorithms, a geometric approach based on grid search called ParallelGridSearchRefine and a combinatorics approach based on prefix-based hash join called ParallelHashJoin-Refine, similar to sequential algorithms discussed in Huang et al. [3]. We now introduce the two algorithms below.

Geometric approach: The geometric approach generate an instance table of a size $k + 1$ pattern based on the instance table of a size k pattern. For example, when generating the instance table of pattern (A, B, C), it starts from the instance table of (A, B) and joins each row of the table with instances of the last feature type C. In order to reduce redundant computation, we utilize the grid index and only check the instances of the last feature type within neighboring cells. Algorithm 3 provides details of the proposed grid-based refinement algorithm. Step 2 is kernel assignment. Each kernel thread first finds out all neighboring cells of the size k row instance $rIns$ (step 3). Then, for each neighboring cell, the kernel thread finds out every instance ins of the last feature type in the cell. It joins ins with size k instance $rIns$ to create a size $k + 1$ pattern instance $rInsC$ if they are spatial neighbors (steps 4 to 7). The new pattern instance $rInsC$ is inserted into the final instance table $I_{k+1}.c$, and a bitmap is updated to mark the instances that participate into the colocation pattern (steps 8 to 10). Finally, the participation ratio and participation index are computed (steps 11 to 13).

Algorithm 3. *ParallelGridSearchRefine*

Input: I_k, instance table of patterns of size k
Input: *CellInstances*, feature instances for each cell
Input: *BitMap*, bitmap for participating instances from different features
Output: $I_{k+1}.c$, instance table of colocation c (size $k+1$) if prevalent
Output: PI, participation index of pattern c

1: $//I_k.(c[1..k])$ is instance table of sub-pattern of c with first k features
2:
3: Initialize $I_{k+1}.c \leftarrow \emptyset$
4: **for each** row instance $rIns \in I_k.(c[1..k])$ **do in parallel**
5:      get *neighborhood* cells of first feature instance in $rIns$
6:      **for each** cell i in neighborhood **do**
7:          **for each** instance ins of feature type $c[k+1]$ in cell i **do**
8:              **if** ins is neighbor of all feature instances in $rIns$ **then**
9:                  Create new row instance of c, $rInsC = < rIns, ins >$
10:                 $I_{k+1}.c = I_{k+1}.c \cup rInsC$ //add new instance into c's instance table
11:                 **for each** feature $f \in c$ **do**
12:                     $BitMap[f][rInsC[f]] =$ true
13: **for each** feature $f \in c$ **do**
14:     $PR[f] =$ PARALLELSUM$(BitMap[\ ][f])/|I_1.f|$
15: $PI =$ MIN(PR)
16: **return** $I_{k+1}.c$, PI

Figure 3 shows an example. The input data is the same as Fig. 1. Assume that the candidate pattern is (A, B, C). A kernel thread is assigned to each row instance of table (A, B). Thread 1 is assigned to instance (A_1, B_1), and it scans all neighboring cells of A_1 (cells $0, 1, 4, 5$). Based on the cell to instance index, the kernel thread checks all instances of feature C (C_1, C_3, C_2) in these cells, and conducts a spatial join operation. The final output size $k+1$ instances from this thread are (A_1, B_1, C_1), (A_1, B_1, C_3), and (A_1, B_1, C_2).

One issue in GPU implementation is that we need to allocate memory for an output instance table, and specify the specific memory location to which each kernel thread writes its results. For example, in the output instance table of pattern (A, B, C) in Fig. 3, the first kernel thread generates 3 row instances, so

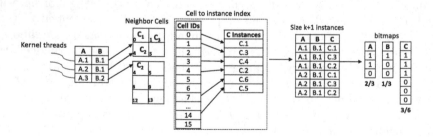

Fig. 3. Illustrative execution trace for grid-based refinement

the second kernel thread has to start with the $4th$ row when writing its instances. It is hard to predetermine the total required memory and enforcing memory coalesce when threads are writing results. Thus, we use a two-run strategy in which in the first run we can calculate the exact size of output instance table as well as slot counts of the number of row instances generated by each kernel thread. In the second run, we allocate memory for output instance table, and use the slot counts to guide which row a kernel thread needs to start from when writing results. Similar to the grid-based refinement, we use two-run strategy to allocate memory and enforce memory coalesce.

Algorithm 4. $ParallelHashJoinRefine$

Input: I_k, instance table of patterns of size k
Input: $BitMap$, bitmap for instances of different features
Output: $I_{k+1}.c$, instance table of colocation c if prevalent
Output: PI, participation index of pattern c
1: $//I_k.c1$ and $I_k.c2$ instance tables of $c1 = c[1..k]$ and $c2 = c[1..k-1, k+1]$
2: **for each** row instance $rIns1$ in $I_k.c1$ **do in parallel**
3:    **for each** row instance $rIns2$ in $I_k.c2$ starting with $rIns1[1]$ **do**
4:      **if** $rIns1$ and $rIns2$ forms an instance of c **then**
5:        Create new instance $rInsC$ by merging $rIns1$ and $rIns2$
6:        $I_{k+1}.c \leftarrow I_{k+1}.c \cup rInsC$
7:        **for each** feature $f \in c$ **do**
8:           $BitMap[f].[rInsC[f]] = $ true
9: **for each** feature $f \in c$ **do**
10:   $PR[f] = \text{PARALLELSUM}(BitMap[\ ][f])/|I_1.f|$
11: $PI = \text{MIN}(PR)$
12: **return** $I_{k+1}.c, PI$

Prefix-based hash Join based refinement: Another option is to generate size $k+1$ instance table by a combinatorics approach. For example, when generating instance table of pattern (A, B, C), we can join rows in instance tables of (A, B) and (A, C). The join condition is that the first k instances from the two tables should be the same, and the last instances from two tables should be spatial neighbors. For example, when joining a row (A_1, B_1) with another row (A_1, C_1), we check that the first instance is the same (A_1), and the last instances B_1 and C_1 are spatial neighbors. So these two rows are joined to form a new row instance (A_1, B_1, C_1). In sequential implementations [3], the join process can be done efficiently through sort-merge join. However, for GPU algorithm, sort merge is difficult due to the order, dependency and multi-attribute keys. We choose to use hash join instead. A prefix-based hash index is built on the second table based on instances of the first spatial feature. Details are shown in Algorithm 4. A kernel thread is allocated to each row in the first size k instance table $I_k.c1$ (step 2). The kernel thread then scans all rows in the second size k instance table $I_k.c2$ that has the same first feature instance. For example, if the row in the first table is (A_1, B_1), then the thread only scans rows starting with A_1 in the instance table of (A, C). If the two rows satisfy the join condition (sharing the same first k instances, and having last instances as neighbors), a size $k + 1$ instance is created and inserted into output size $k + 1$ table (steps 5 to 8). Finally, the participation index is computed (steps 9 to 11). It is worth noting that when generating instance tables of size $k = 2$ patterns, we use the grid-based method since hash-index cannot be created in that case.

An illustrative execution trace is shown in Fig. 4. The raw input data is still the same. Each kernel thread is allocated to a row in instance table (A, B). For example, thread 1 works on pattern instance (A_1, B_1), and scans instance table (A, C). Based on the hash index on A instances, the thread only needs to check (A_1, C_1), (A_1, C_3) and (A_1, C_2). It turns out that B_1 is a neighbor for all C_1, C_2 and C_3. So these instances are inserted to the final output instance table (A, B, C).

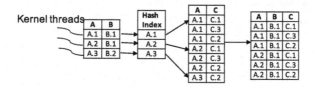

Fig. 4. Illustrative execution trace for hash-join-based refinement

4   Evaluation

The goals of our evaluation are to:

- Evaluate the speedup of GPU colocation algorithms against a CPU algorithm.
- Compare cell-aggregate-based filter with multi-resolution filter on GPU.
- Compare grid-based refinement with hash-join-based refinement on GPU.
- Test the sensitivity of GPU algorithms to different factors.

Experiment Setup: As shown in Fig. 5, we implemented four GPU colocation mining algorithms with two filter options (**M** for multi-resolution filter and **C** for cell-aggregate based filter) and two refinement options (**G** for grid-based and **H** for hash-join based). We also implemented a CPU colocation mining algorithms by Huang et al. [3] (multi-resolution filter, grid-based instance table generation for size $k = 2$, and sort-merge based instance table generation for size $k > 2$). We only compared computational performance since all methods produce the same patterns. For each experiment, we measured the time cost of one run for CPU algorithm, and averaged time cost of 10 runs for GPU algorithms. Algorithms were implemented in C++ and CUDA, and run on a Dell workstation with Intel(R) Xeon(R) CPU E5-2687w v4 @ 3.00 GHz, 64 GB main memory, and a Nvidia Quadro K6000 GPU with 2880 cores and 12 GB memory.

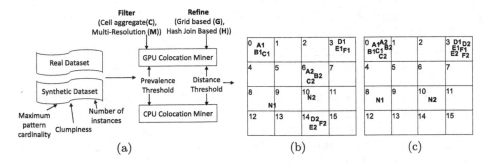

Fig. 5. Experiment setup (a) Experiment design with different candidate approaches; (b) An example of synthetic dataset generated with 2 maximal patterns (A, B, C) and (D, E, F), each pattern with 2 instances with a clumpiness of 1, 2 noise instances N_1 and N_2; (c) Another synthetic dataset similar to (b) but with a clumpiness of 2.

Dataset description: The real dataset contains 13 crime types and 165,000 crime event instances from Seattle in 2012 [22]. The synthetic data is generated similarly to [3]. Figure 5(b–c) provide illustrative examples. We first chose a study area size of 10000×10000, a neighborhood distance threshold (also the size of a grid cell) of 10, a maximal pattern cardinality of 5, and the number of maximal colocation patterns as 2. The total number of features was 12 (5×2 plus 2 additional noise features). We then generated a *number of instances* for each maximal colocation pattern. Their locations were randomly distributed to different cells according to the **clumpiness** (i.e., the number of overlaying colocation instances within the same neighborhood, higher clumpiness means larger instance tables). In our experiments, we varied the number of instances and clumpiness to test sensitivity.

Evaluation metric: We used the speedup of proposed GPU algorithms over the CPU algorithm on computational time.

4.1   Results on Synthetic Data

Effect of the Number of Instances. We conducted this experiment with two different parameter settings. For both settings, the minimum participation index threshold was 0.5. In the first setting, we set the clumpiness to 1 (very low clumpiness), and varied the number of feature instances as 250,000, 500,000, 1,000,000, 1,500,000 and 2,000,000. Results are summarized in Fig. 6(a). GPU algorithms in the plot are based on grid-based filtering. We can see that the speedup of both GPU algorithms increases with the number of feature instances. The grid-based refinement gradually becomes superior over the hash join based refinement in GPU algorithms as the number of instances increases. The reason can be that the cell-instance index in grid-based refinement is done once and for all, while the prefix-based hash index in hash-join based refinement needs to be created repeatedly for each new instance table. The comparison of two approaches with 250,000 instances (the first two points in the curve) may be less conclusive since the running time for both approaches is too small (far below one second).

In the second setting, we set the clumpiness value as 20, and varied the number of feature instances as 50,000, 100,000, 150,000, 200,000, and 250,000. The number of feature instances were set smaller in this setting due to the fact that given the same number of feature instances, a higher clumpiness value results in far more colocation pattern instances (see Fig. 5(b) versus (c)) but we only have limited memory. The results are summarized in Fig. 6(b). We can see that the grid based refinement is persistently better than hash-join based refinement (around 30 versus 5). The reason is that when the clumpiness is high, there are a large number of pattern instances being formed combinatorially. Many of them share the same prefix (i.e., first feature instance). Thus, each GPU kernel thread in prefix-based hash-join refinement was loaded with heavy computation when doing the join operation, impacting the parallel performance. In contrast, in the grid-based refinement, each GPU kernel thread only scans a limited number of instances within neighboring cells.

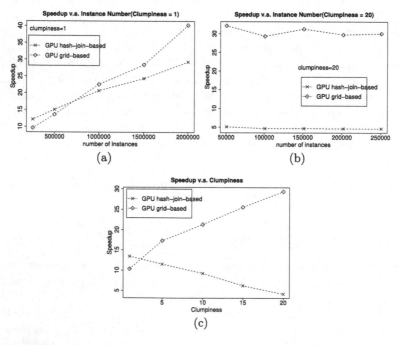

Fig. 6. Results on synthetic datasets: (a) effect of the number of instances with clumpiness as 1 (b) effect of the number of instances with clumpiness as 20 (c) effect of clumpiness with the number of instances as 250 k

Effect of Clumpiness. We set the number of instances to 250 k, and the prevalence threshold to 0.5. Grid-based filtering was used for GPU algorithms. We varied the clumpiness value as 1, 5, 10, 15, and 20. Results in Fig. 6(c) confirm with our analysis above that when clumpiness is higher, the performance of grid-based refinement gets better while the performance of hash-join based refinement gets worse.

Table 1. Comparison of filter and refinement on synthetic data (time in secs)

Clumpiness	Approaches	Filter time	Refine time	Total time	Speedup
1	CPU Baseline	15.3	18.8	34.1	-
	GPU-Filter:M, Refine:H	0.8	1.1	1.9	17.9x
	GPU-Filter:C, Refine:G	**0.2**	**0.7**	**0.9**	**37.9x**
	GPU-Filter:C, Refine:H	**0.2**	**1.0**	**1.2**	**28.4x**
20	CPU Baseline	0.9	407.5	408.4	-
	GPU-Filter:M, Refine:H	**0.1**	97.3	97.4	4.2x
	GPU-Filter:C, Refine:G	**0.1**	**13.8**	**13.9**	**29.4x**
	GPU-Filter:C, Refine:H	**0.1**	96.9	97	4.2x

Comparison on Filter and Refinement. We also compared the computational time of filter and refinement phases of GPU algorithms in the above experiments for the cases with the largest number of instances. Details are summarized in Table 1. When clumpiness is 1, the grid-based filter is much faster than the multi-resolution filter in GPU algorithms (0.2 s versus 0.8 s), making the overall GPU speedup better (37.9 and 28.4 times versus 17.9 times). The reason is that a low clumpiness significantly impacts the multi-resolution filter (coarse scale instance tables cannot be much smaller than true instance tables), while the grid-based filter was less sensitive to clumpiness (more robust) since the time cost of grid-based filtering does not depend on instance distribution. When clumpiness is 20, the refinement phase becomes the bottleneck. The grid-based refinement has a significantly higher speedup than the hash-join refinement (29.4 times versus 4.2 times).

4.2   Results on Real World Dataset

Effect of Minimum Participation Index Threshold. We fixed the distance threshold as 10 meters and varied the prevalence thresholds from 0.3 to 0.9 (we did not chose thresholds lower than 0.3, because there would be too many instance tables exceeding our memory capacity). The clumpiness of the real dataset was high due to a large density of points. Results are summarized in Fig. 7. As we can see, as the prevalence threshold gets higher, the pruning ratio (candidate patterns being pruned out) gets improved (Fig. 7(a)). The GPU algorithm with grid based refinement is much better than the GPU algorithm

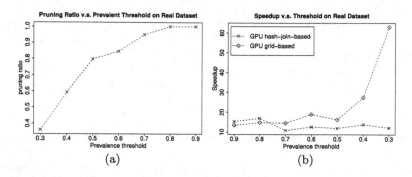

Fig. 7. Results on real world dataset: (a) pruning ratio versus prevalence thresholds (b) speedup versus prevalence thresholds

with hash join based refinement. This is consistent with the results on synthetic datasets when the clumpiness is high.

Comparison of Filter and Refinement. We also compared the detailed computational time in the filter and refinement phases. The distance threshold was 10 meters, and the prevalent threshold was 0.3. Results are summarized in Table 2. Due to a high clumpiness, the refinement phase is the bottleneck, and the grid-based refinement is better than the hash-join based refinement (63.2 times overall speedup versus 12.7 times overall speedup).

Table 2. Comparison of filter and refinement on real dataset (time in secs)

Approaches	Filter time	Refine time	Total time	Speedup
CPU baseline	6.5	340.9	347.4	-
GPU-Filter:M, Refine:H	0.2	26	26.8	13x
GPU-Filter:C, Refine:G	**0.5**	**5.0**	**5.5**	**63.2x**
GPU-Filter:C, Refine:H	**0.5**	26.9	27.4	12.7x

5   Discussion

Preliminary results show that GPU algorithms are promising for the colocation mining problem. One limitation of the proposed GPU algorithm is memory bottleneck. Our algorithm generates instance tables of candidate colocation patterns in GPU. When spatial points are dense and the number of points is large (e.g., millions), such instance tables can reach gigabytes in size. In our implementation, we generate one instance table each time in GPU global memory, and transfer the results to the host in pinned memory. Results showed that the time cost of memory copy was significantly lower (due to pinned memory) than the time cost of GPU computation. In case that the GPU global memory is insufficient for a very large instance table, we can slice it into smaller pieces, compute one piece each time, and transfer results to the host memory. We need to store all relevant instance tables of cardinality k in host memory when computing instance tables of cardinality $k + 1$. Thus, the host memory size can also be a bottleneck. This may be less a concern in future when the main memory price gets lower.

6   Conclusion and Future Work

This paper investigates GPU colocation mining algorithms. We propose a novel cell-aggregate-based upper bound filter, which is easier to implement on GPU and less sensitive to data clumpiness compared with the existing multi-resolution filter. We also design two GPU refinement algorithms, based on grid-based search and prefix based hash-join. We provide theoretical analysis on the correctness and completeness of proposed algorithms. Preliminary results on both real world data and synthetic data on various parameter settings show that proposed GPU algorithms are promising.

In future work, we will explore further refinements on GPU implementation to achieve higher speedup, e.g., avoid redundant distance computation in instance table generation. We will also explore other computational pruning methods.

Acknowledgement. We would like to thank Dr. Simin You for helpful comments and suggestions.

References

1. Koperski, K., Han, J.: Discovery of spatial association rules in geographic information databases. In: Egenhofer, M.J., Herring, J.R. (eds.) SSD 1995. LNCS, vol. 951, pp. 47–66. Springer, Heidelberg (1995). doi:10.1007/3-540-60159-7_4
2. Morimoto, Y.: Mining frequent neighboring class sets in spatial databases. In: Proceedings of the Seventh ACM SIGKDD International Conference on Knowledge Discovery and Data Mining, pp. 353–358. ACM (2001)
3. Huang, Y., Shekhar, S., Xiong, H.: Discovering colocation patterns from spatial data sets: a general approach. IEEE Trans. Knowl. Data Eng. **16**(12), 1472–1485 (2004)
4. Yoo, J.S., Shekhar, S., Smith, J., Kumquat, J.P.: A partial join approach for mining co-location patterns. In: Proceedings of the 12th Annual ACM International Workshop on Geographic Information Systems, pp. 241–249. ACM (2004)
5. Yoo, J.S., Shekhar, S.: A joinless approach for mining spatial colocation patterns. IEEE Trans. Knowl. Data Eng. **18**(10), 1323–1337 (2006)
6. Boinski, P., Zakrzewicz, M.: Collocation pattern mining in a limited memory environment using materialized iCPI-tree. In: International Conference on Data Warehousing and Knowledge Discovery, pp. 279–290. Springer, Heidelberg (2012)
7. Mohan, P., Shekhar, S., Shine, J.A., Rogers, J.P., Jiang, Z., Wayant, N.: A neighborhood graph based approach to regional co-location pattern discovery: A summary of results. In: Proceedings of the 19th ACM SIGSPATIAL International Conference on Advances in Geographic Information Systems, pp. 122–132. ACM (2011)

8. Wang, S., Huang, Y., Wang, X.S.: Regional co-locations of arbitrary shapes. In: Nascimento, M.A., Sellis, T., Cheng, R., Sander, J., Zheng, Y., Kriegel, H.-P., Renz, M., Sengstock, C. (eds.) SSTD 2013. LNCS, vol. 8098, pp. 19–37. Springer, Heidelberg (2013). doi:10.1007/978-3-642-40235-7_2

9. Liu, B., Chen, L., Liu, C., Zhang, C., Qiu, W.: RCP mining: towards the summarization of spatial co-location patterns. In: Claramunt, C., Schneider, M., Wong, R.C.-W., Xiong, L., Loh, W.-K., Shahabi, C., Li, K.-J. (eds.) SSTD 2015. LNCS, vol. 9239, pp. 451–469. Springer, Cham (2015). doi:10.1007/978-3-319-22363-6_24

10. Celik, M., Kang, J.M., Shekhar, S.: Zonal co-location pattern discovery with dynamic parameters. In: Seventh IEEE International Conference on Data Mining. ICDM 2007, pp. 433–438. IEEE (2007)

11. Barua, S., Sander, J.: SSCP: mining statistically significant co-location patterns. In: Pfoser, D., Tao, Y., Mouratidis, K., Nascimento, M.A., Mokbel, M., Shekhar, S., Huang, Y. (eds.) SSTD 2011. LNCS, vol. 6849, pp. 2–20. Springer, Heidelberg (2011). doi:10.1007/978-3-642-22922-0_2

12. Barua, S., Sander, J.: Mining statistically significant co-location and segregation patterns. IEEE Trans. Knowl. Data Eng. **26**(5), 1185–1199 (2014)

13. Barua, S., Sander, J.: Statistically significant co-location pattern mining (2015)

14. Yoo, J.S., Bow, M.: Mining top-k closed co-location patterns. In: 2011 IEEE International Conference on Spatial Data Mining and Geographical Knowledge Services (ICSDM), pp. 100–105. IEEE (2011)

15. Huang, Y., Xiong, H., Shekhar, S., Pei, J.: Mining confident co-location rules without a support threshold. In: Proceedings of the 2003 ACM Symposium on Applied Computing, pp. 497–501. ACM (2003)

16. Yoo, J.S., Boulware, D., Kimmey, D.: A parallel spatial co-location mining algorithm based on mapreduce. In: 2014 IEEE International Congress on Big Data (BigData Congress), pp. 25–31. IEEE (2014)

17. Andrzejewski, W., Boinski, P.: GPU-accelerated collocation pattern discovery. In: Catania, B., Guerrini, G., Pokorný, J. (eds.) ADBIS 2013. LNCS, vol. 8133, pp. 302–315. Springer, Heidelberg (2013). doi:10.1007/978-3-642-40683-6_23

18. Andrzejewski, W., Boinski, P.: A parallel algorithm for building iCPI-trees. In: Manolopoulos, Y., Trajcevski, G., Kon-Popovska, M. (eds.) ADBIS 2014. LNCS, vol. 8716, pp. 276–289. Springer, Cham (2014). doi:10.1007/978-3-319-10933-6_21

19. Yoo, J.S., Boulware, D.: Incremental and parallel spatial association mining. In: 2014 IEEE International Conference on Big Data (Big Data), pp. 75–76. IEEE (2014)

20. Andrzejewski, W., Boinski, P.: Parallel GPU-based plane-sweep algorithm for construction of iCPI-trees. J. Database Manage. (JDM) **26**(3), 1–20 (2015)

21. Agrawal, R., Srikant, R., et al.: Fast algorithms for mining association rules. In: Proceedings of 20th International Conference on Very Large Databases, VLDB. vol. 1215, pp. 487–499 (1994)

22. City of Seattle, Department of Information Technology, Seattle Police Department: Seattle Police Department 911 Incident Response. https://data.seattle.gov/Public-Safety/Seattle-Police-Department-911-Incident-Response/3k2p-39jp

Detecting Isodistance Hotspots on Spatial Networks: A Summary of Results

Xun Tang$^{(\boxtimes)}$, Emre Eftelioglu, and Shashi Shekhar

Department of Computer Science and Engineering,
University of Minnesota, Minneapolis, MN 55455, USA
xuntang@cs.umn.edu

Abstract. Spatial hotspot detection aims to find regions of interest with statistically significant high concentration of activities. In recent years, it has presented significant value in many critical application domains such as epidemiology, criminology and transportation engineering. However, existing spatial hotspot detection approaches focus on either on Euclidean space or are unable to find the entire set of hotspots. In this paper, we first formulate the problem of Network Isodistance Hotspot Detection (NIHD) as finding all sub-networks whose nodes and edges are reachable from a activity center and have significantly high concentration of activities. Then, we propose a novel algorithm based on network partitioning and pruning (NPP) which overcomes the computational challenges due to the high costs from candidate enumeration and statistical significance test based on randomization. Theoretical and experimental analysis show that NPP substantially improves the scalability over the baseline approach while keeping the results correct and complete. Moreover, case studies on real crime datasets show that NPP detects hotspots with higher accuracy and is able to reveal the hotspots that are missed by existing approaches.

1 Introduction

Spatial hotspot detection aims to find regions of interest with statistically significantly high concentration of activities (e.g., crimes). Different from traditional density based clustering algorithms such as DBScan [1], it utilizes statistical significance test to prevent false positives caused by chance patterns which cannot be tolerated in many critical applications such as criminology, epidemiology, and transportation engineering. For example, a chance crime hotspot may cause unnecessary panic and decrease of housing price. In this paper, we study the problem of Network Isodistant Hotspot Detection (NIHD) which reveals all network isodistance hotspots (NIHs) formed by activities diffused isotropically (i.e., towards all directions with same distance) along the network.

Application Domain: Recently, hotspot detection has been extensively applied in environmental criminology. Being aware of the crime hotspots, police officers are able to deploy their force and suppress the crime rates efficiently. It is stated that most crimes spread from a source point within a travel distance [2,3].

© Springer International Publishing AG 2017
M. Gertz et al. (Eds.): SSTD 2017, LNCS 10411, pp. 281–299, 2017.
DOI: 10.1007/978-3-319-64367-0_15

For example, midnight street robbery cases typically form spreading hotspot patterns from bars. Network isodistance hotspot (NIH), which precisely captures the travel distance, is designed to model such patterns.

In epidemiology and public health, hotspot detection also has been used for analyzing the outbreak and diffusion of diseases. Epidemiology studies show that victims of a disease usually form a spreading pattern from the source of infection [4]. For example, HIV is indicated to be usually spreading along important roads [5]. NIHD models such disease diffusion patterns, thereby helps epidemiology and public health experts detect and control disease outbreaks.

In transportation engineering, one of the most critical problems is how to mitigate traffic congestion. A severe congestion is usually initiated by a "generator" such as the end of a sport event [6], then diffuses along the roads. However, most generators are hard to predict (known as phantom traffic jams) since they may be as simple as an aggressive driver forcing others to make hard brakes. NIHD can be used to capture the diffusion of congestion and reveal the hidden generators, and thus help improve the traffic conditions.

Related Work: Traditionally, density based clustering algorithms [1,7,8] are used to detect regions with high concentration of activities. However, due to lacking a statistical significance test, those may output false positive chance hotspots. For example, given a set of points following complete spatial randomness, DBScan [1] will still output some clusters. In some critical application domains, chance hotspots may lead to unacceptable societal and economic consequences (e.g., a crime hotspot may cause panic and decrease of housing price).

In order to eliminate chance pattern, approaches with statistical significance test detect hotspots on Euclidean space with a variety of shapes (e.g., circular and elliptical [9,10], rectangular [11], and ring-shaped [12]). However, using Eulidean distance to model the distance between two points is inaccurate and causes bias since human activities usually diffuse along road networks [13], especially in urban areas.

Recently, in order to overcome the limitation using Euclidean distance, an emerging collection of hotspot detection approaches which leverage the spatial networks were proposed. Network clumping method [14] finds clusters of activities on a network whose distances to each other are smaller than a user-specified global threshold. However, the activities in a hotspot are not necessarily closer than a threshold, also a global threshold does not apply to the whole dataset since the activities are always distributed inhomogenously. Some approaches are able to find significant sub-networks with or without shape constraints (e.g., tree) [15,16]. However, they return only one hotspot that has the maximum score of concentration, making them incapable for handling multiple hotspots existed in the study area. Also, there are approaches that focus on finding the complete set of hotspots. However, the results are limited to linear paths [17] (i.e., individual paths or road segments with significantly high concentrations of activities), which does not capture the hotspots formed by isotropically diffusing towards multiple directions.

Challenges: NIHD is a computationally challenging task due to the following aspects. First, number of isodistance sub-networks are quadratic to the numbers of nodes and edges in the network. In addition, monotonicity does not hold for test statistic (e.g., log likelihood ratio) in this problem. Therefore, monotonicity based pruning techniques (e.g., Apriori [18]) are inapplicable for this problem. Futhermore, the edges in an isodistance sub-network are not necessarily within any shortest paths from the diffusion center. Therefore, searching algorithms (e.g., breadth-first-search, depth-first-search) on shortest path tree cannot be directly applied to identify the edges in each isodistance sub-network. Finally, Monte Carlo simulations is needed for the statistical significance test, which multiplies the cost.

Contributions: This paper formulates the network isodistance hotspots detection (NIHD) problem which is important to many application fields such as criminology, epidemiology and transportation engineering. To solve this computationally challenging problem effectively and efficiently, this paper proposes a novel network partition and pruning algorithm (NPP). Both the theoretical and experimental analysis show that NPP substantially improves scalability over the baseline approach. A case study on real world crime data demonstrates that NPP is able to detect crime hotspots that are missed by existing approaches.

Scope and organization of the paper: This paper only considers isodistance hotspots on the spatial network (e.g., road network). Isodistance polygons or curves are out of the scope. The thresholds of test statistics are set by domain experts and are out of the scope. Also, hotspots found by the proposed approach are to be interpreted by domain experts and the analysis from the domain perspective is beyond the scope.

The rest of this paper is organized as follows: Sect. 2 introduces the basic concepts and formulates the NIHD problem. Sections 3 and 4 propose a baseline algorithm (BaseNIHD) and an advanced algorithm (NPP), respectively. Section 5 provides theoretical proofs and asymptotic cost analysis of the proposed algorithms. Case studies on real datasets are discussed in Sect. 6. In Sect. 7, experiments were conducted to evaluate the proposed algorithms in terms of scalability. Section 8 concludes the paper and previews the future work.

2   Problem Statement

In this section, we formulate the problem of Network Isodistance Hotspot Detection (NIHD). We start with the definitions and examples of some basic concepts, and then give the formal definition of NIHD.

2.1   Basic Concepts

Definition 1. *A **spatial network** $G = (N, E)$ consists of a node set N and an edge set $E \in N \times N$. Each node n in N is associated with a 2-D spatial location. Each element e_{n_A, n_B} in E connects node n_A and node n_B. The example spatial network shown in Fig. 1 consists of 16 nodes and 21 edges, modeling roads and their intersections.*

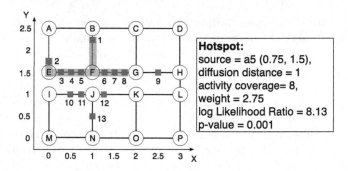

Fig. 1. An illustrative example of NIH

Definition 2. *An **activity set** A is a collection of activities (e.g., crime, disease cases) located on spatial network G. Each activity $a \in A$ is an object of interest that is associated with one edge $e \in E$ and a two-dimensional spatial location. In Fig. 1, activity a_1 at $(1, 2.25)$ is associated with edge e_{n_B, n_F}.*

Definition 3. *An **isodistance sub-network** $ISN(s, d)$ consists of the nodes, edges and sub-edges on the spatial network G that are reachable from a source point s within a network diffusion distance. For example in Fig. 1, with a source point of a_5 at $(0.75, 1.5)$, a diffusion distance $d = 1$, an isodistance sub-network $ISN(a_5, 1)$ consists of nodes n_E, n_F, edge e_{n_E, n_F}, and sub-edges e_{n_E, a_2}, e_{n_F, a_1}, and e_{n_F, a_8}.*

Definition 4. *The **activity coverage** of an isodistance sub-network ac_{ISN} refers to the number of activities on the sub-network. For example in Fig. 1, the activity coverage of $ISN(a_5, 1)$ is 8 since $a_1, a_2, a_3, a_4, a_5, a_6, a_7$, and a_8 is on the sub-network.*

Definition 5. *A **weight** of an isodistance sub-network w_{ISN} means the sum of weights (length) of all edges and sub-edges in ISN. For example in Fig. 1,*
$$w_{ISN(a_5, 1)} = dis_{a_5, a_2} + dis_{a_5, a_f} + dis_{a_f, a_1} + dis_{a_f, a_8} = 2.75.$$

Following the framework of spatial scan statistics [9], we use a hypothesis test to determine if an enumerated isodistance sub-network (ISN) is a hotspot. The null hypothesis H_0 states that the concentration of activities inside the isodistance sub-network is equal to the concentration outside, whereas the alternative hypothesis H_1 states that the concentration inside is higher than outside. The log likelihood ratio [9] measures the ratio of how likely H_1 is accepted rather than H_0, in other words, how likely the tested ISN is a hotspot. Then, a p-value measures the statistical significance the log likelihood ratio. If the p-value of an ISN is smaller than or equal to a threshold (e.g., 0.05 by convention), the ISN is detected as a hotspot.

Definition 6. *The **log likelihood ratio** [9] $logLR\,(ISN)$ which measures how likely an ISN is a hotspot is formally defined by the following formula:*

$$logLR_{ISN} = log\left(\left(\frac{ac_{ISN}}{b_{ISN}}\right)^{ac_{ISN}} \times \left(\frac{|A| - ac_{ISN}}{|A| - b_{ISN}}\right)^{(|A| - ac_{ISN})} \times I()\right) \quad (1)$$

where $|A|$ is the total number of activities on the spatial network; and $b_{ISN} = \frac{w_{ISN}}{w_G} \times |A|$ is the expected number of activities inside ISN, where w_G is the weight of G. $I() = \begin{cases} 1, & \text{if } ac_{ISN} > b_{ISN} \\ 0, & \text{otherwise} \end{cases}$ is an indicator function to filter out the cases where $ac_{ISN} \le b_{ISN}$ since we are only interested in ISN with log likelihood ratio higher than the expectation.

The log likelihood ratio of the hotspot in Fig. 1 can be computed using (1) as follow: $LogLR\,(ISN\,(a_5, 1)) = log\left(\left(\frac{8}{1.74}\right)^8 \times \left(\frac{13-8}{13-1.74}\right)^{(13-8)} \times 1\right) = 8.13$

Definition 7. *Monte Carlo Simulation* *is a randomization technique used to determine the p-values of isodistance sub-networks since the distribution of the log likelihood ratio is unknown beforehand. The number of simulations, m, depends on the desired precision of p-value (precision 0.001 requires $m \ge 999$). The hotspot in Fig. 1 has a p-value of 0.001, which means that its log likelihood ratio is greater than the highest log likelihood ratio for all 999 simulated datasets.*

2.2 Problem Formulation

Network isodistance hotspot detection problem is formulated as follows:

Given:

1. A spatial network $G = (N, E)$ that consists of a node set N and an edge set E,
2. An activity set A on G
3. A log likelihood ratio threshold θ_{LR},
4. A p-value threshold θ_p,

Find: All network isodistance hotspots (NIH) on G which are ISN with $logLR \ge \theta_{LR}$ and p-value$\le \theta_p$.

Objective: Computational efficiency.

Constraints:

1. The source point of each NIH is an activity,
2. The diffusion distance of an NIH is the distance from the source to an activity,
3. Correctness and completeness.

The enumeration space of ISN is infinite which is computationally intractable. However, we only enumerate ISN whose source point is an activity since hotspots are usually initialized by diffusion from a "root" case (e.g., disease diffused from first infected patient). In addition, the diffusion distance of ISN are constrained to distance from the source activity to each of the other activities. This constraint avoids the redundant enumerations of concentric sub-networks having the same activity coverage but different weights. For example in Fig. 1, when activity a_5 is the source, dis_{a_5,a_4}, dis_{a_5,a_3} are two of the diffusion distances considered, but not any point between them.

3   BaseNIHD: A Baseline Algorithm Using Known Algorithmic Refinements

In this section, we introduce BaseNIHD, a baseline algorithm using some known algorithmic refinements. BaseNIHD exhaustively enumerate and then compute the logLR for all the possible isodistance sub-networks (ISN). Then, it computes the p-values for all NIH candidates that pass the logLR threshold.

Step 1: Compute all node-pair distances
First, BaseNIHD computes all-pairs shortest paths between nodes using Dijkstra's algorithm. These distances between nodes are stored in a hashmap. Step 1 in Fig. 2 shows three rows of the hashmap, which are distances between three pairs of nodes, namely $\langle n_A, n_B \rangle$, $\langle n_A, n_C \rangle$, and $\langle n_A, n_D \rangle$.

Step 2: Compute all activity-pair distances
In this step, if two activities are on the same edge, their distance is computed by their spatial locations directly. If two activities are on different edges, their distance is computed by stitching together three parts: the distances from each activity to one of its neighbor nodes respectively and the distance between the two nodes. Since each activity has 2 neighbor nodes, the distance between two activities is the smallest among the distance of 4 stitched paths. This method re-uses the fixed spatial network to compute the distance between a pair of activities in constant time. Step 2 in Fig. 2 shows the distances between three pairs of activities, respectively.

Step 3: Enumerate ISN and Compute logLR
The pseudocode of Step 3 is presented in Algorithm 1. First, for each activity a_i that is enumerated as the source, all activities are sorted in ascending order based on their distance to a_i (Line 3). Next, the distance between the source activity and each of the other activities is enumerated as the diffusion distance, and an ISN is defined accordingly. Figure 2 shows an example of the sorted activities for each source and the ISN enumrated when a_5 is the source.

In order to compute the log likelihood ratio (logLR) of an enumerated isodistance sub-network $ISN_{ij} = ISN\left(a_i, dis_{a_i,a_j}\right)$, we need its activity coverage $ac_{ISN_{ij}}$ and weight $w_{ISN_{ij}}$. The activity coverage $ac_{ISN_{ij}}$ is simply the rank of a_j in the sorted list of source a_i with considering the activities with the same rank.

Algorithm 1. ISN enumeration and logLR computation based on edge stitching

---

1: NIH candidate set is empty: $NIHCandidates \leftarrow \emptyset$
2: **For each** source activity a_i **do**
3:     Sort all activities by their network distance to a_i
4:     **For each** target activity a_j, where $i \neq j$ **do**
5:         Denote $ISN\left(a_i, dis_{a_i, a_j}\right)$ as ISN_{ij}
6:         Compute the activity coverage: $ac_{ISN_{ij}}$
7:         $w_{ISN_{i,j}} = 0$
8:         **For each** edge $e_{n_x, n_y} \in E$ **do**
9:             $w_{ISN_{i,j}} = w_{ISN_{i,j}} + min(max(dis_{a_i, a_j} - dis_{a_i, n_x}, 0) + max(dis_{a_i, a_j} - dis_{a_i, n_y}, 0), dis_{n_x, n_y})$
10:        compute $logLR\left(ISN_{ij}\right)$ using $ac_{ISN_{ij}}$ and $w_{ISN_{ij}}$
11:        **if** $logLR\left(ISN_{ij}\right) \geq \theta_{LR}$ **then**
12:            $NIHCandidates \leftarrow ISN_{ij}$

For example, Step 3 in Fig. 2 shows that $ac_{ISN_{5,1}} = 8$. Lines 7–9 in Algorithm 1 show the edge traversing approach used to compute $w_{ISN_{ij}}$. For each enumerated isodistance sub-network ISN_{ij}, all edges in E are examined if they are in ISN_{ij}. Note that it is possible that an edge is only partially in ISN_{ij}. For example, $ISN_{5,1}$ contains edge e_{n_E, n_F}, sub-edges e_{n_F, a_2}, e_{n_F, a_1} and e_{n_F, a_8}, so that $w_{ISN_{5,1}} = 1 + 0.25 + 0.75 + 0.75 = 2.75$.

With the activity coverage $ac_{ISN_{ij}}$ and weight $w_{ISN_{ij}}$, $logLR\left(ISN_{ij}\right)$ is computed using Eq. 1. If $logLR\left(ISN_{ij}\right) \geq \theta_{LR}$, ISN_{ij} is a network isodistance hotspot (NIH) candidate (Lines 11–12). For example, $logLR\left(ISN_{5,1}\right) = 8.13$ as shown in Fig. 2. Suppose $\theta_p = 8$, $ISN_{5,1}$ is an NIH candidate.

Step 4: Compute p-values for each NIH candidates using Monte Carlo Simulation

Lastly, multiple Monte Carlo simulations are run for computing the p-value for each NIH candidate. The NIH candidates with p-values $\leq$ the p-value threshold θ_p are network isodistance hotspots. For example, suppose $\theta_p = 0.001$, $ISN_{5,1}$ is an NIH since $logLR\left(ISN_{5,1}\right)$ is higher than the highest logLR in all 999 simulations.

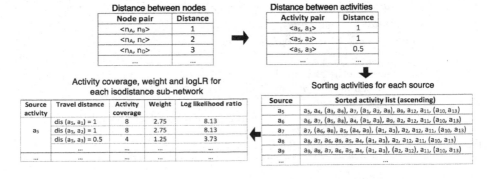

Fig. 2. An illustrative execution trace of BaseNIHD

4 NPP: An Algorithm Based on Network Partitioning and Upper-Bound Pruning

The baseline approach BaseNIHD provides a solution to the network isodistance hotspots detection (NIHD) problem. In this section, we introduce NPP which is a novel algorithm substantially improving the scalability over BaseNIHD. The main idea of NPP is to reduce the high cost of isodistance sub-network (ISN) enumeration and log likelihood ratio (logLR) computation by a filter-and-refine approaches. In the filter phase, NPP determines an upper-bound logLR of a collection of ISN in one shot, then eliminates those ISN without computing their exact logLR. It is realized via partitioning the entire network into grids, by which the ISN defined by the activities from each pair of grids can be examined together. NPP is independent from how the spatial network partitioning is defined. For making the illustration clearly, we use a simple rectangular grid partitioning in this paper. However, many other more sophisticated partitioning methods also apply to NPP, such as Quadtree [19], KD-tree [20], METIS [21], Minimum Cut [22], and Natural Cut [23]. In the refine phase, for each ISN that survives the filtering, NPP provides another speedup by excluding the edges that are guaranteed outside the ISN and only examines a subset of edges which.

NPP begins by computing the all-pairs network distances between nodes and activities as in Step 1 and Step 2 in BaseNIHD. In Step 3, a fast ISN enumeration and logLR computation algorithm is used to obtain the network isodistance hotspot (NIH) which replaces Step 3 in BaseNIHD.

Algorithm 2. ISN enumeration and logLR computation in NPP

1: NIH candidate set is empty: $NIHCandidates \leftarrow \emptyset$
2: Partition the spatial network G into $|P_x| \times |P_y|$ partition: $P_1, P_2, \ldots, P_{|g_x||g_y|}$
3: **For each** activity a_i **do**
4:     Sort all activities by their distances to a_i
5:     Store closest and farthest activities in each partition
6: **For each** source partition P_j **do**
7:     **For each** target partition P_k **do**
8:         **For each** activity a_{jp} in P_j **do**
9:             $rank_{jp} = $ the rank of farthest activity to a_{jp} in P_j
10:         $highestRank = \max rank_{jp}$
11:         Compute $\widehat{ac_{jk}}$, w_{jk} and $\widehat{logLR}_{jk}$
12:         **if** $\widehat{logLR}_{jk} \geq \theta_{LR}$ **then**
13:             Determine the examined edge set E_{jk}
14:             **For each** activity a_{jp} in P_j **do**
15:                 **For each** activity a_{kq} in P_k **do**
16:                     Compute $ac_{ISN_{jpkq}}$, $w_{ISN_{jpkq}}$ and $logLR(ISN_{jpkq})$
17:                     **if** $logLR(ISN_{jpkq}) \geq \theta_{LR}$ **then**
18:                         $NIHCandidates \leftarrow ISN_{jpkq}$

The pseudocode of this algorithm is presented in Algorithm 2, and the execution examples are shown in Fig. 3. We use a simple rectangular grid partitioning here for the purpose of illustration. First, the entire spatial network G is partitioned into two-dimensional rectangular partitions using given a desired number of partitions in each dimension (Line 2). The partitioning starts by assigning the nodes. Assuming the entire study area has the size of $|S_x| \times |S_y|$, and is partitioned into $|P_x| \times |P_y|$ partitions, the size of each partition is equally initialized as $\frac{|S_x|}{|P_x|} \times \frac{|S_y|}{|P_y|}$. The partitions are indexed in rows and each partition contains the nodes that are within its range. Edges and activities are assigned to the partition that contains the edges' neighbor nodes. If the two neighbor nodes of an edge belong to two different partitions, the edge is always assigned to the partition with the node that has the higher index. Activities are assigned with their associated edges.

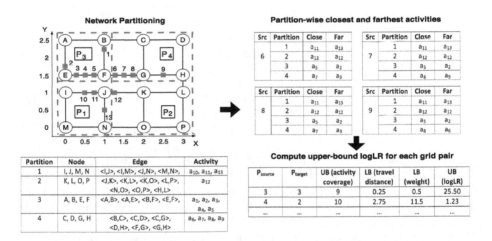

Fig. 3. An illustrative execution trace of ISN enumeration and logLR computation in NPP

Figure 3 shows a study area in the size of 3×3 partitioned into $2 \times 2 = 4$ partitions indexed by rows; the first row includes P_1 and P_2, and the second row includes P_3 and P_4. Each partition is assigned 4 nodes according to their spatial locations: P_1 contains n_I, n_J, n_M, n_N; P_2 contains n_K, n_L, n_O, n_P; P_3 contains n_A, n_B, n_E, n_F; and P_4 contains n_C, n_D, n_G, n_H. Edges and their associated activities are assigned together with their neighbor nodes. For example, e_{n_E,n_F} and its associated activities a_3, a_4 and a_5 are assigned to P_3 since the edge's neighbor nodes n_E and n_F are in P_3. The blue lines in the figure visualize the nodes, edges and activities contained in each partition.

Next, for each activity a_i, NPP generates a sorted list as in Algorithm 1 (Line 4). Then, in a new step, it generates another partition-wise list that stores the closest and farthest activities to a_i in each partition (Line 5). Note that if more than one activity are equally the closest or farthest, only one of them is stored. This list is generated by a traverse of the global list without another sorting process. For example in Fig. 3, the partition-wise list of activity a_5 stores the activities closest and farthest to a_5 in each partition.

Lines 6–11 present the upper-bound logLR ($\widehat{logLR_{jk}}$) computation for each partition pair. Given a source partition P_j and a target partition P_k, $\widehat{logLR_{jk}}$ bounds the largest possible logLR of all the isodistance sub-networks (ISN) that are defined by the source of any activities in P_j and the network distance between source and any target activities in P_k.

$\widehat{logLR_{jk}}$ is computed from the upper-bound activity coverage ($\widehat{ac_{jk}}$) and the lower-bound weight w_{jk} using the Eq. 2.

$$logLR_{jk} = log\left(\left(\frac{\widehat{ac_{jk}}}{b_{jk}}\right)^{\widehat{ac_{jk}}} \times \left(\frac{|A| - \widehat{ac_{jk}}}{|A| - b_{jk}}\right)^{(|A| - \widehat{ac_{jk}})} \times I()\right) \qquad (2)$$

where $b_{jk} = \frac{w_{jk} \times |A|}{w_G}$.

To compute $\widehat{ac_{jk}}$, NPP traverses all activities a_{jp} in P_j, finding the highest global rank of the farthest activity in P_k to a_{jp}:

$$\widehat{ac_{jk}} = \max rank_{a_{jp}}, where \qquad (3)$$

$$j = \underset{h}{\operatorname{argmax}}\, dist_{hp,lk}, l = \underset{i}{\operatorname{argmax}}\, dis_{jp,ik} \qquad (4)$$

For example in Fig. 3, if the source partition is P_4 and the target partition is P_2, NPP traverses all activities in P_4 and determines the farthest activities in P_2 from these activities in P_4 are equally a_{12}, and the highest global rank is 10 since a_{12} ranks 10 in the global sorted list of source a_9 shown in Fig. 2. $\widehat{ac_{jk}}$ is the activity coverage of the $ISN_{6,12} = 10$.

To compute the lower-bound weight w_{jk}, NPP computes the weight of ISN with the source of each activity a_{jp} in P_j and a shortest distance which is the shortest among the network distances between a_{jp} and their closest activities in P_k:

$$w_{jk} = \underset{j}{\min}\, w_{ISN_{jp,lk}}, l = \underset{i}{\operatorname{argmin}}\, dis_{jp,ik} \qquad (5)$$

For example in Fig. 3, if the source partition is P_4 and the target partition is P_2, the shortest distance is $min\{dis_{a_6,a_{12}}, dis_{a_7,a_{12}}, dis_{a_8,a_{12}}, dis_{a_9,a_{12}}\} = 2.75$. Using this distance, the $w_{jk} = min\{w_{ISN_{6,12}}, w_{ISN_{7,12}}, w_{ISN_{8,12}}, w_{ISN_{9,12}}\} = 11.5$. Therefore, $\widehat{logLR_{4,2}} = 1.23$. Since this value is smaller than the logLR threshold $\theta_{LR} = 8$, none of the ISN with the source activities in P_4 and target activities in P_3 can be a network isodistance hotspot (NIH) candidate. If the upper-bound logLR of source P_j and target $P_k \geq \theta_{LR}$, the exact logLR will be computed for each ISN with a source activity from P_j and target activity from P_k in the refine phase (Lines 12–17). As shown in Fig. 3, when the source and target partitions are both P_3, the upper-bound logLR is 25.50 > 8. Therefore, the exact logLR for each ISN from this partition pair is computed. However, instead of checking the entire edge set E to determine the weight for each ISN, NPP only checks the edges in the partitions that are reachable in Euclidean space within the radius of each ISN. The reason is that the Euclidean distance between two points is the lower-bound of their network distance. An alternative approach that provides speedup on computing weights is searching the shortest path tree to determine the nodes within the radius, then only examine the edges that are neighbors to these nodes. Both of these two approaches examine a portion of edges and provide comparable amount of acceleration over BaseNIHD.

Each Monte Carlo simulation repeats all the steps except computing the all-pairs distances between nodes, finding all the NIH by computing the p-values.

5   Theoretical Analysis

In this section, we formally prove that the NPP guarantees the correctness and completeness. Also, we analyze the asymptotic complexities of BaseNIHD and NPP.

Lemma 1.   *(1) Given a fixed expected number of activities, the logLR increases monotonically with the number of activities. (2) Given a fixed number of activities, the logLR increase anti-monotonically with the expected number of activities.*

Proof. We show the proof of (1) and leave the proof of (2) since they are very similar. Given a number of activities ac_{ISN}, total number of activities $|A|$, and an expected number of activities b_{ISN}. We prove that $L_{diff} = L_1 - L_2 \geq 0$, where

$$L_1 = (ac_{ISN} + 1)log\left(\frac{ac_{ISN}+1}{b_{ISN}}\right) + (|A| - ac_{ISN} - 1)log\left(\frac{|A|-ac_{ISN}-1}{|A|-b_{ISN}}\right),$$

$$L_2 = ac_{ISN}log\left(\frac{ac_{ISN}}{b_{ISN}}\right) + (|A| - ac_{ISN})log\left(\frac{|A|-ac_{ISN}}{|A|-b_{ISN}}\right), ac_{ISN} \geq b_{ISN}.$$

Since $ac_{ISN} \geq b_{ISN}$, we know $log\left(\frac{|A|-ac_{ISN}}{|A|-b_{ISN}}\right) \leq 0$, and $log\frac{ac_{ISN}}{b_{ISN}} \geq 0$. Therefore,

$$L_2 \leq L_2^{new} = (ac_{ISN} + 1)log\frac{ac_{ISN}}{b_{ISN}} + (|A| - ac_{ISN} - 1)log\left(\frac{|A|-ac_{ISN}}{|A|-b_{ISN}}\right)$$

$$\Rightarrow L_{diff} \geq L_{diff}^{new} = L_1 - L_2^{new}$$

$$= \underbrace{(ac_{ISN} + 1)log\left(\frac{ac_{ISN}+1}{ac_{ISN}}\right)}_{L_{first}} + \underbrace{(|A| - ac_{ISN} - 1)log\left(\frac{|A| - ac_{ISN} - 1}{|A| - ac_{ISN}}\right)}_{L_{second}}.$$

$\frac{\partial L_{first}}{\partial ac_{ISN}} = log\left(1 + \frac{1}{ac_{ISN}}\right) - \frac{1}{ac_{ISN}} < 0$

$\Rightarrow L_{first}$ monotonically decreases with $a \Rightarrow L_{first} \geq \lim_{a\to\infty} L_{first} = 1$

Let $t = |A| - ac_{ISN}$, then $\frac{\partial L_{second}}{\partial t} = log\left(1 - \frac{1}{t}\right) + \frac{1}{t} < 0$

$\Rightarrow L_{second}$ monotonically decreases with $t \Rightarrow L_{second} \geq \lim_{t\to\infty} L_{second} = -1$

$\Rightarrow$ the minimum value of $L_{diff}^{new} = 1 + (-1) = 0 \Rightarrow L_{diff} \geq 0$

Theorem 1.   *The upper-bound pruning is a correct pruning.*

Proof. Given j and k, $\widehat{ac_{jk}} = \max ac_{jk}$ and $\underline{w_{jk}} = \min w_{jk}$. According to Lemma 1, $logLR(ISN) \geq \theta_{LR}$. Therefore, the upper-bound pruning is correct since it does not prune any ISN that $logLR(ISN) \geq \theta_{LR}$.

Theorem 2. *NPP is complete and correct.*

Proof. According to Theorem 1, all ISN with $logLR\,(ISN) \geq \theta_{LR}$ are sent to the refine phase. Thus, the solution of NPP is complete. Also, since the refine phase computes the exact logLR for each candidate, the solution set is correct.

Time complexity of BaseNIHD: Suppose the input spatial network contains $|N|$ nodes and $|E|$ edges. $|A|$ denotes the total number of activities. m denotes the number of Monte Carlo simulations required. Since only a small number of road segments can connect at an intersection, $|N| \approx |E|$. For simplicity of the analysis, the numbers of nodes and edges are both denoted by $|N|$. The cost of computing all-pairs shortest paths (Step 1) is $O\left(|N|^2 log\,(|N|)\right)$ using Dijkstra's algorithm with a priority queue. This cost will be dominated by the following steps. The cost of Step 2 is $O\left(|A|^2\right)$ since there are $|A|^2$ activity pairs. The cost of Step 3 consists of three parts: (1) activity sorting for each source, (2) activity coverage computing for each enumerated isodistance sub-network (ISN), and (3) weight computing for each ISN. Since there are $|A|$ source activities, and each needs a sorting over $|A|$ activities, the cost for sorting is $O\left(|A|^2 log\,(|A|)\right)$. There are $|A|^2$ enumerated ISN, and each takes $O\,(1)$ to compute the activity coverage. Therefore, the total cost of activity coverage computation is $O\left(|A|^2\right)$. Finally, traversing all edges for each enumerated ISN costs $O\,(|N|)$, making the total cost for weight computation $O\left(|A|^2|N|\right)$. Computing the weight of an ISN, therefore, dominates the cost in activity coverage computation. Together, Step 1, 2 and 3 of BaseNIHD cost $O\left(|A|^2 log\,(|A|) + |A|^2|N|\right)$. The Monte Carlo Simulation (Step 4) multiplies the cost of Steps 2 and 3 for m times. Therefore the total cost of BaseNIHD is $O\left(m \times \left(|A|^2 log\,(|A|) + |A|^2|N|\right)\right)$.

Time complexity of NPP: The costs for computing all-pairs distances between nodes and activities are the same in NPP as BaseNIHD, which are dominated. The network partitioning step costs $O\,(|N| + |A|)$ which is dominated by the steps later. For each source activity, the global sorting costs $O\,(|A| log\,(|A|))$. After that, finding the closest and farthest activities to the source in each grid costs $O\,(|A|)$ which can also be ignored since only a traverse over the globally sorted list is needed. Suppose the spatial network is partitioned into $|P|$ partitions, and the $i - th$ partition P_i contains $|A|_i$ activities, where $i = 1 \ldots |P|$ and $\sum_i |A|_i = |A|$, and the upper-bound logLR computation costs $O\left(\sum_i \sum_j |A|_i|N|\right) = O\,(|P||A||N|)$. In the worst case, no ISN is pruned by the filter phase, the exact logLR computation costs $O\left(|A|^2|N|\right)$, which is the same as in BaseNIHD and dominates the upper-bound logLR computation. With Monte Carlo Simulation, the total cost of NPP is $O\left(m \times \left(|A|^2 log\,(|A|) + |A|^2|N|\right)\right)$.

Analysis: It can be concluded that BaseNIHD and NPP have the same asymptotic computational cost in the worst case, and the dominating term for both is $O\left(|A|^2|N|\right)$. However, if the upper-bound logLR prunes some partitions, the actual running time of NPP will be shorter. In addition, when computing the weights of ISN, NPP only examines a subset of edges, providing another speedup compared to BaseNIHD. Therefore, the complexity of NPP can be written as $O\left(m \times \left(|A|^2 log\left(|A|\right) + f_1 \times f_2 \times |A|^2 |N|\right)\right)$, where f_1 and f_2 represent the speedup from the upper-bound logLR pruning and the reduced number of examined edges, respectively. The experimental results in Sect. 7 show that NPP provides substantial speedup over BaseNIHD in practice.

The number of partitions $|P|$ does not appear in the final costs of NPP. However, it affects the effectiveness of the upper-bound pruning. A small $|P|$ gives loose upper-bound, thus less pruning will happen. A large $|P|$ causes large overhead of upper-bound computation which compromises the speedup from pruning. In extreme cases when $|P| = 1$ or $|P| = |A|$, NPP provides no speedup over BaseNIHD. In Sect. 7, the factor of $|P|$ will be analyzed through experiments.

6   Case Studies on Real World Crime Data

To evaluate the power of the proposed approach in detecting hotspots, we conducted case studies on two real world crime datasets. SaTScan (www.satscan. org), a popular hotspots detection approaches based on Euclidean distance is used as the baseline approach. The proposed approach finds the complete set of hotspots which may overlap each other. To reduce the visual clutter and redundancy, we used the same post-processing as used in SaTScan: if NIH_i and NIH_j are partially overlapped, only the one with the higher log likelihood ratio is shown. The results showed that the proposed approach outperformed the baseline approach both visually and quantitatively. The datasets were provided by "Pinellas County Crime Viewer" (http://egis.pinellascounty.org/) and "City of Seattle Open Data" (https://data.seattle.gov), respectively. The roadmaps were extracted OpenStreetMap (https://www.openstreetmap.org).

6.1   Robberies Occurred in Pinellas County, Florida

We compared the results from SaTScan [24] and the proposed approach on a relatively small dataset which contained 32 robberies (blue dots) from 8/31/11 to 8/31/16 in the Pinellas County, Florida, shown in Fig. 4.

With a p-value threshold of 0.001, as shown in Fig. 4(a), SaTScan [24] found a circular hotspot at bottom-left of the map (i.e., City of Sunset Beach) with a log likelihood ratio (logLR) of 27.31. NPP also detected a hotspot in this region with a higher logLR of 41.22 shown in Fig. 5(b). It can be seen that the circular hotspot found by SaTScan covers a large portion of empty area, whereas the NPP was much better in capturing the actual streets where robberies actually happened. More importantly, NPP detected two more hotspots at upper-right of the map (i.e., City of South Pasadena) which were completely missed by SaTScan. The existence of these two hotspots was demonstrated by a public safety report [25] which states that this region suffers much higher robbery index (i.e., 180) than the average index over the United States (i.e., ≈ 110).

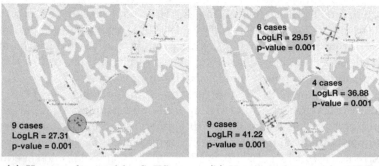

(a) Hotspot detected by SaTScan   (b) Hotspots detected by NPP

Fig. 4. Hotspots detected by SaTScan and NPP on robberies in Pinellas County, Florida

6.2 Assaults Occurred in Fremont, Washington

Moreover, we compared the baseline and proposed approach on a relatively larger dataset which contains 371 assaults occurred in Fremont, Washington State in the year of 2012. With a p-value threshold of 0.05, SaTScan found a hotspot in the middle of the study area that covered both banks of the river

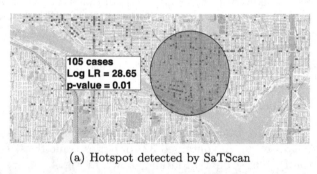

(a) Hotspot detected by SaTScan

(b) Hotspots detected by FastNIHD

Fig. 5. Comparison of outputs by SaTScan and FastNIHD on assault data in Fremont, Washington (best in color)

shown in Fig. 5(a). However, since the two banks are actually far from each other due to the limited connection by bridges, they should not be covered in the same hotspot. In Fig. 5(b), NPP also found a hotspot in the middle which was much more focused and covered only the upper bank of the river. This hotspot was validated by the fact that the upper bank was a commercial area with a few bars which caused more crimes affected by alcohol, but the lower bank was a residential area with a lower crime rate which should be excluded from the hotspot. Another hotspot detected by NPP covered the upper-left region with visually high density of activities. The major reason why SaTScan failed to detect this region was that the activities were dense only in the upper bank of the river, not showing a regular shape in Euclidean space (e.g., circular and elliptical). Another area along a major road (right) was also detected by NPP but was completely missed by SaTScan.

7   Experimental Evaluation

We conducted experiments to evaluate the proposed approaches in terms of scalability by comparing BaseNIHD with NPP under a variety of parameters.

7.1   Experimental Setup

The spatial network used in the experiments was from Pinellas County, Florida containing about 2000 nodes and 4300 edges (the same network as in the second case study). The activity datasets were synthetic datasets that contain hotspots generated following Gaussian distribution plus noise data generated following complete spatial randomness. In order to evaluate how different parameters affect the performance, only one of the following six parameters was varied in each experiment: (1) number of activities (default = 2000), (2) number of nodes (default = 2000), (3) log likelihood ratio threshold (default = 50), (4) number of partitions (default = 16 × 8 = 32), (5) percentage of noise data (default = 50%), and (6) number of hotspots (default = 2) while the others were set to default values. In order to rule out misleading factors and show the effects of different parameters clearly, all the execution times excluded the cost for all-pairs distances between nodes computation and the Monte Carlo simulations. All experiments were implemented in Java and performed on a laptop with an Intel Core i7 2.5 GHz CPU and 16 GB memory.

7.2   Experimental Results

Effect of number of activities: In this experiment, the numbers of activities were varied as 2000, 4000, 8000, and 16000. The results in Fig. 6(a) show that the costs of BaseNIHD increased faster than NPP as numbers of activities increased. The ratios of costs between two algorithms also increased, indicating that the speedup provided by NPP increased as number of activities increased.

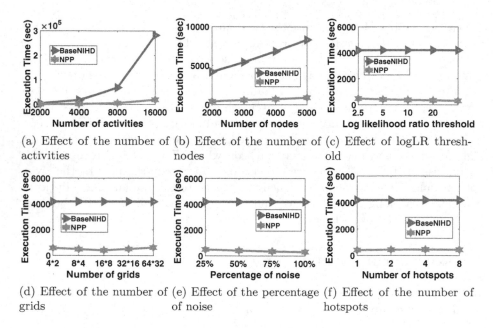

(a) Effect of the number of activities

(b) Effect of the number of nodes

(c) Effect of logLR threshold

(d) Effect of the number of grids

(e) Effect of the percentage of noise

(f) Effect of the number of hotspots

Fig. 6. Comparing execution times under different parameters

Effect of number of nodes: In this experiment, the numbers of nodes were varied as 2000, 3000, 4000 and 5000 by putting synthetic nodes on random edges of the original spatial network, so the number of edges also increased to 5300, 6300 and 7300 accordingly. The results in Fig. 6(b) show that the gap between the costs of BaseNIHD and NPP increased fast as the numbers of nodes increased, and the ratios between costs kept similar. It showed that NPP was able to provide consistent speedup over BaseNIHD with the increase of nodes and edges.

Effect of log likelihood ratio threshold: In this experiment, the log likelihood ratio thresholds (θ_{LG}) were varied as 25, 50, 100 and 200. The results in Fig. 6(c) show that the costs of BaseNIHD was not affected. As θ_{LG} increased, the costs of NPP kept decreasing since the upper-bound filter could prune more and more grid pairs.

Effect of number of grids: In this experiment, the numbers of grids were varied as 4×2, 8×4, 16×8 and 32×16. The results in Fig. 6(d) show that the costs of BaseNIHD were not affected and the costs of NPP kept much smaller than BaseNIHD. As the numbers of grids increased from 4×2 to 16×8, the costs of NPP decreased since the upper-bound logLR became tighter and the number of examined edges became smaller. After that, when numbers of grids increased from 16×8 to 64×32, the costs of NPP increased since the overhead of the pruning phase compromised the speedup.

Effect of percentage of noise: In this experiment, the percentage of noise data were varied as 25%, 50%, 75%, and 100%. The results in Fig. 6(e) show that the costs of BaseNIHD were not affected whereas the costs of NPP decreased when there are more noise in the datasets. The reason behind is that more data are pruned since the number of grids with high activity density decreased.

Effect of number of hotspots: In this experiment, the number of hotspots were varied as 1, 2, 4, and 8. The results in Fig. 6(f) show that the costs of BaseNIHD were not affected whereas the costs of NPP slightly increased at the beginning and then decreased when there were 8 hotspots. This is because that as the numbers of hotspots increased, more grids tended to have higher density of points and thereby survived the pruning. However, as the numbers of hotspots kept increasing, the activity density inside each hotspot decreased, and thus more grids are pruned.

Overall, NPP achieved a substantial speedup over BaseNIHD. With the increase of number of activities, size of the networks, logLR threshold, and the percentage of noise data, the speedup kept increasing. With increased numbers of grids, NPP performed better at first since the upper-bound logLR became tighter. Then, the performance decreased due to the increase of the overhead brought by the pruning phase. As the increase of number of hotspots, NPP became little slower at the beginning and then got faster because that more grids had a decreased activity density.

8 Conclusion and Future Work

This paper formulates the problem of network isodistance hotspots detection (NIHD) in relation to important application domains such as criminology, epidemiology and transportation engineering. We propose a novel algorithm based on network partitioning and upper-bound pruning (NPP) which substantially saves the computational cost without loss of result quality in terms of correctness and completeness. Two case studies on real crime datasets showed that the proposed approach was able to detect hotspots that are missed by existing approaches. The existence of detected hotspots are demonstrated by a report regarding public safety. Both theoretical analysis and experimental evaluation results show that with a simple rectangular grid partitioning, our proposed approach yields substantial computational savings while keeping the completeness and correctness of the solution.

In the future, we plan to consider more factors that potentially affect the formation of hotspots such as population and other demographic information. We also plan to investigate how different network partitioning approaches [19–23] affect the performance of the proposed approaches. In addition, we plan to explore hotspot detection based on travel time rather than travel distance. It may give different hotspots in different time of a day (e.g., rush hour vs. midnight), providing more knowledge to help understand the life cycle of hotspots. Futhermore, we plan to study the parallelization of NPP, starting by investigating the independency across the logLR computation of different partitions.

Acknowledgement. This material is based upon work supported by the National Science Foundation under Grant No. 1029711, IIS-1320580, 0940818, IIS-1218168, and IIS-1541876, the USDOD under Grant No. HM0210-13-1-0005, the ARPA-E under Grand No. DE-AR0000795, and the University of Minnesota under the OVPR U-Spatial. We are particularly grateful to Kim Koffolt and the members of the University of Minnesota Spatial Computing Research Group for their valuable comments.

References

1. Ester, M., Kriegel, H.P., Sander, J., Xu, X.: A density-based algorithm for discovering clusters in large spatial databases with noise. In: KDD 1996, pp. 226–231 (1996)
2. Brantingham, P.J., Brantingham, P.L.: Environmental Criminology. Sage Publications, Beverly Hills (1981)
3. Guerette, R.T.: Analyzing crime displacement and diffusion. US Department of Justice, Office of Community Oriented Policing Services, Washington, DC (2009)
4. Merrill, R.M.: Introduction to Epidemiology. Jones and Bartlett Publishers (2013)
5. Tatem, A.J., Rogers, D.J., Hay, S.: Global transport networks and infectious disease spread. Adv. Parasitol. **62**, 293–343 (2006)
6. U.S. Department of Transportation: Traffic congestion and reliability: Trends and advanced strategies for congestion mitigation, http://www.ops.fhwa.dot.gov/
7. Yiu, M.L., Mamoulis, N.: Clustering Objects on a Spatial Network. In: Proceedings of the 2004 ACM SIGMOD International Conference on Management of Data, pp. 443–454. ACM (2004)
8. Levine, N.: Crimestat iii. Houston (TX): Ned Levine & Associates. National Institute of Justice, Washington, DC (2004)
9. Kulldorff, M.: A spatial scan statistic. Commun. Stat. Theory Methods **26**(6), 1481–1496 (1997)
10. Tang, X., Eftelioglu, E., Shekhar, S.: Elliptical hotspot detection: a summary of results. In: Proceedings of the 4th International ACM SIGSPATIAL Workshop on Analytics for Big Geospatial Data, pp. 15–24. ACM (2015)
11. Neill, D.B., Moore, A.W.: Rapid detection of significant spatial clusters. In: Proceedings of the 10th ACM SIGKDD, pp. 256–265. ACM (2004)
12. Eftelioglu, E., Shekhar, S., Kang, J.M., Farah, C.C.: Ring-shaped hotspot detection. IEEE Trans. Knowl. Data Eng. **28**(12), 3367–3381 (2016)
13. Beavon, D.J., Brantingham, P.L., Brantingham, P.J.: The influence of street networks on the patterning of property offenses. Crime Prev. Stud. **2**, 115–148 (1994)
14. Shiode, S., Shiode, N.: Detection of multi-scale clusters in network space. Int. J. Geogr. Inf. Sci. **23**(1), 75–92 (2009)
15. Rozenshtein, P., Anagnostopoulos, A., Gionis, A., Tatti, N.: Event detection in activity networks. In: Proceedings of the 20th ACM SIGKDD International Conference on Knowledge Discovery and Data Mining, pp. 1176–1185. ACM (2014)
16. Cadena, J., Chen, F., Vullikanti, A.: Near-optimal and practical algorithms for graph scan statistics (2017)
17. Oliver, D., Shekhar, S., Zhou, X., Eftelioglu, E., Evans, M., Zhuang, Q., Kang, J., Laubscher, R., Farah, C.: Significant route discovery: a summary of results. In: Proceedings of Geographic Information Science International Conference, pp. 284–300 (2014)

18. Inokuchi, A., Washio, T., Motoda, H.: An apriori-based algorithm for mining frequent substructures from graph data. In: Zighed, D.A., Komorowski, J., Żytkow, J. (eds.) PKDD 2000. LNCS, vol. 1910, pp. 13–23. Springer, Heidelberg (2000). doi:10.1007/3-540-45372-5_2
19. Samet, H.: The quadtree and related hierarchical data structures. ACM Comput. Surv. (CSUR) 16(2), 187–260 (1984)
20. Bentley, J.L.: Multidimensional binary search trees used for associative searching. Commun. ACM 18(9), 509–517 (1975)
21. Karypis, G., Kumar, V.: Metis-unstructured graph partitioning and sparse matrix ordering system, version 2.0. (1995)
22. Karger, D.R.: Global min-cuts in RNC, and other ramifications of a simple min-cut algorithm. In: SODA 1993, pp. 21–30 (1993)
23. Delling, D., Goldberg, A.V., Razenshteyn, I., Werneck, R.F.: Graph partitioning with natural cuts. In: 2011 IEEE International Parallel & Distributed Processing Symposium (IPDPS), pp. 1135–1146. IEEE (2011)
24. Kulldorff, M.: Satscan user guide for version 9.4, http://www.satscan.org/
25. Tampa Bay Business Journal: Pinellas county's highest crime areas, http://www.bizjournals.com/tampabay/news/2013/10/14/pinellas-countys-highest-crime-areas.html

Detection and Prediction of Natural Hazards Using Large-Scale Environmental Data

Nina Hubig[1], Philip Fengler[1], Andreas Züfle[2(✉)], Ruixin Yang[2],
and Stephan Günnemann[1]

[1] Technical University of Munich, Munich, Germany
{hubig,fengler,guennemann}@in.tum.de
[2] George Mason University, Fairfax, VA, USA
{azufle,ryang}@gmu.edu

Abstract. Recent developments in remote sensing have made it possible to instrument and sense the physical world with high resolution and fidelity. Consequently, very large spatio-temporal environmental data sets, have become available to the research community. Such data consists of time-series, starting as early as 1973, monitoring up to thousands of environmental parameters, for each spatial region of a resolution as low as $0.5' \times 0.5'$. To make this flood of data actionable, in this work, we employ a data driven approach to detect and predict natural hazards. Our supervised learning approach learns from labeled historic events. We describe each event by a three-mode tensor, covering space, time and environmental parameters. Due to the very large number of environmental parameters, and the possibility of latent features hidden within these parameters, we employ a tensor factorization approach to learn latent factors. As the corresponding tensors can grow very large, we propose to employ an outlier-score for sparsification, thus explicitly modeling interesting (location, time, parameter) triples only. In our experimental evaluation, we apply our data-driven learning approach to the use-case of predicting the rapid-intensification of tropical storms. Learning from past tropical storms, we show that our approach is able to predict the future rapid-intesification of tropical storms with high accuracy, matching the accuracy of domain specific solutions, yet without using any domain knowledge.

1 Introduction

Modern advances in sensor technology, including weather stations, satellite imagery, ground and aerial LIDAR, weather radar, and citizen-supplied observations, have coined the notion of *Next Generation Sensor Networks and Environmental Science* [9]. With the recent advances in the collection and integration of such data sources, as, e.g., offered by the state-of-the-art GEOS-5 data assimilation system [17], it is now possible to integrate these data sources into single data sets available publicly. An example of such large scale data set is the MERRA data, provided by NASA [18]. It covers the whole time period of the modern era of remotely sensed data, from 1979 until today, and contains a large variety of

M. Gertz et al. (Eds.): SSTD 2017, LNCS 10411, pp. 300–316, 2017.
DOI: 10.1007/978-3-319-64367-0_16

environmental parameters such as temperature, humidity and precipitation, on a spatial resolution of 0.5° latitude times 0.67° longitude produced at one-hour intervals. With huge environmental data like MERRA available, we are suddenly able to access many Terabytes of historic environmental data.

In this work, we propose a framework to make this data actionable for classifying and predicting environmental events. An environmental event is a point in space and time enriched by a label describing the event, such as a storm or an earthquake. Given such an event in region s at time t, we propose to map this event into the MERRA data set as a spatio-temporal environmental database to obtain information about the change of environmental attributes, i.e., their time-series, in spatial regions around s, and during the time around t.

To detect events, our first step is to prune areas in space and time that are non-interesting. Therefore, we propose an unsupervised learning approach, which detects outlying events by comparing time-series of individual regions to the corresponding time-series of adjacent regions. Following Tobler's first law of geography, time-series of a region should be highly correlated to time-series in its close vicinity. Accordingly, we can argue that any region that is not correlated to its vicinity indicates an outlier event. Second, extending the first approach, we propose a supervised learning principle, which assumes that historical points in time and space are labeled with event information, such as the existence of a tropical storm, or whether a tropical storm is increasing or decreasing. Tensor factorization [16] creates latent factors as explanatory variables used for predicting the natural hazards.

To summarize, our contribution of this work are

- Using labeled event data, we propose two tensor factorization approaches to learn the latent factor to classify future events. We propose a simple 4-mode tensor approach, considering different labeled instances as a fourth mode, and we propose an approach using coupled tensor-tensor factorization.
- To reduce the density of these tensors, we propose an Environmental Extremeness Measure (EEM), which maps each point in space and time to a score value describing its local extremeness. For any point in space and time where this measure is below a threshold, we set it to zero, thus indicating that nothing interesting is happening.
- In our experiments, we apply our tensor factorization based learning approach to a current problem in geo-information science: The problem of predicting rapid intensification of tropical cyclones. We show that our approach, by exploiting a spatio-temporal environmental database, is able to predict whether a tropical cyclone will rapidly intensify its wind-speed by at least 30 knots in the next 24 h. Our approach outperforms current literature on this problem.

1.1 Roadmap

We start with our supervised classification of spatio-temporal events in Sect. 2. Here, we assume that historical events are labeled with ground-truth information

Table 1. Table of Symbols and Acronyms.

Symbols	Definitions
Cor	(Pearson) Correlation
$CTTF$	Coupled Tensor-Tensor Factorization
$4MTF$	Four-Mode Tensor Factorization
$\mathcal{D}$	An environmental spatio-temporal data set
$\mathcal{S}$	Set of spatial regions
$s \in \mathcal{S}$	Spatial region
$\mathcal{T}$	Set of all time points
$T \subseteq \mathcal{T}$	Time interval
$t \in T$	A time point
$\mathcal{A}$	Set of attribute domains
$A \in \mathcal{A}$	Attribute domains
$a \in A$	A attribute value
$TS_{s,A,T}$	Time series for attribute A in region s in T
r	Radius for surrounding areas
τ	Threshold, computed for every domain A
$X(s, T, \mathcal{A})$	A 3-mode tensor with modes $s, T, \mathcal{A}$
$\mathcal{X}(s, T, \mathcal{A}, ID)$	A 4-mode tensor with modes $s, T, \mathcal{A}, ID$
$lock(\mathcal{A})$	For CTTF: coupled attribute mode
$lock(MODES)$	For 4MTF: all modes coupled *except ID*

that can be used for model training to similar events in the future. In Sect. 3 we improve the run-time of this approach by pruning areas in space and time which show non-interesting attribute values compared to their spatial vicinity. The experimental evaluation presented in Sect. 4 shows our results in predicting whether a tropical cyclone will increase at least 30 knots in 24 h, showing that our approach is capable of matching the prediction accuracy of current state-of-the-art which uses domain-specific expert knowledge. Then we bring our work in the context of related work in Sect. 5 and conclude in Sect. 6. An overview of notations and variables used and defined in the remainder of this work is given in Table 1.

2   Framework for Natural Hazards Detection

To predict spatio-temporal events and hazards, we first give a formal definition of the type of data that we are considering.

Definition 1 (Spatio-Temporal Database). *Let $\mathcal{T}$ be a time domain, let $\mathcal{S}$ be a set of spatial regions, and let $\mathcal{A}$ be a set of attribute domains. A spatio-*

temporal database $\mathcal{D}$ is a collection of tuples (t, s, a), where $t \in \mathcal{T}$ is a point of time, $s \in \mathcal{S}$ is a spatial region, and $a \in A$ is an attribute value of domain $A \in \mathcal{A}$.

Definition 2 (Attribute Time Series). *Let $T \subseteq \mathcal{T}$ be a time interval, let $s \in \mathcal{S}$ be a spatial region, and let $A \in \mathcal{A}$ be an attribute domain, then $TS_{s,A,T}$ is a time series, i.e. a function mapping each point of time $t \in T$ to an attribute value $a \in A$ for a region $s \in \mathcal{S}$. Thus, each region $s \in \mathcal{S}$ represents a specific time series:*

$$TS_{(s \in \mathcal{S}, A \in \mathcal{A}, T \subseteq \mathcal{T})} : T \to A$$

Our supervised classification approach assumes labeled ground-truth hazard information, such as the time and location of historical tropical cyclones. These can be obtained either from a natural hazard database such as the NOAA natural hazard database[1] or by employing the unsupervised outlier detection approach presented in Sect. 3. Thus, in the following, we assume a set GT of labeled ground-truth events, such that each element $(s, \mathcal{A}, T) \in GT$ describes a hazard at a spatial region s, during a time interval T, using a set of attributes $\mathcal{A}$. In the following we will discuss the second main building block of our framework: the tensor factorization and classification.

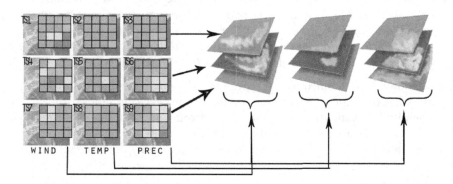

Fig. 1. Creating a tensor from precipitation, temperature and wind speed data.

2.1 Environmental Tensor Factorization

For our supervised framework the exact position (latitude, longitude) of possible labeled outliers and their vicinity are taken to form an environmental tensor X, formally:

Definition 3 (Environmental Tensor). *Let $(s, \mathcal{A}, T) \in GT$ be an event, and let $S = \{s_1, ..., s_n\}$ be a set of spatial regions around the event region s. Then we use the set of time series $TS(s \in S, T, A \in \mathcal{A})$ to construct a space-time-attribute tensor as follows:*

$$X(1 \le i \le |S|, 1 \le j \le |\mathcal{A}|, 1 \le k \le |T|) = TS(s_i, A_j, T)(k)$$

---

[1] https://www.ngdc.noaa.gov/hazard/.

Intuitively, $X(i, j, k)$ is the k'th value of the time series of attribute A_j in region s_i. For a specific point in region s, a set of spatial regions $S = \{s_1, ..., s_n\}$ in spatial vicinity of s, a time interval T, and attribute domains $\mathcal{A} = \{A_1, ..., A_{|\mathcal{A}|}\}$, the set of time series $TS_{(s \in S, A \in \mathcal{A}, T)}$ constitutes the three dimensional tensor X.

Figure 1 depicts the construction of *one* environmental tensor. For each of the three considered attributes $A \in \mathcal{A}$, for each of the 16 considered spatial regions $s \in S$, we obtain a time series. On the left of Fig. 1, we show three points of time for each attribute. A 3-mode tensor X is created from these time series by, concatenating the time series of each spatial location, yielding a two dimensional space-time matrix, which are concatenated over all attributes, yielding a space-time-attribute tensor corresponding to the tensor defined in Definition 3.

To derive latent features from such an environmental tensor, we employ a tensor factorization.

Definition 4 (3-Mode Tensor Factorization). *A classical CANDECOMP/ PARAFAC* [10, 16] *tensor factorization decomposes a tensor $X(S, T, \mathcal{A})$ as follows*

$$X(S, T, \mathcal{A}) = U \circ V \circ W + error$$

where $\circ$ denotes the 3-mode tensor product ([15]), U is a $|S| \times K$ matrix, V is a $|T| \times K$ matrix and W is a $|\mathcal{A}| \times K$ matrix. The parameter K is a parameter of the tensor factorization, controlling the number of latent features extracted for each mode. Furthermore, error corresponds to the loss of information incurred by factorizing a large three mode tensor into three small matrices.

The 3-mode tensor defined above represents the environmental information of a single labeled region. For creating a model of all existing labeled regions and their vicinity we want to decompose *all* environmental tensors simultaneously. This can be done in two ways: First is making use of the special case that all environmental tensors have all three modes in common by applying a Four-Mode Tensor Factorization:

Definition 5 (Environmental Four-Mode Tensor). *For n environmental tensors $X_1(s_1, t_1, \mathcal{A}_1), ..., X_n(s_n, t_n, \mathcal{A}_n)$, the environmental four-mode tensor $\mathcal{X}$ is defined as:*

$$\mathcal{X}(1 \leq i \leq |S|, 1 \leq j \leq |\mathcal{A}|, 1 \leq k \leq |T|, 1 \leq l \leq n) = X_l(i, j, k)$$

having

$$\mathcal{X}(U, V, W, ID) = U \circ V \circ W \circ ID$$

as the four-mode tensor factorization (4MTF) with U, V, W defined as in Definition 4, $\circ$ denoting the 4-mode tensor product, and ID is a $n \times K$ matrix describing latent features of each environmental tensor, and thus, of each environmental 3-mode tensor stored in the 4-mode tensor $\mathcal{X}$.

Besides the 4MTF where all three modes of the environmental tensors are coupled, it is possible to couple only one dimension of these tensors. This is done by coupled tensor-tensor factorization (CTTF), which works similar for this case as the coupled matrix-tensor factorization (CMTF) [1].

Definition 6 (Coupled Tensor-Tensor Factorization). *Given* n *environmental tensors* $X(s, T, \mathcal{A})$ *and let them be sharing only the attribute mode* $\mathcal{A}$, *their decomposition optimizes:*

$$\frac{1}{2}\|s - (A_1 \cdot A_2 \ldots \cdot A_n)\|^2 \|T - (A_1 \cdot A_2 \ldots A_n)\|^2$$

where s *and* T *are the variable modes of the environmental tensors and* $A_1 \ldots A_n$ *is the number of fixed matrices over all environmental tensors for each attribute domain* A. *More details on coupled factorization can be found in* [1].

For the collective factorization of matrices, coupling is crucial in at least one mode. Else the latent factors (concepts) of the environmental tensors would be independent and thus incomparable. Semantically it is different to lock either only one mode (CTTF) or three modes as we will evaluate in the experimental Sect. 4. The overall goal of the environmental tensor factorization is:

- finding discriminative latent attributes describing the environment around labeled environmental hazards,
- reducing the dimensionality of the environmental tensors,
- finding similarities among different spatial regions using matrix U, among different time intervals using matrix V, among different environmental attributes using matrix W, and among different environmental events using matrix ID.

In the next subsection, we employ the two tensor factorization approaches 4MTF (Definition 5) and CTTF (Definition 6) to classify new unlabeled environmental hazards given a set of labeled environmental hazards, using the environmental tensors extracted with knowledge about time and location of the labeled hazards.

2.2   Classifying Natural Hazards

After the environmental tensor factorization, we employ a black box of classifiers to model historic natural hazards that can predict future occurrences of such hazards basing on the historic data. We note that these classification algorithms are not the main focus of this work. Depending on which tensor factorization is used we have two different inputs for a given classifier:

- CTTF: all uncoupled modes, in this case the space s and time T.
- 4MTF: in the 4-mode case, the modes U, V, W describe the *common* aspects between *different* tensors regarding their attributes/time/space. The ID, in contrast, describes for *each* tensor its *individual* aspects. Thus, using these individual aspects might lead to a discrimination between the tensors and can thus be used as input features to a classifier. Only the fourth mode ID is uncoupled and as such used as input for the classification.

These latent features can be fed to traditional classification algorithms. The quality of these algorithms on our factorized latent features will be explored in our experimental evaluation in Sect. 4.

The main computation drawback of our event classification approach is that an environmental 3-mode tensor, comprising information about space, time and environmental attributes of an event is highly dense, thus incurring high computational cost in the tensor factorization. To solve this bottleneck, the aim of the next section is to reduce the density of such tensors, by setting non-interesting attribute-location-time triples to zero. For this purpose, we correlate the attribute time-series at a given location to time-series in the vicinity. At points in time where this correlation is extremely high, we argue that nothing interesting is happening in this location, thus that we do not need to explicitly model these parts of the tensor.

3   Spatio-Temporal Tensor Sparsification

The aim of this section is to find regions whose attribute information, such as temperature, differ significantly from other regions in the vicinity. This approach is used as a pre-processing step for our classification presented in Sect. 2. This step has two purposes: first, it can be used to reduce the data tensor to prune (set to zero) time-series of locations and attributes that are strongly correlated to their surroundings. This indicates that nothing interesting is happening for the given location, attribute pair. Second, by finding extremely low correlations, we can use this approach to identify and label anomalous environmental events in cases where these are not given by a authoritative ground-truth.

For detecting natural hazards finding outliers without prior knowledge of the given data set is the first crucial task. Our spatio-temporal outlier detection finds extreme regions using a calculated environmental extremeness measure (EEM). We define this measure by using Pearson correlation of a time-series to other time-series of regions in spatially close vicinity. In general consider the example in Fig. 2, where the time series $TS_{s,A,T}$ of the attribute $A =$ temperature is shown for 17 different regions s for some time interval T. Most of these curves show a similar behavior in terms of temperature change. Only the highlighted green time series shows little correlation to the other time series. As such it is regarded an outlier. To formalize such outliers, we define the notion of our EEM for spatio-temporal data as:

Definition 7 (Environmental Extremeness Measure). *Let T be a time interval, let $A \in \mathcal{A}$ be an attribute domain, and let $s_1, s_2 \in S$ be two spatial regions, then*

$$Cor(TS_{(s_1,A,T)}, TS_{(s_2,A,T)}) :=$$

$$\frac{\sum_{t \in T}(TS_{(s_1,A,T)}(t) - \overline{(TS_{(s_1,A,T)}(t))})(TS_{(s_2,A,T)}(t) - \overline{(TS_{(s_2,A,T)}(t))})}{\sqrt{\sum_{t \in T}(TS_{(s_1,A,T)}(t) - \overline{(TS_{(s_1,A,T)}(t))})^2 \sum_t (TS_{(s_2,A,T)}(t) - \overline{(TS_{(s_2,A,T)}(t))})^2}}$$

denotes the correlation of the time-series of regions s_1 and s_2 in attribute A during time T, where $\overline{(TS_{(s_1,A,T)}(t))}$ denotes the mean of all values, respective for region s_1. For a set of spatial regions $S \subseteq \mathcal{S}$, we define our environmental extremeness score of a region s as the average correlation to all time-series of S, formally:

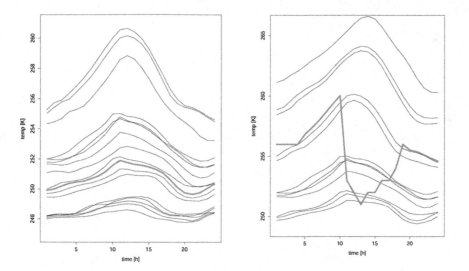

Fig. 2. Example temperature data for 17 neighboring geo-locations. Every line corresponds to one location. The green line corresponds on the left hand to a closely correlated region compared to the others, on the right hand it is an extreme outlier with even negative correlation. (Color figure online)

$$EEM(s, S, A, T) = \sum_{s' \in S} Cor(TS_{(s,A,T)}, TS_{(s',A,T)})/|S|$$

The current literature defines an outlier and its difference from other points in terms of distance and density [6, 14]. For spatio-temporal objects, we need to adapt this definition. Clearly, we could employ a time-series distance to measure the similarity of two spatio-temporal regions (s_1, t_1) and (s_2, t_2), such as Euclidean distance or Dynamic Time Warping [2]. However, this might not be possible as two regions may have similar trends of contextual information over time, but their absolute values may be different as illustrated in the following example.

Example 1. Given in Fig. 3 are two time series A, B showing the temperature (in Kelvin) to different time frames. A and B in Fig. 3(left) are showing the same functional behavior for different regions: A is for example the temperature measured in a valley while B is showing the temperature on a mountain. Thus, their actual base temperatures do not share any similarity. However in Fig. 3(right), time series B starts at around the same temperature level as time series A but does show a quite different altitude over time. Now, Dynamic Time Warping and Edit distance would both choose the two time series in Fig. 3(right) as the most similar time series as their temperature values differ less between and the space between A and B (yellow) is minimized. But regarding Definition 7 we

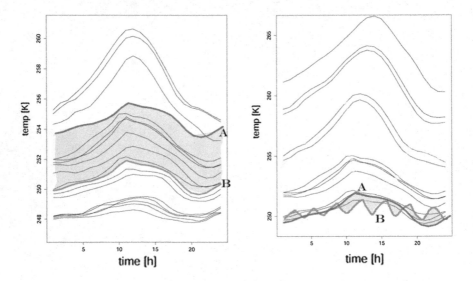

Fig. 3. Correlation vs. DTW and Edit Distance. Correlating the red and green temperature curves would regard the curves in the left plot more similar than the ones on the right. Using DTW or Edit Distance the curves on the right are more similar than the ones on the left since the covered area (yellow) is smaller. (Color figure online)

would want our measure to regard the temperature curves in Fig. 3(left) as similar (even as exact match) because the amount the temperature *changes* between these two time series are the same. Correlation reasonably considers the similarity of two time series matching our intuition and the application-specific requirements from geospatial data. Moreover, correlation is efficient to compute.

Following Tobler's first law of geography, two regions are expected to have a similar context (attribute values) if they are spatially close. This allows a straightforward extension of the general concept for spatio-temporal outliers:

Intuitively, a region $s \in S$ is called a spatio-temporal outlier, if the attribute A in region s during time interval T, is so different from the attribute at time T in regions spatially close to s, such that the suspicion arises, that it appears to be generated by a different process. More formally, we define a spatio-temporal outlier as follows.

Definition 8 (Spatio-Temporal Outlier). *Let $\mathcal{D}$ be a spatio-temporal database, let τ be a real value, let $s \in S$ be a spatial region and let $S \subseteq \mathcal{S}$ be a set of spatial regions close to s. Region s is called a spatio-temporal outlier for attribute A during time T if*

$$EEM(s, S, A, T) \leq \tau.$$

According to Definition 8, a spatio-temporal outlier is a region whose average EEM value is below a user specified threshold τ. To reduce the density of an environmental 3-mode tensor X, when a time series is not considered as an outlier at a time, the corresponding EEM value is set to zero. Parameter τ is chosen individually for each attribute domain $A \in \mathcal{A}$. In our experimental evaluation, we chose this parameter empirically.

To illustrate this approach, reconsider Fig. 1, and assume that the color intensities of cells correspond to their EEM score. By setting all green values to zero, we can reduce the density of the tensor, while still keeping information on the most relevant time periods.

3.1   Algorithmic procedure

Our overall goal is to find spatio-temporal outliers for a specific region. Applying Toblers Law of Geography [24], we intend to find such regions by comparing the correlation between this region to the other surrounding regions by using the formula in Definition 7 on each of them. This correlation calculated with knowledge of the vicinity of the region is called an environmental extremeness measure (EEM). To restrict the areas of these surrounding regions we have to define a radius in which we regard the regions as "connected":

Definition 9 (Spatial Radius). *Let* $r \in \mathbb{N}$. *Let* s *be a region, then* $Radius(s, r)$ *is the set of spatial regions having a distance to* s *that is smaller than* r.

With this definition we are able to compute for a given time interval T, a given attribute A, and a given spatial region s, the average correlation between s and the set of its spatial neighbors $Radius(s, r)$ – and accordingly the set of outliers by collecting all regions whose EEM score is below τ.

4   Experimental Evaluation

Our experimental evaluation includes two steps: (a) an evaluation of our unsupervised spatio-temporal outlier detection method using data where we synthetically added outliers to a sample set of real MERRA data and (b) an evaluation of our full framework applying unsupervised approaches like the outlier detection from (a) as well as supervised approaches like tensor factorization and classification. The full framework is not only capable of finding outliers but works also as a multi-solution tool for attribute selection and mining massive data. We start by introducing the used large real data set.

4.1   Global Climate Data

The open source data collection MERRA, used in this paper, is provided by NASA [17] and consists of more than six gigabytes of spatio-temporal environmental data per day, having collected more than 70 terabytes of data in the last years.[2] The MERRA time period covers the modern era of remotely sensed data, from 1979 until today (as of February 2017 for this evaluation), and covers

---

[2] https://gmao.gsfc.NASA.gov/products/documents/Merra_File_Specification.pdf.

a large variety of hundreds of environmental parameters on a spatial resolution of 0.5° latitude times 0.67° longitude produced at one-hour intervals.

For our supervised classification experiments, we used the *Statistical Hurricane Intensity Prediction Scheme (SHIPS)* database [5,26]. This database contains location and time of 800 tropical cyclones (TCs) from 1984–2011. Each TC is recorded every six hours. Each observation of a TC is also enriched with many attributes of the TC, such as wind-speed and air-pressure.

4.2   Finding Synthetic Spatio-Temporal Outliers

The first step for proving our concept for spatio-temporal outlier detection is showing whether we are generally able to find given outliers. Since outliers, i.e., triples of space s, time t and attribute A, cannot be verified easily without traveling back in time to t, we generate synthetic outliers. For this purpose, we randomly *distort* real time series, by changing the corresponding attribute values. We distort a time-series in two ways: (i) the number of values that are changed in the time series, ranging from a single changed value up to 20 changed values; and (ii) the magnitude v of the change itself, i.e. how strong is the original value altered, ranging from 10 to 100 of the original value. The effect of this change is of course also regarded to the given measurement scale of the three attributes. Temperature is in Kelvin, wind speed in km/h and precipitation in mm/h. The radius for the size of the vicinity is set to a manhattan distance of 1.

Figure 4 shows the result of our outlier detection for three attributes *Wind Speed, Temperature* and *Precipitation*. We measure the fraction of distorted time-location pairs for which we distorted the corresponding time series. As expected, we observe that a larger magnitude v of distortion increases the fraction of distortions found as outliers. At the same time, an increased duration of distortion, specified by the number of values changed, also improves the detection rate. Overall, we can see that our unsupervised outlier detection approach is able to detect environmental time series, whose attribute values are sufficiently distorted in terms of the number of time series values changed and in terms of magnitude of this change, with high accuracy. However, the more applicative problem of supervised classification of environmental events, given a set of labeled events, is evaluated in the next section.

4.3   Finding Natural Hazards on Real Data

Our experimental evaluation for the supervised approach aims at classifying Rapid Intensification (RI) of tropical cyclones [11] in the Atlantic ocean using Merra data. For each TC in the SHIPS dataset, at time t and location s, we obtained an environmental tensor (as defined in Definition 3) from the Merra dataset as follows: We queried the MERRA dataset for the 24-hour time interval before t, in all 25 spatial regions having a manhattan distance of at most two to s, using all 387 environmental attributes available in Merra. This way, we were able to link each TC available in the SHIPS data set, to a size $25 \times 24 \times 387$ tensor of environmental measurements. We labeled each TC as "RI" if the wind

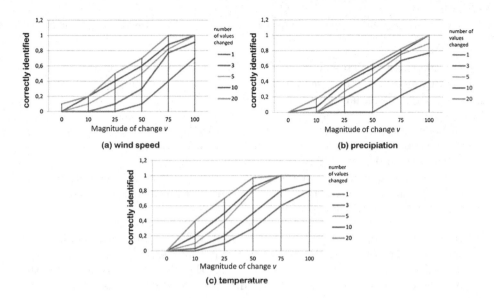

Fig. 4. Finding synthetic outliers in real world data of MERRA for the attributes precipitation, temperature and wind speed. The parameter v denotes the changed value added to the original value. The vertical axis shows the percentage of correctly identified synthetic outliers.

speed of the TC increased by at least 30 knots within the next 24 h, or "non-RI" otherwise. About 5% of TCs were labeled as "RI" this way. The task is to predict this label. We evaluated the quality of our proposed algorithms using F1 measure on the two classes "RI" and "non-RI". For reference, according to [12], the rapid intensification forecasting from the National Hurricane Center (NHC) official forecast at 24 h lead time for the Atlantic Basin reached only a 10% probability of detection with more than a 30% false alarm rate. We evaluate the following algorithms.

- Competitor [26]: A data mining approach recently presented in [26], which uses only data in the SHIPS data set, which includes less than two hundred parameter per TC, and uses standard classification algorithms on these features. According to the results of [26], this approach had an F1 measure of no more than 60% for both classes in any experiment.
- Baseline: The baseline is provided by putting the raw data, without any dimensionality reduction (4MTF or CTTF) into the classifiers.
- Orig-CTTF and Orig-4MTF: These two algorithms apply our two tensor factorization approaches (either CTTF or 4MTF) on the raw data tensors, and use the resulting latent features for classification.
- EEM-CTTF and EEM-4MTF: These two algorithms use the full framework, thus applying the tensor sparsification using our environmental extremeness measure (EEM). Then, tensor factorization (CTTF or 4MTF) is applied to

this pre-processed tensor and the resulting latent features are used for classification.

For the classification step, we use the following standard Matlab implementations, using 10-fold cross validation. If the parameter setting is not provided, it is kept at the standard given by Matlab:

- SVM support vector machine with linear kernel.
- SVM support vector machine with polynomial kernel.
- Decision Tree. Decision Tree classification.
- k-nearest neighbors (knn) classification using $k = 10$ nearest neighbors.
- kNN Ensemble using random subspaces and $k = 10$ in each subspace.
- Random Forrest Ensemble using RUSBoost.

In all cases except EEM-CTTF and EEM-4MTF, the data was standardized before using the classifiers. For these two experiments this was not necessary as the EEM already normalizes the input for the classifiers. In addition, the algorithms were set up to take the prior distribution of the labels into account.

Results. Our tropical cyclone rapid intensification classification results are shown in Fig. 5, depicting the F1 measure for the two classes "RI" and "non-RI". Overall, we can see that all proposed approaches using the full information contained in the spatio-temporal tensor of a TC, including the baseline, yield a higher classification quality than the classification of features of the SHIPS dataset in [26], which showed a F1-measure of no more than 60% in any experiment. This promising result proves that our concept, of joining a natural hazard database (such as the SHIPS database) with a spatio-temporal environmental database (such as MERRA), has the potential to improve the classification and forecast of natural hazards beyond the state-of-the-art of domain-specific solutions, by exploiting the wealth of public environmental data. We want to discuss this promising result at SSTDM, in order to discuss potential new applications, of joining a set of known historic events, with MERRA.

Comparing the five approaches using spatio-temporal tensors obtained from MERRA, we note that our baseline solution performs very well. In fact, our solutions using tensor factorization are not able to outperform our baseline classification for many classification algorithms in many settings. This indicates that our four tensor factorization based algorithms still allow room for future research

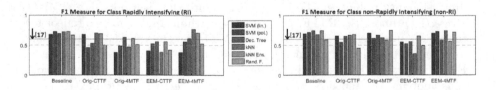

Fig. 5. Tropical Cyclone classification results.

and improvement. Still, see that our most sophisticated approach *EEM-CTTF*, which employs our environmental extremeness measure (cf. Definition 8) and the coupled Tensor-Tensor factorization (c.f. Definition 6) is able to achieve a much higher classification rate using Random Forests, than any other algorithm can achieve using the baseline features. This improvement is quite significant, and shows that our approach of factorizing space, time and environmental attributes can indeed be used to learn and classify complex environmental phenomena, such as the prediction of whether a TC will rapidly intensify in wind speed.

5   Related Work

5.1   Spatio-Temporal Outlier Detection

Large spatio-temporal environmental data has created an enormous interest in findings patterns in earth climatic changes. In [8], the authors introduce climate challenges to the data mining community and compare and contrast climate data mining to spatial and spatio-temporal data mining. The authors of [7] also did a similar study and present the challenges that spatio-temporal data mining (STDM) faces within climate data. The authors also apply STDM pattern mining to identify mesoscale ocean eddies from satellite data and provides concrete examples of challenges faced when mining climate data and how effectively analyzing the data's spatio-temporal context may improve existing methods' accuracy, interpretability, and scalability. However, none of these works are able to find spatio-temporal environmental outliers efficiently.

Steinbach *et al.* [22] argue to apply principal components analysis (PCA) and singular value decomposition (SVD), to discover climate indices. Our outlier detection algorithm is closest related to [3], where a neighborhood-based outlier detection algorithm is provided that computes similarity score for a region represented by a $n \times m$ matrix by considering its r-point neighborhood. Unlike us, the authors detect outliers over space and time only *separately* to explain some of the extreme events like drought and severe rainfall at specific locations on earth. For the use case of finding tropical cyclones, [27], Yang *et al.* used association rule mining to look for combinations of persistent and synoptic conditions which provide improved RI (Rapidly Intensifying) probability estimates. Similar in [28], the authors apply association rule mining on the analysis of intensity changes of North Atlantic tropical cyclones.

5.2   Classification and Prediction

As mentioned earlier, in addition to outlier detection, classification algorithms can be useful when trying to find, characterize and especially predict natural hazards. For instance, [19] provides examples for classification being used in combination with spatio-temporal data. More specifically, these works employ spatio-temporal data obtained from satellite images to classify types of vegetation. Clearly, identifying vegetation from environmental time series data using our framework will be an interesting new case study. The authors use Support Vector Machines (SVM) in order to identify cartographic regions with similar behaviours from Satellite Image Time Series. In [4], data from the weather forecast model Eta was used to identify patterns which can be associated with unusual weather activity. The results of the study showed good classification performance for regions in Brazil during 2007, yet similar to the approaches of Réjichi and Chaabane [19,20] they can only help to identify natural hazards in past data, but do not offer functionality to predict such extreme events. Wang and Ding [25] use classification in their three step approach to build a forecasting model for extreme rainfalls using location-based patterns. This approach can be used alternatively to our EEM-measure to find occurrences of spatio-temporal outliers on an unlabeled dataset. A comparative study, using different environmental outlier detection approaches is planned for our future work. The survey in [15] gives an overview over implementations of tensor factorizations used in this work.

5.3   Tensor Factorization

Like matrix factorizations, tensor decompositions have been used in a wide range of applications and for similar wide range of purposes. In [13], the authors apply both, nonnegative matrix factorization as well as nonnegative tensor factorization to extract features from sea ice SAR images and compare the results. In addition, the extracted features are utilized in classification algorithms. They were able to show that both methods lead to meaningful results and that both can be used in classification. The study conducted on structural magnetic resonance images of Alzheimer patients in [29] draws a comparable conclusion. Yang and Cui also used tensor factorization in combination with a support vector machine to classify the patient data. Those findings are of interest as in our work, we likewise apply tensor factorization to extract features for classification. In general, tensor decompositions are implemented in two ways: CANDECOMP/ PARAFAC (CP), which decomposes a tensor into the sum of rank-one tensors, and the Tucker decomposition, which can be considered as a higher-order form of principal component analysis (PCA). The survey in [15] gives an overview over the two and more implementations, including advantages. Since the MERRA data set is extremely massive, scalability is a critical factor for our project. In [23], an algorithm implementing non-negative tensor factorization is introduced. The authors use auxiliary tensors to speed up the decomposition of highly sparse tensors. Shin and Kang [21] propose a distributed approach to calculating the decomposition of a tensor. Two algorithms are introduced, both of which work based on MapReduce. Likewise, the algorithm HaTen2, which is described in [10], uses MapReduce in combination with both, Tucker and PARAFAC decompositions. Another interesting approach is presented in [16]. The algorithm Par-Cube is well suited for sparse tensors, scales to truly large datasets and is highly parallelizable.

6   Conclusion

Concluding, we presented a framework to solve several hypotheses regarding spatio-temporal environmental data at once with high accuracy. For that we proposed an Environmental Extremeness Measure (EEM) score to find local outliers in the space-, time- and attribute-space. On top of this unsupervised learning approach, we created a model and predictor consisting of three building blocks: sparsifying the data based on the EEM score, creating an environmental tensor and use CTTF or 4MTF on it and classify or predict this data. Our experimental evaluation shows that this supervised approach can be successfully applied to problems in environmental sciences, such as the prediction of rapid intensification of tropical cyclones.

Acknowledgments. This research was supported by the Technical University of Munich - Institute for Advanced Study, funded by the German Excellence Initiative and the European Union Seventh Framework Programme under grant agreement no. 291763, co-funded by the European Union.

References

1. Acar, E., Kolda, T.G., Dunlavy, D.M.: All-at-once optimization for coupled matrix and tensor factorizations. arXiv preprint arXiv:1105.3422 (2011)
2. Berndt, D., Clifford, J.: Using dynamic time warping to find patterns in time series. In: KDD Workshop, pp. 359–370 (1994)
3. Das, M., Parthasarathy, S.: Anomaly detection and Spatio-temporal analysis of global climate system. In: SensorKDD 2009 (2009)
4. de Lima, G.R.T., Stephany, S.: A new classification approach for detecting severe weather patterns. Comput. Geosci. **57**, 158–165 (2013)
5. DeMaria, M., Kaplan, J.: A statistical hurricane intensity prediction scheme (ships) for the atlantic basin. Weather Forecast. **9**(2), 209–220 (1994)
6. Ester, M., Kriegel, H.-P., Sander, J., Xu, X.: A density-based algorithm for discovering clusters in large spatial databases with noise, pp. 226–231 (1996)
7. Faghmous, J.H., Kumar, V.: Spatio-temporal data mining for climate data: advances, challenges, and opportunities. In: Chu, W. (ed.) Data Mining and Knowledge Discovery for Big Data. Studies in Big Data, vol. 1, pp. 83–116. Springer, Heidelberg (2014). doi:10.1007/978-3-642-40837-3_3
8. Ganguly, A.R., Steinhaeuser, K.: Data mining for climate change and impacts. In: ICDMW, pp. 385–394 (2008)
9. Hey, T., Tansley, S., Tolle, K.M., et al.: The Fourth Paradigm: Data-Intensive Scientific Discovery, vol. 1. Microsoft research Redmond, WA (2009)
10. Jeon, I., Papalexakis, E.E., Kang, U., Faloutsos, C.: Haten2: billion-scale tensor decompositions. In: ICDE (2015)
11. Kaplan, J., DeMaria, M.: Large-scale characteristics of rapidly intensifying tropical cyclones in the North Atlantic Basin. Weather Forecast. **18**(6), 1093–1108 (2003)
12. Kaplan, J., et al.: Evaluating environmental impacts on tropical cyclone rapid intensification predictability utilizing statistical models. Weather Forecast. **30**(5), 1374–1396 (2015)

13. Karvonen, J., Kaarna, A.: Sea ice SAR feature extraction by non-negative matrix and tensor factorization. In: 2008 IEEE International Geoscience and Remote Sensing Symposium (IGARSS 2008). vol. 4, pp. IV-1093 (2008)
14. Knorr, E.M., Ng, R.T.: Algorithms for mining distance-based outliers in large datasets, pp. 392–403 (1998)
15. Kolda, T.G., Bader, B.W.: Tensor decompositions and applications. In: SIAM, pp. 455–500 (2009)
16. Papalexakis, E.E., Faloutsos, C., Sidiropoulos, N.D.: ParCube: sparse parallelizable tensor decompositions. In: Flach, P.A., Bie, T., Cristianini, N. (eds.) ECML PKDD 2012. LNCS, vol. 7523, pp. 521–536. Springer, Heidelberg (2012). doi:10.1007/978-3-642-33460-3_39
17. Rienecker, M., Suarez, M., Todling, R., Bacmeister, J., Takacs, L., Liu, H., Gu, W., Sienkiewicz, M., Koster, R., Gelaro, R., et al.: The GEOs-5 data assimilation system-documentation of versions 5.0.1, 5.1.0, and 5.2.0. NASA Technical Memorandum 104606(27):2008 (2008)
18. Rienecker, M.M., Suarez, M.J., Gelaro, R., Todling, R., Bacmeister, J., Liu, E., Bosilovich, M.G., Schubert, S.D., Takacs, L., Kim, G.-K., et al.: Merra: Nasa's modern-era retrospective analysis for research and applications. J. Clim. **24**(14), 3624–3648 (2011)
19. Réjichi, S., Chaabane, F.: Feature extraction using PCA for VHR satellite image time series Spatio-temporal classification. In: IGARSS, pp. 485–488 (2015)
20. Réjichi, S., Chaâbane, F.: SVM Spatio-temporal vegetation classification using HR satellite images, vol. 8176 (2011)
21. Shin, K., Kang, U.: Distributed methods for high-dimensional and large-scale tensor factorization. In: 2014 IEEE International Conference on Data Mining, pp. 989–994, December 2014
22. Steinbach, M., Tan, P.-N., Kumar, V., Klooster, S., Potter, C.: Discovery of climate indices using clustering. In: KDD, pp. 446–455 (2003)
23. Takeuchi, K., Tomioka, R., Ishiguro, K., Kimura, A., Sawada, H.: Non-negative multiple tensor factorization. In: 2013 IEEE 13th International Conference on Data Mining, pp. 1199–1204, December 2013
24. Tobler, W.R.: A computer movie simulating urban growth in the detroit region. Econ. Geogr. **46**, 234–240 (1970)
25. Wang, D., Ding, W.: A hierarchical pattern learning framework for forecasting extreme weather events. In: ICDM (2015)
26. Yang, R.: A systematic classification investigation of rapid intensification of atlantic tropical cyclones with the ships database. Weather Forecast. **31**(2), 495–513 (2016)
27. Yang, R., Tang, J., Kafatos, M.: Mining "optimal" conditions for rapid intensifications of tropical cyclones. In: Program of the 19th Conference on Probability and Statistics. American Meteorological Society (2008)
28. Yang, R., Tang, J., Sun, D.: Association rule data mining applications for atlantic tropical cyclone intensity changes. Weather Forecast. **26**(3), 337–353 (2011)
29. Yang, W., Cui, Y.: Tensor factorization-based classification of Alzheimer's disease vs healthy controls. In: 2012 5th International Conference on Biomedical Engineering and Informatics (BMEI), pp. 371–374. IEEE (2012)

Localization and Spatial Allocation

FF-SA: Fragmentation-Free Spatial Allocation

Yiqun Xie$^{(\boxtimes)}$ and Shashi Shekhar

Department of Computer Science and Engineering,
University of Minnesota, Twin Cities, USA
{xiexx347,shekhar}@umn.edu

Abstract. Given a grid G containing a set of orthogonal grid cells and a list L of choices on each grid cell (e.g., land use), the fragmentation-free spatial allocation (FF-SA) aims to find a tile-partition of G and choice assignment on each tile so that the overall benefit is maximized under a cost constraint as well as spatial geometric constraints (e.g., minimum tile area, shape). The spatial constraints are necessary to avoid fragmentation and maintain practicality of the spatial allocation result. The application domains include agricultural landscape design (a.k.a., Geodesign), urban land-use planning, building floor zoning, etc. The problem is computationally challenging as an APX-hard problem. Existing spatial allocation techniques either do not consider spatial constraints during space-tiling or are very limited in enumeration space. We propose a Hierarchical Fragmentation Elimination (HFE) algorithm to address the fragmentation issue and significantly increase the enumeration space of tiling schemes. The new algorithm was evaluated through a detailed case study on spatial allocation of agricultural lands in mid-western US, and the results showed improved solution quality compared to the existing work.

1   Introduction

Given a $m \times n$ grid G containing a set of orthogonal grid cells g_{ij} and a list L of choices, where each choice on g_{ij} has a benefit value b_{ij} and cost c_{ij}, the fragmentation-free spatial allocation (FF-SA) aims to find a tile-partition of G and a choice for each tile to maximize the overall benefit under a cost constraint and spatial constraints (e.g., minimum tile area, shape) which enforce spatial contiguity and regularity. FF-SA is related to both optimization and computational geometry.

FF-SA is an important problem in many application domains. In agricultural landscape design (a.k.a., Geodesign) [1,2], land covers and management practices need to be determined for each spatial location in order to improve environmental or economic objectives. For example, maximizing food production under water quality constraint. The allocated land cover patches (e.g., continuous area of corn or wheat) must be geometrically large enough and regular in shape to allow efficient use of modern farm equipment (e.g., commonly used 40–60 feet wide combine harvester heads in mid-western US). Without the spatial constraints, a spatial allocation result often contains land fragmentation

© Springer International Publishing AG 2017
M. Gertz et al. (Eds.): SSTD 2017, LNCS 10411, pp. 319–338, 2017.
DOI: 10.1007/978-3-319-64367-0_17

(e.g., tiny land cover patches), which can significantly reduce farming efficiency [3,4]. Similarly, in urban land-use planning and building floor zoning [5,6], zones need also satisfy the spatial constraints to avoid undesired fragmentation, such as tiny nested zones of different types of land-uses (e.g., residential, industrial). FF-SA addresses these fragmentation issues using spatial geometric constraints.

FF-SA is approximation-hard (APX-hard), which implies NP-hardness. The APX-hardness shows that FF-SA has no polynomial time approximation scheme (PTAS) unless P = NP. The problem is computationally challenging given the large number of decision variables and constraints needed for real-world applications (e.g., million, trillion). Due to the hardness and scale of the problem, standard MIP solvers cannot efficiently solve FF-SA. Without modeling spatial constraints, conventional integer programming formulations have also been used for spatial allocation problems [2,5,6]. However, grid cells are treated as independent variables during optimization and this leads to fragmentation in the final results. In Geodesign Optimization [7], a dynamic growth tiling framework (DGTF) was used to avoid fragmentation during tiling scheme generation. The limitation of DGTF is that it only enumerates one full tiling scheme in the heuristic optimization process, which means its enumeration space of tiling scheme is very small. The new approach targets improving the solution quality of FF-SA by significantly increase the enumeration space of tiling schemes during optimization. Details are discussed in Sects. 4 and 5.

In this work, we prove that the hardness class of FF-SA is at least APX-hard beyond NP-hard, which further encourages research on heuristic solutions instead of exact or approximation algorithms. Then, we propose a new algorithm based on dynamic programming, namely Hierarchical Fragmentation Eliminator (HFE), to better approach an upper bound on optimal solution quality by significantly increasing the enumeration space of tiling schemes. An acceleration algorithm based on multi-layer integral image is also proposed to improve the performance of the new approach. The computational time complexity of the new algorithms are analyzed in details.

A detailed case study in agricultural spatial allocation was performed to evaluate the proposed algorithms. The case study was carried out at Seven Mile Creek watershed (Minnesota, US) which has an area of 25,000 acres. Experiment results show that: (1) the HFE algorithm was able to eliminate fragmentation existed in conventional integer programming (no spatial constraints) solution; (2) HFE consistently produced better spatial allocation solutions compared to DGTF [7] in the study area; (3) the acceleration algorithms significantly improved the computational performance of HFE.

2   Problem Statement: Fragmentation-Free Spatial Allocation

The problem formulation for FF-SA inherits the formulation of Geodesign Optimization problem [7] and is generalized to a broader domain. Some key concepts are defined as follows: **(1) Grid:** A grid partition contains m rows and n columns.

Each grid cell is identified by its row i and column j. Each grid cell g_{ij} has d choices, and each choice k has a benefit b_{ijk} and cost c_{ijk}. The choices for each grid cell are the same, but the benefit and cost values for each choice vary across grid cells. **(2) Tile:** A tile is rectangular in shape. Each tile $tile_t = (i_0, j_0, i_1, j_1)$ is identified by its top-left grid cell (i_0, j_0) and bottom-right grid cell (i_1, j_1). The choice made on all grid cells within a tile is the same. For choice k on a tile $tile_t$, the cost c_{tk} and benefit b_{tk} are the sum of b_{ijk} and c_{ijk} of the grid cells g_{ij} inside $tile_t$. Given N grid cells, N^2 tiles can be uniquely defined. **(3) Tiling scheme:** A tiling scheme is a tile-partition of the study area. Tiles in a tiling scheme must not have overlaps. FF-SA is formulated as:

Inputs:

- A grid G and a list L of choices;
- A list Z contains all combinations of tiles and choices with the corresponding benefit and cost: $\{tile_t : (i_0, j_0, i_1, j_1), choice : k \in L, b_{tk}, c_{tk}\}$;
- A limit on total cost c_{tot}, minimum tile area α, minimum tile width (also height) β.

Output: A tiling scheme with choice assignment, which is a subset Z' of Z.

Objective: Maximize the total benefit: $\sum_{p=1}^{|Z|} (Z_p.b) \cdot s_p$, where $Z_p.b$ is the benefit value of Z_p, and s_p is 1 if $Z_p \in Z'$ and otherwise 0.

Constraints:

- Binary value constraint on s_p: $s_p \in \{0, 1\}$
- Each tile can only have one choice: $\sum_{p \in V} s_p = 1, V = \{p | Z_p.choice = k\}, \forall k$
- Total cost is less than or equal to c_{tot}: $\sum_{p=1}^{|Z|} (Z_p.c) \cdot s_p \leq c_{tot}$
- Each element Z'_p in Z' must satisfy a minimum area α and width β:

$$i_1 - i_0 + 1 \geqslant \beta, i_1, i_0 \in Z'_p \tag{1}$$

$$j_1 - j_0 + 1 \geqslant \beta, j_1, j_0 \in Z'_p \tag{2}$$

$$(i_1 - i_0 + 1) \cdot (j_1 - j_0 + 1) \geqslant \alpha, i_1, i_0, j_0, j_1 \in Z'_p \tag{3}$$

- There is no spatial overlap among elements in Z':

$$\forall Z'_1, Z'_2 \in Z', Z'_1.tile \cap Z_2.tile = \phi \tag{4}$$

Constraints (1) to (4) are spatial constraints imposed on tiles to avoid fragmentation. Constraints (1) to (3) are single-tile constraints. For implementation, these constraints can be imposed by preprocessing, which removes Z_p in Z if $Z_p.tile$ does not satisfy the area and width constraints. Constraints (4) is a pair-wise constraint. A concrete mathematical formulation of (4) requires adding constraints (5) and (6) for all pairs of elements in Z (either (5) **or** (6) needs to be satisfied). Two elements Z_x and Z_y in Z are spatially disjoint only if none of the vertices in Z_x's tile falls into Z_y's tile. This indicates that each vertex (i, j)

of Z_x's tile must satisfy the following non-linear constraints when $Z_x \in Z'$ and $Z_y \in Z'$:

$$(i - i_0) \cdot (i - i_1) > 0, \ i_1, i_0 \in Z_y \ or \ s_x = 0 \ or \ s_y = 0 \tag{5}$$

$$\mathbf{or} \ (j - j_0) \cdot (j - j_1) > 0, \ j_1, j_0 \in Z_y \ or \ s_x = 0 \ or \ s_y = 0 \tag{6}$$

Scope of illustration: In the general FF-SA problem, the minimum area and width constraints for each choice may vary. For example, in agricultural spatial allocation, several land cover types (e.g., alfalfa) may not require large-size farm equipment operation while the others (e.g., corn) still require it. Thus, α and β can be varied for each choice. Since the proposed HFE algorithm can be trivially generalized to the case where different α and β are used for different choices, for simplicity and clarity of illustration, we assume α and β are the same for each choice in later sections to avoid overloaded details.

3 Challenges

APX-hardness. The FF-SA problem is APX-hard, and the APX-hardness implies NP-hardness. The hardness proof is based on reduction from the MAX-3-Satisfiability (MAX-3SAT) problem [8]. We show FF-SA is more general than the MAX-3SAT problem, which is APX-hard and does not admit a polynomial-time

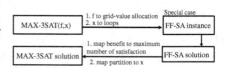

Fig. 1. Reduction graph.

approximation scheme (PTAS). We propose an extension of the proof in [9], which shows 3-Satisfiability (3SAT, a variant of MAX-3SAT) problem is a special case of box packing (BOX-PACK). In general, it was shown that for any 3SAT instance, a box packing instance can be constructed with a special design of box-allocation, in polynomial-time, and the optimal solution of box packing can be converted to the solution of the 3SAT problem in polynomial time. We generalize the proof in [9] to show that MAX-3SAT is a special case of FF-SA. Since the proof is non-trivial and lengthy, it is provided in Appendix A. The general reduction graph is shown in Fig. 1.

4 Related Work and Limitations

Without spatial constraints, there is no need to define the tiles and the spatial allocation problem becomes a 0-1 Multiple-Choice Knapsack Problem (MCKP) [10]. In MCKP formulation, the spatial relationship among grid cells is ignored and each grid cell is considered as an independent variable. There are multiple choices for each variable and each choice has a benefit and cost value. The optimization goal in MCKP is to maximize the benefit under a cost constraint. The 0-1 integer programming problem in MCKP is a well studied problem [10] and can be solved by dynamic programming with a large space cost. However, without modeling of spatial constraints, the allocation results often show large amount of fragmentation, which prohibits its use in domain applications.

In computational geometry, rectangular partitioning problems have been studied [11,12]. These problems (e.g., R-TILE, R-PACK) only have one fixed choice for each grid cell and there are no multiple choices to choose from. Thus, R-TILE and R-PACK problem do not concern with spatial contiguity issue and fragmentation is not modeled since there are no different choices to make for adjacent grid cells. The techniques in R-TILE and R-PACK were based on this single-choice property, and thus cannot be trivially applied to FF-SA. The spatially-constrained integer programming formulation of FF-SA can only be exactly solved for very small size problems (e.g., a 10 by 10 grid) given its computational challenge (e.g., APX-hardness result in Sect. 3). In this formulation, denote the number of grid cells as N, and the number of choices as d. Then the length of list Z is in $O(N^2 \cdot d)$, and the non-linear pair-wise constraints needed is $O((N^2 \cdot d)^2)$. This makes a 500 by 500 grid with 5 choices require at least 3.125×10^{11} elements (decisions) in Z and 9×10^{22} non-linear pair-wise constraints. [7] proposes a dynamic-growth tiling framework (DGTF) to find a heuristic solution of spatial allocation problem that honors spatial constraints. DGTF focuses on generating a single tiling scheme that satisfy minimum area and width constraints. It has a large potential search space but the algorithm only enumerates one complete tiling scheme within the search space. Thus, it has a very limited enumeration space. In addition, the tile-level choice-assignment phase in [7] is based on heuristic local-search. In this paper, we targets improving solution quality of FF-SA using a significantly larger enumeration space of tiling schemes, and an exact global optimizer for the tile-level choice assignment phase.

5   HFE: Hierarchical Fragmentation Eliminator

The APX-hardness result encourages the design of heuristic algorithms. We propose a Hierarchical Fragmentation Eliminator (HFE), to solve FF-SA with two-phases: (1) space tiling and (2) choice-assignment on tiles. Comparing to [7], the new approach aims to significantly increase the volume of enumeration space of solutions to improve the solution quality of FF-SA.

Search Space vs. Enumeration Space. In general, there exists three types of space tiling frameworks [7,11], namely (1) arbitrary; (2) hierarchical; and (3) $p \times q$. Arbitrary tiling framework considers all possible tiling schemes. Hierarchical tiling framework uses a straight-line to partition the current rectangle into two sub-rectangles at each step. A $p \times q$ tiling framework considers all schemes with p rows and q columns. A tiling algorithm generates tiling schemes based on these frameworks. We define the **search space** of a tiling framework as the space that contains all possible tiling schemes out of the framework. In contrast, **enumeration space** contains all tiling schemes that are actually enumerated by a tiling algorithm. Since arbitrary tiling framework contains all possible tiling schemes, it has the largest search space. Hierarchical tiling framework includes tiling schemes following the hierarchical bi-partition structure, so its search space is a subset of that of arbitrary tiling framework. $p \times q$ tiling framework is a special case of hierarchical tiling framework, which means its search space is a smaller subset. Although the tiling algorithm proposed in [7] considers a search space defined by arbitrary tiling scheme, it has a very limited enumeration space containing only a single complete tiling scheme. The number of tiling schemes covered by the new HFE algorithm is exponential to the number of grid cells in the input. HFE aims to use this enlarged enumeration space to explore new opportunities to improve solution quality.

5.1   Phase-1: Space Tiling

In phase-1, we propose a Hierarchical Fragmentation Eliminator (HFE) to find a tiling scheme of FF-SA. As discussed in Sect. 4, spatial allocation (SA) can be solved using conventional integer programming (IP) formulation without spatial constraints, and this approach is denoted as "IP-SA". IP-SA solution S_{IP}^* is able to maximize benefit under a given cost constraint but contains fragmentation. HFE uses the IP-SA solution S_{IP}^* as an input, and finds a hierarchical tiling scheme to eliminate the land fragmentation in S_{IP}^* with minimum number of choice-changes on grid cells.

The solution quality of IP-SA solution S_{IP}^* can be considered as an upper bound on FF-SA solution S_{FF}^* because the spatial constraints are relaxed. For each grid cell, consider the choice made by S_{IP}^* as the optimal choice, adding spatial constraints will force some non-optimal choices in order to maintain the minimum tile area and width.

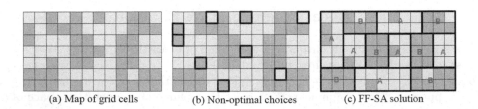

(a) Map of grid cells   (b) Non-optimal choices   (c) FF-SA solution

Fig. 2. A toy example of the spatial allocation problem (best in color).

Figure 2 shows a toy example of the spatial allocation problem to illustrate the intuition of HFE. In the example, the FF-SA problem aims to maximize the benefit under a cost constraint. There are 84 grid cells and 2 choices

Table 1. Benefit and cost values

Grid cell	Choice A	Choice B
Yellow	b = 2, c = 1	b = 1, c = 2
Green	b = 1, c = 2	b = 2, c = 1

b is benefit, c is cost.

"A" and "B" for each cell. For simplicity of demonstration, the cells are colored by green and yellow (to represent spatial variation), and the grid cells of the same color have the same benefit and cost values for each choice. The values are shown in Table 1. In this simplified example, we can see "A" is an optimal (dominant) choice for all yellow grid cells since it has a higher benefit of 2 and lower cost of 1; for green cells, "B" is an optimal choice. Suppose the cost constraint is 100. Without any spatial constraints, the optimal integer programming solution S_{IP}^* will always choose "A" for all yellow grid cells and "B" for green, and this gives a benefit of 168 and cost of 84. If we choose a non-optimal choice at a grid cell, the total benefit will decrease by 1 and cost will increase by 1. Suppose we impose two **spatial constraints** to construct the fragmentation-free spatial allocation (FF-SA) problem: minimum tile area 6 and minimum tile width 2 (unit: grid cell). With the constraints, we have to make non-optimal choices on some grid cells in order to satisfy the spatial constraints (in other words, avoid fragmentation). Here "non-optimal" means non-optimal in the IP-SA solution S_{IP}^*. For example, some grid cells are highlighted in Fig. 2(b). For each of the highlighted grid cells, we have to make non-optimal choices either on itself or on its neighbor; otherwise it is not possible to have a valid tile containing it since each tile must have a homogeneous choice. Thus, an optimal solution of FF-SA at least makes non-optimal choices at 9 grid cells. Based on this observation, we can see Fig. 2(c) gives an optimal FF-SA solution S_{FF}^* with a benefit of 159 and cost of 93.

Minimization problem of HFE. Within the search space containing all hierarchical tiling schemes, the newly proposed HFE finds a tiling scheme that can globally minimize the total number of choice-changes needed on grid cells to eliminate the fragmentation in the IP-SA solution S_{IP}^* (e.g., Fig. 2(b) and (c): 9 changes needed).

The total number of hierarchical tiling schemes can be exponential to the number of grid cells (e.g., when minimum area and width constraints are small). To avoid a brute-force search of all schemes, dynamic programming is used in [11] to minimize the total number of tiles in a hierarchical tiling scheme under a constraint on heft-function values. We show that the minimization problem of HFE also processes the two key properties needed for dynamic programming, namely optimal substructure property and overlapping sub-problems property.

Optimal substructure property: At each level, the hierarchical tiling framework partitions each of the rectangles at this level into two sub-rectangles using a straight line. Denote the range of each rectangle as (i_0, j_0, i_1, j_1), where (i_0, j_0) is top-left grid cell and (i_1, j_1) is the bottom-right. Denote the minimum number of choice-changes needed to eliminate fragmentation in range (i_0, j_0, i_1, j_1) as $M(i_0, j_0, i_1, j_1)$, the minimum tile area in FF-SA as α, and the minimum width as β. Since there are two directions to split a rectangular region into two sub-regions, we have:

$$M(i_0, j_0, i_1, j_1) = \min(M_{hor}, M_{ver}) \tag{7}$$

$$M_{hor} = \min_{x_{min} \leq x \leq x_{max}} [M(i_0, j_0, x, j_1) + M(x+1, j_0, i_1, j_1)] \tag{8}$$

$$\begin{cases} x_{min} = \max(i_0 + \beta - 1, i_0 + \lceil \frac{\alpha}{j_1 - j_0 + 1} \rceil - 1) \\ x_{max} = \min(i_1 - \beta, i_1 - \lceil \frac{\alpha}{j_1 - j_0 + 1} \rceil) \end{cases} \tag{9}$$

$$M_{ver} = \min_{y_{min} \leq y \leq y_{max}} [M(i_0, j_0, i_1, y) + M(i_0, y + 1, i_1, j_1)] \tag{10}$$

$$\begin{cases} y_{min} = \max(j_0 + \beta - 1, j_0 + \lceil \frac{\alpha}{i_1 - i_0 + 1} \rceil - 1) \\ y_{max} = \min(j_1 - \beta, j_1 - \lceil \frac{\alpha}{i_1 - i_0 + 1} \rceil) \end{cases} \tag{11}$$

In Eq. 7, the minimum of $M(i_0, j_0, i_1, j_1)$ is achieved either with a horizontal split or vertical split. Since the minimum number of choice-changes needed in each sub-rectangle is independent from the other, the minimum of a split is a direct sum of the minimums of the two sub-rectangles as shown in Eqs. 8 and 10. This shows the optimal substructure property of the minimization problem in HFE, that is, the solution of the original problem can be efficiently constructed using solutions of the sub-problems. In this case, the construction takes $O(m+n)$ time, where m, n are the number of rows and columns. The feasible locations of a horizontal and vertical split are defined by $[x_{min}, x_{max}]$ and $[y_{min}, y_{max}]$, respectively. The ranges guarantee that the two sub-rectangles generated by the split satisfy the minimum area α and width β.

Overlapping sub-problems property: A rectangular region (i_0, j_0, i_1, j_1) in the study area can be heavily shared by many hierarchical tiling schemes. An example is shown in Fig. 3, where the region "a" shows up in all three tiling schemes. In the minimization problem of HFE, the minimum achieved in a region will be re-used by many tiling schemes. HFE uses a **four dimensional storage matrix** R to save the minimums obtained in the sub-problems, and each of the four dimensions corresponds to a coordinate in (i_0, j_0, i_1, j_1) which defines the rectangular region of the sub-problem. During hierarchical tiling, when the minimum of a rectangular region (i_0, j_0, i_1, j_1) is needed, HFE queries $R(i_0, j_0, i_1, j_1)$ for the minimum. If $R(i_0, j_0, i_1, j_1)$ is null, the minimum will be computed based on hierarchical tiling according to Eqs. 7–11; otherwise the value of $R(i_0, j_0, i_1, j_1)$ will be directly used.

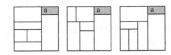

Fig. 3. Region shared by schemes.

The pseudo-code of HFE consists of two recursive functions: (1) find_min: finds the minimum for each rectangular region within the study area to complete matrix R (Algorithm 1); (2) find_tile: uses the computed matrix R to find all tiles belonging to the optimal solution (Algorithm 2). The reason to split the work into two functions is to avoid extra storage needed to save all split points and directions during hierarchical tiling. It is also worth-noting that the process in Algorithm 1 is different than a k-d tree, which is a one-way process without recursive functions. In addition, in line 17–19 of Algorithm 1, the minimum is computed directly since the rectangular region given is not splittable under the spatial constraints. In this non-splittable case, no sub-problems can be further generated and we need to directly compute the minimum number of choice-changes needed. As defined in Sect. 2, all grid cells in a tile must share the same choice in a FF-SA solution. Thus, the minimum is achieved by assigning all grid cells with the majority choice in S_{IP}^* in this region, and the minimum value is the number of grid cells in the region that are not originally assigned with this majority choice.

Algorithm 1. find_min($i_0, j_0, i_1, j_1, S_{IP}^*, \alpha, \beta, R$)

Require: 1. rectangular region: (i_0, j_0, i_1, j_1); **2.** IP-SA solution matrix S_{IP}^*; **3.** Minimum area α and width β; **4.** Storage matrix R of sub-problem minimums
1: Initialization: $minRect = +\infty$; $splittable$ =FALSE
2: **if** $R(i_0, j_0, i_1, j_1)$! =NULL **then return; end if**
3: $x_{min} = \max(i_0 + \beta - 1, i_0 + \lceil (\alpha/(j_1 - j_0 + 1)) \rceil - 1)$
4: $x_{max} = \max(i_0 - \beta, i_0 - \lceil (\alpha/(j_1 - j_0 + 1)) \rceil)$
5: **for** $x = x_{min} : x_{max}$ **do**
6:    $splittable$ = TRUE
7:    find_min($i_0, j_0, x, j_1, S, R, \alpha, \beta$); find_min($x + 1, j_0, i_1, j_1, S, R, \alpha, \beta$)
8:    $minRect = \min(minRect, R(i_0, j_0, x, j_1) + R(x + 1, j_0, i_1, j_1))$
9: **end for**
10: $y_{min} = \max(j_0 + \beta - 1, j_0 + \lceil (\alpha/(i_1 - i_0 + 1)) \rceil - 1)$
11: $y_{max} = \max(j_0 - \beta, j_0 - \lceil (\alpha/(i_1 - i_0 + 1)) \rceil)$
12: **for** $y = y_{min} : y_{max}$ **do**
13:    $splittable$ = TRUE
14:    find_min($i_0, j_0, i_1, y, S, R, \alpha, \beta$); find_min($i_0, y + 1, i_1, j_1, S, R, \alpha, \beta$)
15:    $minRect = \min(minRect, R(i_0, j_0, i_1, y) + R(i_0, y + 1, i_1, j_1))$
16: **end for**
17: **if** $splittable$ == FALSE **then**
18:    $minRect = area(i_0, j_0, i_1, j_1)$ - majority($S_{IP}^*, i_0, j_0, i_1, j_1$).count()
19: **end if**
20: $R(i_0, j_0, i_1, j_1) = minRect$

To find the optimal hierarchical tiling scheme for the minimization problem of HFE, Algorithms 1 and 2 are executed sequentially with inputs $(i_0, j_0, i_1, j_1) = (1, 1, i_{max}, j_{max})$, IP-SA solution S_{IP}^*, minimum area α, minimum width β, R initialized with all NULL values and an empty list $list$ (Algorithm 2 only). The output is a complete list $list$ containing all tiles in the final tiling scheme.

5.2   Phase-2: Choice Assignment

Phase-1 of FF-SA gives an initial solution for spatial allocation. It also finalizes the tiling scheme of FF-SA in the proposed algorithm. A fixed tiling scheme simplifies the FF-SA problem since the only task left is to find an optimal choice assignment.

Preprocessing: This step combines all grid cells in each tile into one single variable. Given the spatial range of each tile, the benefit value and cost value of the combined variable can be computed for each choice by summing up all benefit or cost values of the grid cells inside.

Algorithm 2. find_tile($i_0, j_0, i_1, j_1, R, \alpha, \beta, list$)

Require: 1. rectangular region: (i_0, j_0, i_1, j_1); **2.** Storage matrix R of sub-problem minimums; **3.** Minimum area α and width β; **4.** List of tiles found: $list$

1: $x_{min} = \max(i_0 + \beta - 1, i_0 + \lceil (\alpha/(j_1 - j_0 + 1)) \rceil - 1)$
2: $x_{max} = \max(i_0 - \beta, i_0 - \lceil (\alpha/(j_1 - j_0 + 1)) \rceil)$
3: **for** $x = x_{min} : x_{max}$ **do**
4:    **if** $R(i_0, j_0, x, j_1) + R(x + 1, j_0, i_1, j_1) == R(i_0, j_0, i_1, j_1)$ **then**
5:       find_tile($i_0, j_0, x, j_1, R, \alpha, \beta, list$); find_tile($x + 1, j_0, i_1, j_1, R, \alpha, \beta, list$)
6:       **return**
7:    **end if**
8: **end for**
9: $y_{min} = \max(j_0 + \beta - 1, j_0 + \lceil (\alpha/(i_1 - i_0 + 1)) \rceil - 1)$
10: $y_{max} = \max(j_0 - \beta, j_0 - \lceil (\alpha/(i_1 - i_0 + 1)) \rceil)$
11: **for** $y = y_{min} : y_{max}$ **do**
12:    **if** $R(i_0, j_0, i_1, y) + R(i_0, y + 1, i_1, j_1) == R(i_0, j_0, i_1, j_1)$ **then**
13:       find_tile($i_0, j_0, i_1, y, R, \alpha, \beta, list$); find_tile($i_0, y + 1, i_1, j_1, R, \alpha, \beta, list$)
14:       **return**
15:    **end if**
16: **end for**
17: $list.insert(i_0, j_0, i_1, j_1)$

Optimization: Since the tiling scheme from HFE in phase-1 does not contain fragmentation, we just need to optimize the choices on each tile to maximize the total benefit under the cost constraint for the given tiling scheme. This problem's structure is the same as IP-SA (integer programming formulation without spatial constraints). As discussed in Sect. 4, the 0-1 integer programming problem in IP-SA is a well-studied problem and can be formulated into 0-1 multiple-choice knapsack problem [13]. This paper does not attempt to improve the performance of conventional integer programming solvers, and thus we use standard CPLEX solvers [14] in phase-2 to optimize the choice-assignment.

5.3 Algorithm Acceleration

The focus of this section is to reduce the time complexity on computing the minimum of a non-splittable tile (i_0, j_0, i_1, j_1) (line 18 in Algorithm 1). The goal is to count the minimum number of grid cells that need a choice-change to remove the fragmentation and get homogeneous choice assignment inside the tile. This minimum number $minRect$ is the number of grid cells that are not assigned with the majority choice in this tile (Sect. 5.1).

Naive algorithm: The naive algorithm performs a single scan of all grid cells in this region, and counts the number grid cells of each choice. The maximum count $maxRect$ can be tracked and updated in the process. After the scan, the minimum $minRect$ is achieved by $minRect = (i_1 - i_0 + 1) * (j_1 - j_0 + 1) - maxRect$. The time complexity of the naive algorithm is $O(r)$, where r is the number of grid cells in the tile.

Accelerated algorithm: In this algorithm, an integral image [15] of **the whole** S_{IP}^{*} is pre-computed prior to the start of Algorithm 1 to reduce the time complexity. An IP-SA solution S_{IP}^{*} can be considered a two-dimensional image $I^{m \times n}$ where the value on each pixel is a choice ID assigned on the corresponding grid cell (each grid cell becomes a pixel here). Suppose there are d choices in total. The accelerated algorithm first converts the two-dimensional image into a **multi-layer representation** with dimensions $m \times n \times d$, where d is the number of layers and each layer corresponds to a different choice. Each two-dimensional layer is a binary image: for layer k, if the kth choice is assigned to the pixel, the value is set to 1; otherwise 0. Then, an integral image $I_{int}^{m \times n}$ is computed for each layer. An **integral image** is defined as: Given an image I, its integral image I_{int} has the same dimension, and each pixel value $I_{int}(p, q) = \sum_{i=1}^{p} \sum_{j=1}^{q} I(i, j)$.

Algorithm 3. $minRect$=compute_min($i_0, j_0, i_1, j_1, I_{int}[d]$)

Require: 1. Region: (i_0, j_0, i_1, j_1); **2.** Multi-layer integral image $I_{int}[d]$ (d layers)

1:   $maxRect = -\infty$
2:   **for** $k = 1 : d$ **do**
3:     $count_k = I_{int}[k](i_1, j_1) - I_{int}[k](i_0 - 1, j_1) - I_{int}[k](i_1, j_0 - 1) + I_{int}[k](i_0 - 1, j_0 - 1)$
4:     $maxRect = \max(maxRect, count_k)$
5:   **end for**
6:   $minRect = (i_1 - i_0 + 1) * (j_1 - j_0 + 1) - maxRect$

Integral image I_{int} can be constructed with two linear scans of the original image I. The first linear scan is performed row-wise. For row i, each element is sequentially updated as: $I_{int}(i, 1) = I(i, 1)$, and $I_{int}(i, j) = I_{int}(i, j - 1) + I(i, j)$, $j = 2, ..., n_{col}$. Thus, after the row-wise linear scan, each $I_{int}(i, j) = \sum_{j=1}^{n} I(i, j)$, $\forall i$. To get the final integral image, a similar column-wise linear scan is performed. For column j, each element is sequentially updated as: $I_{int}(1, j) = I(1, j)$, and $I_{int}(i, j) = I_{int}(i - 1, j) + I(i, j)$, $i = 2, ..., m_{row}$. Using I_{int}, the sum of values in a rectangular region (i_0, j_0, i_1, j_1) of the original image I can be computed by $\sum_{i=i_0}^{i_1} \sum_{j=j_0}^{j_1} I(i, j) = I_{int}(i_1, j_1) - I_{int}(i_0 - 1, j_1) - I_{int}(i_1, j_0 - 1) + I_{int}(i_0 - 1, j_0 - 1)$ [15] in O(1) time.

With each layer of the multi-layer IP-SA solution converted to an integral image, the minimum $minRect$ can be computed in O(d) time using Algorithm 3, where d is the total number of choices.

Matlab implementation is available: https://github.com/yqthanks/ffsa.

5.4 Computational Time Analysis

For the computational time analysis, this section mainly focuses on deriving an upper bound for the proposed Minimum Fragmentation Eliminator (MFE) algorithm. The 0-1 integer programming problem in IP-SA is a well studied problem [13] and in general is less difficult than FF-SA since it admits a fully polynomial time approximation scheme whereas FF-SA does not (APX-hard). Thus, in this paper we assume that IP-SA problem can be solved in reasonable time using the state-of-the-art solvers (e.g., CPLEX [14]), and this is a prerequisite of our proposed HFE algorithm.

Time complexity of HFE: Suppose the input grid of the study area has m rows and n columns ($N = mn$ grid cells), each grid cell has d choices, and the minimum area is α and minimum width is β (unit of α, β is grid cell). In addition, denote the total number of non-splittable rectangular regions as N' (line 17, Algorithm 1). N' is determined by m, n, α and β. In order for a region to be non-splittable, both x and y in Eqs. 8 and 10 must not be feasible. In other words, $x_{min} > x_{max}$ and $y_{min} > y_{max}$. Since the relationship between x_{min} and x_{max} as well as y_{min} and y_{max} are difficult to determine without specific values of (i_0, j_0, i_1, j_1) according to Eqs. 9 and 11, here we use N' to represent the number of non-splittable regions instead of a close-form expression using m, n, α and β.

Theorem 1. *The upper bound on computational time complexity of HFE (accelerated) is $O((N^2 - N')(m + n))$. The upper bound on computational time complexity of HFE (naive) is $O((N^2 - N')(m + n) + \alpha N')$.*

Proof. Since each rectangular region can be uniquely identified by two grid cells (e.g., top-left and bottom-right), there are at most N^2 different rectangular regions to be enumerated. Among these rectangles, we need to enumerate split points for $N^2 - N'$ of them. For each splittable rectangular region, there exist at most $m + n$ possible split points. Thus, to get the solution, the total number of queries on the rectangles is at most $(N^2 - N') \times (m + n)$. For each non-splittable rectangle, HFE uses the multi-layer integral image to compute its minimum in $O(d)$ time by enumerating d layers. Thus, the total time complexity of Algorithm 1 is upper bounded by $O((N^2 - N')(m + n) + dN')$. In order to use the accelerated algorithm, the multi-layer integral image needs to be computed prior to the execution of Algorithm 1. Since only two linear scans of each layer is needed, the multi-layer integral image can be computed in $O(dN)$ time. Thus, the total time complexity for multi-layer image computation and Algorithm 1 is $O((N^2 - N')(m + n) + dN' + dN)$. In FF-SA the total number of choices on each cell is assumed to be much smaller than $(m + n)$ (e.g., 5 vs. 500), so we have $d << (m + n)$. In addition, N is dominated by N^2. Thus, the upper bound is simplified to $O((N^2 - N')(m + n))$. Algorithm 2 finds all the tiles using a top-down and one-way search. For each rectangle at a certain level in the hierarchy, we enumerate at most $m + n$ split points. Since all the minimum values $minRect$ of the rectangular regions have been computed and stored in R, we only need to check which one of the split point gives that the sum of the minimums from the two sub-rectangles is equal to the minimum of the current rectangle. After one split point satisfies this criterion, there is no need to further check other splits and this split must belong to an optimal tiling scheme. In an optimal tiling scheme, there are at most $\lfloor \frac{N}{\alpha} \rfloor$ tiles given the minimum area constraint. Since the hierarchical tiling framework partitions each rectangular region into two sub-regions at each step, we can consider the final hierarchy as a full binary tree.

Thus, given that the tree has at most $\lfloor \frac{N}{\alpha} \rfloor$ leaf nodes, there are at most $2\lfloor \frac{N}{\alpha} \rfloor$ nodes in the full binary tree. Thus, Algorithm 2 requires at most $2\lfloor \frac{N}{\alpha} \rfloor (m+n)$ enumerations of split points, and the result of each split can be evaluated in $O(1)$ time (e.g., line 4 and 12 in Algorithm 2). This shows the time complexity of Algorithm 2 is bounded by $O(\lfloor \frac{N}{\alpha} \rfloor (m+n))$. Thus, the total time complexity of HFE is $O((N^2 - N' + \lfloor \frac{N}{\alpha} \rfloor)(m+n)) = O((N^2 - N')(m+n))$.

For the naive algorithm, the only difference lies in the computation of minimums for non-splittable rectangular regions. The accelerated algorithm needs $O(d)$ time for the computation whereas the naive algorithm requires $O(\alpha)$ time since it needs to enumerate all grid cells in each non-splittable region. As analyzed in the proof above, the time complexity of the accelerated version can be written as $O((N^2 - N')(m+n) + dN')$ where the second term represents the time needed for computing minimums for non-splittable regions. Since d is subsumed by $m+n$, the final time complexity is $O((N^2 - N')(m+n))$. In the naive algorithm case, the time is $O((N^2 - N')(m+n) + \alpha N')$. The difference is that α may not be subsumed by $m+n$ and in fact α can be much greater than $m+n$ (e.g., when α is a portion of $N = mn$). Thus, the final time complexity is written as $O((N^2 - N')(m+n) + \alpha N')$.

Theorem 2. *The space complexity of HFE is $O(N^2)$.*

Proof. In HFE, there are in total N^2 sub-problems as shown in the proof of Theorem 1. Thus, storing the minimums of all sub-problem requires at most $O(N^2)$ space.

6   Validation: Case Study in Agricultural Land Allocation

We evaluate the performance of the proposed HFE algorithm through a case study on agricultural land allocation at the Seven Mile Creek watershed in Minnesota, US. The study area is discretized into a 455×477 grid with 30 m by 30 m grid cells. In this case study, the choices are 5 land management practices, namely conventional tillage, conservation tillage (with corn stover), low-fertilizer application, prairie grass and switch grass. The **benefit** is the amount of sediment reduction in water and the **cost** is the investment needed for changing current land management practices. The current land management practice in the study area is homogeneously conventional tillage.

Since the original shape of the study area is irregular while the rectangular version of FF-SA assumes a rectangular study area, an orthogonal rectangular bounding box (ORBB) of the study area was used to generate the grid. All grid cells outside the original study area in the ORBB are assigned with 0 benefit and 0 cost for all choices so that they do not impact the final solution. In addition, there are some unchangeable landscape in the study area (e.g., roads). The grid cells at these places are also assigned with zero benefit and cost values. In the current work, the minimum size and shape constraints are relaxed at the irregular boundaries of both the study area and unchangeable landscapes.

Figure 4(a) shows the IP-SA solution (integer programming without spatial constraints) at a local zoom-in window overlaid with the tiling scheme generated by the HFE algorithm. In the maps, each color represents a choice of land management practice. There are some regions (white) cut out from the

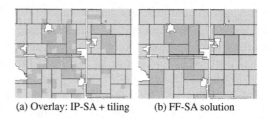

(a) Overlay: IP-SA + tiling   (b) FF-SA solution

Fig. 4. Fragmentation elimination.

tiling scheme, and those regions are the unchangeable landscape inside the study area as discussed above. We can see fragmentation in the IP-SA solution, such as irregular shapes and tiny area of patches (e.g., standing alone grid cells). The land fragmentation prohibits efficient use of modern farm equipments (e.g., combine harvesters). The fragmentation is removed by the HFE algorithm with each tile being constrained by a minimum area and width. Based on the hierarchical tiling scheme, the HFE algorithm minimizes the number of choice-changes needed to remove the fragmentation in a IP-SA solution. Figure 4(b) shows the final output of FF-SA at this local window with both a tiling scheme and choice assignment.

6.1 Solution Quality Comparison

In this section, we compare the solution quality of the proposed HFE algorithm and the dynamic-growth tiling framework (DGTF) in [7]. The goal of this comparison is to evaluate the impact of tiling schemes on final solution quality (e.g., total benefit). Thus, we use the same choice-assignment algorithm in phase-2 after a tiling scheme is generated from either HFE or DGTF to eliminate the impact of choice-assignment algorithm. Since the original choice-assignment algorithm used for DGTF in [7] is a heuristic algorithm whereas the proposed phase-2 algorithm in this work is an exact algorithm (global optimizer), using the new choice-assignment algorithm for DGTF will strictly improve the solution quality of DGTF in [7].

Experiment setup: We consider two resolutions of the input grid. The first grid is the default grid where the cell-size is 30m. The second grid reduces the resolution to half and has a cell-size of 60m. For each of the grid, we evaluate 5 different combinations of (minimum area, minimum width): (8,2), (50,5), (200,10), (800,20) and (1800,30), where the unit is grid cell. The budget limit used is $100,000 as suggested by domain experts. As a reference, the smallest minimum area used is about the size of two standard American football fields (playing field).

Comparison: Figure 5(a) and (b) shows comparisons of solution quality of HFE and DGTF. The vertical axis shows sediment reduction (tons/year) and horizontal axis shows minimum area (grid cell) imposed on tiles. The first trend in Fig. 5 is that the solution quality of both algorithms decrease as the minimum

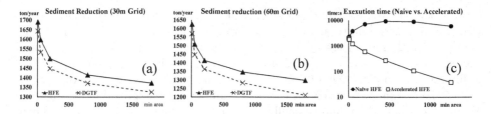

Fig. 5. Experiment results on solution quality and execution-time.

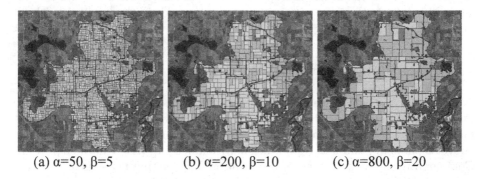

(a) α=50, β=5   (b) α=200, β=10   (c) α=800, β=20

Fig. 6. Maps of agricultural land allocations under different spatial constraints.

area increases. As the minimum area increases, space becomes more contiguous with larger tile area. However, the algorithms are also forced to make more non-optimal choices (compared to IP-SA solution) at places in order to maintain a higher level of contiguity and regularity. The second trend is that the solution quality of HFE is consistently better than DGTF in the experiment. The result is expected since the enumeration space of HFE is much larger than that of DGTF and it provides potential opportunities to identify better solutions during optimization. The average difference between solution quality of HFE and DGTF is about 51 tons/year in Fig. 5(a) and 64 tons/year in (b). Figure 6 shows the maps of land allocation achieved by HFE with three different minimum area constraints (unchangeable landscape removed). The grid cell size is 30 m (default).

6.2   Execution-Time Analysis

Runtime is measured on a 64-bit Windows 8 laptop with Core i7 and 8 GB of RAM. Since large N leads to very expensive time cost for the naive algorithm (Sect. 5.4), we used a grid with 100 m cell-size to compare the average run-time of the naive and the accelerated algorithm. The chart in Fig. 5(c) shows the performance improvement gained by algorithm acceleration proposed in Sect. 5.3. The time is shown in seconds using a log scale (Y-axis). For the accelerated HFE, the execution time decreases fast as the minimum area α increases, because in general the hierarchical split can stop earlier for larger minimum area α. The performance of the naive algorithm is more complicated. When α is small, the computational time increases sharply as α increases. However, as α becomes larger, the time begins to decrease as α increases. This is because the naive algorithm's time complexity is $O((N^2 - N')(m + n) + \alpha N')$, where α shows up as one of the dominant term (N' is the total number of non-splittable rectangles). When α is small, N' remains small. For example, when $\alpha = 1$, $N' = N$. Then as α increases, N' also temporarily increases. However, when α gets very large, N' becomes small again. For example, an extreme case is when $\alpha = N$, $N' = 1$. Thus, this reflects the increasing and decreasing trends in the naive algorithm's execution time.

7   Conclusion and Future Work

We propose a Hierarchical Fragmentation Eliminator (HFE) to solve the Fragmentation-Free Spatial Allocation (FF-SA) problem. FF-SA is important for many societal applications such as agricultural landscape design and urban land-use planning, and it is computationally challenging (APX-hard as proved). Compared to the related work in optimization and computational geometry, the proposed HFE algorithm addresses the fragmentation-issue while having a significantly larger enumeration space, which helps improve solution quality. The evaluation results show that the HFE algorithm indeed provided higher solution quality compared to existing algorithms and fixed the fragmentation issue. Future work will explore: (1) penalty function formulations of spatial constraints and (2) a variety of regular-shape constraints.

Acknowledgment. This material is based upon work supported by the National Science Foundation under Grants No. 1541876, 1029711, IIS-1320580, 0940818 and IIS-1218168, the USDOD under Grants No. HM1582-08-1-0017 and HM0210-13-1-0005, ARPA-E under Grant No. DE-AR0000795, the OVPR U-Spatial and Minnesota Supercomputing Institute. We would like to thank Dr. David Mulla and Brent Dalzell for providing the dataset.

Appendix A   Proof on APX-hardness

We show FF-SA is more general than the MAX-3SAT problem [8], which is APX-hard.

Definition 1 (3SAT and MAX-3SAT). *Given a Boolean formula:* $(x_1 \vee x_2 \vee x_3) \wedge (\neg x_2 \vee x_4 \vee x_5) \wedge (\neg x_5 \vee x_6 \neg x_7) \wedge ...$, *which is composed by m clauses where each clause has 3 binary variables,* ***3SAT*** *determines whether there exists a 0-1 assignment of all variables such that the output of the Boolean formula is 1 (satisfiable).* ***MAX-3SAT*** *is a variation of 3SAT. Instead of determining whether the formula is satisfiable, MAX-3SAT finds a 0-1 variable assignment to maximize the number of satisfied clauses.*

Definition 2 (BOX-PACK and MAX-BOX-PACK). *Given a set S of axis-parallel square boxes with integer height h_i, width w_i and coordinates (locations fixed),* **BOX-PACK** *determines whether there exists a subset of B non-overlapping boxes in S.* **MAX-BOX-PACK**, *instead of determining this existence, finds the subset that has the maximum number of non-overlapping boxes.*

The proof in [9] reduces 3SAT to BOX-PACK. It was shown given a 3SAT instance, a specific two-dimensional allocation of boxes can be constructed in polynomial time in BOX-PACK, and 3SAT can be satisfied (return 1) if and only if a subset of B non-overlapping boxes can be found. Every unsatisfied clause in 3SAT will cause a deduction of 1 from B. The proof needs to be extended in two ways to reduce MAX-3SAT to FF-SA: (1) extend from "3SAT $\Longrightarrow$ BOX-PACK" to "MAX-3SAT $\Longrightarrow$ MAX-BOX-PACK"; and (2) re-formulate the box-allocation in MAX-BOX-PACK to the grid representation of FF-SA by assigning specific grid cell values.

Polynomial-time reduction from BOX-PACK to MAX-BOX-PACK: (1) If BOX-PACK can be determined in polynomial time, then we show there exists a simple algorithm to find the maximum number B^* in MAX-BOX-PACK. Suppose the total number of boxes in the instance is N. Then the algorithm can simply enumerate through all integers in $[1, N]$ using BOX-PACK and find B^* in polynomial time. (2) If MAX-BOX-PACK can be solved in polynomial time, then for any number B in BOX-PACK, we can compare it to B^* and get the decision in polynomial time. Thus, based on (1) and (2), BOX-PACK and MAX-BOX-PACK can be mutually transformed in polynomial time.

MAX-3SAT is a special case of MAX-BOX-PACK: We show MAX-3-SAT can be reduced to MAX-BOX-PACK in polynomial time. So far we have shown 3-SAT can be reduced to MAX-BOX-PACK. Suppose we have a 3SAT instance I_{3SAT} and a corresponding BOX-PACK instance I_{BP}, and B is the largest number of non-overlapping boxes achievable in I_{BP} when I_{3SAT} is satisfiable. [9] shows that every unsatisfied clause in I_{3SAT} is equivalent to a deduction of one in B of I_{BP}. Since for I_{3SAT} we can create a corresponding MAX-BOX-PACK instance I_{MBP} with exactly the same input (box-allocation) as I_{BP}. Let B^* denote the maximum number of non-overlapping boxes achievable in I_{MBP}, and $B^* = B$ when I_{3SAT} is satisfiable. Similarly, we construct a MAX-3SAT instance I_{M3SAT} using the same input of I_{3SAT}. Given B, B^*, and m (total number of clauses in I_{M3SAT}), the maximum number of satisfied clauses in I_{M3SAT} can be computed as $M^* = m - (B - B^*)$. Thus, I_{MBP} can be constructed in polynomial time given I_{M3SAT}, and the solution M^* of I_{M3SAT} can be achieved in polynomial time given B^*.

MAX-3SAT is a special case of FF-SA: To prove FF-SA is APX-hard, we use the above-mentioned special case (specific input box-allocation) of MAX-BOX-PACK to build the corresponding instance of FF-SA. First, we reformulate the special case of box-allocation into a grid representation. Since the coordinates of boxes in MAX-BOX-PACK are all integers, it is straightforward to map the boxes to a grid with cell-size 1. Next, we enrich all the grid cells with values (e.g., benefits) to create the corresponding instance of FF-SA. The goal is to make sure that for the special case of box-allocation, the optimal solution of FF-SA gives the optimal solution of MAX-BOX-PACK in polynomial time.

Overview of the construction of box-allocation [9]: Generally, for each variable in MAX-3SAT, there is a loop of equal-size boxes with side-length 2 and the number of boxes is even. Half of the boxes are labeled "1" and the others are "0", where the boxes representing "1" and boxes representing "0" are interleaving (Fig. 7(a)), so that each box with "0" (resp. "1") intersects with two boxes with "1" (resp. "0"). Given this allocation, the largest subset of non-overlapping boxes for each loop is either all boxes with "1" or all with "0", which corresponds to the binary choice of each variable in MAX-3SAT. For a given clause with three variables, there will be a clause region, where the three box-loops corresponding to the three variables approach each other. The space available in the clause region is determined by the 0-1 choices made at the three loops. The allocation is designed in a way that if any loop chooses "1" boxes, then there will be enough space to include one and only one more box in the clause region into the solution (a subset of all boxes) without overlapping any other boxes. Given a solution containing a subset of boxes in MAX-BOX-PACK, the number of satisfied clauses in MAX-3SAT can be inferred by counting how many boxes are added from all the clause regions. [9] shows it is sufficient to discuss three general cases appearing in the box-allocation. To avoid heavy redundancy, we will focus on the three cases without proving how exactly they translates to MAX-3SAT, as this omitted proof is in [9].

In the following discussion of FF-SA instance I_{FF}, (1) we relax the cost constraint by choosing a $+\infty$ budget; (2) there are only two choices for each grid cell, namely "A" and "B". A's benefit values are shown as the grid cell values, and B's benefit values are always 0; and (3) for choice A, the minimum area is 4 and minimum width is 2; for choice B, the minimum area and width are 1 (discussed in scope, Sect. 2).

Case 1: A fragment of a single loop and this fragment does not intersect with any other loop (Fig. 7(a), left). It is easy to see we can either choose all boxes with "1" or all with "0" for a maximum non-overlapping subset. To achieve the same result in FF-SA, a benefit-value assignment is proposed in Fig. 7(a) (right), where γ denotes a very large positive value (e.g., 9999). Since tile with choice "A" cannot overlap and must satisfy minimum area 4 and width 2, the optimal solution must be choice "A" on either all 2×2 grid boxes corresponding to "1" or "0", which is the same as MAX-BOX-PACK. The rest of the cells will just choose choice "B" with 0 benefit.

Case 2: An intersection of two different loops (Fig. 7(b), left). An intersection is a special case in this box-allocation. A square region (dashed) of 3×3 is added to the intersection, which contains 4 mutually overlapping 2×2-boxes (not drawn) and the four boxes do not belong to either loop. This special modeling guarantees that whatever combinations of "1" and "0" x_1 and x_2 (Fig. 7(b)) choose, we can always add one and only one 2×2-box from the dashed region

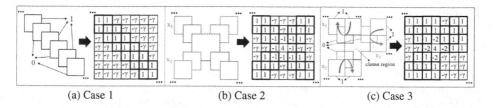

(a) Case 1	(b) Case 2	(c) Case 3

Fig. 7. Three cases of box-allocation: from MAX-BOX-PACK to FF-SA. (best in color)

into the optimal non-overlapping subset. Thus, the intersection has no effect on the decision in choosing the optimal subset. A corresponding benefit-value assignment is proposed in Fig. 7(b)(right). For both loops, choosing boxes with either "1" or "0" will include one and only one cell with value of -1 at the intersection. Same as MAX-BOX-PACK, it does not matter whether "0" or "1" boxes are chosen by the two loops to choose choice "A", since we can always add one and only one 2×2 box with choice "A" at the intersection and increase benefit by 1. The cells left in the intersecting region will choose choice "B" (0 benefit value).

Case 3: This case is the only one that affects the final optimal solution of MAX-BOX-PACK and FF-SA. As shown in Fig. 7(c), there are three loops of boxes intersecting one dashed region at their turns. This dashed region is called a "clause region". The dashed region has an up-side-down "L" shape and it has three mutually overlapping 2×2-boxes. Each "clause region" in the box allocation corresponds to a clause in the MAX-3SAT problem. The loops are constructed such that only a "0" box intersects the "clause region". In the clause region, if there exists at least one loop that chooses boxes with "1", then we can add one and only one more 2×2-box to the final subset. If none of the three loops choose boxes with "1", then we cannot add any of the 2×2-box due to overlaps. Thus, for the MAX-BOX-PACK problem, finding a maximum non-overlapping subset is equivalent to finding the maximum number of satisfied clauses in MAX-3SAT. For FF-SA instance I_{FF}, a benefit-assignment is given in Fig. 7(c) (right). Similarly, any cell with negative γ is prohibitive for choice "A" and needs to be assigned with choice "B". As with MAX-BOX-PACK, if at least one of the three loops chooses the "1" boxes with choice of "A", then we can add one and only one more 2×2 "A" box into the FF-SA solution. Due to the spatial constraints, the rest of the cells in the clause region must choose "B". In addition, we cannot allocate a 3×2 tile in this region because it will cause a decrease in total benefit.

Thus, with the optimal solution of I_{FF}, all 2×2 boxes with choice "A" represent exactly the optimal subset of boxes chosen in MAX-BOX-PACK instance I_{MBP}, which then gives the optimal solution of the MAX-3SAT instance I_{M3SAT} in polynomial time.

References

1. Slotterback, C.S., Runck, B.C., Pitt, D., Kne, L., Jordan, N.: Collaborative geodesign to advance multifunctional landscapes. Landscape Urban Plann. **156**, 71–80 (2016)
2. Kumari, P., Reddy, G., Krishna, T.: Optimum allocation of agricultural land to the vegetable crops under uncertain profits using fuzzy multiobjective linear programming. IOSR J. Agric. Vet. Sci. **7**(12), 19–28 (2014)
3. Van Dijk, T.: Scenarios of central european land fragmentation. Land Use Policy **20**(2), 149–158 (2003)
4. Niroula, G.S., Thapa, G.B.: Impacts and causes of land fragmentation, and lessons learned from land consolidation in South Asia. Land Use Policy **22**(4), 358–372 (2005)
5. Cao, K., Batty, M., Huang, B., Liu, Y., Yu, L., Chen, J.: Spatial multi-objective land use optimization: extensions to the non-dominated sorting genetic algorithm-II. Int. J. Geogr. Inf. Sci. **25**(12), 1949–1969 (2011)
6. Yansui, L., Jianzhou, G., Wenli, C.: Optimal land use allocation of urban fringe in Guangzhou. China Postdoctoral Sci. Found. **22**, 179–191 (2012)
7. Xie, Y., Yang, K., Shekhar, S., Dalzell, B., Mulla, D.: Geodesign optimization (GOP) towards improving agricultural watershed sustainability. In: Thirty-First AAAI Conference on Artificial Intelligence, Workshop on AI and OR for Social Good, AAAI-17, pp. 57–63 (2017)
8. Håstad, J.: Some optimal inapproximability results. J. ACM **48**(4), 798–859 (2001)
9. Fowler, R.J., Paterson, M.S., Tanimoto, S.L.: Optimal packing and covering in the plane are NP-complete. Inf. Process. Lett. **12**(3), 133–137 (1981)
10. Lust, T., Teghem, J.: The multiobjective multidimensional knapsack problem: a survey and a new approach. Int. Trans. Oper. Res. **19**(4), 495–520 (2012)
11. Muthukrishnan, S., Poosala, V., Suel, T.: On rectangular partitionings in two dimensions: algorithms, complexity and applications. In: Beeri, C., Buneman, P. (eds.) ICDT 1999. LNCS, vol. 1540, pp. 236–256. Springer, Heidelberg (1999). doi:10.1007/3-540-49257-7_16
12. Berman, P., DasGupta, B., Muthukrishnan, S., Ramaswami, S.: Improved approximation algorithms for rectangle tiling and packing. In: Proceedings of the Twelfth Annual ACM-SIAM Symposium on Discrete Algorithms, SODA 2001, pp. 427–436 (2001)
13. Kellerer, H., Pferschy, U., Pisinger, D.: The multiple-choice knapsack problem. Knapsack Problems, pp. 317–347. Springer, Heidelberg (2004). doi:10.1007/978-3-540-24777-7_11
14. Using CPLEX in AMPL. http://www.ampl.com/BOOKLETS/ampl-cplex3.pdf
15. Viola, P., Jones, M.J.: Robust real-time face detection. Int. J. Comput. Vis. **57**(2), 137–154 (2004)

Collective-k Optimal Location Selection

Fangshu Chen[1], Huaizhong Lin[1(✉)], Jianzhong Qi[2], Pengfei Li[1],
and Yunjun Gao[1]

[1] College of Computer Science and Technology,
Zhejiang University, Hangzhou, People's Republic of China
youyou_chen@foxmail.com, {linhz,gaoyj}@zju.edu.cn, lpff1218@sina.com
[2] School of Computing and Information Systems,
University of Melbourne, Melbourne, Australia
jianzhong.qi@unimelb.edu.au

Abstract. We study a novel location optimization problem, the *Collective-k Optimal Location Selection* (CkOLS) problem. This problem finds k regions such that setting up k service sites, one in each region, collectively attracts the maximum number of customers by proximity. The problem generalizes the traditional influence maximizing location selection problem from searching for one optimal region to k regions. This increases the complexity of the problem. We prove that the CkOLS problem is NP-hard, and propose both precise and approximate algorithms to solve this problem. The precise algorithm uses a brute-force search with a pruning strategy. The approximate algorithm adopts a clustering-based approach, which restricts the combinational search space within representative regions of clusters, and has a bounded approximation ratio. Extensive experiments show that the approximate algorithm is effective and efficient, and its output results are close to optimal.

Keywords: BRNN · k optimal location · Clustering

1 Introduction

Given a set of service sites P and a set of customer points O, for a service site $p \in P$, a *Bichromatic Reverse Nearest Neighbor* (BRNN) query finds all customer points $o \in O$ whose nearest neighbor in P is p. The set of BRNN points of p is called the *influence set* of p, and the cardinality of this set is called the *influence value* of p. Influence maximization has been a common goal in traditional location optimization problems. A basic problem is the MaxBRNN problem [2–6,13–15], which finds a region S in which all the points have the maximum influence value, i.e., setting up a new service site within this region will attract the maximum number of customers by proximity. Here, a point is said to attract a customer if a service site set up at this point will become the new nearest neighbor of the customer.

This work is partially done when Fangshu is visiting the University of Melbourne.

M. Gertz et al. (Eds.): SSTD 2017, LNCS 10411, pp. 339–356, 2017.
DOI: 10.1007/978-3-319-64367-0_18

We generalize the MaxBRNN problem to find k regions such that setting up k new service sites, one in each region, collectively attract the maximum number of customers by proximity. We call our generalized problem the *Collective-k Optimal Location Selection* (CkOLS) problem. The distinctive feature of this problem is that it optimizes the combined impact of k selected points rather than the impact of any single point. As a result, we need to check every combination of k regions to find the optimal result, which leads to a much higher computation complexity than that of the traditional MaxBRNN problem. Although there are existing algorithms for MaxBRNN [4–6, 13–15], a straightforward adaptation of these solutions is inefficient for CkOLS because there are too many region combinations to be considered.

Meanwhile, CkOLS has many real life applications in service site planning and emergency scheduling. For example, consider a food truck chain which wishes to deploy a number of food trucks in the city CBD. It is preferable to deploy the food trucks within the regions which collectively attract the largest number of customers possible.

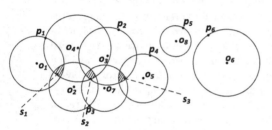

Fig. 1. An example of CkOLS

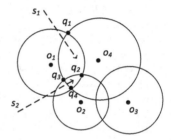

Fig. 2. An example of regions covered by NLCs

To define and compute the optimal regions, we make use of a concept called the *Nearest Location Circle* (NLC) [14]. An NLC is a circle centered at a customer point o with the radius being the distance from o to its nearest service site. For any customer point o, if we build a new service site p outside the NLC of o, p cannot attract o for that there already exists a service site closer to o. As a result, to attract more customer points, the new service site should be located within as many NLCs as possible. The objective of CkOLS then becomes finding k regions covered by the maximum number of NLCs.

For example, in Fig. 1, there are 8 customer points denoted by $o_1, o_2, ..., o_8$ and 6 existing service sites denoted by $p_1, p_2, ..., p_6$. The NLC of each customer is represented by the circle enclosing the customer point. The NLCs partition the space into regions. Three shaded regions s_1, s_2, and s_3 are covered by multiple NLCs. Region s_2 is covered by 4 NLCs. A new service site in s_2 will attract the 4 corresponding customers $\{o_2, o_3, o_4, o_7\}$. Similarly, each of regions s_1 and s_3 are covered by 3 NLCs and a new service site in them will attract 3 customers.

Assuming that we want to build 2 new service sites in 2 different regions. The optimal result in this case is $\{s_1, s_3\}$ which attracts 6 customer points together: $\{o_1, o_2, o_3, o_4, o_5, o_7\}$. In comparison, the combination $\{s_1, s_2\}$ attracts 5 customer points, $\{o_1, o_2, o_3, o_4, o_7\}$; and $\{s_2, s_3\}$ also attracts 5 customer points. As we can see from Fig. 1, the single optimal region is s_2 (attracts 4 customer points which is more than any other regions). This region, however, is not part of the optimal regions for the CkOLS problem when $k = 2$.

To solve the CkOLS problem, we propose two algorithms. The first algorithm is a precise algorithm. It computes all intersection points of NLCs, and obtain the influence value of each intersection point (i.e., number of customers whose NLCs encloses the intersection point). Then, these intersection points are used to represent the corresponding intersection regions. The algorithm sorts the intersection points in descending order of their influence value, and enumerate all combinations of k intersection points to find the optimal combination. We choose the point with largest influence value first to form a combination. During the process of checking all combinations, we can prune the combinations that are unpromising. The second algorithm adopts an approximate strategy to reduce the computation complexity. The algorithm also computes all the intersection points first, but then adopts a clustering method to find a small set of candidate representative points (one from each cluster). At last, it chooses the optimal combination of k representative points to form the final result. Our contributions are summarized as follows:

1. We propose and formulate the CkOLS problem, and prove that the problem is NP-hard.
2. We propose both a precise and an approximate algorithm to solve the CkOLS problem. The precise algorithm uses a brute-force search with a pruning strategy. The approximate algorithm adopts a clustering-based approach, which restricts the combinational search space within representative regions of clusters, and has a bounded approximation ratio.
3. We conduct extensive experiments over both real and synthetic datasets to evaluate the proposed algorithms. Experimental results verify the effectiveness and efficiency of the proposed algorithms.

The rest of the paper is organized as follows. Section 2 discusses related work. Problem statement is presented in Sect. 3. Sections 4 and 5 describe the proposed algorithms and analyze their complexity. Section 6 presents experimental results and Sect. 7 concludes the paper.

2   Related Work

The MaxBRNN problem was first introduced by Cabello et al. in [2,3]. They call it MaxCOV and propose an algorithm in two-dimensional Euclidean space.

Wong et al. propose the MaxOverlap algorithm to solve the MaxBRNN problem [14]. MaxOverlap uses region-to-point transformation to reduce the optimal region search problem into an optimal intersection point search problem.

The intersection points are computed by the NLCs of customer points, and the optimal intersection points are those covered by the largest number of NLCs. In another paper [13], Wong et al. extend the MaxOverlap algorithm to L_p-norm in two- and three-dimensional space. An algorithm called MaxFirst is presented by Zhou et al. [15] to solve the MaxBRNN problem. MaxFirst partitions the data space into quadrants, computes the upper and lower bounds of each quadrant's BRNN, and recursively partitions the quadrants with the largest upper bound until the upper bound and lower bound of some quadrants come to the same value. Liu et al. propose an algorithm called MaxSegment [5] for the MaxBRNN problem where the L_p-norm is used. Lin et al. present an algorithm called OptRegion [6] to solve the MaxBRNN problem. The algorithm employs the sweep line technique and a pruning strategy based on upper bound estimation to improve the search performance.

The maximum coverage problem is a similar problem, which computes a group of locations $p \in P$ (where P is a given candidate service site set) that maximizes the total weight of the clients attracted [10,12]. This problem differs from ours in that, it dose not consider the existing facilities, and hence the solutions are not suitable to our problem.

Another similar problem is the Top-k MaxBRNN query [5,11]. The Top-k MaxBRNN problem simply ranks the regions by the number of BRNNs attracted and returns the Top-k regions. Although it also returns a group of k regions such that setting up a new service site within each region attracts a large number of customers, the Top-k regions may shared many customers attracted, which is different from our CkOLS problem. However, the algorithm proposed for Top-k MaxBRNN can also be considered as an approximate method for CkOLS. We will compare the proposed approximate algorithm with Top-k MaxBRNN query in the experiment section. Qi et al. conducted a series of studies [7–9] on the Mindist location selection problems. They find a service site location to minimize the average distance between the service sites and the customers they are serving. This is a different optimization goal and the studies will not be discussed further.

3 Preliminaries and Problem Statement

3.1 Problem Statement

Given a set of customer points O and a set of service sites P, each customer point o is associated with a weight $w(o)$, which indicates the number of customers at o. For a point $p \in P$, $BRNN(p, O, P)$ represents the set of customer points that take p as their nearest neighbor in P. For a point $s \notin P$, $BRNN(s, O, P \cup \{s\})$ represents the set of points that take s as their nearest neighbor if s is added into P. In order to describe the attracted customer points of a new service site, we define *influence set, influence list* and *influence value* in Definitions 1 and 2.

Definition 1 (*Influence set/value of a single service site*). *Given a service site s, we define the influence set of s to be $BRNN(s, O, P \cup \{s\})$. The influence value of s is equal to $\sum_{o \in BRNN(s,O,P \cup \{s\})} w(o)$.*

Definition 2 (*Influence list/value of k points*). *Let $\mathcal{S}$ be a set of k new service sites. The influence list of $\mathcal{S}$ is the union of BRNN(s, O, P) for every service site $s \in \mathcal{S}$. We denote this list by U, $U = \cup_{s \in \mathcal{S}} BRNN(s, O, P \cup \mathcal{S})$. The influence value of $\mathcal{S}$ is the sum of the weights of the customers in U, i.e., $\sum_{o \in U} w(o)$.*

CkOLS: Given a set of customer points O and a set of service sites P in a two-dimensional Euclidean space, the CkOLS problem finds k optimal regions, such that setting up k new service sites, one in each region, will collectively attract the maximum number of customers.

3.2   Region-to-Point Transformation

Wong et al. transform finding optimal regions to finding optimal points for the MaxBRNN problem [14]. We also use this transformation for the CkOLS problem. We use the intersection points of NLCs (intersection points, for short) to represent regions. We say that a region is *determined by* the intersection points of the NLC arcs enclosing the region.

Notice that, the influence set of q_1 is $\{o_1, o_4\}$ (as shown in Fig. 2), which is different from that of q_2 and q_3, while q_2, q_3, q_4 share the same influence set $\{o_1, o_2, o_4\}$, and region s_1 is not an optimal region of CkOLS, since s_2 attracts more customers. We can see that, if a region is determined by intersection points with different influence sets, it cannot be the optimal region. This is because, a region's influence value is always bounded by the intersection point with the lowest influence value among all the intersection points that determine this region. Next, we introduce Theorem 1 to help us find the optimal region efficiently.

Theorem 1. *Any optimal region for CkOLS must be determined by the intersection points with the same influence set.*

Proof. For any region s that is determined by intersection points $p_1, p_2, ..., p_n$ with different influence list, we assume that p is the one with largest influence value in $p_1, p_2, ..., p_n$. Thus there must exist some other intersection points (not in $p_1, p_2, ..., p_n$) with the same influence list as p. The region s_0, which is determined by p and the intersection points with the same influence list as p, must be a better region than s for the CkOLS problem (s_0 attracts more customer points than s). This means that the region determined by intersection points with different influence list cannot be the optimal region of CkOLS. Therefore, it can be deduced that the optimal region of CkOLS problem must be determined by the intersection points with the same influence list.

According to Theorem 1, we can prune a large number of regions which cannot be the optimal regions of CkOLS. We only need to deal with the regions determined by intersection points with the same influence set. Since any point inside s has the same influence set, we use just one intersection point p_i in $p_1, p_2, ..., p_n$ to represent s. Adopting this region-to-point transformation, the CkOLS problem becomes finding k intersection points, with the objective that the influence value of the k intersection points is maximum.

3.3   NP-hardness of C*k*OLS

Next, we show that the C*k*OLS problem is NP-hard by reducing an existing NP-complete problem called the *Maximum Coverage Problem* to C*k*OLS in polynomial time.

Maximum Coverage Problem: Given a number t and a collection of sets $\mathcal{S} = \{S_1, S_2, S_3, ..., S_m\}$, the objective is to find a subset $\mathcal{S}' \subset \mathcal{S}$ such that $|\mathcal{S}'| \leq t$ and the number of covered elements $|\cup_{S_i \in \mathcal{S}'} S_i|$ is maximum. In the weighted version, every element has a weight, and the objective is to find a collection of coverage sets which has the maximum sum of weight.

We reduce the Maximum Coverage Problem to C*k*OLS as follows. Sets $S_1, S_2, S_3, ..., S_m$ correspond to the influence sets of the intersection points. The collection of sets $\mathcal{S}$ is equal to the collection of the influence sets of all intersection points, and t is equal to k. It is easy to see that this transformation can be constructed in polynomial time, and it can be easily verified that when the problem is solved in the transformed C*k*OLS problem, the original Maximum Coverage Problem is also solved. Since the Maximum Coverage Problem is an NP-complete problem, C*k*OLS is NP-hard.

Theorem 2. *The Ck OLS problem is NP-hard.*

4   Precise Algorithm

In this section, we propose a precise algorithm which is a brute-force enumeration based algorithm. We call the algorithm C*k*OLS-Enumerate. Using the intersection points to represent regions, we compute all the intersection points and check the total combinations of k intersection points to obtain the optimal result of C*k*OLS. We divide the precise algorithm into two phases.

Phase 1. We compute all intersection points of NLCs, and obtain the influence set of each intersection point. In order to obtain the NLC of each customer point, we use kd-tree [15] to perform nearest neighbor query. We build a kd-tree of service sites and use the algorithm ANN [1] to find the nearest service sites over the kd-tree. The influence set of each intersection point is the set of customer points whose NLCs cover the intersection point.

Phase 2. We sort the intersection points in descending order of their influence values, and the point with the largest influence value is processed first. We enumerate to check all combinations of k intersection points to get the optimal result of C*k*OLS. The algorithm is summarized in Algorithm 1.

4.1   Enumeration Algorithm

We adopt a recursive algorithm in Algorithm 1. The main idea is to choose one point from all intersection points and choose $m - 1$ (m is the number of points need to be added into one combination, and initially $m = k$) points from the remaining intersection points. During the process of enumeration, we add the intersection point one by one into a combination of k points. Lines 7–9 check whether we have added k points into one combination. Lines 12–14 prune the combinations which are unnecessary to check. We will elaborate this pruning strategy in the next subsection. Keeping the intersection points in a sorted list, lines 15–17 choose the next point to add into one combination, and iteratively choose $m - 1$ points from the remaining $n - i - 1$ (n is the number of intersection points) intersection points. After having added k points into one combination, lines 18–23 verify whether the influence value of this combination is larger than the maximum influence value ever found, and update the maximum influence value if necessary. Finally, we obtain the optimal result after checking all combinations of k intersection points.

Algorithm 1. $CkOLS$-Enumerate algorithm

Input: k: number of optimal regions, IP: set of sorted intersection points
Output: k optimal points
1:   n: number of intersection points; $i = 0$
2:   $m := k$ /*number of points to be added into one combination */
3:   $maxValue := 0$ /* the largest combination influence value already found */
4:   $optCombination := \emptyset$ /* the optimal combination */
5:   $cValue := 0$ /* combination influence value of added points in one combination */
6:   **procedure** COMBINATION(n, IP, m)
7:      **if** $m == 0$ **then**
8:         return $optCombination$
9:      **end if**
10:    **while** $i < n - m$ **do**
11:       $value :=$ influence value of the ith point
12:       **if** $m \cdot value + cValue < maxValue$ **then**
13:          break
14:       **end if**
15:       Add i into this combination
16:       $m := m - 1$, update $cValue$
17:       Combination($n - i - 1$,IP,m)
18:       **if** $m == 0$ **then**
19:         **if** $cValue > maxValue$ **then**
20:            $maxValue := cValue$
21:            update $optCombination$
22:         **end if**
23:       **end if**
24:       $i := i + 1$
25:    **end while**
26: **end procedure**

4.2 Pruning Strategy

In this subsection, we introduce our pruning strategy exploited by the enumeration algorithm (Lines 12–14 in Algorithm 1). Using the sorted intersection point list IP, we choose the point with larger influence value earlier to add into one combination. During the process of checking any combination $\mathcal{C}$, we assume that there are m points to be added into $\mathcal{C}$. The influence value of the $k - m$ added points of $\mathcal{C}$ is $cValue$. The point with the largest influence value left in IP is $maxP$ (which is the very first point left in IP), and the influence value of $maxP$ is $maxPValue$. The influence value of set $\mathcal{C}$ cannot be larger than the sum of the influence value of each intersection point. Therefore, after adding the remaining m points into $\mathcal{C}$, the influence value of $\mathcal{C}$ cannot be larger than $m \cdot maxPValue + cValue$. Let $maxValue$ be the maximal influence value of combinations already found. As long as $m \cdot maxPValue + cValue < maxValue$, combination $\mathcal{C}$ can be pruned.

4.3 Time Complexity of CkOLS-Enumerate

First, we discuss the time complexity to compute all intersection points. In order to compute the NLC of each customer point, we use the All Nearest Neighbor algorithm [1], which requires $O(\log |P|)$ time. The time complexity of the construction of a kd-tree for P is $O(|P| \log |P|)$. Then, it takes $O(|P| \log |P| + |O| \log |P|)$ time to get NLCs of all customer points in O. After obtaining NLCs, we use MBR of NLCs to compute the intersection points of NLCs. Suppose that, for each NLC, its MBR intersects with at most d MBRs. Then it takes $O(d)$ time to process each NLC. Therefore, the total time complexity to compute the intersection points is $O(|P| \log |P| + |O| (\log |P| + d))$. In addition, we need to sort the intersection points IP, which requires $O(|IP| \log |IP|)$ time.

Second, as we can see that the enumeration algorithm is a brute-force algorithm. In the worst case, let N be the number of intersection points, the enumeration algorithm will check $\binom{k}{N}$ combinations to get the optimal result. Let m be the maximum size of influence set of points in the intersection points list. The time complexity of checking each combination is $O(k^2 \cdot m)$. Thus, the worst-case time complexity of the enumeration algorithm is $O(k^2 \cdot m \cdot \binom{k}{N})$. With the increase of the data volumes, the enumeration algorithm degrades rapidly.

5 Approximate Algorithm

In the enumeration algorithm, with the growth of the cardinality of intersection points, the combinatorial number of k intersection points would be very large. As a result, it would take a long running time. In order to reduce the cost of enumeration, we aim to reduce the number of intersection points to be checked. Towards this aim, we propose a clustering based approach.

As we can see, there are many intersection points whose influence sets are similar to each other. For two intersection points with largely overlapping influence sets, the influence value of the combination of the two intersection points may not be much larger than the influence value of either point. Putting the two points in a combination is less desired. Hence, we partition the intersection points into clusters according to the similarity of their influence sets. We choose a *representative point* of each cluster, and use the representative point to represent all the other points inside the cluster. Afterwards, we only need to check the combinations of the representative points to obtain the approximate optimal result. Since the number of representative points is much smaller than that of the entire set of intersection points, the enumeration cost can be greatly reduced. We call the clustering-based approximate algorithm CkOLS-Approximate. The approximate algorithm also runs in two phases. The first phase clusters the intersection points and compute the representative points. The second phase checks the combination of the representative points in the same way as the precise algorithm. We omit the full pseudo-code of CkOLS-Approximate, and only describe *Clustering algorithm* here.

5.1   Clustering Algorithm

In order to cluster the intersection points according to the similarity between their influence sets, we propose the concept of *Discrepancy*. For ease of discussion, we first define a new operation in Definition 3.

Definition 3 (*Operation* $\langle\rangle$). *Given an influence set* N, $\langle N \rangle = \sum_{o \in N} w(o)$. *Point* o *is a customer point in* N, *and* $w(o)$ *is the weight of* o.

Definition 4 (*Discrepancy*). *Given two service sites* s_1 *and* s_2, *let* N_1 *and* N_2 *denote the influence sets of* s_1 *and* s_2. *We define the discrepancy of* s_2 *to* s_1 *as* $d(s_1, s_2)$, $d(s_1, s_2) = \frac{\langle N_2 - N_1 \rangle}{\langle N_1 \rangle}$. *Notice that* $d(s_2, s_1)$, *which denotes the discrepancy of* s_1 *to* s_2, *is not equal to* $d(s_1, s_2)$.

We cluster the intersection points as follows. First, we sort the intersection point list L in decreasing order of influence value. Second, we pick the point p with the largest influence value in L as the representative point of a new cluster cl, and remove p from L. Third, after building a new cluster, we need to add the points with discrepancy to p less than α into cl, and remove these points from L as well. We say that the points in one cluster are represented by the representative point of this cluster. We repeat the second and third steps until L becomes empty. When finishing clustering, the influence value of any representative point is larger than that of any other point in the same cluster.

As described in Algorithm 2, lines 3–5 get the point p with the largest influence value in L to build a new cluster cl. Point p is the representative point of cl. Since L is sorted by decreasing order of influence value, the top point of L is just the one with the largest influence value. We say that, an intersection point cp is represented by p if $d(p, cp) < \alpha$. Lines 6–12 get the candidate points that could probably be represented by p, check all these points, and add these points into cl. The procedure stops when every intersection point has been partitioned into a cluster.

Algorithm 2. Clustering algorithm

Input: L: sorted intersection point list
Output: C: set of clusters of intersection points
1: **procedure** CLUSTERING(L)
2:     **while** L is not empty **do**
3:         $p :=$ top point of L
4:         remove p from L
5:         build a new cluster cl
6:         $cl.clusterCenter := p$
7:         $Cand :=$ GetCandidate(p)
8:         **for** $cp \in Cand$ **do**
9:             **if** $d(p, cp) < \alpha$ **then**
10:                 add cp into cl
11:                 remove cp from L
12:             **end if**
13:         **end for**
14:     **end while**
15: **end procedure**
16:
17: **procedure** GETCANDIDATE(p)
18:     $set :=$ influence set of p
19:     $maxRadius :=$ get maximum NLC radius of point in set
20:     $Cand :=$ the points with distance to p smaller than $2 \cdot maxRadius$
21:     return $Cand$
22: **end procedure**

Table 1 shows a part of intersection points and clustering result of Fig. 1. Points $q_1, ..., q_{10}$ are intersection points of NLCs of $o_1, ..., o_4$. The clustering algorithm partitions $q_1, ..., q_{10}$ into clusters cl_1 and cl_2, and $cl_1 = \{q_1, q_2, q_3, q_7, q_8\}$, $cl_2 = \{q_4, q_5, q_6, q_9, q_{10}\}$. Point q_1 (q_4) is the representative point of cl_1 (cl_2). The discrepancies of the other points in cl_1 (cl_2) to q_1 (q_4) are all 0. The discrepancy of q_4 to q_1 is 0.33. As a result, q_4 cannot be partitioned into the cluster of q_1 if α equals 0.2. We list some examples of discrepancy in Table 2.

5.2   Getting Candidate Represented Points

In the process of getting represented points of any representative point, it will take much time if we check all points in the intersection point list. Considering the principle of locality, given an intersection point p, the intersection points that are closer to p are more likely to be represented by p. Next, we will describe how to get the candidate represented points in Lines 17–22 of Algorithm 2 by the explanation of Theorem 3.

Theorem 3. *Given two intersection points p and q, let $maxRadius$ be the maximum NLC radius of points in p's influence set; let N_1 and N_2 be the influence set of p and q. If $dist(p, q)$ (distance from p to q) is larger than $2 \cdot maxRadius$, and $\langle N_2 \rangle > \langle N_1 \rangle \cdot \alpha$, then the discrepancy of q to p must be larger than α.*

Table 1. An example of intersection points

Intersection point	Influence set	Cluster
q_1	$\{o_1, o_2, o_4\}$	cl_1
q_2	$\{o_1, o_2, o_4\}$	cl_1
q_3	$\{o_1, o_2, o_4\}$	cl_1
q_4	$\{o_2, o_3, o_4\}$	cl_2
q_5	$\{o_2, o_3, o_4\}$	cl_2
q_6	$\{o_2, o_3, o_4\}$	cl_2
q_7	$\{o_1, o_2\}$	cl_1
q_8	$\{o_1, o_4\}$	cl_1
q_9	$\{o_2, o_3\}$	cl_2
q_{10}	$\{o_3, o_4\}$	cl_2

Table 2. Discrepancy value

Discrepancy	Value
(q_1, q_2)	0
(q_1, q_3)	0
(q_1, q_4)	0.33
(q_1, q_7)	0
(q_1, q_9)	0.33

Proof. Since $dist(p, q) > 2 \cdot maxRadius$, p and q cannot be covered by any NLC simultaneously. If there is an NLC c that covers p and q at the same time, let r be the radius of c, then $dist(p, q)$ must be smaller than $2r$. As we have discussed in Sect. 3, the influence set of p equals the set of points whose NLC covers s. Thus r must be smaller than $maxRadius$, and we get the contradiction that $dist(p, q) < 2 \cdot maxRadius$. Therefore, there is no intersection of p and q's influence sets. This means that $N_2 - N_1 = N_2$. Then we can get that $d(p, q) = \frac{\langle N_2 \rangle}{\langle N_1 \rangle}$. Under the condition that $\langle N_2 \rangle > \langle N_1 \rangle \cdot \alpha$, we can prove that $d(p, q) > \alpha$.

Let c be the circle centered at p with radius equals $2 \cdot maxRadius$, then the candidate represented points of p are all the intersection points inside c. According to Theorem 3, we can get limited candidate represented points of p instead of checking every point in L, and this helps saving lots of time while clustering the intersection points. In this paper, we traverse the influence set of the representative points to get $maxRadius$, and construct a kd-tree [15] of intersection points to help us get the candidate represented points. Given an representative point p, we do a range query on kd-tree to obtain the other intersection points that are within the distance of $2 \cdot maxRadius$ to p.

As shown in Fig. 3, given four intersection points $q_1, ..., q_4$, $c_1, ..., c_4$ are the NLCs of points in q_1's influence set. $MaxRadius$ is the maximum NLC radius of q_1's influence set, which is the radius of c_3 in this example. Since $dist(q_1, q_2) > 2 \cdot maxRadius$, there cannot be any NLC that covers q_1 and q_2 simultaneously. According to Theorem 3, we can easily get the candidate represented points set of q_1, which is $\{q_3, q_4\}$. During the second phase of the approximate algorithm, we adopt the same enumeration process as the $CkOLS$-Enumerate algorithm. We simply check the overall combinations of representative points, and finally obtain an approximate optimal k points.

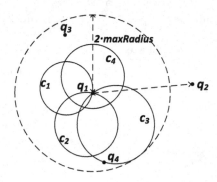

Fig. 3. An example of Theorem 3

5.3   Accuracy of C*k*OLS-Approximate

Next, we will prove the accuracy of algorithm C*k*OLS-Approximate.

Theorem 4. *The accuracy of C*k*OLS-Approximate is at least* $\frac{1+\alpha}{1+k\cdot\alpha}$.

Proof. We suppose that the optimal result of C*k*OLS is $\{a_1, a_2, a_3, ..., a_k\}$, each a_i indicates the intersection point. The corresponding representative point of a_i is b_i. For convenience, we use A_i and B_i to indicate the influence set of a_i and b_i. Next, we will prove that $\frac{\langle B_1 \cup B_2 \cup B_3 \cup, ..., \cup B_k \rangle}{\langle A_1 \cup A_2 \cup A_3 \cup, ..., \cup A_k \rangle} > \frac{1+\alpha}{1+k\cdot\alpha}$.

Let $\Delta B_i = A_i - B_i$, we can easily see that $A_i \subset B_i \cup \Delta B_i$, then

$$
\frac{\langle B_1 \cup B_2 \cup B_3 \cup, ..., \cup B_k \rangle}{\langle A_1 \cup A_2 \cup A_3 \cup, ..., \cup A_k \rangle}
$$

$$
\geq \frac{\langle B_1 \cup B_2 \cup B_3 \cup, ..., \cup B_k \rangle}{\langle (B_1 \cup \Delta B_1) \cup (B_2 \cup \Delta B_2) \cup (B_3 \cup \Delta B_3) \cup, ..., \cup (B_k \cup \Delta B_k) \rangle}
$$

$$
\geq \frac{\langle B_1 \cup B_2 \cup B_3 \cup, ..., \cup B_k \rangle}{\langle B_1 \cup B_2 \cup B_3 \cup, ..., \cup B_k \rangle + \langle \Delta B_1 \cup \Delta B_2 \cup \Delta B_3 \cup, ..., \cup \Delta B_k \rangle} \tag{1}
$$

$$
= \frac{1}{1 + \frac{\langle \Delta B_1 \cup \Delta B_2 \cup \Delta B_3 \cup, ..., \cup \Delta B_k \rangle}{\langle B_1 \cup B_2 \cup B_3 \cup, ..., \cup B_k \rangle}}
$$

With the precondition that the discrepancy of a_i to b_i is less than $1 - \alpha$, ΔB_i must be less than $\langle B_i \rangle \cdot \alpha$. Supposing that b_j is the point with the largest influence value in $\{b_1, b_2, b_3, ..., b_k\}$, then

$$
\langle \Delta B_1 \cup \Delta B_2 \cup \Delta B_3 \cup, ..., \cup \Delta B_k \rangle
$$

$$
\leq \langle \Delta B_1 \rangle + \langle \Delta B_2 \rangle + \langle \Delta B_3 \rangle + ... + \langle \Delta B_k \rangle
$$

$$
\leq \langle B_1 \rangle \cdot \alpha + \langle B_2 \rangle \cdot \alpha + \langle B_3 \rangle \cdot \alpha + ... + \langle B_k \rangle \cdot \alpha \tag{2}
$$

$$
\leq \langle B_j \rangle \cdot \alpha \cdot k
$$

Knowing that b_j has the largest influence value, there must exist a certain point b_m in $\{b_1, b_2, b_3, ..., b_k\}$, that satisfies $d(b_m, b_j) > \alpha(b_i \neq b_m)$. If not, it means that all of $\{b_1, b_2, b_3, ..., b_k\}$ can be partitioned into the same cluster as b_j, which means that $\{b_1, b_2, b_3, ..., b_k\}$ cannot be the result of the approximate algorithm. As a result, $\langle B_j \cup B_m \rangle = \langle B_j \rangle + \langle B_m - B_j \rangle > \langle B_j \rangle \cdot (1+\alpha)$. Therefore, $\langle B_1 \cup B_2 \cup B_3 \cup, ..., \cup B_k \rangle$ must be larger than $\langle B_j \rangle \cdot (1 + \alpha)$, then

$$\frac{\langle \Delta B_1 \cup \Delta B_2 \cup \Delta B_3 \cup, ..., \cup \Delta B_k \rangle}{\langle B_1 \cup B_2 \cup B_3 \cup, ..., \cup B_k \rangle} \leq \frac{\langle B_j \rangle \cdot \alpha \cdot k}{\langle B_j \rangle \cdot (1+\alpha)} = \frac{k\alpha}{1+\alpha} \tag{3}$$

$$\frac{\langle B_1 \cup B_2 \cup B_3 \cup, ..., \cup B_k \rangle}{\langle A_1 \cup A_2 \cup A_3 \cup, ..., \cup A_k \rangle} \geq \frac{1}{1 + \frac{k\alpha}{1+\alpha}} \geq \frac{1+\alpha}{1+k \cdot \alpha} \tag{4}$$

Since the approximate algorithm makes combination of the representative points, the output result of the approximate algorithm must be larger than (or at least equal to) $\langle B_1 \cup B_2 \cup B_3 \cup, ..., \cup B_k \rangle$. Through the derivation above, we can conclude that the accuracy of the approximate algorithm is at least $\frac{1+\alpha}{1+k \cdot \alpha}$.

5.4   Time Complexity of CkOLS-Approximate

We partition the time complexity into three parts. First, we calculate all intersection points and the time complexity is the same as we have discussed in Sect. 4.3. We omit it here. Second, we analyze the time complexity of the clustering algorithm. We use heap sort to sort the intersection points IP, which requires $O(|IP| \log |IP|)$ time. Then we build a kd-tree of IP, which requires $O(|IP| \log |IP|)$ time. Let m be the maximum size of influence sets of intersection points in IP, n be the maximum size of candidate represented points of the representative points. Assuming that there are t clusters after partition of the intersection points, it would take $O(t(\log |P| + m))$ time to get the candidate represented points. For the convenience of discrepancy computation, we sort the influence set of each point in IP, which requires $O(|IP| \cdot m \log m)$ time. The time complexity of discrepancy computation in one cluster is $O(m \cdot n)$. Totally, the time complexity of clustering algorithm is $O(|IP|(m \log m + \log |IP|) + t(\log |IP| + m \cdot n))$. Third, in the worst case, we have to check $\binom{k}{0.1 \cdot |IP|}$ combinations to get the approximate optimal result, which is far fewer than the precise algorithm. The worst query time complexity is $O(k^2 \cdot m \cdot \binom{k}{0.1 \cdot |IP|})$.

The total time complexity of the approximate algorithm is the sum of the above three parts.

6   Performance Study

We have conducted extensive experiments to evaluate the proposed algorithms using both synthetic and real datasets. The algorithms are implemented in C++. All experiments are carried out on a Linux Machine with an Intel(R) Core i5-4590 3.30 GHz CPU and 8.00 GB memory. The synthetic datasets follow Gaussian and Uniform distribution, and the data volume ranges from 1 K to 100 K. The customer dataset and service dataset follow the same distribution. As for real datasets, we use LB and CA, which contain 2D points representing geometric locations in Long Beach Country and California respectively[1]. To keep consistency, we partition the real datasets into two parts to ensure that the customer points and service sites share the same distribution. Considering that the number of service sites is usually much fewer than that of customer points, we set the cardinality of P to be half of the cardinality of O for all datasets.

Section 6.1 tests effect of clustering parameter α. Section 6.2 compares the running time and accuracy of the proposed algorithms. Section 6.3 tests the algorithms' scalability. Section 6.4 compares the effectiveness of our CkOLS query with Top-k MaxBRNN query.

6.1   Effect of Clustering Parameter

The clustering algorithm outputs the representative points of the intersection points. Instead of making combination of the total intersection points, we only use the representative points to get the optimal combination. As a result, the ratio of the number of representative points to the number of entire intersection points is essential for the efficiency of the approximate algorithm. For simplicity, we call the ratio *clustering efficiency*. A small *clustering efficiency* means a small number of representative points.

Table 3. Effect of α

α	*clustering efficiency*	Accuracy	Running time
0.05	0.22	0.98	$3s$
0.1	0.22	0.95	$2.9s$
0.2	0.1	0.9	$1.2s$
0.3	0.04	0.83	$0.8s$
0.4	0.02	0.8	$0.6s$

The experiments are conducted on synthetic datasets with 1 K to 100 K customer points, and we get the average clustering efficiency when α varies from 0.05 to 0.4. As we can see from Table 3, the *clustering efficiency* decreases with the increase of α. Besides impacting the clustering efficiency, α also decides the effectiveness of the approximate algorithm. As shown in Table 3, the larger α is, the worse approximate result we would get with shorter running time. For the balance of effectiveness and clustering efficiency, we conduct experiments with $\alpha = 0.2$ in the following experiments.

6.2   Comparison of Algorithms

The results are given in Figs. 4 and 5 over datasets with different cardinality and distribution. We compare the running time of the precise algorithm with

---

[1] http://www.rtreeportal.org/spatial.html.

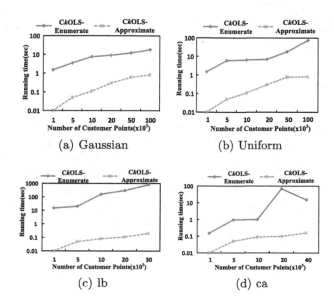

Fig. 4. Running time of proposed algorithms

approximate algorithm with $k = 10$ and $\alpha = 0.2$, and set the weight of all customer points to 1. We only consider the running time without the time of calculating the intersection points of NLCs.

From Fig. 4 we can see, as the cardinality of customer points increases, the query time of CkOLS-Approximate is at least one order of magnitude faster than CkOLS-Enumerate. The efficiency of CkOLS-Enumerate is unstable on different datasets, and the query time of dataset with larger cardinality is possibly less than the same dataset with smaller cardinality. For example, in the experiments on CA dataset, the query time of CkOLS-Enumerate over 40 K customer points is less than that over 20 K customer points. The reason behind is that the efficiency of enumeration algorithm is decided by how fast the optimal combination is found during the enumeration of combination. For the convenience of comparison, we set k to 5 in the experiments of CA dataset for the reason that it takes too long for the enumeration algorithm to get the optimal result with $k = 10$. Figure 5 shows the accuracy of CkOLS-Approximate algorithm. For the synthetic datasets, the

Fig. 5. Accuracy of CkOLS-Approximate algorithm

accuracies of different datasets are almost 1. While for the real datasets, the average accuracy is about 0.9 to 0.95.

6.3   Scalability

Figure 6 shows the scalability of proposed algorithms. We set $\alpha = 0.2$ and $k = 10$ as their default values. We vary the number of customer points from 100 K to 500 K, and use Uniform dataset here, other distribution can get similar results. From Fig. 6(a), we can see that, as the increase of the number of customer points, the running time of approximate algorithm keeps almost stable, while the precise algorithm degrades a lot, especially when the dataset is 500 K, the precise algorithm is very slow. As shown in Fig. 6(b), compare to the accuracy result on small scale datasets, the accuracy of approximate algorithm decreases, however, the worst case is still beyond 0.8.

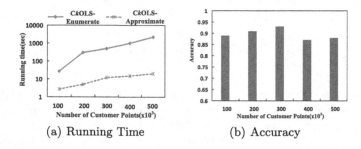

(a) Running Time   (b) Accuracy

Fig. 6. Scalability of proposed algorithms

6.4   Comparison to Top-k MaxBRNN

We know that the Top-k MaxBRNN query can also return a group of k regions to attract large number of customers. And it can be regarded as an approximate method. We compare the proposed algorithm CkOLS-Approximate with Top-k MaxBRNN query to find out which can collectively attract the maximum number of customers by proximity. Figure 7 shows that, CkOLS query can always get the larger number of attracted customer points. And it is almost three times the number of Top-k MaxBRNN query when using Gaussian data set with $|O| = 100$ K. From Fig. 7(c), we can see that the result on real data set is similar to the synthetic data set. We also compare the running time of the two methods. As shown in Fig. 8, the two methods almost have the same running time no matter on which data distribution. However, as we have discussed, the Top-k MaxBRNN query has much worse accuracy.

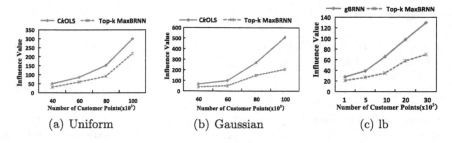

Fig. 7. Comparison to Top-k MaxBRNN query (accuracy)

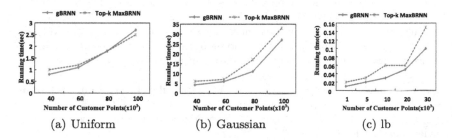

Fig. 8. Comparison to Top-k MaxBRNN query (running time)

7   Conclusion

We studied the CkOLS problem and proposed a precise and an approximate algorithm to solve the problem. The precise algorithm uses a brute-force search with a pruning strategy. The approximate algorithm adopts a clustering-based approach, which restricts the combinational search space within representative regions of clusters, and has a bounded approximation ratio. Experiment results show that the approximate algorithm is effective and efficient. Its output is very close to that of the precise algorithm, with an approximation ratio of up to 0.99 on all the datasets. The algorithm is also efficient. It outperforms the precise algorithm by up to three orders of magnitudes in terms of the running time.

For future work, we are interested in adapting the proposed algorithms to solve collective-k location selection problems with other optimization goals such as min-dist [7–9].

Acknowledgements. This work was partly supported by The University of Melbourne Early Career Researcher Grant (project number 603049), National Science and Technology Supporting plan (2015BAH45F00), the public key plan of Zhejiang Province (2014C23005), the cultural relic protection science and technology project of Zhejiang Province.

References

1. Arya, S., Mount, D.-M., Netanyahu, N.-S., Silverman, R., Wu, A.-Y.: An optimal algorithm for approximate nearest neighbor searching in fixed dimensions. JACM **45**(6), 891–923 (1998)
2. Cabello, S., Díaz-Báñez, J.M., Langerman, S., Seara, C., Ventura, I.: Reverse facility location problems. In: CCCG, pp. 1–17 (2005)
3. Cabello, S., Díaz-Báñez, J.M., Langerman, S., Seara, C., Ventura, I.: Facility location problems in the plane based on reverse nearest neighbor queries. Eur. J. Oper. Res. **202**(1), 99–106 (2010)
4. Chen, Z., Liu, Y., Wong, C., Xiong, J., Mai, G., Long, C.: Efficient algorithms for optimal location queries in road networks. In: SIGMOD, pp. 123–134 (2014)
5. Liu, Y.-B., Wong, R.-C., Wang, K., Li, Z.-J., Chen, C.: A new approach for maximizing bichromatic reverse nearest neighbor search. In: KAIS, pp. 23–58 (2013)
6. Lin, H., Chen, F., Gao, Y., Lu, D.: Finding optimal region for bichromatic reverse nearest neighbor in two- and three-dimensional spaces. Geoinformatica **20**(3), 351–384 (2016)
7. Qi, J., Zhang, R., Wang, Y., Xue, Y., Yu, G., Kulik, L.: The min-dist location selection and facility replacement queries. WWWJ **17**(6), 1261–1293 (2014)
8. Qi, J., Zhang, R., Kulik, L., Lin, D., Xue, Y.: The Min-dist Location Selection Query. In: ICDE, pp. 366–377 (2012)
9. Qi, J., Zhang, R., Xue, Y., Wen, Z.: A branch and bound method for min-dist location selection queries. In: ADC, pp. 51–60 (2012)
10. Sakai, K., Sun, M.-T., Ku, W.-S., Lai, T.H., Vasilakos, A.V.: A framework for the optimal-coverage deployment patterns of wireless sensors. IEEE Sens. J. **15**(12), 7273–7283 (2015)
11. Sun, Y., Huang, J., Chen, Y., Du, X., Zhang, R.: Top-k most incremental location selection with capacity constraint. In: WAIM, pp. 165–171 (2012)
12. Sun, Y., Qi, J., Zhang, R., Chen, Y., Du, X.: MapReduce based location selection algorithm for utility maximization with capacity constraints. Computing **97**(4), 403–423 (2015)
13. Wong, R.-C., Tamer Özsu, M., Fu, A.-W., Yu, P.-S., Liu, L., Liu, Y.: Maximizing bichromatic reverse nearest neighbor for LP-norm in two- and three-dimensional space. In: VLDB, pp. 893–919 (2011)
14. Wong, R.-C., Özsu, M.T., Yu, P.-S., Fu, A.-W., Liu, L.: Efficient method for maximizing bichromatic reverse nearest neighbor. In: VLDB, pp. 1126–1137 (2009)
15. Zhou, Z., Wu, W., Li, X., Lee, M.-L.: Wynne Hsu: MaxFirst for MaxBRkNN. In: ICDE, pp. 828–839 (2011)

Inherent-Cost Aware Collective Spatial Keyword Queries

Harry Kai-Ho Chan[1(✉)], Cheng Long[2], and Raymond Chi-Wing Wong[1]

[1] The Hong Kong University of Science and Technology,
Clear Water Bay, Hong Kong
{khchanak,raywong}@cse.ust.hk
[2] Queen's University Belfast, Belfast, UK
cheng.long@qub.ac.uk

Abstract. With the proliferation of spatial-textual data such as location-based services and geo-tagged websites, spatial keyword queries become popular in the literature. One example of these queries is the *collective spatial keyword query* (CoSKQ) which is to find a set of objects in the database such that it *covers* a given set of query keywords collectively and has the smallest *cost*. Some existing cost functions were proposed in the literature, which capture different aspects of the distances among the objects in the set and the query. However, we observe that in some applications, each object has an inherent cost (e.g., workers have monetary costs) which are not captured by any of the existing cost functions. Motivated by this, in this paper, we propose a new cost function called the *maximum dot size cost* which captures both the distances among objects in a set and a query as existing cost functions do and the inherent costs of the objects. We prove that the CoSKQ problem with the new cost function is NP-hard and develop two algorithms for the problem. One is an exact algorithm which is based on a novel search strategy and employs a few pruning techniques and the other is an approximate algorithm which provides a $\ln |q.\psi|$ approximation factor, where $|q.\psi|$ denotes the number of query keywords. We conducted extensive experiments based on both real datasets and synthetic datasets, which verified our theoretical results and efficiency of our algorithms.

1 Introduction

Nowadays, geo-textual data which refers to data with both spatial and textual information is ubiquitous. Some examples of geo-textual data include the spatial points of interest with textual description, geo-tagged web objects (e.g., webpages and photos at Flicker), and also geo-social networking data (e.g., users of FourSquare have their check-in histories which are spatial and also profiles which are textual).

Collective Spatial Keyword Query (CoSKQ) [2,3,18] is a query type recently proposed on geo-textual data which is described as follows. Let $\mathcal{O}$ be a set of objects. Each object $o \in \mathcal{O}$ is associated with a spatial location, denoted by $o.\lambda$, and a set of keywords, denoted by $o.\psi$. Given a query q with a location $q.\lambda$ and

© Springer International Publishing AG 2017
M. Gertz et al. (Eds.): SSTD 2017, LNCS 10411, pp. 357–375, 2017.
DOI: 10.1007/978-3-319-64367-0_19

a set of keywords $q.\psi$, CoSKQ is to find a set S of objects such that S covers $q.\psi$, i.e., $q.\psi \subseteq \cup_{o \in S} o.\psi$, and the *cost* of S, denoted by $cost(S)$, is minimized. CoSKQ is useful in applications where a user wants to find a set of objects to collectively satisfy his/her needs. One example is that a tourist wants to find some points-of-interest to do sight-seeing, shopping and dining, where the user's needs could be captured by the query keywords of a CoSKQ query. Another example is that a manager wants to finds some workers, collectively offering a set of skills, to set up a project.

One key component of the CoSKQ problem is the cost function that measures the cost of a set of objects S wrt the query q, i.e., $cost(S)$. In the literature, five different cost functions have been proposed for $cost(S)$ [2,3,18]. Each of these cost functions is based on one or a few of the following distances: (D1) the *sum* of the distances between the objects in S and q, (D2) the *min* of the distances between the objects in S and q, (D3) the *max* of the distances between the objects in S and q, and (D4) the *max* of the distances between two objects in S. Specifically, (1) the *sum cost function* [2,3] defines $cost(S)$ as D1, (2) the *maximum sum cost function* [2,3,18] defines $cost(S)$ as D3 + D4, (3) the *diameter cost function* [18] defines $cost(S)$ as max{D1, D2}, (4) the *sum max cost function* [2] defines $cost(S)$ as D1 + D4, and (5) the *min max cost function* [2] defines $cost(S)$ as D2 + D4.

These cost functions are suitable in applications where the cost of a set of objects could be captured well by the spatial distances among the objects and the query *only*. For example, in the application where a tourist wants to find a set of points-of-interest, the cost is due to the distances to travel among the points-of-interest and the query. However, in some other applications, each object has an inherent cost and thus the cost of a set of objects would be better captured by both the distances among the objects and the query and the inherent costs of the objects. In the application where a manager wants to find a group of workers, each worker is associated with some monetary cost. Another example is that a tourist wants to visit some POIs (e.g., museums, parks), each POI is associated with an admission fee. Motivated by this, in this paper, we propose a new cost function called *maximum dot size function* which captures both some spatial distances between objects and a query and the inherent costs of the objects. Specifically, the *maximum dot size function* defines the cost of a set S of objects, denoted by $cost_{MaxDotSize}(S)$, as the multiplication between the maximum distance between the objects in S and q and the sum of the inherent costs of objects in S.

The CoSKQ problem with the maximum dot size function is proven to be NP-hard and an exact algorithm would run in exponential time where the exponent is equal to the number of query keywords. We design an exact algorithm called *MaxDotSize-E*, which adopts a novel strategy of traversing possible sets of objects so that only a small fraction of the search space is traversed (which is achieved by designing effective pruning techniques which are made possible due to the search strategy). For better efficiency, we also design an approximate algorithm called *MaxDotSize-A* for the problem. Specifically, our main contribution is summarized as follows.

- Firstly, we propose a new cost function $cost_{MaxDotSize}$, which captures both the spatial distances between the objects and the query, and the inherent costs of the objects.
- Secondly, we prove the NP-hardness of *MaxDotSize-CoSKQ* and design two algorithms, namely *MaxDotSize-E* and *MaxDotSize-A*. *MaxDotSize-E* is an exact algorithm and runs faster than an adapted algorithm in the literature [3]. *MaxDotSize-A* gives a solution set S with a $\ln |q.\psi|$-factor approximation. In particular, if $|S| \leq 3$ it is guaranteed that S is an optimal solution.
- Thirdly, we conducted extensive experiments on both real and synthetic datasets, which verified our theoretical results and the efficiency of our algorithms.

The rest of this paper is organized as follows. Section 2 gives the related work. Section 3 defines the problem and discusses its hardness. Section 4 presents our proposed algorithms. Section 5 gives the empirical study. Section 6 concludes the paper.

2   Related Work

Many existing studies on spatial keyword queries focus on retrieving a *single object* that is close to the query location and relevant to the query keywords.

A *boolean kNN query* [5,13,25,30] finds a list of k objects each covering all specified query keywords. The objects in the list are ranked based on their spatial proximity to the query location.

A *top-k kNN query* [9,10,16,19–21,26] adopts the ranking function considering both the spatial proximity and the textual relevance of the objects and returns top-k objects based on the ranking function. This type of queries has been studied on Euclidean space [9,16,19], road network databases [20], trajectory databases [10,21] and moving object databases [26]. Usually, the methods for this kind of queries adopt an index structure called the *IR-tree* [9,24] capturing both the spatial proximity and the textual information of the objects to speed up the keyword-based nearest neighbor (NN) queries and range queries. In this paper, we also adopt the *IR-tree* as an index structure.

Some other studies on spatial keyword queries focus on finding an *object set* as a solution. Among them, some [2,3,18] studied the *collective spatial keyword queries* (CoSKQ). These studies on CoSKQ adopted a few cost functions which capture the spatial distances among the objects and the query only and thus they are not suitable in applications where objects are associated with inherent costs.

Another query that is similar to CoSKQ is the *mCK query* [15,28,29] which takes a set of m keywords as input and finds m objects with the minimum *diameter* that cover the m keywords specified in the query. In the existing studies of *mCK* queries, it is usually assumed that each object contains a single keyword. There are some variants of the *mCK* query, including the *SK-COVER* [8] and the *BKC query* [11]. These queries are similar to the CoSKQ problem in that they also return an object set that covers the query keywords, but they only take a set of keywords as input.

There are also some other studies on spatial keyword queries, including [22] which finds top-k groups of objects with the ranking function considering the spatial proximity and textual relevance of the groups, [17] which takes a set of keywords and a clue as inputs, and returns k objects with highest similarities against the clue, [14,23] which finds an object set in the road network, [4,12] which finds a *region* as a solution and [1,27] which finds a *route* as a solution.

3   Problem Definition

Let $\mathcal{O}$ be a set of objects. Each object $o \in \mathcal{O}$ is associated with a location denoted by $o.\lambda$, a set of keywords denoted by $o.\psi$ and an inherent cost denoted by $o.cost$. Given two objects o and o', we denote by $d(o, o')$ the Euclidean distance between $o.\lambda$ and $o'.\lambda$. Given a query q which consists of a location $q.\lambda$ and a set of keywords $q.\psi$, an object is said to be **relevant** if it contains at least one keyword in $q.\psi$, and we denote by $\mathcal{O}_q$ the set containing all relevant objects and say a set of objects is **feasible** if it covers $q.\psi$.

Problem Definition [3]. Given a query $q = (q.\lambda, q.\psi)$, the *Collective Spatial Keyword Query* (CoSKQ) problem is to find a set S of objects in $\mathcal{O}$ such that S covers $q.\psi$ and the *cost* of S is minimized.

In this paper, we propose a new cost function called *maximum dot size* which defines the cost of a set S of objects, denoted by $cost_{MaxDotSize}(S)$, as the multiplication of the maximum distance between objects in S and q and the sum of the inherent costs of objects in S, i.e.,

$$cost_{MaxDotSize}(S) = \max_{o \in S} d(o, q) \cdot \sum_{o \in S} o.cost \tag{1}$$

For simplicity, we assume that each object has a unit cost and thus the overall inherent cost of a set of objects corresponds to the size of this set and $cost_{MaxDotSize}(S)$ corresponds to $\max_{o \in S} d(o, q) \cdot |S|$. However, the exact algorithm and the approximate algorithm developed in this paper could also be applied to the general case with arbitrary costs, while all theoretical results (e.g., approximation ratio) remain applicable (details could be found in Sect. 4.4).

We define the maximum dot size function with distance D3 (i.e., $\max_{o \in S} d(o, q)$) because D3 is the distance traveled to the most far-away object, and it is able to capture other distances (e.g., $2\max_{o \in S} d(o, q) \leq \max_{o_1, o_2 \in S} d(o_1, o_2)$). We use a simple product to combine the two factors (distance and inherent cost) such that it remains applicable and meaningful when the object inherent costs are unavailable. Note that we discarded the normalization term $\max_{o \in \mathcal{O}} d(o, q) \cdot |q.\psi| \cdot \max_{o \in \mathcal{O}} o.cost$, which does not affect the applicability of the cost function.

The CoSKQ problem with the maximum dot size function is denoted as **MaxDotSize-CoSKQ**. In the following, if there is no ambiguity, we write $cost_{MaxDotSize}(\cdot)$ as $cost(\cdot)$ for simplicity.

Intractability. The following lemma shows the NP-hardness of MaxDotSize-CoSKQ.

Lemma 1. *MaxDotSize-CoSKQ is NP-hard.*   □

Proof: We prove by transforming the *set cover* problem which is known to be NP-Complete to the MaxDotSize-CoSKQ problem. The description of the MaxDotSize-CoSKQ problem is given as follows. Given a set $\mathcal{O}$ of spatial objects, each $o \in \mathcal{O}$ is associated with a location $o.\lambda$ and a set of keywords $o.\psi$, a query q consisting of a query location $q.\lambda$ and a set of query keywords $q.\psi$, and a real number C, the problem is to determine whether there exists a set S of objects in $\mathcal{O}$ such that S covers the query keywords and $cost(S)$ is at most C.

The description of the set cover problem is given as follows. Given a universe set $U = \{e_1, e_2, ..., e_n\}$ of n elements and a collection of m subsets of U, $V = \{V_1, V_2, ..., V_m\}$, and a number k, the problem is to determine whether there exists a set $T \subseteq V$ such that T covers all the elements in U and $|T| \leq k$.

We transform a set cover problem instance to a MaxDotSize-CoSKQ problem instance as follows. We construct a query q by setting $q.\lambda$ to be an arbitrary location in the space and $q.\psi$ to be a set of n keywords each corresponding to an element in U. We construct a set $\mathcal{O}$ such that $\mathcal{O}$ contains m objects each corresponding to a subset in V. For each object o in $\mathcal{O}$, we set $o.\lambda$ to be any location at the boundary of the disk which is centered at $q.\lambda$ and has the radius equal to 1, set $o.\psi$ be a set of keywords corresponding to the elements in the subset in V that o is corresponding to and set $o.cost = 1$. Note that for any $o \in \mathcal{O}$, we have $d(o, q) = 1$ and the MaxDotSize cost of any set of objects is exactly equal to the size of the set. Besides, we set C to be equal to k. Clearly, the above transformation can be done in polynomial time.

The equivalence between the set cover problem instance and the corresponding MaxDotSize-CoSKQ problem instance could be verified easily since if there exists a set T such that T cover all the elements in U and $|T| \leq k$, the set of objects corresponding the subsets in T cover all the query keywords in $q.\psi$ and has the MaxDotSize cost at most C and vice versa.

4   Algorithms for MaxDotSize-CoSKQ

4.1   An Exact Algorithm

In this section, we present our exact method called *MaxDotSize-E* for MaxDotSize-CoSKQ. Before we present the algorithm, we first introduce some notations. Given a query q and a non-negative real number r, we denote the circle or the disk centered at $q.\lambda$ with radius r by $D(q, r)$. Given a feasible set S, the **query distance owner** of S is defined to be the object $o \in S$ that is most far away from $q.\lambda$ (i.e., $o = \arg\max_{o \in S} d(o, q)$). Given a query q and a keyword t, the **t-keyword nearest neighbor** of q, denoted by $NN(q, t)$, is defined to be the nearest neighbor (NN) of q containing keyword t. Besides, we define the **nearest neighbor set** of q, denoted by $N(q)$, to be the set containing q's

Algorithm 1. Algorithm *MaxDotSize-E*

Input: A query q and a set $\mathcal{O}$ of objects
Output: A feasible set S with the smallest cost
1: $S \leftarrow N(q)$
2: // Step 1 (Query Distance Owner Finding)
3: **for** each relevant object o in ascending order of $d(o, q)$ **do**
4:     // Step 2 (Feasible Set Construction)
5:     $S' \leftarrow$ findBestFeasibleSet(o)
6:     // Step 3 (Optimal Set Updating)
7:     **if** $S' \neq \emptyset$ and $cost(S') < cost(S)$ **then**
8:         $S \leftarrow S'$
9: **return** S

---

t-keyword nearest neighbor for each $t \in q.\psi$, i.e., $N(q) = \cup_{t \in q.\psi} NN(q, t)$. Note that $N(q)$ is a feasible set.

The *MaxDotSize-E* algorithm is presented in Algorithm 1. At the beginning, it maintains an object set S for storing the best-known solution found so far, which is initialized to $N(q)$. Then, it performs an iterative process where each iteration involves three steps.

1. **Step 1 (Query Distance Owner Finding):** It picks one relevant object o following an ascending order of $d(o, q)$.
2. **Step 2 (Feasible Set Construction):** It constructs the best feasible set S' with object o as the query distance owner via a procedure called "findBest-FeasibleSet".
3. **Step 3 (Optimal Set Updating):** It then updates S to S' if $cost(S') < cost(S)$.

The iterative process continues with the next relevant object until all relevant objects have been processed.

One remaining issue in Algorithm 1 is the "findBestFeasibleSet" procedure which is shown in Algorithm 2.

First, it maintains a variable ψ, denoting the set of keywords in $q.\psi$ not covered by o yet, which is initialized as $q.\psi - o.\psi$. If $\psi = \emptyset$, then it returns $\{o\}$ immediately since we know that it is the best feasible set. Otherwise, it proceeds to retrieve the set $\mathcal{O}'$ of all relevant objects in $D(q, d(o, q))$. If $\mathcal{O}'$ does not cover ψ, it returns $\emptyset$ immediately. Otherwise, it enumerates each possible subset S'' of $\mathcal{O}'$ with size at most $\min\{|\psi|, \frac{cost(S)}{d(o,q)} - 1\}$ in ascending order of $|S''|$ (by utilizing the inverted lists maintained for each keyword in ψ), and checks whether S'' covers ψ (note that $|S''|$ is at most $\frac{cost(S)}{d(o,q)} - 1$ because otherwise it cannot contribute to a better solution). If yes, it returns $S'' \cup \{o\}$ immediately since it is the best feasible set in $\mathcal{O}'$. Otherwise, it checks the next subset of $\mathcal{O}'$. When all subsets of $\mathcal{O}'$ have been processed and there is no feasible set found, it returns $\emptyset$.

To further improve the efficiency of the *MaxDotSize-E* algorithm, we develop some pruning techniques in both Step 1 and Step 2.

Algorithm 2. Algorithm for finding the best feasible set with object o as the query distance owner (findBestFeasibleSet(o))

Input: An object o
Output: The best feasible set with o as the query distance owner (if any)
 1: $\psi \leftarrow q.\psi - o.\psi$
 2: **if** $\psi = \emptyset$ **then**
 3:     **return** $\{o\}$
 4: $\mathcal{O}' \leftarrow$ a set of all relevant objects in $D(q, d(o, q))$
 5: **if** $\mathcal{O}'$ does not cover ψ **then**
 6:     **return** $\emptyset$
 7: **for** each subset S'' of $\mathcal{O}'$ with size at most $\min\{|\psi|, \frac{cost(S)}{d(o,q)} - 1\}$ in ascending order of $|S''|$ **do**
 8:     **if** S'' covers ψ **then**
 9:         **return** $S'' \cup \{o\}$
10: **return** $\emptyset$

4.1.1 Pruning in Step 1

The major idea is that not each object $o \in \mathcal{O}_q$ is necessary to be considered as a query distance owner of S' to be constructed and thus some can be pruned. Specifically, we have the following lemmas.

Lemma 2 (Distance Constraint). *Let S_o be the optimal set and o be the query distance owner of S_o. Then, we have $d_{LB} \leq d(o, q) \leq d_{UB}$, where $d_{LB} = \max_{o \in N(q)} d(o, q)$ and $d_{UB} = cost(S)$, where S is an arbitrary feasible set.* □

Proof: First, we prove $d_{LB} \leq d(o, q)$ by contradiction. Assume $d(o, q) < d_{LB}$. Let o_f be the farthest object from q in $N(q)$, i.e., $d_{LB} = d(q, o_f)$. There exists a keyword $t_f \in o_f.\psi \cap q.\psi$ such that t_f is not contained by any object that is closer to q than o_f since otherwise $o_f \notin N(q)$. This leads to a contradiction since there exists an object $o' \in S_o$ which covers t_f and $d(o', q) \leq d(o, q) < d_{LB}$.

Second, we prove $d(o, q) \leq d_{UB}$ also by contradiction. Assume $d(o, q) > d_{UB}$. Then $cost(S_o) = d(o, q) \cdot |S_o| > d_{UB} \cdot |S_o| > cost(S)$ which contradict the fact that S_o is the optimal set.

Figure 1 shows the distance constraint. We only need to consider the relevant objects inside the gray area (i.e., o_2, o_3 and o_5) to be the query distance owners.

Lemma 3 (Keyword Constraint). *Let o be the query distance owner of the set S' to be constructed. If $d(o, q) > d_{LB}$ and $|o.\psi \cap q.\psi| < 2$, there exist a feasible set S'' s.t. $cost(S'') \leq cost(S')$.* □

Proof: Given the set S', we can construct the feasible set S'' as follows. Consider the following two cases. *Case 1.* $d(o, q) > d_{LB}$ and $|o.\psi \cap q.\psi| = 0$. In this case, we can construct $S'' = S' \setminus \{o\}$ be the feasible set with lower cost. *Case 2.* $d(o, q) > d_{LB}$ and $|o.\psi \cap q.\psi| = 1$. Let the keyword $t = o.\psi \cap q.\psi$. We know that $o \notin N(q)$ because $d(o, q) > d_{LB}$. Note that there exist an object $o' \in N(q)$ that contains t. Thus, we can construct $S'' = S' \setminus \{o\} \cup \{o'\}$ be the feasible set with $cost(S'') \leq cost(S')$. □

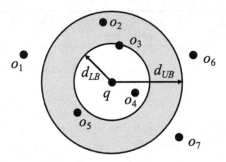

Fig. 1. Distance constraint for query distance owners

The above two lemmas suggest that an object o could be pruned if any of the following two conditions is not satisfied.

1. (Condition 1): $d_{LB} \le d(o, q) \le d_{UB}$; and
2. (Condition 2): $(d(o, q) = d_{LB})$ or $(d(o, q) > d_{LB}$ and $|o.\psi \cap q.\psi| \ge 2)$

4.1.2   Pruning in Step 2

The major idea is that not each object $o' \in \mathcal{O}'$ is necessary to be enumerated when finding the best feasible set S' with the query distance owner o. We first introduce a concept called **dominance**. Given a set of keywords ψ, a set of objects $\mathcal{O}'$, two objects o_1 and o_2 in $\mathcal{O}'$, we say that o_1 dominates o_2 wrt ψ if all keywords in ψ that are covered by o_2 can be covered by o_1 and there exists a keyword in ψ that is covered by o_1 but not by o_2 (i.e., $o_2.\psi \cap \psi \subset o_1.\psi \cap \psi$). An object that is not dominated by another object is said to be a **dominant** object. Then, it could be verified easily that only those objects that are dominant ones need to be considered in Step 2.

4.1.3   Correctness and Time Complexity

Based on the description of the *MaxDotSize-E* algorithm, it is easy to verify that *MaxDotSize-E* is an exact algorithm.

Theorem 1. *MaxDotSize-E returns a feasible set with the smallest cost for MaxDotSize-CoSKQ.*   □

Time Complexity. We use the IR-tree built on $\mathcal{O}$ to support the range query operations involved in the algorithm.

Let n_1 be the number of iterations (lines 3–8) *MaxDotSize-E* and β be the cost of executing one iteration. The time complexity of *MaxDotSize-E* is $O(n_1 \cdot \beta)$. In practice, we have $n_1 \ll |\mathcal{O}_q|$ since n_1 is equal to the number of relevant objects satisfying the two conditions.

Consider β. It is dominated by the cost of executing the "findBestFeasible Set" procedure with Algorithm 2 (line 5 of Algorithm 1). We analyze the cost of Algorithm 2 as follows. The cost of lines 1–3 is dominated by that of the remaining parts in the algorithm. Line 4 could be finished by performing a range query with an additional constraint that the object is relevant, which incurs the cost of $O(\log |\mathcal{O}| + |\mathcal{O}'|)$ [7]. Note that $|\mathcal{O}'|$ corresponds to the number of objects returned by the range query. Lines 5–6 could be finished simply by traversing for each object in $\mathcal{O}'$ the set of keywords associated with it and thus the cost is $O(\sum_{o \in \mathcal{O}'} |o.\psi|)$. Lines 7–9 could be finished by enumerating all possible subsets of $\mathcal{O}'$ with size at most $\min\{|\psi|, \frac{cost(S)}{d(o,q)} - 1\}$ in ascending order of size and for each subset, checking whether it covers ψ, and thus the cost is bounded by $O(|\mathcal{O}'|^{|\psi|} \sum_{o \in \mathcal{O}'} |o.\psi|)$ (since there are at most $|\mathcal{O}'|^{|\psi|}$ subsets to be checked and the cost of checking each subset is bounded by $\sum_{o \in \mathcal{O}'} |o.\psi|$). Overall, we know that the time complexity of $MaxDotSize\text{-}E$ is $O(n_1 \cdot (\log |\mathcal{O}| + |\mathcal{O}'| + \sum_{o \in \mathcal{O}'} |o.\psi| + |\mathcal{O}'|^{|\psi|} \sum_{o \in \mathcal{O}'} |o.\psi|)) = O(n_1 \cdot (\log |\mathcal{O}| + |\mathcal{O}'|^{|\psi|} \sum_{o \in \mathcal{O}'} |o.\psi|))$, where we have $n_1 \ll |\mathcal{O}_q|$, $|\mathcal{O}'| \leq |\mathcal{O}_q|$, and $|\psi| < |q.\psi|$.

4.2   An Approximate Algorithm

In this section, we introduce an approximate algorithm called $MaxDotSize\text{-}A$ for MaxDotSize-CoSKQ, which gives a $\ln |q.\psi|$-factor approximation.

$MaxDotSize\text{-}A$ is exactly the same as the $MaxDotSize\text{-}E$ except that it replaces the "findBestFeasibleSet" procedure which is an expensive exhaustive search process with a greedy process which is much cheaper. Specifically, the greedy process first initializes the set to be returned to $\{o\}$ and then iteratively selects an object in the disk $D(q, d(o, q))$ which has the greatest number of uncovered keywords and inserts it into the set until all keywords are covered.

Theoretical Analysis. Although the set S returned by the $MaxDotSize\text{-}A$ algorithm might have a larger cost than the optimal set S_o, the difference is bounded.

Theorem 2. *MaxDotSize-A gives a $\ln |q.\psi|$-factor approximation for the MaxDotSize-CoSKQ problem. In particular, if the solution returned by MaxDotSize-A has the size at most 3, the solution is an exact solution.*   □

Proof: Let S_o be the optimal solution and S_a be the solution returned by $MaxDotSize\text{-}A$. Let o be the query distance owner of S_o. There exists an iteration that the algorithm processes o and the greedy process construct the set S' by selects the objects in $D(q, d(o, q))$.

In the following, we show that $cost(S') \leq \ln |q.\psi| \cdot cost(S_o)$ which immediately implies the correctness of the theorem since $cost(S_a) \leq cost(S')$.

Let $S' = \{o, o'_1, o'_2, ..., o'_a\}$ and $S_o = \{o, o_1, o_2, ..., o_b\}$. That is, $|S'| = 1 + a$ and $|S_o| = 1 + b$. Without loss of generality, assume that after o is included in S', $o'_1, o'_2, ..., o'_a$ are included in S' sequentially in the greedy procedure.

Let ψ_t denote the set of keywords in $q.\psi$ that are not covered by S' after o'_{t-1} (if any) is included in S' but before o'_t is included in S' for $t = 1, 2, ..., a$. For example, $\psi_1 = q.\psi - o.\psi$.

In the case that $|S'| = 1$, i.e., $S' = \{o\}$ and $a = 0$, we know that o covers all the keywords in $q.\psi$ and $S_o = \{o\} = S'$. Therefore, S' is an exact solution. In the case that $|S'| = 2$, i.e., $S' = \{o, o_1'\}$ and $a = 1$, S_o also involves exactly two objects and thus $cost(S_o) = d(o, q) \cdot 2 = cost(S')$. Again, S' is an exact solution. In the case that $|S'| = 3$, i.e., $S' = \{o, o_1', o_2'\}$ and $a = 2$, it could be verified that S_o involves exactly three objects (this is because (1) S_o involves no more than three objects by definition and (2) S_o involves at least three objects since otherwise S' involves less than three objects which leads to a contradiction) and thus $cost(S_o) = d(o, q) \cdot 3 = cost(S')$. Still, S' is an exact solution. In the case that $|S'| \geq 4$, i.e., $a \geq 3$, we continue to prove as follows. First, it could be verified that $b \geq 2$. Second, it could be verified that $|\psi_1| \geq 3$ since $a \geq 3$. Third, in the case that $|\psi_1| \in [3, 7]$, we verify that $\frac{cost(S')}{cost(S_o)} \leq \ln |q.\psi|$ and the details of this step could be found in [6]. Therefore, in the following, we focus on the case that $|\psi_1| \geq 8$.

First, we have

$$|o_t' \cap \psi_t| \geq \frac{|\psi_t|}{b} \tag{2}$$

which could be verified by as follows. Consider the moment when o_{t-1}' (if any) has been included in S' but o_t' has not. By the greedy nature of the process, o_t' is the object that covers the greatest number of keywords that are not covered yet, i.e., $o_t' = \arg\max_{o \in \mathcal{O'}}\{|o.\psi \cap \psi_t|\}$. Therefore, we know $|o_t'.\psi \cap \psi_t| \geq \max_{1 \leq i \leq b}\{|o_i.\psi \cap \psi_t|\} \geq \frac{\sum_{1 \leq i \leq b} |o_i.\psi \cap \psi_t|}{b} \geq \frac{|\psi_t|}{b}$.

Based on Eq. (2), we have

$$|\psi_{t+1}| = |\psi_t| - |o_t'.\psi \cap \psi_t|$$
$$\leq |\psi_t| - \frac{|\psi_t|}{b} = (1 - \frac{1}{b})|\psi_t| \tag{3}$$

Based on Eq. (3), we further deduce that

$$|\psi_a| \leq (1 - \frac{1}{b})|\psi_{a-1}| \leq (1 - \frac{1}{b})^{a-1}|\psi_1| \tag{4}$$

Note that $|\psi_a| \geq 1$ since otherwise o_a' would not be included in S'. As a result, we know

$$(1 - \frac{1}{b})^{a-1}|\psi_1| \geq |\psi_a| \geq 1 \tag{5}$$

Based on Eq. (5), we know that

$$a \leq \frac{\ln |\psi_1|}{-\ln(1 - \frac{1}{b})} + 1 \leq \frac{\ln |\psi_1|}{\frac{1}{b}} + 1 \tag{6}$$
$$\leq \ln |\psi_1| \cdot b + 1 \tag{7}$$

The correctness of Eq. (6) is based on the fact that $\ln(1 + x) \leq x$ for $x \in (-1, 0]$ (to illustrate, consider $f(x) = \ln(1 + x) - x$. We have $f'(x) = \frac{1}{1+x} - 1 \geq 0$ for $x \in (-1, 0]$ and $f(0) = 0$).

As a result, we know

$$\frac{cost(S')}{cost(S_o)} = \frac{d(o,q) \cdot |S'|}{d(o,q) \cdot |S_o|} = \frac{a+1}{b+1}$$

$$\leq \frac{\ln |\psi_1| \cdot b + 2}{b+1} \leq \ln |\psi_1| + \frac{2 - \ln |\psi_1|}{b+1}$$

$$\leq \ln |\psi_1| + \frac{2 - \ln 8}{b+1} < \ln |q.\psi| \tag{8}$$

which immediately implies the correctness of the theorem since $cost(S_a) \leq cost(S')$. □

Time Complexity. We use the IR-tree built on $\mathcal{O}$ to support the range query operations involved in the algorithm.

Let n_1 be the number of iterations in *MaxDotSize-A*. The time complexity of *MaxDotSize-A* is $O(n_1 \cdot \gamma)$, where γ is the cost of the greedy process. The cost of the greedy process is $O(|\psi| \cdot |\psi||\mathcal{O}'|)$ since it involves at most $|\psi|$ iterations and for each iteration, it checks for at most $|\mathcal{O}'|$ objects the number of keywords in ψ newly covered by the object being checked, which could be done in $O(|\psi| \cdot |\mathcal{O}'|)$ time with the help of an inverted list of $|\mathcal{O}'|$ based on ψ (note that the cost of building the inverted list is simply $O(\sum_{o \in \mathcal{O}'} |o.\psi|)$).

Therefore, the time complexity of *MaxDotSize-A* is $O(n_1 \cdot (\sum_{o \in \mathcal{O}'} |o.\psi| + |\psi|^2 |\mathcal{O}'|))$, where $n_1 << |\mathcal{O}_q|$, $|\mathcal{O}'| \leq |\mathcal{O}_q| << |\mathcal{O}|$, and $|\psi| < |q.\psi|$.

4.3   Adaptations of Existing Algorithms

In this section, we adapt the existing algorithms in [2,3,18], which are originally designed for CoSKQ problem with other cost functions, for MaxDotSize-CoSKQ.

Cao-E. *Cao-E* is an exact algorithm proposed in [3] for CoSKQ problem with $cost_{MaxSum}$. It can be adapted to MaxDotSize-CoSKQ problem directly by replacing the cost function from $cost_{MaxSum}$ to $cost_{MaxDotSize}$, because it is a best-first search algorithm which is independent of the cost function used in the problem.

Other Exact Algorithms. Some other exact algorithms were proposed in the literature for CoSKQ problem with different cost functions, namely *Cao-Sum-E* [2,3] (for $cost_{Sum}$), *Cao-E-New* [2] (for either $cost_{MaxSum}$ or $cost_{MinMax}$) and *Long-E* [18] (for either $cost_{MaxSum}$ or $cost_{Dia}$). They cannot be adapted to CoSKQ problem with $cost_{MaxDotSize}$ because they all rely on the property of their original cost functions. Consider *Long-E* as an example. The core of *Long-E* relies on an important property of the distance owner group (containing three objects) that different sets of objects with the same distance owner group have the same cost for the cost function of either $cost_{MaxSum}$ or $cost_{Dia}$ studied in [18]. However, this important property could not be applied to our cost function studied in this paper, i.e., $cost_{MaxDotSize}$. In fact, it is possible that two sets of objects with the same distance owner group have different costs for the cost function of $cost_{MaxDotSize}$.

Cao-A1. *Cao-A1* is an approximate algorithm proposed in [2,3] for CoSKQ problem with either $cost_{MaxSum}$ or $cost_{MinMax}$. In [6], we prove that *Cao-A1* gives $|q.\psi|$-factor approximation for MaxDotSize-CoSKQ.

Cao-A2. *Cao-A2* is an approximate algorithm proposed in [2] for CoSKQ problem with $cost_{MaxSum}$. In [6], we prove that *Cao-A2* gives $\frac{|q.\psi|}{2}$-factor approximation for MaxDotSize-CoSKQ.

Cao-A3. *Cao-A3* is an approximate algorithm proposed in [2,3] for CoSKQ problem with $cost_{Sum}$. In [6], we prove that *Cao-A3* gives $|q.\psi|$-factor approximation for MaxDotSize-CoSKQ.

Long-A. *Long-A* is an approximate algorithm proposed in [18] for CoSKQ problem with either $cost_{MaxSum}$ or $cost_{Dia}$. In [6], we prove that *Long-A* gives $\frac{|q.\psi|}{2}$-factor approximation for MaxDotSize-CoSKQ.

Table 1. Approximation factors for MaxDotSize-CoSKQ

Approximate algorithm	Approximation factor
Cao-A1 [2,3]	$|q.\psi|$
Cao-A2 [2]	$|q.\psi|/2$
Cao-A3 [2,3]	$|q.\psi|$
Long-A [18]	$|q.\psi|/2$
MaxDotSize-A (this paper)	$\ln |q.\psi|$

Table 1 shows the approximation factors of the above adaptations of existing approximate algorithms and also the approximate algorithm *MaxDotSize-A* in this paper. Among all approximate algorithms, our *MaxDotSize-A* provides the best approximation factor for MaxDotSize-CoSKQ.

4.4   Extension to Arbitrary Inherent Costs

Our algorithms can also be applied to the general case with arbitrary object costs, with the following small changes in both *MaxDotSize-E* and *MaxDotSize-A*.

Specifically, for *MaxDotSize-E*, we do not return the solution immediately after we found a set S'' covers ψ (i.e., line 9 in Algorithm 2). Instead, we enumerate all subsets and find the one with the minimum cost. Note that the distance constraint pruning (Lemma 2) is still applicable. Also, we adjust the definition of dominance as follows. Given a set of keywords ψ, a set of objects $\mathcal{O}'$, two objects o_1 and o_2 in $\mathcal{O}'$, we say that o_1 dominates o_2 wrt ψ if all keywords in ψ that are covered by o_2 can be covered by o_1, there exists a keyword in ψ that is covered by o_1 but not by o_2, and $o_1.cost \leq o_2.cost$. Then, the pruning based on dominance remains applicable. It is easy to see that the above changes do not affect the correctness and time complexity of the algorithm.

For *MaxDotSize-A*, we need to change the selection criteria in the greedy process as follows. The greedy process first initializes the set to be returned to $\{o\}$ and then iteratively selects an object in the disk $D(q, d(o, q))$ which has the greatest ratio of (number of uncovered keywords covered by the object)/(object inherent cost) and inserts it into the set until all keywords are covered. The following theorem shows the approximation ratio of *MaxDotSize-A*.

Theorem 3. *MaxDotSize-A gives a $(\ln |q.\psi| + 1)$-factor approximation for the MaxDotSize-CoSKQ problem, when objects have arbitrary inherent costs. In particular, if the solution returned by MaxDotSize-A has the size at most 2, the solution is an exact solution.*

Proof: Let $c = o.cost$, $a = \sum_{o \in S'} o.cost - c$ and $b = \sum_{o \in S_o} o.cost - c$. Let ψ_t denote the set of keywords in $q.\psi$ that are not covered by S' after o'_{t-1} (if any) is included in S' but before o'_t is included in S' for $t = 1, 2, ...$ For example, $\psi_1 = q.\psi - o.\psi$.

In the case that $|S'| = 1$, i.e., $S' = \{o\}$, we know that o covers all the keywords in $q.\psi$ and $S_o = \{o\} = S'$. Therefore, S' is an exact solution. In the case that $|S'| = 2$, i.e., $S' = \{o, o'_1\}$, S_o also involves exactly o and o'_1 and thus $cost(S_o) = cost(S')$. Again, S' is an exact solution.

In the case that $|S'| \geq 3$, we verify $a \leq b \cdot (\ln |\psi_1| + 1)$ as follows. Consider an object $o_i \in S_o$. Let $o_i.\psi = \{t_k, t_{k-1}, ..., t_1\}$, where the algorithm covers the keywords in o_i in the order $t_k, t_{k-1}, ..., t_1$. There exist an iteration in our algorithm that pick an object o'_t that covers a keyword t_j of o_i. In that iteration, at least i keywords in o_i remain uncovered. Thus, if the algorithm were to pick o_i in that iteration, the cost per keyword at most $o_i.cost/i$. Summing over the keywords in o_i, the total amount charged to keywords in o_i is at most $o_i.cost \cdot (1 + 1/2 + 1/3 + ..1/k) \leq o_i.cost \cdot (\ln k + 1) \leq o_i.cost \cdot (\ln |\psi_1| + 1)$. Summing over the objects in S_o and noting that every keywords in ψ_1 is covered by some objects in S_o, we get

$$a = \sum_{o_i \in S_o \setminus o} o_i.cost \cdot (\ln |\psi_1| + 1)$$

$$= b \cdot (\ln |\psi_1| + 1) \tag{9}$$

Therefore, we know

$$\frac{cost(S')}{cost(S_o)} = \frac{d(o, q) \cdot \sum_{o \in S'} o.cost}{d(o, q) \cdot \sum_{o \in S_o} o.cost} = \frac{a + c}{b + c}$$

$$\leq \frac{b \cdot (\ln |\psi_1| + 1) + c}{b + c}$$

$$\leq \ln |\psi_1| + 1 - \frac{c \cdot (\ln |\psi_1| + 1)}{b + c} < \ln |q.\psi| + 1 \tag{10}$$

which immediately implies the correctness of the theorem since $cost(S_a) \leq cost(S')$. $\square$

It is easy to see that changing the object selection criteria does not affect the time complexity of the algorithm.

5   Empirical Studies

5.1   Experimental Set-Up

Datasets. We used three real datasets adopted in [2,3,18], namely Hotel, GN and Web. Dataset Hotel contains a set of hotels in the U.S. (www.allstays.com), each of which has a spatial location and a set of words that describe the hotel (e.g., restaurant, pool). Dataset GN was collected from the U.S. Board on Geographic Names (geonames.usgs.gov), where each object has a location and also a set of descriptive keywords (e.g., a geographic name such as valley). Dataset Web was generated by merging two real datasets. One is a spatial dataset called TigerCensusBlock[1], which contains a set of census blocks in Iowa, Kansas, Missouri and Nebraska. The other is WEBSPAM-UK2007[2], which consists of a set of web documents. Table 2 shows the statistics of the datasets. We set the inherent costs of the objects to 1 by default.

Table 2. Datasets used in the experiments

	Hotel	GN	Web
Number of objects	20,790	1,868,821	579,727
Number of unique words	602	222,409	2,899,175
Number of words	80,645	18,374,228	249,132,883

Query Generation. Let $\mathcal{O}$ be a dataset of objects. Given an integer k, we generate a query q with k query keywords similarly as [3,18] did. Specifically, to generate $q.\lambda$, we randomly pick a location from the MBR of the objects in $\mathcal{O}$, and to generate $q.\psi$, we first rank all the keywords that are associated with objects in $\mathcal{O}$ in descending order of their frequencies and then randomly pick k keywords in the percentile range of $[10, 40]$. In this way, each query keyword has a relatively high frequency.

Algorithms. We studied our *MaxDotSize-E*, *MaxDotSize-A* and adapted algorithms as mentioned in Sect. 4.3. Specifically, we consider two exact algorithms, namely *MaxDotSize-E* and *Cao-E* [3] (the adaption), and five approximate algorithms, namely *MaxDotSize-A*, *Cao-A1* [2,3], *Cao-A2* [2], *Cao-A3* [2,3] and *Long-A* [18].

All algorithms were implemented in C++ and all experiments were conducted on a Linux platform with a 2.66 GHz machine and 32GB RAM. The IR-tree index structure is memory resident.

---

[1] http://www.rtreeportal.org.
[2] http://barcelona.research.yahoo.net/webspam/datasets/uk2007.

5.2   Experimental Results

We used the running time and the approximation ratio (for approximate algorithms only) as measurements. For each set of settings, we generated 50 queries, ran the algorithms with each of these 50 queries. The averaged measurements are reported.

5.2.1   Effect of $|q.\psi|$

Following the existing studies [3, 18], we vary the number of query keywords (i.e., $|q.\psi|$) from $\{3, 6, 9, 12, 15\}$.

Experiment on Dataset Hotel. Figure 2 shows the results on dataset Hotel. According to Fig. 2(a), the running time increases with the query size. Our *MaxDotSize-E* is faster than *Cao-E* by 1–3 orders of magnitude, and the order of magnitude increases with the query size. It is because *Cao-E* has to enumerate all sets while *MaxDotSize-E* has more effective pruning strategies. According to Fig. 2(b), *MaxDotSize-A*, *Cao-A2*, *Cao-A3* and *Long-A* have comparable running time. The running time of *Cao-A1* is the smallest but as shown in Fig. 2(c), however, the empirical approximation ratio of *Cao-A1* is the greatest. Our *MaxDotSize-A* has the best performance with an approximate ratio close to 1, which shows that *MaxDotSize-A* achieves a high accuracy in practice. This is also consistent with our theoretical results that *MaxDotSize-A* gives the best approximation factor among all approximate algorithms.

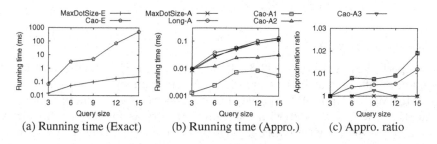

(a) Running time (Exact)   (b) Running time (Appro.)   (c) Appro. ratio

Fig. 2. Effect of $|q.\psi|$ (Hotel)

Experiment on Dataset GN. Figure 3 shows the results on dataset GN. The results of *Cao-E* with the query size 9, 12, 15 are not shown because it took more than 3 days or ran out of memory. According to Fig. 3(a), our *MaxDotSize-E* is faster than *Cao-E*. Note that when query size increases, the running time of *MaxDotSize-E* only increases slightly because the pruning strategy based on dominant objects reduces the search space effectively. According to Fig. 3(b), *Long-A* runs the slowest, *Cao-A1* runs the fastest, and all other approximate algorithms including *MaxDotSize-A* run comparably fast. As shown in Fig. 3(c), all approximate algorithms have approximate ratios 1. We found that this is

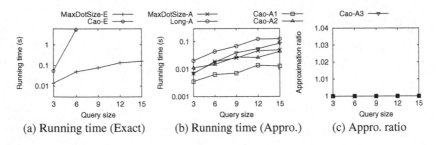

(a) Running time (Exact)   (b) Running time (Appro.)   (c) Appro. ratio

Fig. 3. Effect of $|q.\psi|$ (GN)

because each object in the optimal solution S_o only contains one query keyword and thus the size of S_o is equivalent to the number of query keywords (i.e., $|S_o| = |q.\psi|$). In this case, $S_o = N(q)$ which is used as a starting point in each of the approximate algorithms.

Experiment on Dataset Web. Figure 4 shows the results on dataset Web, which gives similar clues.

In the following, we do not show the results of *Cao-E* since it is not scalable and consistently dominated by our *MaxDotSize-E* algorithm.

5.2.2   Effect of Average $|o.\psi|$

Following the existing studies [3,18], we conduced experiments on average $|o.\psi|$. The details could be found in [6].

5.2.3   Scalability Test

Following the existing studies [2,3,18], we generated 5 synthetic datasets for the experiments of scalability test, in which the numbers of objects used are 2M, 4M, 6M, 8M and 10M. Specifically, we generated a synthetic dataset by augmenting the GN dataset with additional objects as follows. Each time, we create a new object o with $o.\lambda$ set to be a random location from the original GN dataset by following the distribution and $o.\psi$ set to be a random document from GN and then add it into the GN dataset. We vary the number of objects from $\{2M, 4M, 6M, 8M, 10M\}$, and the query size $|q.\psi|$ is set to 6.

Figures 5 shows the results for the scalability test. According to Fig. 5(a), our *MaxDotSize-E* is scalable wrt the number of objects in the datasets, e.g., it ran within 10 s on a dataset with 10M objects. Besides, according to Fig. 5(b), our *MaxDotSize-A* runs consistently faster than *Cao-A2* and *Long-A* and it is scalable, e.g., it ran within 1 s on a dataset with 10M objects. According to Fig. 5(c), all approximate algorithms can achieve approximation ratios close to 1.

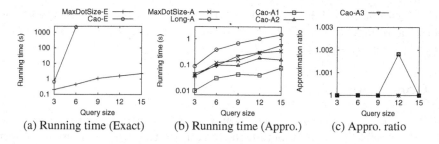

(a) Running time (Exact)   (b) Running time (Appro.)   (c) Appro. ratio

Fig. 4. Effect of $|q.\psi|$ (Web)

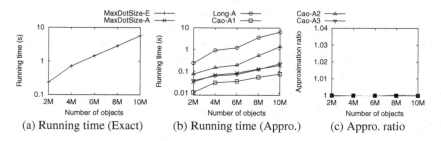

(a) Running time (Exact)   (b) Running time (Appro.)   (c) Appro. ratio

Fig. 5. Scalability test

5.2.4 Objects with Inherent Cost

We further generated a dataset based on the Hotel dataset, where each object is associated with an inherent cost. For each object, we assign an integer inherent cost in the range $[1, 5]$ randomly.

Figure 6 shows the results. As shown in Fig. 6(a) and (b), the running time of the algorithms are similar to the case without object inherent cost (i.e., Fig. 2(a) and (b)). According to Fig. 6(c), the approximation ratio of our *MaxDotSize-A* is near to 1, which shows that the accuracy of *MaxDotSize-A* is high in practice. The approximation ratio shown in the figure was computed based on the average of 50 queries. We found that the approximation ratio of *MaxDotSize-A* is exactly 1 for most queries (e.g., more than 47). Therefore, the averaged approximation

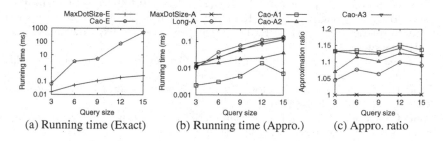

(a) Running time (Exact)   (b) Running time (Appro.)   (c) Appro. ratio

Fig. 6. Effect of $|q.\psi|$ (Inherent cost)

ratio of *MaxDotSize-A* is always near to 1. This is also consistent with our theoretical results that *MaxDotSize-A* gives the best approximation factor among all approximate algorithms.

6   Conclusion

In this paper, we proposed a new cost function, *maximum dot size cost* for CoSKQ problem. The cost function captures both the distances among objects and the inherent costs of objects. We proved the NP-hardness of MaxDotSize-CoSKQ and designed an exact algorithm and an approximate algorithm with a theoretical error guarantee. Extensive experiments were conducted which verified our theoretical findings.

There are several interesting future research directions. One direction is to penalize objects with too much keywords such that the results would not always favour these objects (this could be used to make it more difficult to cheat an algorithm for the CoSKQ problem by associating an object with many keywords). It is interesting to see how to make a good balance between objects with many keywords and the number of objects.

Acknowledgements. We are grateful to the anonymous reviewers for their constructive comments on this paper. The research of Harry Kai-Ho Chan and Raymond Chi-Wing Wong is supported by HKRGC GRF 16219816.

References

1. Cao, X., Chen, L., Cong, G., Xiao, X.: Keyword-aware optimal route search. PVLDB **5**(11), 1136–1147 (2012)
2. Cao, X., Cong, G., Guo, T., Jensen, C.S., Ooi, B.C.: Efficient processing of spatial group keyword queries. TODS **40**(2), 13 (2015)
3. Cao, X., Cong, G., Jensen, C.S., Ooi, B.C.: Collective spatial keyword querying. In: SIGMOD, pp. 373–384. ACM (2011)
4. Cao, X., Cong, G., Jensen, C.S., Yiu, M.L.: Retrieving regions of intersect for user exploration. PVLDB **7**(9), 733–744 (2014)
5. Cary, A., Wolfson, O., Rishe, N.: Efficient and scalable method for processing top-k spatial boolean queries. In: Gertz, M., Ludäscher, B. (eds.) SSDBM 2010. LNCS, vol. 6187, pp. 87–95. Springer, Heidelberg (2010). doi:10.1007/978-3-642-13818-8_8
6. Chan, H.K.-H., Long, C., Wong, R.C.-W.: Inherent-cost aware collective spatial keyword queries (full version) (2017). http://www.cse.ust.hk/~khchanak/paper/sstd17-coskq-full.pdf
7. Chazelle, B., Cole, R., Preparata, F.P., Yap, C.: New upper bounds for neighbor searching. Inf. Control **68**(1), 105–124 (1986)
8. Choi, D.-W., Pei, J., Lin, X.: Finding the minimum spatial keyword cover. In: ICDE, pp. 685–696. IEEE (2016)
9. Cong, G., Jensen, C.S., Wu, D.: Efficient retrieval of the top-k most relevant spatial web objects. PVLDB **2**(1), 337–348 (2009)
10. Cong, G., Lu, H., Ooi, B.C., Zhang, D., Zhang, M.: Efficient spatial keyword search in trajectory databases. Arxiv preprint arXiv:1205.2880 (2012)

11. Deng, K., Li, X., Lu, J., Zhou, X.: Best keyword cover search. TKDE **27**(1), 61–73 (2015)
12. Fan, J., Li, G., Chen, L.Z.S., Hu, J.: Seal: spatio-textual similarity search. PVLDB **5**(9), 824–835 (2012)
13. Felipe, I.D., Hristidis, V., Rishe, N.: Keyword search on spatial databases. In: ICDE, pp. 656–665. IEEE (2008)
14. Gao, Y., Zhao, J., Zheng, B., Chen, G.: Efficient collective spatial keyword query processing on road networks. ITS **17**(2), 469–480 (2016)
15. Guo, T., Cao, X., Cong, G.: Efficient algorithms for answering the m-closest keywords query. In: SIGMOD. ACM (2015)
16. Li, Z., Lee, K., Zheng, B., Lee, W., Lee, D., Wang, X.: IR-tree: an efficient index for geographic document search. TKDE **23**(4), 585–599 (2011)
17. Liu, J., Deng, K., Sun, H., Ge, Y., Zhou, X., Jensen, C.: Clue-based spatio-textual query. PVLDB **10**(5), 529–540 (2017)
18. Long, C., Wong, R.C.-W., Wang, K., Fu, A.W.-C.: Collective spatial keyword queries:a distance owner-driven approach. In: SIGMOD, pp. 689–700. ACM (2013)
19. Rocha-Junior, J.B., Gkorgkas, O., Jonassen, S., Nørvåg, K.: Efficient processing of top-k spatial keyword queries. In: Pfoser, D., Tao, Y., Mouratidis, K., Nascimento, M.A., Mokbel, M., Shekhar, S., Huang, Y. (eds.) SSTD 2011. LNCS, vol. 6849, pp. 205–222. Springer, Heidelberg (2011). doi:10.1007/978-3-642-22922-0_13
20. Rocha-Junior, J.B., Nørvåg, K.: Top-k spatial keyword queries on road networks. In: EDBT, pp. 168–179. ACM (2012)
21. Shang, S., Ding, R., Yuan, B., Xie, K., Zheng, K., Kalnis, P.: User oriented trajectory search for trip recommendation. In: EDBT, pp. 156–167. ACM (2012)
22. Skovsgaard, A., Jensen, C.S.: Finding top-k relevant groups of spatial web objects. VLDBJ **24**(4), 537–555 (2015)
23. Su, S., Zhao, S., Cheng, X., Bi, R., Cao, X., Wang, J.: Group-based collective keyword querying in road networks. Inf. Process. Lett. **118**, 83–90 (2017)
24. Wu, D., Cong, G., Jensen, C.: A framework for efficient spatial web object retrieval. VLDBJ **21**(6), 797–822 (2012)
25. Wu, D., Yiu, M., Cong, G., Jensen, C.: Joint top-k spatial keyword query processing. TKDE **24**(10), 1889–1903 (2012)
26. Wu, D., Yiu, M.L., Jensen, C.S., Cong, G.: Efficient continuously moving top-k spatial keyword query processing. In: ICDE, pp. 541–552. IEEE (2011)
27. Zeng, Y., Chen, X., Cao, X., Qin, S., Cavazza, M., Xiang, Y.: Optimal route search with the coverage of users' preferences. In: IJCAI, pp. 2118–2124 (2015)
28. Zhang, D., Chee, Y.M., Mondal, A., Tung, A., Kitsuregawa, M.: Keyword search in spatial databases: towards searching by document. In: ICDE, pp. 688–699. IEEE (2009)
29. Zhang, D., Ooi, B.C., Tung, A.K.H.: Locating mapped resources in web 2.0. In: ICDE, pp. 521–532. IEEE (2010)
30. Zhang, D., Tan, K.-L., Tung, A.K.H.: Scalable top-k spatial keyword search. In: EDBT/ICDT, pp. 359–370. ACM (2013)

Vision/Challenge Papers

Towards a Unified Spatial Crowdsourcing Platform

Christopher Jonathan$^{(\boxtimes)}$ and Mohamed F. Mokbel

Department of Computer Science and Engineering,
University of Minnesota, Minneapolis, MN, USA
{cjonathan,mokbel}@cs.umn.edu

Abstract. This paper provides the vision of a unified *spatial crowdsourcing* platform that is designed to efficiently tackle different types of *spatial tasks* which have been gaining a lot of popularity in recent years. Several examples of *spatial tasks* are ride-sharing services, delivery services, translation tasks, and crowd-sensing tasks. While existing crowdsourcing platforms, such as Amazon Mechanical Turk and Upwork, are widely used to solve lots of general tasks, e.g., image labeling; using these marketplaces to solve spatial tasks results in low quality results. This paper identifies a set of characteristics for a unified *spatial crowdsourcing* environment and provides the core components of the platform that are required to empower the capability in solving different types of *spatial tasks*.

1 Introduction

In recent years, crowdsourcing has been gaining a lot of popularity due to its capability in solving various computer-hard tasks, e.g., data labeling [1] and image sorting [2]. The popularity of crowdsourcing can be seen by the existence of several famous commercial crowdsourcing marketplaces, including Amazon Mechanical Turk[1] and Upwork[2]. While these marketplaces are widely used to solve lots of general tasks, many tasks were born to be spatially oriented, namely *spatial tasks*. In such tasks, the location of the worker plays an important role in solving the tasks efficiently. For example, ride-sharing and delivery services can only be done by workers who are located within the area of the tasks, image geotagging tasks will be done more accurately by workers who live near the location of the image, and rating a restaurant would be preferred to be done by local workers. Using the general crowdsourcing marketplaces to solve *spatial tasks* results in low quality results as they are not spatially-aware by randomly selecting workers to solve the tasks regardless of the tasks' location.

The popularity of *spatial tasks*, hindered by the limitations of existing general crowdsourcing marketplaces to support them, urges both industry and academia

This work is partially supported by the National Science Foundation, USA, under Grants IIS-1525953, CNS-1512877, IIS-1218168, and IIS-0952977.

[1] https://www.mturk.com/.
[2] https://www.upwork.com/.

© Springer International Publishing AG 2017
M. Gertz et al. (Eds.): SSTD 2017, LNCS 10411, pp. 379–383, 2017.
DOI: 10.1007/978-3-319-64367-0_20

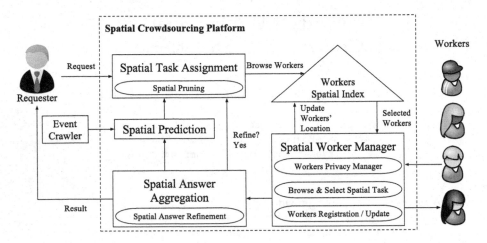

Fig. 1. *Spatial Crowdsourcing* Platform

to provide several *spatial crowdsourcing* solutions where each solution is designed
to tackle a specific type of *spatial tasks*. This includes specific solutions for ride
sharing [3,4], crowd sensing [5], and asking workers to go to a certain location
to do a task [6,7]. For surveys of such tasks, see [8,9]. Despite all such efforts
on providing solutions for different kinds of *spatial tasks*, these approaches are
not scalable from a system point of view. Currently, to create a new solution for
a *spatial task*, developers need to rebuild the same components that have been
built by other *spatial task's* solutions. Thus, identifying the core components
that can be shared among different *spatial tasks* will allow us to have a single
holistic framework that can facilitate and take the burden of recreating the same
components for different *spatial tasks* away from the developers.

In this paper, we envision a unified *spatial crowdsourcing* platform that provides a one-stop solution for any *spatial tasks*. Our envisioned platform will
be equivalent to Amazon Mechanical Turk, however, tailored for *spatial crowdsourcing*. In particular, a *spatial crowdsourcing* platform will provide the basic
modules that are required to efficiently solve *spatial tasks*. Requesters of *spatial
tasks* do not need to worry about the details of each internal decision of the *spatial crowdsourcing* process. For example, Uber[3] can use the *spatial crowdsourcing*
platform for ride-sharing while other users can use it to translate a document.

2   The Spatial Crowdsourcing Platform

Figure 1 gives the system architecture of our vision for a unified *spatial crowdsourcing* platform. There are two types of users that are using the platform,
namely, a requester and a worker. The requester provides a *spatial task* along
with the budget that she is willing to pay to the workers (i.e., the crowd) to

---

[3] https://www.uber.com/.

the platform. Then, the platform reports back the result to the requester. The worker will provide her current location when she is available to do some tasks and the platform will provide the tasks that she is qualified for which depends on her current location.

Similar to general crowdsourcing marketplaces, a *spatial crowdsourcing* platform would have three main components, namely, *worker manager, task assignment*, and *answer aggregation*. However, all three modules need to be modified to be spatially-aware. In addition, the *spatial crowdsourcing* platform introduces two new components, namely a *spatial index* and a *spatial prediction* module. The details of each component is described briefly below:

Spatial Worker Manager. In general crowdsourcing marketplaces, this module is responsible on maintaining workers' information, broadcasting available tasks to qualified workers, allowing workers to browse available tasks, and reporting workers' answers. In *spatial crowdsourcing* platform, this module will also store the current location information of every worker. It will also be equipped with a *worker privacy manager* submodule to maintain the worker's location privacy.

Workers Spatial Index. Most *spatial tasks* need to be assigned to workers within spatial proximity of the task. Thus, *spatial crowdsourcing* platform incorporates a *spatial index* to manage its workers' locations for fast retrieval. The index can be any types of spatial index, e.g., a quadtree, an R-tree, or a grid index.

Spatial Task Assignment. In general crowdsourcing marketplaces, this module receives a request from the requester and runs a task assignment algorithm to find qualified workers, e.g., workers within a certain age or gender. In *spatial crowdsourcing* platform, this module is modified to be spatially-aware by providing the capability of assigning workers within spatial proximity of the task by exploiting the spatial index. Furthermore, this module is also equipped with a *spatial pruning* submodule that tries to prune out workers that do not need to be checked by the *spatial task assignment* module for a given task, even when the workers are within spatial proximity of the task. Consider an example of a ride-sharing task where the requester provides both pick-up and drop-off locations to the platform. The *spatial pruning* submodule may prune out workers who are currently not traveling towards the drop-off location of the task even when they are currently located near the pick-up location of the task.

Spatial Answer Aggregation. In general crowdsourcing marketplaces, this module receives the results from workers and tries to deduce the final answer and returns it to the requester. In *spatial crowdsourcing* platform, this module is modified to be spatially-aware by taking into consideration the worker's location in evaluating each worker's answer. The closer the distance of a worker to the task, the higher the confidence of the worker in providing the answer is. This module is also equipped with a *spatial answer refinement* submodule to decide whether another crowdsourcing cycle is needed to further spatially refine the result, e.g., to assign workers within a finer spatial granularity in the next crowdsourcing cycle, thus, results in a better quality result.

Spatial Prediction. This module receives a stream of information, e.g., traffic, weather, and event information, in order to help the *spatial task assignment* module to assign more *spatial tasks*. For example, when this module receives an information about a football game that is currently happening at a certain location, it will tell the *spatial task assignment* module to send drivers to the location of the game before the game is finished as there might be many ride-sharing tasks available later.

3   Case Studies

This section discusses three *spatial tasks*: ride-sharing, taking picture of a land-mark, and image geotagging, as case studies to show the functionality of the platform.

Case Study 1: Ride-Sharing. Requester provides the pick-up location and the drop-off location to the platform and she will receive the driver identification. The *spatial task assignment* module uses the spatial index to find the closest k workers to the pick-up location and the *spatial pruning* submodule prunes out workers who are currently not traveling in the direction of the drop-off location. The *spatial worker manager* module receives the information about the driver who accepts the task. Then, the *spatial answer aggregation* module returns this information to the requester.

Case Study 2: Taking Picture of a Landmark. Requester provides a land-mark with its location to the platform and she will receive a picture of the landmark. The *spatial task assignment* uses the spatial index to find the closest worker to the landmark. The *spatial worker manager* module receives the image from the worker who accepts the task. Then, the *spatial answer aggregation* module returns the image to the requester.

Case Study 3: Image Geotagging. Requester provides an image to the *spatial crowdsourcing* platform and it will return the location of the image. The *spatial task assignment* module uniformly selects workers from around the world. The *spatial worker manager* module receives locations of the image from the workers who accept the task. Then, the *spatial answer aggregation* module assigns a different weight to each worker's answer depending on the distance of her answer to her location. Then, it aggregates the results to find the location of the image and return it to the requester. The *spatial answer refinement* submodule may run another crowdsourcing cycle, however, this time with a more refined location based on the current result, if it is unable to find the location of the image.

References

1. Haas, D., Wang, J., Wu, E., Franklin, M.J.: Clamshell: Speeding up crowds for low-latency data labeling. In: PVLDB (2015)
2. Marcus, A., Wu, E., Karger, D., Madden, S., Miller, R.: Human-powered sorts and joins. In: PVLDB (2011)
3. Asghari, M., Deng, D., Shahabi, C., Demiryurek, U., Li, Y.: Price-aware real-time ride-sharing at scale: an auction-based approach. In: SIGSPATIAL (2016)
4. Cici, B., Markopoulou, A., Laoutaris, N.: Designing an on-line ride-sharing system. In: SIGSPATIAL (2015)
5. Ganti, R.K., Ye, F., Lei, H.: Mobile crowdsensing: current state and future challenges. IEEE Commun. Mag. **49**(11), 32–39 (2011)
6. Kazemi, L., Shahabi, C.: Geocrowd: enabling query answering with spatial crowdsourcing. In: SIGSPATIAL (2012)
7. Chen, Z., Fu, R., Zhao, Z., Liu, Z., Xia, L., Chen, L., Cheng, P., Cao, C.C., Tong, Y., Zhang, C.J.: gMission: a general spatial crowdsourcing platform. In: PVLDB (2014)
8. Chen, L., Shahabi, C.: Spatial crowdsourcing: challenges and opportunities. IEEE Data Eng. Bull. **39**(4), 14–25 (2016)
9. Zhao, Y., Han, Q.: Spatial crowdsourcing: current state and future directions. IEEE Commun. Mag. **54**(7), 102–107 (2016)

On Designing a GeoViz-Aware Database System - Challenges and Opportunities

Mohamed Sarwat[1]([✉]) and Arnab Nandi[2]

[1] Arizona State University, 699 S. Mill Ave, Tempe, AZ, USA
`msarwat@asu.edu`
[2] Ohio State University, 2015 Neil Ave, Columbus, OH, USA
`arnab@cse.osu.edu`

1 Motivation and Challenges

The human's ability to perceive, consume, and interact with the data is in-fact limited. One key observation is worth considering – that Geospatial data is typically consumed as aggregate visualizations. e.g., Heatmap, Choropleth map (Fig. 2), Cartogram. For instance, Fig. 1 shows a heatmap of the drop-off locations of 1.1 billion NYC Taxi trips. Also, interactions are performed in a manner constrained by the user map interface. Thus, leveraging this observation allows us to aspire towards large data sizes, while keeping the user-facing outputs constant. GeoVisual analytics, *abbr. GeoViz*, is the science of analytical reasoning assisted by interactive GeoVisual map interfaces. While there

Fig. 1. NYC taxi trip heatmap

Fig. 2. Tweets Choropleth map

exists decades of research in spatial data management, enabling practitioners in non-Computer Science fields to perform highly interactive GeoViz over large-scale spatial data remains challenging. Off-the-shelf visualization and data exploration tools focus on business domains, while existing GIS products focus on data management and exploration. We recognize a number of challenges in analysis of spatiotemporal data: *(1) Scalability:* The massive-scale of spatial data hinders visualizing it using traditional GIS tools. Many map services such as MapBox and GIS tools (e.g., QGIS) allow users to visualize a fairly small amount of spatial data. The problem becomes more challenging when the GIS tool has to load, render, and visualize terabytes of geospatial data. *(2) Interactivity:* Currently, an expert can easily produce a one-shot GeoViz analysis in spatial data, producing a static visualization. However, sifting through different attributes to interactively inspect a map view of the data is a challenging task, due to the latency involved in regenerating the visualization. Since interactivity impacts the ability to derive insights at the speed of thought, it is important to reduce roundtrip latency at all points of the stack.

© Springer International Publishing AG 2017
M. Gertz et al. (Eds.): SSTD 2017, LNCS 10411, pp. 384–387, 2017.
DOI: 10.1007/978-3-319-64367-0_21

2   Vision: A GeoViz-Aware Database System

The straightforward approach to interactively visualizing spatial data completely decouples the GIS (Geographic Information System) application and the spatial database system such as PostGIS. In this approach, the GIS tool runs at the client side and the DBMS runs at the server side; each performs its task independently from the other. When the user performs a spatial data visualization task, this approach first loads the spatial data that lie within the visualization window from the database into a format understood by the GIS tool. The visualization tool, in turn, visualizes the retrieved spatial data on the map. Analysis of spatial data often begins without a specific intent as an agnostic search, and blends through browsing into concrete user intent and precise querying. An expert interacts with the data by constructing consecutive queries using insights gained from the query session. Such interactions are enabled by interactive dashboards, which provide instantaneous response, helping the user quickly discover insights such as trends and patterns. However, the straightforward approach issues a new spatial range query to the database system for each user interaction with the map. The user will not tolerate delays introduced by the underlying spatial database system to execute a new spatial query for every single change in the viewport. Instead, the user needs to visualize useful information quickly and interactively change her visualization (e.g., zoom in/out and pan) if necessary.

Expressing GeoViz in Database Systems: We envision a new approach, namely GeoVizDB, that injects interactive GeoViz map exploration awareness inside the spatial database system. The user can issue a GeoViz query to the system. Once the initial query is exhibited, the expert interacts with data. She can continue studying the preliminary GeoViz query result, generate new visualizations, or replace the default one with the analysis and measure of his choice, which will utilize the already generated materialized view. Note that the initial query is independent of the user's intent, and hence can be stored as a materialized view in the database. Also, note that any change to the underlying spatial data or the visualization window will result in the materialized view being updated, and resulting queries and visualizations being regenerated. Following [10], GeoVizDB needs to recognize following categories of interactions: *(a) Navigational Interactions:* They consist of actions such as zoom in/out or panning on the map. Navigational interactions can be formulated as *brushings* (a brush on any visual component immediately updates other components) on the visual components. Brushing is an *incremental query*, and results in a union or deletion of one of the current spatial filters and the subsetted data. As a result, an incremental query does not need to be fully materialized – only differential data subset (and result in the case of algebraic and distributive measures [9]), needs to be retrieved. A brushing action is composed of three components: *action* (union, subtraction), *affected boundary* (lower, upper bounds), and *attribute*. *(b) Informational Interactions:* These actions provide more details for a selected point. An example of this kind of interaction is the *"what's here"* operation in Google Maps. *(c) Fusion Interactions:* Such interactions enable merging current data with external sources, which is necessary to discover the extrinsic causalities of current spatial observations. Note that a pair of spatial data points can be joined using their latitude, longitude, and time. An expert often goes through a sequence of navigational interactions in an *exploratory context* and then switches to *investigation context* by using informational interactions to request complementary details, or by using fusion interactions to explain observations.

Prefetching and Caching Spatial Data: Spatial data is likely to be accessed again in the near future due to the temporal locality principle, hence the system needs to cache results of recently executed queries in memory. If a subsequent request is a subset of one of the previous queries (e.g. zoom), the database systems stays untouched – results can be returned to the analyst almost immediately from the cache. Assume the user visualizes spatial objects within a specific rectangular range R and then decided to slightly expand, shrink, or move the original viewport R to R'. If the system could predict R', it might speculatively pre-fetch the answer to R' so that the user gets the answer to R' very fast when needed. To achieve that, the system needs to employ a smart speculation algorithm that is able to predict what kind of interactions the user might issue in her GeoViz session. This task is challenging for the following reasons: (1) The spatial query variations might be endless and hence speculatively computing the answer to all possible variations leads to huge system overhead. (2) Even if the number of speculative queries is reasonable, the user might wind up not using any of the speculatively calculated answers and hence the amount of work spent on speculation and data pre-fetching would be a waste. With cache being a limited resource, the system needs to provide principled caching strategies to improve the cache-hit rate. Three have been several research efforts on predicting and preloading possible upcoming data chunks. ATLAS [4] and ImMens [8] employ simple user movement prediction approaches such as Momentum and Hotspot. Existing work [3] leverages Markov chain to further improve the prediction accuracy. Researchers [6] in vehicle navigation community also use Markov chain to predict traffic trajectory which is similar to user movement. DICE [5] considers a cube traversal-based model to shrink the prefetching space while ForeCache [1] argues that the prediction model should consider not only user movement but also the data chunk features (e.g. color histogram). However, these systems do not provide native support for general spatial objects, e.g., points, polygons. Also, these systems are crafted for a custom visualization tool and cannot be easily plugged into generic GIS tools. GeoVizDB must, on the other hand, support GeoViz-aware spatial data caching/pre-fetching as a generic middleware between the GIS application and the spatial database system.

Sampling Spatial Data: Achieving real-time performance for GeoViz applications is quite challenging even when employing high performance computing and modern hardware infrastructure. For instance, the NYC heatmap in Fig. 1 requires the retrieval of billions of spatial objects from the database, which may take so long to run. The problem is further amplified when more spatial objects need to be loaded in response to the user's interactions with the GeoViz map. Given that a heatmap (same for other GeoViz) represent an aggregate view of the data, there is room for trading interactive performance for accuracy. To achieve that, data sampling techniques may scale the GeoViz process by getting rid of overly-detailed spatial objects. Random sampling and stratified sampling are two widely-used simple approaches when people want to only pick the most representative objects. Nano Cube [7] and Hashed Cube [11] maintain compressed aggregates of the spatial data to scale the GeoViz process. RS-Tree [13] augments the R-tree data structure to retrieve just a sample of the spatial data that lie within the query range. ScalaR [2] and VAS [12] store precomputed multiple resolution aggregates of the data using a database system to achieve interactive performance. Even though spatial data sampling/compression techniques allow users to visualize the spatial sample using de-facto GIS applications. This is due to the fact that each spatial query returns compact version of the spatial data that the GIS tool is able to efficiently visualize. Nonetheless, such sampling/compression techniques face the following challenges: (a) fail to provide high quality GeoViz map images for the user, (b) are not tailored to handle geospatial map visualizations and hence may highly compromise the GeoViz accuracy, (c) cannot easily support the streaming nature of geospatial data.

References

1. Battle, L., Chang, R., Stonebraker, M.: Dynamic prefetching of data tiles for interactive visualization. In: SIGMOD. ACM (2016)
2. Battle, L., Stonebraker, M., Chang, R.: Dynamic reduction of query result sets for interactive visualizaton. In: Big Data, pp. 1–8. IEEE (2013)
3. Cetintemel, U., Cherniack, M., DeBrabant, J., Diao, Y., Dimitriadou, K., Kalinin, A., Papaemmanouil, O., Zdonik, S.B.: Query steering for interactive data exploration. In: CIDR (2013)
4. Chan, S.-M., Xiao, L., Gerth, J., Hanrahan, P.: Maintaining interactivity while exploring massive time series. In: Symposium on Visual Analytics Science and Technology, pp. 59–66. IEEE (2008)
5. Kamat, N., Jayachandran, P., Tunga, K., Nandi, A.: Distributed interactive cube exploration. In: ICDE (2014)
6. Krumm, J.: A Markov model for driver turn prediction. In SAE (2008)
7. Lins, L., Klosowski, J.T., Scheidegger, C.: Nanocubes for real-time exploration of spatiotemporal datasets. TVCG 19(12), 2456–2465 (2013)
8. Liu, Z., Jiang, B., Heer, J.: imMens: real-time visual querying of big data. In: Computer Graphics Forum, vol. 32, pp. 421–430. Wiley Online Library (2013)
9. Nandi, A., Yu, C., Bohannon, P., Ramakrishnan, R.: Data cube materialization and mining over mapreduce. IEEE TKDE 24(10), 1747–1759 (2012)
10. Omidvar-Tehrani, B., Amer-Yahia, S., Termier, A.: Interactive user group analysis. In: CIKM, pp. 403–412. ACM (2015)
11. Pahins, C., Stephens, S., Scheidegger, C., Comba, J.: Hashedcubes: simple, low memory, real-time visual exploration of big data. TVCG 23, 671–680 (2016)
12. Park, Y., Cafarella, M., Mozafari, B.: Visualization-aware sampling for very large databases. In: ICDE. IEEE (2016)
13. Wang, L., Christensen, R., Li, F., Yi, K.: Spatial online sampling and aggregation. In: VLDB, vol. 9, pp. 84–95. VLDB Endowment (2015)

Predicting the Evolution of Narratives
in Social Media

Klaus Arthur Schmid[1], Andreas Züfle[2(✉)], Dieter Pfoser[2], Andrew Crooks[2],
Arie Croitoru[2], and Anthony Stefanidis[2]

[1] Institute of Informatics, Ludwig-Maximilians-Universität München,
Munich, Germany
schmid@dbs.ifi.lmu.de
[2] Department for Geography and Geoinformation Science,
Center for Geospatial Intelligence, George Mason University, Fairfax, USA
{azufle,dpfoser,acrooks2,acroitor,astefani}@gmu.edu

Abstract. The emergence of global networking capabilities (e.g. social media) has provided newfound mechanisms and avenues for information to be generated, disseminated, shaped, and consumed. The spread and evolution of online information represents a unique narrative ecosystem that is facilitated by cyberspace but operates at the nexus of three dimensions: the social network, the contextual, and the spatial. Current approaches to predict patterns of information spread across social media primarily focus on the social network dimension of the problem. The novel challenge formulated in this work is to blend the social, spatial, and contextual dimensions of online narratives in order to support high fidelity simulations that are contextually informed by past events, and support the multi-granular, reconfigural and dynamic prediction of the dissemination of a new narrative.

1   Introduction

Information technology of the 20th century allows members of social media communities to spread updates to friends and followers in real-time all over the world. Through this process information reaches a broader community while at the same time being re-shaped (i.e., altered, refined, or complemented). This represents a newfound example of communally curated narratives, reflecting the 21st century ethos of public sharing of information. We have observed this in numerous instances and at various scales, from the use of social media by the general public to provide timely information in the aftermath of the terrorist attacks in Boston and Paris, to the evolving narrative regarding election campaigns across the world, and the dissemination of information, hopes, and fears that follows the emergence of global epidemics like Zika. The spread and evolution of online information in such instances represent a unique narrative ecosystem that is facilitated by cyberspace but operates at the nexus of three dimensions:

- the social network dimension, as defined by the social networks that are formed between individual nodes and serve as potential dissemination routes,

M. Gertz et al. (Eds.): SSTD 2017, LNCS 10411, pp. 388–392, 2017.
DOI: 10.1007/978-3-319-64367-0_22

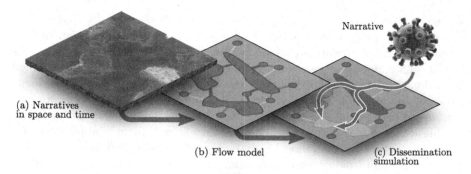

(a) Narratives
in space and time

(b) Flow model

(c) Dissemination
simulation

Narrative

Fig. 1. Abstraction from narrative occurrences to a flow model and simulation.

- the spatial dimension, reflecting communication in the physical world, and
- the contextual dimension of the particular interests and opinions of these
  networks on diverse topics, providing context that discerns event responses.

Current approaches to study this narrative ecosystem, e.g. [1–4,6], have primarily
focused on the social network dimension of the problem. As a result, we have
seen the development of complex network-based solutions to study, model, and
compute information dissemination patterns as a function of the structure of the
corresponding social networks. However, these approaches do not perform well
when attempting to predict the dissemination of a narrative in the real-world.
This is due to their lack of an understanding of the other dimensions of the
communication process. The framework proposed in [8] considers the geo-spatial
component, but without consideration of the underlying social network structure.
To the best of our knowledge, no solutions exist that consider the contextual
dimension to predict the dissemination of a narrative in space and time, thus
learning how various narrative types tend to spread and evolve differently.

In order to develop more powerful models to predict the spread and evolution
of narratives, we need to blend the social, spatial and contextual dimensions of
online narratives in order to support high fidelity simulations that are contextually informed by past events, and support the multi-granular, reconfigural and
dynamic nature of these networks. An overview of this vision is shown in Fig. 1:
First, Fig. 1(a) illustrates the raw data, i.e., occurrences of a given narrative in
space and time. For such narrative, a flow-model can be constructed by reducing the space of individuals relevant for a specific topic, as shown in Fig. 1(b).
Doing this for a large number of historic narratives yields a library of narrative
dissemination models. This library can be used to predict the dissemination of
a new narrative, by searching for similar historic narratives and using these to
predict future dissemination.

2   State-of-the-Art

Traditionally, two types of models have been established to model diffusion
and dissemination of information in networks: differential equation models [4]

and agent based models [3]. An overview of these models can be found in [6]. A weakness of models using differential equations is the aggregation of individual agents into a relatively small number of compartments or populations. Within each population, people are assumed to be homogeneous and well mixed. Transitions among compartments are modeled as their expected value, losing important information about individual influencers and gate-keepers. In contrast, agent based models are able to capture heterogeneity between individuals, thus allowing to exploit the network structure, as well as individual attributes in the information diffusion model to improve diffusion prediction accuracy. Yet, such models suffer from a high computational complexity, scaling up to hundreds of thousands of agents [3,5]. Several models have been proposed to incorporate social and spatial information to detect current events [7,9,11], however, such work often lack the temporal component necessary to follow a narrative in time. Therefore, the ability to predict future information propagation patterns using past data remains a substantial scientific challenge.

3   Proposed Direction

In our work we aim to combine the high efficiency of differential equation models with the high modeling power of agent based models, by aggregating individuals to compartments/groups only as necessary. Towards such a hybrid model, we can reduce the problem complexity in two ways:

Reduction of the topic space: Different narratives share similar diffusion patterns in space as time, as we were able to show in preliminary work [8]. For instance, different topics related to entertainment disseminate in a similar way, whereas topics related to politics exhibit different dissemination patterns. Such clusters of similar narratives can be organized hierarchically, for example, a broader topic may be health, and under it we may have several subtopics, such as infectious diseases, or chronic diseases. These subgroups can be generated through the analysis of historical data (such as Twitter data). The resulting *library of narrative groups* represents abstractions of the information dissemination process, and can be fine-tuned in terms of its thematic resolution (moving from broader categories to more specific categories), in terms of its network resolution (moving from broader clusters of nodes to finer ones, as we will see,in the next section). This allows us to balance computational performance and fidelity as desired for future simulations.

Reduction of the agent space: For a single specific topic, many users may have similar opinions, and can be aggregated into a population without significant loss of information. For instance, one user might be a vocal influencer and gate-keeper for the information dissemination of topics related to a specific type of sports, whereas this user may, at the same time, be oblivious to topics related to politics. Thus, for politics related topics, we may not need to model this user by an individual agent. This further observation allows to further reduce the space of users that need to be modelled. It allows to group of individuals that are oblivious to the given narrative on an aggregated level, while giving full detail to individuals that the model identifies to be trend-setters and vocal. We can model the information dissemination given (conditioned to) a specific topic this topic-archetype and apply an attributed graph clustering algorithm [10] to find communities of users which share a similar opinion towards the topic. This layer of abstraction, which is illustrated by the transition from Fig. 1(a) to Fig. 1(b), yields a set of abstracted information dissemination models for each narrative group.

Simulation and Prediction of Narrative Dissemination: Once we have a library of abstracted information dissemination models, these will be used for grounding an agent based simulation. In this simulation, an agent will be a person or a whole population, as dictated by the dissemination model. To start a simulation, a new narrative is injected into the system. The first step of this grounding is to identify the narrative group of a new narrative. This identification task can be formulated as a supervised classification problem, mapping the new narrative to the most fitting narrative group in the narrative group library. Depending on the available information about the new narrative, this classification may be more or less detailed, thus allowing to dive more or less deep into the narrative group hierarchy, and thus, yielding a more or less detailed dissemination model. Intuitively, the longer we observe a new narrative, the more confident our model fitting will become.

4   Conclusion

The proposed framework would result in models that extend their power beyond the mere structure of the underlying social network. By learning the dissemination of past narratives, latent forms of spreading and evolving a narrative are also captured by this model: While we can not directly observe individuals sharing ideas physically, we can observe the consequence of both individuals frequently sharing the same ideas. This enables us to implicitly capture forms on information dissemination, and allow to learn features of individuals that are more likely, for the given narrative group, to pass on the ideas of others for further dissemination. In doing so we will have the potential to predict the dissemination of new narratives as they emerge.

References

1. Bakshy, E., Rosenn, I., Marlow, C., Adamic, L.: The role of social networks in information diffusion. In: WWW, pp. 519–528. ACM (2012)
2. Cha, M., Mislove, A., Gummadi, K.P.: A measurement-driven analysis of information propagation in the Flickr social network. In: WWW, pp. 721–730. ACM (2009)
3. Lymperopoulos, I.N., Ioannou, G.D.: Online social contagion modeling through the dynamics of integrate-and-fire neurons. Inf. Sci. **320**, 26–61 (2015)
4. Mahajan, V., Muller, E., Wind, Y.: New-Product Diffusion Models, vol. 11. Springer, New York (2000)

5. Pires, B., Crooks, A.T.: Modeling the emergence of riots: a geosimulation approach. Comput. Environ. Urban Syst. **61**(Part A), 66–80 (2017)
6. Rahmandad, H., Sterman, J.: Heterogeneity and network structure in the dynamics of diffusion: comparing agent-based and differential equation models. Manag. Sci. **54**(5), 998–1014 (2008)
7. Sakaki, T., Okazaki, M., Matsuo, Y.: Earthquake shakes twitter users: real-time event detection by social sensors. In: Proceedings of the 19th International Conference on World Wide Web, pp. 851–860. ACM (2010)
8. Schmid, K.A., Frey, C., Peng, F., Weiler, M., Züfle, A., Chen, L., Renz, M.: Trend-Tracker: modelling the motion of trends in space and time. In: SSTDM@ICDM Workshop, pp. 1145–1152 (2016)
9. Unankard, S., Li, X., Sharaf, M.A.: Emerging event detection in social networks with location sensitivity. World Wide Web **18**(5), 1393–1417 (2015)
10. Xu, Z., Ke, Y., Wang, Y., Cheng, H., Cheng, J.: A model-based approach to attributed graph clustering. In: SIGMOD, pp. 505–516. ACM (2012)
11. Zhou, X., Chen, L.: Event detection over twitter social media streams. VLDB J. **23**(3), 381–400 (2014)

A Unified Framework to Predict Movement

Olga Gkountouna[1], Dieter Pfoser[1], Carola Wenk[2], and Andreas Züfle[1(✉)]

[1] George Mason University, Fairfax, USA
{ogkounto,dpfoser,azufle}@gmu.edu
[2] Tulane University, New Orleans, USA
cwenk@tulane.edu

Abstract. In the current data-centered era, there are many highly diverse data sources that provide information about movement on networks, such as GPS trajectories, traffic flow measurements, farecard data, pedestrian cameras, bike-share data and even geo-social movement trajectories. The challenge identified in this vision paper is to create a unified framework for aggregating and analyzing such diverse and uncertain movement data on networks. This requires probabilistic models to capture flow/volume and movement probabilities on a network over time. Novel algorithms are required to train these models from datasets with varying levels of uncertainty. By combining information from different networks, immediate applications of such a unifying movement model include optimal site planning, map construction, traffic management, and emergency management.

1 Introduction

Numerous data sources exist that *capture movement* on (movement) networks relating to the same geographic area. The most popular movement data sources are GPS trajectories, which are readily available from a myriad of smartphone apps. In the case of social media data, movement is often captured implicitly, as trajectories are derived from geocoded tweets, Flickr images, or Foursquare check-ins. Such data is considerably less frequently sampled, i.e., the time between samples may range from minutes to hours. Additional movement data sources include farecard, bikeshare, and taxi trip data. Figure 1 shows various movement data sources for the greater Washington DC metropolitan area including GPS traces uploaded by users to Openstreetmap[1], a colored road network to symbolize traffic conditions derived from measuring stations, public transit fare card data, traces of Twitter users, and bikeshare data. Existing solutions to predict traffic in networks [1,6–9,11] focus on a single network, and assume that positional data is fairly accurate and frequent, such as GPS data.

Our vision is to *use this multitude of available data sources to generalize traffic prediction towards movement prediction for known and unknown movement networks.* We anticipate that meaningful movement information from unconventional data sources can be extracted by utilizing knowledge from a variety of

---

[1] http://www.openstreetmap.org.

© Springer International Publishing AG 2017
M. Gertz et al. (Eds.): SSTD 2017, LNCS 10411, pp. 393–397, 2017.
DOI: 10.1007/978-3-319-64367-0_23

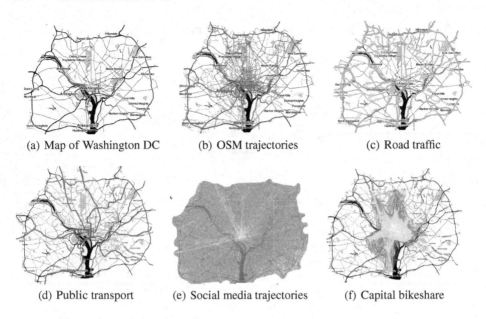

(a) Map of Washington DC   (b) OSM trajectories   (c) Road traffic

(d) Public transport   (e) Social media trajectories   (f) Capital bikeshare

Fig. 1. Heterogenous sources of movement network data.

data sources with varying degrees of uncertainty. The goal is to use *all* of these highly diverse and uncertain movement data sources *together* to enable movement prediction. This will allow us to estimate and to predict movement even in locations where no authoritative data is available, e.g., in the extreme case, to estimate traffic on road networks solely based on social media data.

2   Challenges

To develop a unified model for probabilistic movement prediction among multiple transportation networks, the following three major challenges need to be addressed. In the following, we describe these challenges and their feasibility in more detail.

Making Sense of Extreme Uncertainty: Techniques have been developed [1,7–9] that yield accurate models to probabilistically predict trajectories in transportation networks. These models work well in the case when trajectories are given at high-frequency, thus narrowing down the space of possible locations for a user given their previous location. However, in cases of extreme uncertainty where observations are hours apart from each other, predicting possible locations of individuals becomes futile. The space of possible locations that could be reached in an hour becomes too large to make reliable predictions. Yet, we can use such uncertain observation data to estimate traffic density/volume in the network, rather than estimating the location of individuals. For instance, if we have a large number of users that have a very small probability of being in a specific area, then Poisson's Law of Small Numbers states that the number of individuals in this area is Poisson distributed, and thus, can be predicted accurately.

Correlation of Different Networks: Another research challenge is the problem of matching and correlating data from different movement networks, such as road-traffic loops data, public-transport data and pedestrian GPS data. These datasets stem from different modes of transportation. They may vary in size and may include different attributes (such as vehicular speed, volume, traffic density, etc.). Furthermore, they may be supplied to us in an aggregated form or as raw data, and their measurements may be taken at different sets of timestamps. Finally, they may consist exclusively of observed data points, or may include interpolated values as well.

The correlation of separate movement networks used to be impossible due to the lack of open data sources. Using *all* available data sets discussed in Sect. 1, we are now able to align different networks to identify how individuals move and transition between them. Consider here for example a daily commute that includes driving by car to the metro station, taking the metro downtown, walking to an office building and then inside, directly to the coffee maker. By joining available movement data, we can identify transitions in between road networks, public transportation networks, pedestrian networks, and building floor plans. Considering this correlation between networks of varying degrees of uncertainty, we can further enhance movement flow predictions. For example, an inbound full metro train will lead to increased road traffic around metro parking lots.

Correlation of External Variables: The movement volume estimation can be improved using other parameters such as vehicular speed. If the number of cars observed on an edge is relatively low, but we notice that the traffic moves at a relatively slow speed, then this might be an indicator that the actual traffic volume is higher than the estimated movement flow. Other observable variables that can be used to infer movement flow are weather conditions, day-time, and event information (e.g., a football game). By taking into account such information, we can condition the model to a set of current variables, resulting in a model that is more tailored towards current conditions by exploiting learned implications of these variables. For this purpose, we need to learn the impact of such observable (and publicly available) parameters and correlate them to movement flow, to improve prediction. The main challenge is to assess the significance of each data source, which can be learned empirically from authoritative ground-truth traffic volume data measured by road-side detectors.

3   Applications of Multi-network Movement Flow Prediction

Traffic Management: Given the capability to connect different networks for volume prediction, we can improve efficient decision making in traffic management. By simulating road closures, or closed public transportation lines, critical areas can be identified in other networks, which would be impacted by such an event. Furthermore, for a road closure that has just occurred, our vision will allow online simulation of the consequences, thus quickly predicting areas that will become congested soon, and so providing traffic management systems with decision criteria to reroute traffic.

Emergency Management: In the case of a large scale disaster or emergency such as an earthquake, a forest fire, or a terrorist attack, an optimal deployment of fire fighters and police forces is paramount to saving lives. In order to make the right decisions, emergency management personnel needs accurate information quickly and in the right form. They need predictions about what may happen and information about what has happened and where. In the report to the US Congress on Hurricane Katrina, Secretary of Homeland Security Michael Chertoff emphasized "the importance of having accurate, timely and reliable information about true conditions on the ground" and pointed out that the response efforts during Katrina "were significantly hampered by a lack of information from the ground" [2]. In addition to existing work [3,4,10], which uses only Twitter as an information source, our envisioned framework enables the fusion of streams of location and movement data such as tweets, OSM trajectories, traffic loop data and other sources, to model a traffic and population flow under the new conditions, and thus making this flood of geo-social data actionable for immediate decision making.

Managing the flow of pedestrian movement in the case of an emergency is vital to saving human lives. On 7/24/10, a crowd disaster at the Love Parade electronic dance music festival in Duisburg, Germany caused the death of 21 people from suffocation [5]. To prevent such disasters, we envision a powerful movement prediction model for both prevention and counter-measures. For prevention, we can simulate a disaster, showing potential bottlenecks. To enable counter-measures, we can use real-time social media data to quickly predict large crowd movements. This way, emergency responders can be alerted within minutes before a deadly stampede occurs.

4   Conclusions

Most sources of movement and traffic information are highly uncertain, due to many reasons. Existing work focuses on analyzing these sources individually. In this vision paper we propose to unify these sources into a single movement prediction framework, that is able to learn movement and traffic patterns from *all* data sources and movement networks, including very sparse and uncertain data, simultaneously. Such a framework will not only improve existing applications, but will also inspire entirely new research directions in applications where precise movement data is not available.

Acknowledgements. This research has been supported by National Science Foundation AitF grants CCF-1637576 and CCF-1637541.

References

1. Abadi, A., Rajabioun, T., Ioannou, P.A.: Traffic flow prediction for road transportation networks with limited traffic data. IEEE ITS **16**(2), 653–662 (2015)
2. Chertoff, M.: Statement by homeland security secretary Michael Chertoff before the United States house select committee on hurricane Katrina. Department of Homeland Security, Washington, DC (2005)
3. De Longueville, B., Smith, R.S., Luraschi, G.: OMG, from here, I can see the flames!: a use case of mining location based social networks to acquire spatio-temporal data on forest fires. In: LBSN, pp. 73–80. ACM (2009)
4. Goodchild, M.F.: Citizens as sensors: the world of volunteered geography. Geo J. **69**(4), 211–221 (2007)
5. Helbing, D., Mukerji, P.: Crowd disasters as systemic failures: analysis of the love parade disaster. EPJ Data Sci. **1**(1), 1–40 (2012)
6. Jeung, H., Yiu, M.L., Zhou, X., Jensen, C.S.: Path prediction and predictive range querying in road network databases. VLDB J. **19**(4), 585–602 (2010)
7. Kriegel, H.-P., Renz, M., Schubert, M., Züfle, A.: Statistical density prediction in traffic networks. In: SDM, vol. 8, pp. 200–211. SIAM (2008)
8. Niedermayer, J., Züfle, A., Emrich, T., Renz, M., Mamoulis, N., Chen, L., Kriegel, H.-P.: Probabilistic nearest neighbor queries on uncertain moving object trajectories. Proc. VLDB Endowment **7**(3), 205–216 (2013)
9. Pfoser, D., Jensen, C.S.: Capturing the uncertainty of moving-object representations. In: Güting, R.H., Papadias, D., Lochovsky, F. (eds.) SSD 1999. LNCS, vol. 1651, pp. 111–131. Springer, Heidelberg (1999). doi:10.1007/3-540-48482-5_9
10. Vieweg, S., Hughes, A.L., Starbird, K., Palen, L.: Microblogging during two natural hazards events: what twitter may contribute to situational awareness. In: SIGCHI, pp. 1079–1088. ACM (2010)
11. Zhang, J., Zheng, Y., Qi, D.: Deep spatio-temporal residual networks for citywide crowd flows prediction. In: AAAI 2017, pp. 1655–1661 (2017)

Semantic Understanding of Spatial Trajectories

Zhenhui Li$^{(\boxtimes)}$

College of Information Sciences and Technology, Pennsylvania State University,
University Park, PA, USA
jessieli@ist.psu.edu

1 Trajectory Mining Without Contexts

The advances in location-acquisition technologies and the prevalence of location-based services have generated massive spatial trajectory data, which represent the mobility of a diversity of moving objects, such as people, vehicles, and animals. Such trajectories offer us unprecedented information to understand moving objects and locations that could benefit a broad range of applications. These important applications in turn calls for novel computing technologies for discovering knowledge from trajectory data.

Under the circumstances, trajectory data mining has become an increasingly important research theme in the past decade [10]. Extensive research has been done in the field of trajectory data mining with many interesting patterns have been proposed and studied. However, most existing studies focus on **only trajectory data** and did not consider **rich spatial-temporal contexts** that are associated with trajectories. As a consequence, trajectory patterns detected from existing methods could be trivial. Let's examine an example below.

Example 1. Trajectory outliers or anomalies are usually defined as trajectory segments that are different from other trajectories in terms of some similarity metric. It can also be observations (represented by a collection of trajectories) that do not conform to an expected pattern (e.g., traffic jam). However, such anomalies might actually be expected if we consider the contexts. For example, as shown in Fig. 1, it might be expected that, whenever there is a baseball game in San Francisco, the traffic is heavier than usual and a car not going to the game may choose a longer detour path. Such "anomalies" are actually normal under the condition of contexts (e.g., baseball game in this example).

2 Vision: Semantic Trajectories Understanding

The increasing availability of contextual information (e.g., venue information, local events, weather, and landscape) can potentially lead to a revolution in trajectory data mining. Mining trajectory data should no longer focus on trajectory only, but should also utilize the rich contexts from other data sources to provide a semantic understanding of trajectories.

© Springer International Publishing AG 2017
M. Gertz et al. (Eds.): SSTD 2017, LNCS 10411, pp. 398–401, 2017.
DOI: 10.1007/978-3-319-64367-0_24

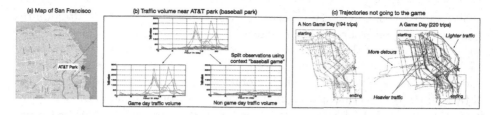

Fig. 1. Taxi trajectories in San Francisco. (a) San Francisco map and the location of AT&T park (a baseball stadium). (b) Recurrence pattern detected if we consider the context of baseball game near AT&T park. (c) Trajectories not going to the game will try to avoid the area near AT&T park. Such "anomalies" (e.g., heavier traffic and detours) are expected during the game time.

Semantic trajectories with contexts could benefit a number of applications: (1) User profiling. Contexts associated with trajectories will provide a more accurate profiling of users' interest, socioeconomic status, and health conditions. For example, a person frequently visits kindergarten and kids-friendly restaurants, he/she may have a young kid and we could recommend kids-related activities on weekend; if a person frequently visits fast food and convenience stores, and rarely visits recreational places (e.g., parks, fitness facilities), the person may live a poor lifestyle and might be at a greater risk of chronic health conditions. (2) Intelligent transportation. A significant change in intelligent transportation system is to use more data collected from a variety of sources. Understanding traffic patterns with contexts, we could better predict the traffic and suggest the best route for drivers under different conditions of time, weather, and events. (3) Ecology. Organism-environment interaction is a fundamental question in ecology that tells us how animals respond to the dynamic changes of environment and helps us predict how environmental change will impact animals' behaviors.

3   Research Challenges

Mining trajectory data with contexts is not simply using all the nearby contextual data or simply extending current data mining techniques with extra context information. To enable the power of contexts in trajectory mining, we need to understand how trajectories are associated with or impacted by the surrounding contexts. There are two key challenges we will face:

Implicit and complicated correlations. Since there are many surrounding contexts near a location, it is ambiguous which context correlates with the trajectory. For example, a person observed at Madison Square Garden (MSG) could be attending a concert in MSG, or could be transitioning at Penn Station which sits below MSG, or could be visiting a restaurant nearby. Moreover, the observed trajectories are impacted by many factors simultaneously, such as daily/weekly regularity, local events, weather, car accidents, and traffic jams. The impacts could be also at different scales from small farmer's market, to big football game, to extreme weather.

Sparse and noisy data. Observations on trajectories and contexts are often quite sparse in real applications. For example, we may only have sporadic observations on individual data if the data collection mechanism requires users to voluntarily contribute data; some trajectory datasets, such as taxi data, only reflect a biased and incomplete version of the overall mobility density. It is also not realistic to obtain all the context information that impact trajectories. In addition, the data we obtained could also be noisy and imprecise. GPS positioning often has errors that vary from a few meters to hundred of meters, depending on the sensing equipments and atmospheric effects. For an event obtained from the news article with a description as "football game at 12 p.m. on Saturday", the game was probably from 12 p.m. to 2 p.m. In order to capture more local events, we may even need to extract the contexts from the noisy raw data (e.g., extracting events geo-tagged tweets).

4   Preliminary Studies and Future Directions

Recent studies have realized the importance of utilizing external context data to enrich the semantics of mobility patterns. However, most of these studies assume that the contexts are already associated with mobility records (e.g., check-in data). To bridge the gap between raw trajectory data and contexts, we need to associate the trajectories with the corresponding contexts. Various methods have been proposed to annotate the mobility records with landmarks [1], landscapes [4], land-use categories [7,8], geo-tagged tweets [6], and POI [5,8]. These methods can be generally classified into two categories. The first category [1–4,6,7] is to consider each mobility record separately and to annotate each record independent of other records. For example, the most common approach is to attach the closest context to a mobility record. The second category [5,8,9] considers the dependency among records. For example, Yan et al. [8,9] propose a hidden Markov model to consider the transition dependency in individual movement. Wu and Li [5] propose to use a Markov Random Field to consider the consistency in individual preference.

However, many challenges have not been well addressed by these preliminary studies. Here we discuss a few potential future research topics.

First, the context data are often messy and ambiguous. For example, there could be multiple duplicate POI entries corresponding to the same POI entity because POI entries are often generated by the crowd. Also, even though events could have a significant impact on trajectories, there is no such a good data source documenting all the events in a city. All the existing studies have been assuming the context data are clean and contain no ambiguity.

Second, depending on the data collection mechanism, the spatial trajectories could be in different forms. Constant GPS tracking may give a complete trajectory, but will require pre-processing to extract the meaningful location records. Data collected by social media or smartphone applications are often very sparse. Such sporadic location data are only collected when users use the applications. In addition, due to privacy concern, sometimes we may only have the crowd information, such as taxi pick-ups and drop-offs without knowing the passenger identities. Different data properties will require different methods for semantic understanding.

Third, it remains challenging how to evaluate the semantic patterns. How do we know whether the annotated venues are the true destination venues of a user? How do we know that an event is the cause of a person visiting a location or a person happens to locate at that venue during the event time? It will be valuable to generate a benchmark dataset that people can evaluate their methods on semantic trajectory mining.

Acknowledgements. This work was supported in part by NSF awards #1618448, #1652525, #1639150, and #1544455. The views and conclusions contained in this paper are those of the author and should not be interpreted as representing any funding agencies.

References

1. Alvares, L.O., Bogorny, V., Kuijpers, B., de Macedo, J.A.F., Moelans, B., Vaisman, A.: A model for enriching trajectories with semantic geographical information. In: Proceedings of the 15th Annual ACM International Symposium on Advances in Geographic Information Systems (GIS 2007), p. 22. ACM (2007)
2. Fileto, R., May, C., Renso, C., Pelekis, N., Klein, D., Theodoridis, Y.: The baquara 2 knowledge-based framework for semantic enrichment and analysis of movement data. Data Knowl. Eng. **98**, 104–122 (2015)
3. Ruback, L., Casanova, M.A., Raffaetà, A., Renso, C., Vidal, V.: Enriching mobility data with linked open data. In: Proceedings of the 20th International Database Engineering & Applications Symposium, pp. 173–182. ACM (2016)
4. Spaccapietra, S., Parent, C., Damiani, M.L., de Macedo, J.A., Porto, F., Vangenot, C.: A conceptual view on trajectories. IEEE Trans. Knowl. Data Eng. (TKDE) **65**(1), 126–146 (2008)
5. Wu, F., Li, Z.: Where did you go: personalized annotation of mobility records. In: Proceedings of the 25th ACM International on Conference on Information and Knowledge Management (CIKM 2016), pp. 589–598. ACM (2016)
6. Wu, F., Li, Z., Lee, W.-C., Wang, H., Huang, Z.: Semantic annotation of mobility data using social media. In: Proceedings of the 24th International Conference on World Wide Web (WWW 2015) (2015)
7. Yan, Z., Chakraborty, D., Parent, C., Spaccapietra, S., Aberer, K.: Semitri: a framework for semantic annotation of heterogeneous trajectories. In: Proceedings of 14th International Conference on Extending Database Technology (EDBT 2011), pp. 259–270. ACM (2011)
8. Yan, Z., Chakraborty, D., Parent, C., Spaccapietra, S., Aberer, K.: Semantic trajectories: mobility data computation and annotation. ACM Trans. Intell. Syst. Technol. (TIST) **4**(3), 49 (2013)
9. Yan, Z., Giatrakos, N., Katsikaros, V., Pelekis, N., Theodoridis, Y.: SeTraStream: semantic-aware trajectory construction over streaming movement data. In: Pfoser, D., Tao, Y., Mouratidis, K., Nascimento, M.A., Mokbel, M., Shekhar, S., Huang, Y. (eds.) SSTD 2011. LNCS, vol. 6849, pp. 367–385. Springer, Heidelberg (2011). doi:10.1007/978-3-642-22922-0_22
10. Zheng, Y.: Trajectory data mining: an overview. ACM Trans. Intell. Syst. Technol. (TIST) **6**(3), 29 (2015)

Demonstrations

An Integrated Solar Database (ISD) with Extended Spatiotemporal Querying Capabilities

Ahmet Kucuk$^{(\boxtimes)}$, Berkay Aydin, Soukaina Filali Boubrahimi,
Dustin Kempton, and Rafal A. Angryk

Department of Computer Science, Georgia State University, Atlanta, USA
akucuk1@cs.gsu.edu

Abstract. Over the last decade, the volume of solar big data have increased immensely. However, the availability and standardization of solar data resources has not received much attention primarily due to the scattered structure among different data providers, lack of consensus on data formats and querying capabilities on metadata. Moreover, there is limited access to the derived solar data such as image parameters extracted either from solar images or tracked solar events. In this paper, we introduce the Integrated Solar Database (ISD), which aims to integrate the heterogeneous solar data sources. In ISD, we store solar event metadata, tracked and interpolated solar events, compressed solar images, and texture parameters extracted from high resolution solar images. ISD offers a rich variety of spatiotemporal and aggregate queries served via a web Application Program Interface (API) and visualized through a web interface.

1 Introduction

One of the key aspects of conducting high quality research is accessible and reliable data sources. For the case of solar astronomy, the accessibility to the data is limited due to several factors such as disintegrated data sources, the scarcity of clean data, and lack of querying capabilities. A number of challenges arise when creating a single platform that addresses such limitations such as data acquisition from heterogeneous data sources, providing spatial and temporal querying capabilities to solar astronomy researchers, and building a user-friendly system.

One of the most prevalent raw data sources for solar physicists is the NASA's Solar Dynamics Observatory (SDO). It captures approximately 60,000 images every day, and generates 0.55 PBs of raster data each year [1]. To process and analyze the SDO's data, NASA selected a consortium called Feature Finding Team (FFT), to produce a comprehensive automated solar event recognition systems for solar images. These systems contains many modules detecting the spatial locations and physical parameters of different types of phenomena from the SDO data [2], which is called solar event metadata.

© Springer International Publishing AG 2017
M. Gertz et al. (Eds.): SSTD 2017, LNCS 10411, pp. 405–410, 2017.
DOI: 10.1007/978-3-319-64367-0_25

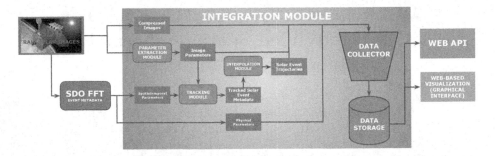

Fig. 1. The illustration of ISD's data pipeline. ISD obtains solar metadata (FFT) and solar images (SDO). Next, metadata is fed to automated tracking and interpolation modules. Solar images are used for extracting the image parameters. Data collector collects, cleans, and integrates the data. The integrated data is served and can be obtained from a web API or visualized online.

A number of existing systems address the integration issues from their own perspectives. For instance, Heliophysics Events Knowledgebase (HEK) provides solar event metadata with basic visualization and querying capabilities [2]. Helioviewer is another tool that provides visualization of solar events whose boundaries overlaid on top of the solar images. Helioviewer also provides an API for retrieving compressed SDO images [3]. While HEK and Helioviewer focus on storing and serving both image and metadata, they are limited with regard to querying capabilities.

There is a growing need for a tool focused on integrating different sources and providing preprocessed, cleaned and reliable data coupled with querying capabilities to accelerate solar astronomy research. To this end, we present the Integrated Solar Database (ISD), a publicly accessible platform that addresses the aforementioned limitations.

Data pipeline of ISD is demonstrated in Fig. 1. ISD integrates the following data sources: solar event metadata, tracked and interpolated event metadata, SDO images, and image parameters. Using these data sources, the ISD enables various queries such as temporal, spatiotemporal, and aggregation queries. In addition, ISD is equipped with a web based visualization tool along with a raw data API. With ISD, we aim to create a system where researchers can query for all these data sources from a single source.

2   Integrating the Solar Data Sources

Solar Event Metadata: Solar events are the natural phenomena appearing on the surface of the Sun. Each event is recorded as a set of spatial, temporal and non-spatiotemporal attributes. The spatial attributes includes the event centroid, bounding box, and the polygon boundary (if available) in different solar coordinate systems. The temporal attributes are the start and end times of the event. Non-spatiotemporal attributes vary by the event type and show various physical measurements. Solar events metadata are produced by FFT modules in a near-real time fashion. To retrieve the solar event metadata, ISD uses HEK's public API. There is a total of 24 event types stored in ISD and each event record contains at least 104 attributes (depending on the event type). ISD contains over 1.3 M event records, and it continuously grows with newly reported events.

Tracking and Spatiotemporal Interpolation: The goal of the tracking module is to link the solar events into sequences representing the trajectory of the solar phenomena. The tracking algorithm [4], firstly links the event instances by projecting a detected object forward using the known differential rotation of the solar surface and searching for the potential detections that overlap with the search area at the next time step. If the algorithm finds multiple paths to choose from, several aspects of visual and motion similarity are compared to produce the most probable path for the object. The resultant paths are again fed into another iteration of the algorithm with larger and larger gaps between detections allowed to account for missed detections in the original metadata.

Though the tracking algorithm generates trajectories that can last over days, there are gaps of reports that we filled using our spatiotemporal interpolation techniques as appeared in [5]. Different interpolation strategies were proposed depending on the event types. The simplest interpolation method is *MBR-Interpolation*, which is designed for events where there are no complex polygon boundaries such as Flares. Complex-Polygon (CP) interpolation is the generic interpolation technique for events with complex polygon boundaries (such as Sunspots). It uses the centroid shape signature along with dynamic time warping alignment to match the boundaries of complex polygons. Filament (FI) Interpolation is an extended complex polygon interpolation method that considers the unique shape characteristics of the filament events.

Compressed Images and Image Parameter Data: The image parameter data consists of ten texture parameters that are designed for use in solar image retrieval [1]. The parameters are calculated by dividing each SDO image into 64 by 64 cells, and performing image parameter extraction in the resultant grid cell. The image parameter extraction process effectively reduces the full-resolution images to ten 64 by 64 matrices, for each image coming from the SDO's Atmospheric Imaging Assembly (AIA) instrument. There are nine wavelengths at a cadence rate of 6 min in the dataset; 94 Å, 131 Å, 171 Å, 193 Å, 211 Å, 304 Å, 335 Å, 1600 Å, and 1700 Å.

3 Data Organization

Our storage structure can be divided into three parts: compressed images, image parameters, event metadata in the form of spatiotemporal trajectories. Images are stored as blobs in two tables storing 2048 by 2048 JP2 images and 256 by 256 thumbnails. To store images parameters, we use image parameter tables, which references the parameters based on the unique identifier of images.

The image parameters are stored as a record containing all the ten parameters for a cell of a given image. Thus, along with the image's identifier, we store the cell location, where the parameters are extracted.

The storage of the solar event metadata is more complex due to the need for spatial and spatiotemporal queries. Currently, ISD supports temporal point and range, spatiotemporal window, and spatiotemporal nearest neighbor queries. To answer these queries efficiently, we store the event metadata in relational tables. We use PostgreSQL database with PostGIS extension [6] to create tables storing the spatial and temporal aspects of the events. PostgreSQL database enables us to index the spatial and temporal attributes of the solar event metadata, which eventually leads to better query performance.

For performance and extensibility reasons, we store the instances of each event type in separate tables. The performance aspect of our choice is related to being able to search the tables in the database separately. While some of the attributes are shared across all the event types, some attributes are available to only a specific event type. For example, Flare table has 12 unique attributes, while Filament table has 11 unique attributes, and they share 106 attributes. To be able to extend our database for possible new event types and perform more efficient spatial and temporal queries, we sliced the events based on their event type.

The spatial locations of the events (region polygons) are indexed with built-in R-tree indexes, while the temporal attributes (start and end times of the events) are indexed using a B-tree on composite time range attributes. In our current settings, joining all event metadata tables could be inefficient when compared to storing all events in the same table; however, considering the solar astronomy use cases such as searching certain regions with spatial or temporal predicates, the separation of the event types provides better query efficiency. Coupled with our data organization schema, the spatial and temporal indexing allows us to perform efficient temporal range, spatiotemporal window queries and nearest neighbor queries.

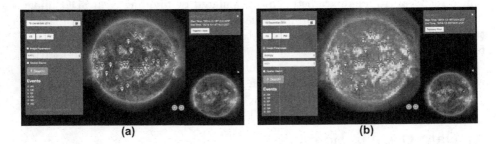

Fig. 2. ISD web-based visualization application screenshots of temporal query result on (a) SDO AIA image and (b) entropy image parameter.

4 Visualization and API

There are two ways of consuming the previously mentioned queries: via the web-based interface[1] or via the web API[2].

For the first option, users can access ISD's web client from our website, and search the database using the graphical interface. In Fig. 2, we demonstrate a snapshot of our web-based interface. Figure 2.a shows the searched events overlaid on top of 171Å image on 2015-06-15 03:26. Figure 2.b shows the same events overlaid on image parameter map (for Entropy parameter from 94Å image). Our interface allows the users to change the background images. In addition to the solar images available in nine different wavelengths (as shown in Fig. 2.a), the users can choose the image parameter maps as background images (as shown in Fig. 2.b). Furthermore, the web interface enables users to pick the events overlaid on images to further show a preview of its trajectory.

The second option to consume the queries is via our web API. This option is more convenient for users who want to have a longer temporal span or higher loads of raw data. All of the web API functions are listed on our API documentation (see footnote 2) and may be consumed by HTTP requests. The response to the request will either be a text in JSON format or a JPEG image. The current version of the web API includes temporal and spatiotemporal search on the event records, trajectory search for a given event identifier, and temporal search for full-disk image and image parameters.

In Fig. 3, we demonstrate a solar physics use-case, where a researcher uses our unique tracking and image parameter data for flare forecasting along with the web visualization tool. We will be presenting various similar use-cases during our demonstration.

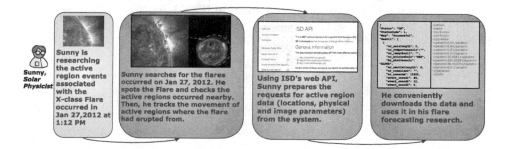

Fig. 3. A use-case scenario showing how to query solar events in ISD.

---

[1] ISD is available at http://isd.dmlab.cs.gsu.edu/.
[2] ISD API is available at http://api.isd.dmlab.cs.gsu.edu/.

5   Conclusion

In this paper, we presented ISD, which is an integrated solar data repository with spatiotemporal querying capabilities. ISD addresses the problem of heterogeneity of the currently available solar data and the growing number of data sources from solar observatories. In addition, ISD comes with a web application that enables a user-friendly visualization experience of the provided data that is fetched using our web APIs. As a future direction, we plan to provide near real-time data pipeline for image parameters, as well as tracked and interpolated solar events. ISD will also greatly benefit from the integration of other solar data sources, as well as adaptation of new queries based on the needs of solar physicists.

References

1. Martens, P.C.H., et al.: Computer vision for the solar dynamics observatory (SDO). In: Chamberlin, P., Pesnell, W.D., Thompson, B. (eds.) The Solar Dynamics Observatory, pp. 79–113. Springer, New York (2011). doi:10.1007/978-1-4614-3673-7_6
2. Hurlburt N., et al.: Heliophysics event knowledgebase for the solar dynamics observatory (SDO) and beyond. In: Chamberlin, P., Pesnell, W.D., Thompson, B. (eds.) The Solar Dynamics Observatory, pp. 67–78. Springer, New York (2010). doi:10.1007/978-1-4614-3673-7_5
3. Hughitt, V.K., Ireland, J., Mueller, D., Dimitoglou, G., Garcia Ortiz, J., Schmidt, L., Wamsler, B., Beck, J., Alexanderian, A., Fleck, B.: Helioviewer.org: browsing very large image archives online using JPEG 2000. In: AGU Fall Meeting Abstracts, December 2009
4. Kempton, D., Angryk, R.: Tracking solar events through iterative refinement. Astron. Comput. **13**, 124–135 (2015)
5. Boubrahimi, S.F., Aydin, B., Kempton, D., Angryk, R.: Spatio-temporal interpolation methods for solar events metadata. In: 2016 IEEE International Conference on Big Data, BigData 2016, pp. 3149–3157, December 2016
6. Strobl, C.: Postgis. In: Encyclopedia of GIS, pp. 891–898 (2008)

HX-MATCH: In-Memory Cross-Matching Algorithm for Astronomical Big Data

Mariem Brahem$^{(\boxtimes)}$, Karine Zeitouni, and Laurent Yeh

DAVID Laboratory, University Versailles St Quentin,
Paris Saclay University, Versailles, France
{mariem.brahem,karine.zeitouni,laurent.yeh}@uvsq.fr

Abstract. The advanced progress in telescope facilities is continuously generating observation images containing billions of objects. Cross-match is a fundamental operation in astronomical data processing which enables astronomers to identify and correlate objects belonging to different observations in order to make new scientific achievements by studying the temporal evolution of the sources or combining physical properties. Comparing such vast amount of astronomical catalogs with low latency is a serious challenge. In this demonstration, we propose HX-MATCH, a new cross-matching algorithm based on Healpix and showcase an in-memory distributed framework where astronomers can compare large datasets.

Keywords: Astronomical survey data management · Big data · Cross-matching · Healpix · In-memory distributed framework

1 Introduction

Sky surveys are producing large datasets using various instruments. Cross-matching information of the same celestial objects from different observations is a computation intensive operation. Thus, generating fast matches is indispensable to the development of the astronomical field. Distributed systems like Spark [4] have become increasingly popular as a cluster computing models for processing large amounts of data in many application domains. Spark performs an in-memory computing, with the objective of outperforming disk-based frameworks such as Hadoop. There have been studies on cross-matching astronomical catalogs [9]. But none of these works provides an in-memory cross-matching algorithm based on Spark. Another category of systems proposing distance join operation (equivalent to cross-match) can be included in our related work [6,8] but only SIMBA [8] is based on Spark. These systems are designed for the Geospatial context that differs from the astronomical context in its data types and

This work has made use of data from the European Space Agency (ESA) mission *Gaia* (https://www.cosmos.esa.int/gaia), processed by the *Gaia* Data Processing and Analysis Consortium (DPAC, https://www.cosmos.esa.int/web/gaia/dpac/consortium). Funding for the DPAC has been provided by national institutions, in particular the institutions participating in the *Gaia* Multilateral Agreement.

© Springer International Publishing AG 2017
M. Gertz et al. (Eds.): SSTD 2017, LNCS 10411, pp. 411–415, 2017.
DOI: 10.1007/978-3-319-64367-0_26

operations. They propose spatial indices and queries that are not suitable to the spherical coordinates. In our context, the sky is projected on a sphere and the coordinates are based on the International Celestial Reference System (ICRS).

In this demo, we present HX-MATCH, an efficient cross-matching algorithm based on our server AstroSpark [5]. HX-MATCH stands for Healpix based cross(X)-match. We showcase its performance step-by-step on real data by using both visualization tools and monitoring interfaces.

2   System Overview

2.1   AstroSpark Architecture

AstroSpark is a distributed data server for Big Data in astronomy. It is based on Spark, a distributed in-memory computing framework, to analyze and query huge volume of astronomical data. AstroSpark in a nutshell adapts data partitioning to efficiently processing astronomical queries. To this end, we apply a spatial-aware data partitioning, and first use linearization with the Healpix library to transform a two dimensional data points (represented by spherical coordinates) into a single dimension value represented by a pixel identifier (Healpix ID). Healpix (Hierarchical Equal Area isoLatitude Pixelization) [7] is a structure for the pixelization of data on the sphere. It is an available library maintained by the NASA. We have adopted the Healpix software for the following reasons:

- Healpix is a mapping technique adapted to the spherical space, with equal areas per cell all over the sky, a unique identifier is associated to each cell of the sky. This id allows us to manipulate cells in our algorithm efficiently.
- Data linearization with Healpix ensures preserving data locality, neighboring points in the two-dimensional space are likely to be close in the corresponding one dimensional space. Data locality helps us to organize spatially close points in the same partition or in consecutive partitions, and thus optimizes query execution by reducing accesses to irrelevant partitions.
- The software for Healpix is easily accessible and contains many functionalities that are useful in our context such as filtering neighbor pixels ot those in a cone.

Queries are expressed in Astronomical Data Query Language (ADQL) [1], a well-know language adapted to query astronomical data. It is a SQL-Like language improved with geometrical functions which allows users to express astronomical queries with alphanumeric properties. The query parser of AstroSpark is extended to translate an ADQL query with astronomical functions and predicates into an internal algebraic representation. Then, the query optimizer will enrich some pre-filtering operators based on our spatial partitioning which make global filtering prune out irrelevant partitions. AstroSpark extends the Spark SQL optimizer called Catalyst by integrating particular logical and physical optimization techniques. The interested reader may refer to [5] for more detail.

2.2   Partitioning Phase

Partitioning is a fundamental component for processing in parallel. It reduces computer resources when only a sub-part of relevant data are involved in a query, and then improve query performances. AstroSpark Partitioner should ensure two main requirements: (1) Data locality: points that are located close to each other should be in the same partition, a partition has to represent a portion of the sky (2) Load balancing: the partitions should be roughly of the same size to efficiently distribute tasks between nodes of the cluster. To achieve the first requirement, a spatial grouping of the data is necessary. Nevertheless, a basic spatial partitioning may lead to imbalanced partitions due to the typical skewness of astronomical data. Therefore, the partitioning should be also adaptive to the data distribution. AstroSpark partitions Spark *dataframes* in a way the partitions are balanced while favoring data locality.

At this end, the two dimensional spherical coordinates are first mapped into a single dimensional ID using the Healpix library, which fulfill the data locality requirement. To achieve the load balancing, we leverage the Spark range partitioner based on this ID, which yields data partitions with roughly equal sizes. Then, the partitions are stored on HDFS. Each partition is saved in a separate sub-directory containing records with the same partition number. AstroSpark divides each partition into buckets. This technique optimizes query execution in a way that makes it efficient to retrieve the contents of a bucket and obviate scanning irrelevant partitions. AstroSpark retrieves also partition boundaries and store them as metadata. Note that in our case, all we need to store are the three values *(n, l, u)* where n is the partition number, l is the first Healpix cell of the partition number n and u is the last Healpix cell of the partition number n. It should be noted that we store the partitioned files (with the metadata) on HDFS and use them for future queries, which amortizes their construction cost (Fig. 1).

2.3   Cross Matching Algorithm

Cross-matching in astronomy takes two datasets R, S and a radius ϵ as inputs and returns all pairs of points (r, s) where r $\in$ R, s $\in$ S, s $\neq$ r and spherical distance between r and s is lower than ϵ. The spherical distance is calculated with the Harvesine formula which is used to calculate the great circle distance between two points on a sphere. Spark SQL uses a cartesian product to execute a join on the two inputs datasets, and then filters the output according to the join condition. This naive approach is ineffective. In contrast, HX-MATCH is a tailored solution, using both partitions and Healpix indexing, and applying the principle of filter and refine to limit the pair-wise distance computations.

HX-MATCH runs as follows. First, we match the partitions of R and S based solely on the metadata (i.e., partition number and healpix interval bounds), and store the result in a matching table along with the interval bounds of their intersection. Then, we start processing each pair of such matching partition. The idea is to go further in filtering the candidates by using the Healpix indices (which is more efficient than the pairwise distance computations).

Basically, we consider that the objects in R and S in the same cell (having the same Healpix index) are candidates to satisfy the distance criteria. However, we need to deal also with the matched objects along the borders, i.e., those having an ϵ-distance but belonging to different (neighbor) partitions. For this purpose, we identify the neighbor cells and replicate the data belonging to these cells. Precisely, we chose one dataset as a reference, let say S, and for each matched partition, we filter S restricted to the intersection, then we augment the data of this partition by replicating the objects belonging to all neighboring cells (which are computed thanks to the Healpix Library). We employ a local filtering, which matches the data in the same cell or in neighboring cells based on their Healpix indices. The trick we use in this regard is to substitute the Healpix index of the replicates by the one of the current cell. Thus, a simple equi-join query suffices to filter all the candidate pairs. At last, the refinement step checks the exact objects' harvesine distance and returns the cross-match results. It is worth noticing that the combination of the local filtering and the refinement relies on Spark SQL, and benefits from its lazy evaluation feature.

3 Experimentation and Demonstration Scenario

3.1 Cross Match Evaluation

Experimental Setup. Experiments were performed over a distributed system composed of 6 nodes and 80 cores. The spark cluster is used in Standalone mode. The main memory reserved for Spark is 180 GB in total.

Datasets. We used the public *Gaia* DR1 dataset [3] which counts more than 1 billion objects (1,142,461,316 records) to perform our tests. Each record contains a sourceID, a two-dimensional coordinate (alpha and delta) representing the star position on the sky and other attributes including magnitude, metalicity among 57 attributes. We perform a cross-matching of samples from *Gaia* DR1 with another catalog called Tycho-2 related to prior surveys. The Tycho-2 catalog contains positions and proper motions for 2,539,893 stars in the sky. The search radius is set to 2 arc-seconds, a value recommended by an astronomer.

Tests. We compare the performance of HX-MATCH, Spark SQL and SIMBA. As shown in Fig. 3, HX-MATCH is very efficient, and outperforms by far SIMBA. The performance shows a linear trend, which makes it scalable. For partitioning, SIMBA uses an STR-Tree based partitioning as illustrated in Fig. 2. We can observe that the bounding boxes become very elongated around the poles. This may hinder the distance join performance, due to the increase of the objects along the border, which entails multiple partition access. When the tests exceeds 111 Millions objects, SIMBA fails due to memory problems. Furthermore, SIMBA only implements the Euclidean distance, which leads to erroneous result when cross-matching. Spark SQL is worse, because it performs a cartesian product. As an example, the execution time of a cross-match between 200,000 records of *Gaia* and Tycho-2 takes 13,6 h.

  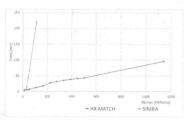

Fig. 1. Partitioning with AstroSpark

Fig. 2. Partitioning with SIMBA

Fig. 3. Cross-Matching time

3.2 Demonstration Scenario

The demonstration will showcase the partitioning and the cross-matching steps. We will illustrate system scalability. AstroSpark GUI is based on Zeppelin, a web-based notebook that enables interactive data analytics. We present a live demo, by running an instance of the cluster described in Sect. 3.1. We will issue a cross-matching query using HX-MATCH and compare it to an equivalent run on Spark SQL and SIMBA. Attendees can look into the progress of the query as it is running. We will also show the execution plan and verify the correctness of the cross-matching result visually with Aladin [2], a tool for viewing astronomical data and acquiring sky maps.

Acknowledgments. This work is partly founded by the CNES (Centre National d'Etudes Spatiales). We would like to thank Frederic Arenou at Paris Observatory, and Veronique Valette at CNES for their cooperation.

References

1. ADQL. http://www.ivoa.net/documents/latest/ADQL.html
2. ALADIN. http://aladin.u-strasbg.fr/
3. GAIA. https://www.cosmos.esa.int/web/gaia/dr1
4. Armbrust, M., et al.: Spark SQL: relational data processing in spark. In: Proceedings of the 2015 SIGMOD International Conference on Management of Data (2015)
5. Brahem, M.A., et al.: AstroSpark: towards a distributed data server for big data in astronomy. SIGSPATIAL Ph.D. Symposium (2016)
6. Eldawy, A., Mokbel, M.F.: Spatialhadoop: a mapreduce framework for spatial data. In: 2015 IEEE 31st International Conference on Data Engineering (ICDE). IEEE (2015)
7. Gorski, K.M., et al.: HEALPix: a framework for high-resolution discretization and fast analysis of data distributed on the sphere. Astrophys. J. **622**(2), 759 (2005)
8. Xie, D., et al.: Simba: efficient in-memory spatial analytics. In: Proceedings of the 2016 International Conference on Management of Data. ACM (2016)
9. Zhao, Q., Sun, J., Yu, C., Cui, C., Lv, L., Xiao, J.: A paralleled large-scale astronomical cross-matching function. In: Hua, A., Chang, S.-L. (eds.) ICA3PP 2009. LNCS, vol. 5574, pp. 604–614. Springer, Heidelberg (2009). doi:10.1007/978-3-642-03095-6_57

pgMemento – A Generic Transaction-Based Audit Trail for Spatial Databases

Felix Kunde$^{(\boxtimes)}$ and Petra Sauer

Beuth University for Applied Sciences, Luxemburger Straße 10,
13353 Berlin, Germany
{fkunde,sauer}@beuth-hochschule.de
https://projekt.beuth-hochschule.de/magda

Abstract. Within the last decades a great amount of location-based data has been integrated into spatial data infrastructures (SDI) on top of spatial databases. One essential but often neglected element for spatial data quality is the lineage, a history about how the data has been created and what transformations have been applied to it. While proprietary spatial databases offer provenance techniques to produce an audit trail for the data, open source alternatives like PostgreSQL leave it to the user to keep track of data changes. Thus, a variety of different approaches has been developed fulfilling this task. However, restore or repair functionalities are often missing or unable to work with data integrity constraints inside the database. pgMemento solves this by providing a transaction-based logging approach that allows for querying and reverting past data transformations more selectively.

Keywords: Audit trail · Versioning · Event trigger · PostgreSQL · JSONB

1 Introduction

The motivation for developing pgMemento[1] arose from the need to access different temporal versions of a 3D city model stored in a PostGIS[2] database [1]. Most cities that have invested in producing a virtual 3D city model care for keeping all objects even if they are outdated due to revision or demolition. This might be useful for visualizing retrospectives of the model [2,3] or different planning alternatives [4]. Changes of environmental parameters like sun exposure, wind circulation or rainfall runoff could be measured against different historical or hypothetical scenarios which can help in decision making for future urban planning projects.

While the SQL:2011 standard proposes a linear history model for data transformations [5], many users of spatial databases favor hierarchical version control

---

[1] https://github.com/pgmemento.
[2] http://postgis.net/.

© Springer International Publishing AG 2017
M. Gertz et al. (Eds.): SSTD 2017, LNCS 10411, pp. 416–420, 2017.
DOI: 10.1007/978-3-319-64367-0_27

solutions to enable GIS-based work flows where collaborative edits are encapsulated in long running transactions [6]. Provenance can be established through additional fields determining the validity period of each tuple (usually with timestamps). Logical changes to the data can easily be handled automatically by database triggers creating different versions at transaction time [7]. Aside from a growing size of the data pedigree, keeping a branching system inside the relational model is challenging in terms of query planning and performance [8]. External tools can be of help here [9].

1.1 Versioning Inside PostgreSQL

PostgreSQL does not provide an in-built audit trail mechanism, but several extensions have been developed over the past years to fill this gap. Most solutions work with triggers that track DML changes and use either two timestamp fields or a corresponding range type. Usually, the historic data is separated from the recent state to facilitate the data management. Schema changes would require an update to the history table and in some cases also to the trigger procedure. In fact, it might be easier to create a new history table and let the trigger function simply point to the new version. To the best of our knowledge there is yet no extension which enables schema versioning in combination with relation history tables. But, some tools use a generic logging approach as a workaround, e.g. CyanAudit[3] or audit_trigger_91plus[4]. pgMemento falls into this category.

2   Concept

pgMemento audits changes in JSONB, a semi-structured data type of PostgreSQL. So far, AFTER row-level triggers track only old versions of a tuple and, in case of an update, only changed fields. For example, when updating a single thematic attribute only one key-value pair will be logged. On the one hand, this can save a lot of disk space if the table contains also complex binary data (e.g. geometry, raster, images), but on the other hand, it requires a full attribute comparison between the OLD and NEW record. Every logged row of a table is referenced to a table event. Multiple table events can be referenced to one transaction (see Fig. 1). Events and transactions are tracked in separate tables by a BEFORE statement-level trigger. To trace the different versions of a tuple a synthetic key (audit_id) is used for each versioned table incremented by a global sequence.

pgMemento provides schema versioning similar to [10]. Within additional log tables temporal information about tables and columns is determined by a range of internal transaction ids. Schema changes are captured with PostgreSQL's event triggers which update these fields and insert new records.

---

[3] http://pgxn.org/dist/cyanaudit/.
[4] https://wiki.postgresql.org/wiki/Audit_trigger_91plus.

Fig. 1. Schema of audit tables in pgMemento

2.1 Restoring Previous Revisions

The main motivation for keeping an audit trail inside a DBMS is the ability to directly query different revisions of the data and analyze who changed what where and when. This can easily be done when using relational history tables with timestamp fields. Generic solutions require either a very different query or a conversion of the schemaless logs back to a relational format. For JSONB fields the *?* operator is used to search for keys and -> is used to retrieve the historic values. When restoring complete tuples pgMemento filters all columns to be found in the DDL log tables and, first, feeds the data to PostgreSQL's *jsonb_build_object* function and, second, converts it to a relational layout with the *jsonb_populate_record* function.

2.2 Rolling Back Transactions

The audit trail could also be used to rollback certain transformations. From the compared tools (see next chapter) this function is only offered by CyanAudit and only for single tables. pgMemento allows for reverting transactions selectively while considering the integrity constraints between multiple tables incl. inheritance or self-references. This is achieved through a correct ordering of events, audit ids and relation dependencies. Because of the columnar data history it is easy to derive SQL statements to revert the changes made by a given transaction.

3 Performance Impact

In this section pgMemento is compared with other open source audit trail solutions in terms of performance impact on DML operations and disk space consumption (temporal_tables[5], tablelog[6], audit_trigger_91plus and CyanAudit). First, a 3D city model generated with Random3Dcity[7] is imported into PostgreSQL using the 3D City Database[8]. Second, one thematic field in one spatial table is updated ten times to measure the update speed and growth in occupied disk space. Third, all buildings are deleted with stored procedures of the 3D City Database.

---

[5] http://pgxn.org/dist/temporal_tables/.
[6] http://andreas.scherbaum.la/writings/tablelog.pdf.
[7] https://github.com/tudelft3d/Random3Dcity.
[8] http://www.3dcitydb.org.

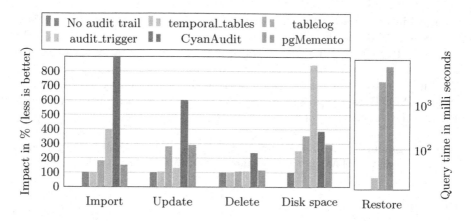

Fig. 2. Performance comparison of different audit trail solutions. Left: Impact on CRUD operations and storage footprint in % compared to unversioned database as baseline (100%). Right: Query time for former database states in milli seconds.

From Fig. 2 we see that most solutions produce a performance overhead to disk writes. Regarding the results for pgMemento this is mostly caused by design choices (JSONB logs, three audit tables, calculating update deltas). The good results for the temporal_tables extension prove that logging complete tuples with relational history tables has the least impact. To improve the performance during disk writes, pgMemento could work with PostgreSQL's logical decoding in the future. This feature has been designed for logical replication but also auditing without the overhead from triggers. An output plugin could parse the Write-Ahead-Log streams into JSON[9] to create the same audit schema proposed in this paper.

When querying for the first database state after the import, again the relational candidates are winning (tablelog only in forward mode). But, pgMemento - as the only generic auditing solution providing an API for restoring previous database states on-the-fly - reaches a comparable performance. It could be a lot faster if we would store complete tuples like the audit_trigger script does. But we think that storing deltas does not only facilitate the interpretation and correction of the audit trail. Branching and merging algorithms are probably easier to develop in the future. In terms of spatial data versioning it would be desirable to log only topological changes which requires a corresponding data model [11]. Finally, keeping the data pedigree in JSON gives us the flexibility to choose alternative backends for various applications.

4   Conclusion

In this demo paper we have presented pgMemento - a generic transaction-based audit trail for PostgreSQL which was designed to enable data and schema

---
[9] https://github.com/sebastian-r-schmidt/logicaldecoding.

versioning in complex spatial databases. Instead of using timestamp fields to reference different database states we decouple audits of transactions, table events and data changes in order to rollback past write operations without violating data integrity. To the best of our knowledge pgMemento is the only open source in-database audit trail software offering this functionality. By using PostgreSQL's JSONB data type and logging only deltas we still achieve comparable performance to relational solutions.

Acknowledgments. The work was supported by the Federal Ministry for Economic Affairs and Energy (BMWi) under grant agreement 01MD15001B (Project: ExCELL).

References

1. Kunde, F.: CityGML in PostGIS. Portability, management and performance analysis using the 3D City Database instance of Berlin. Master thesis, University of Potsdam (in german only) (2013)
2. Chaturvedi, K., Smyth, C.S., Gesquière, G., Kutzner, T., Kolbe, T.H.: Managing versions and history within semantic 3D city models for the next generation of CityGML. In: Abdul-Rahman, A. (ed.) Advances in 3D Geoinformation. LNGC, pp. 191–206. Springer, Cham (2017). doi:10.1007/978-3-319-25691-7_11
3. Redweik, R., Becker, T.: Change detection in CityGML documents. In: Breunig, M., Al-Doori, M., Butwilowski, E., Kuper, P.V., Benner, J., Haefele, K.H. (eds.) 3D Geoinformation Science. LNGC. Springer, Heidelberg (2015). doi:10.1007/978-3-319-12181-9_7
4. Gröger, G., Kolbe, T.H., Schmittwilken, J., Stroh, V., Plümer, L.: Integrating versions, history and levels-of-detail within a 3D geodatabase. In: Gröger, G., Kolbe, T.H. (eds.) Proceedings of the 1st International Workshop on Next Generation 3D City Models, Bonn. EuroSDR Publication 49 (2005)
5. Kulkarni, K., Michels, J.E.: Temporal features in SQL: 2011. ACM SIGMOD Rec. **41**(3), 34 (2012)
6. Bakalov, P., Hoel, E., Menon, S., Tsotras, V.J.: Versioning of network models in a multiuser environment. In: Mamoulis, N., Seidl, T., Pedersen, T.B., Torp, K., Assent, I. (eds.) Advances in Spatial and Temporal Databases. Springer, Heidelberg (2009)
7. Roddick, J.F.: A survey of schema versioning issues for database systems. Inf. Softw. Technol. **37**(7), 383–393 (1996)
8. Stanislav, M.: Framework for managing distinct versions of data in relational databases. In: Ivanović, M., Thalheim, B., Catania, B., Schewe, K.-D., Kirikova, M., Šaloun, P., Dahanayake, A., Cerquitelli, T., Baralis, E., Michiardi, P. (eds.) ADBIS 2016. CCIS, vol. 637, pp. 229–234. Springer, Cham (2016). doi:10.1007/978-3-319-44066-8_24
9. Wieland, M., Pittore, M.: A spatio-temporal building exposure database and information life-cycle management solution. ISPRS Int. J. Geo-Inf. **6**(4), 114 (2017)
10. De Castro, C., Grandi, F., Scala, M.: Schema versioning for multitemporal relational databases. Inf. Syst. **22**(5), 249–290 (1997)
11. van Oosterom, P.: Maintaining Consistent Topology including Historical Data in a Large Spatial Database. Auto Carto 13, Seattle, WA (1997)

ELVIS: Comparing Electric and Conventional Vehicle Energy Consumption and CO_2 Emissions

Ove Andersen[1,2]($\boxtimes$), Benjamin B. Krogh[1], and Kristian Torp[1]

[1] Department of Computer Science, Aalborg University, Aalborg, Denmark
{xcalibur,bkrogh,torp}@cs.aau.dk
[2] FlexDanmark, Aalborg, Denmark
oan@flexdanmark.dk

Abstract. Making the transition from conventional combustion vehicles (CVs) to electric vehicles (EVs) requires the users to be comfortable with the limited range of EVs. We present a system named ELVIS that enables a direct comparison of energy/fuel consumption, CO_2 emissions, and travel-time between CVs and EVs. By letting users enter their everyday driving destinations ELVIS estimates the fuel consumption and whether the users can replace their CV with an EV, optionally by charging the EV at certain stops for a number of minutes. In this demonstration the popular CV Citroën Cactus is compared to the popular EV Nissan Leaf. It is shown that for a typical scenario it is possible reduce CO_2 emissions by 28% when substituting a CV for an EV. ELVIS bases its estimations on 268 million GPS records from 325 CVs and 219 million records from 177 EVs, annotated with fuel/energy consumption data.

Keywords: Electric vehicle · CO_2 · Fuel · kWh · Vehicle range

1 Introduction

To reduce the emissions of CO_2 from the transport sector there is a huge interest in converting from conventional vehicles (CVs), where gasoline or diesel is used as fuel, to electric vehicles (EVs) because EV can be charged with electricity from renewable energy sources. A major challenge in this conversion is that currently the range of EVs is typically much lower than the range of CVs, which may require an EV to be charged on longer trips. Only limited direct comparison of EVs and CVs exist. This paper uses very large sets of GPS and CAN bus data from both EVs and CVs to demonstrate the software prototype named ELVIS (**EL**ectric **V**ehicle **I**tinerary **S**ystem). ELVIS enables a direct comparison of EVs to CVs looking at energy consumption, CO_2 emissions, and travel time. To the best of our knowledge this is the first comparison on EVs to CVs based on very large sets of real-world GPS and CAN bus data.

The main screen of the web-based ELVIS prototype is shown in Fig. 1. At (1) the user can enter the locations that he/she wants to visit in Denmark. In Fig. 1, the user has picked three locations going from Thisted (in the South-West) to

© Springer International Publishing AG 2017
M. Gertz et al. (Eds.): SSTD 2017, LNCS 10411, pp. 421–426, 2017.
DOI: 10.1007/978-3-319-64367-0_28

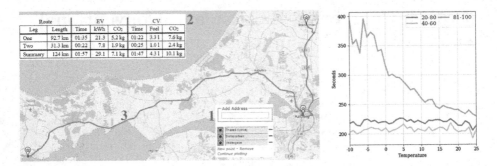

Fig. 1. ELVIS main screen and results

Fig. 2. Temperature impact on charging 1 SoC

Aalborg (in the East) to Brønderslev (in the North-East). ELVIS then computes the energy consumption, CO_2 emissions, and travel time for each leg as shown at (2). Further, the route is displayed as the red line at (3). In (2) it can be seen that the travel time is estimated 10 min longer for an EV compared to a CV but the CO_2 emission is considerable lower 7.1 kg CO_2 for the EV compared to 10.1 kg CO_2 for the CV.

Fuel consumption models have previously been studied and compared with the actual fuel consumption (from CAN bus) of only a small number of vehicles on few routes [1]. In contrast, our fuel estimation uses a large number of vehicles in a country-sized road network. Methods for estimating CO_2 emissions have been studied [2,3]. These methods do generally not support what-if analysis such as those introduced in Sect. 4.

2 Technical Background

The system uses two GPS/CAN bus data sets: one from CVs and another from EVs. 219 million GPS and CAN bus records and 278404 trips are available from EVs in the period 2012 to 2014. 268 million GPS and CAN bus records and 53862 trips are available from CVs in the period 2011 to 2015. Both data sets are one-second logs. The CVs consists of taxis and minibuses and the EVs consist of the three identical vehicles, Citroën C-Zero, Mitsubishi iMiev, and Peugeot Ion. These EVs are in total charged 57511 times. When charging detailed information about the State of Charge, *SoC*, Sect. 3, is reported.

The road network is OpenStreetMap and consists of 0.8 million vertices and 1.7 million directed edges covering all of Denmark. Each edge is associated with four weights: Travel time for EVs and CVs respectively, the fuel consumption of CVs and the energy consumption for EVs (all introduced in Sect. 3).

The data-processing platform is described in [4] and is here extended with functionality for integrating and annotating the road network with fuel-consumption weights. The software stack is based on Python 3.5, pgRouting 2.2, PostGIS 2.3, and PostgreSQL 9.6.

3 Energy and Charging Models

The road network is annotated with weights for normalized average energy consumption ($\mathbf{w}_{kWh}$), fuel consumption ($\mathbf{w}_l$), and travel times for both CVs and EVs. These normalized values are then scaled to the user-provided mileage, km/kWh for EVs and km/l for CVs, when using the ELVIS system, see Sect. 4. We will here focus on the approach for EVs as the approach for CVs is identical, for details see [5]. Equation 1 shows how the average energy consumption on the edge e is computed.

$$\mathbf{w}_{kWhe} = \frac{\sum_{c \in \mathbf{C}_e} \frac{m_c}{m_{agg}} \times f_{e,c,agg}}{|\mathbf{C}_e|} \qquad (1)$$

$\mathbf{C}_e$ is the set of all EVs that have driven on edge e, m_c is the average mileage of vehicle c (from the CAN bus data), m_{agg} is the average mileage of all vehicles (from the CAN bus data), $f_{e,c,agg}$ is the average energy consumption of vehicle c on edge e (from the CAN bus data), and $\mathbf{w}_{kWhe}$ is the estimated normalized fuel consumption for edge e. The normalized energy consumption is scaled according to the user-provided mileage m_c. Equation 2 shows how to estimate the energy consumption $\mathbf{w}_{kWhe,c}$ on a single edge e for the EV c with the user-provided mileage m_c.

$$\mathbf{w}_{kWhe,c} = \frac{m_{agg}}{m_c} \times \mathbf{w}_{kWhe} \qquad (2)$$

If there is an insufficient number of CAN bus measurements on an edge a smoothing technique uses similar neighbor road segments. To convert from kWh to CO_2 emissions information about the g/kWh from the Danish Energy Agency is used. To convert from liters of gasoline/diesel to CO_2 emissions information from the U.S Energy Information Administration is used.

State of Charge (SoC) is the level to which the EVs battery is charged. A SoC of 100 means that the battery is fully charged and a SoC of 0 means the battery is totally drained. To model how the SoC changes when an EV is charging the linear model shown in Eq. 3 is used where SoC_{new} is the SoC after charging, SoC_{old} is the SoC when the charging starts, $Charging_{speed}$ is the charging speed in kWh/hour, $Charging_{time}$ is the charging time in hours, and $Capacity$ is the battery capacity in kWh.

$$SoC_{new} = SoC_{old} + \frac{Charging_{speed} \times Charging_{time} \times 100}{Capacity} \qquad (3)$$

As an example, if SoC_{old} is 10 kWh, $Charging_{speed}$ is 30 kWh/h, $Charging_{time}$ is 0.5 h, and $Capacity$ is 60 kWh then $SoC_{old} = 10 + (30 \text{ kWh/h} \times 0.5 \text{ h} \times 100)/ 60 \text{ kWh} = 10 + 15 = 25$. This charging model is quite accurate as shown in Fig. 2 showing the time in seconds to charge one SoC for the 219 million EVs. The EVs are charged 57511 times at 3.7 kWh home-charging stations. The yellow line shows how long it takes to charge one SoC as the outdoor temperature changes and the SoC of the battery is between 40 and 60. It is mainly the last SoC s that are impacted by temperature, shown by the gray line.

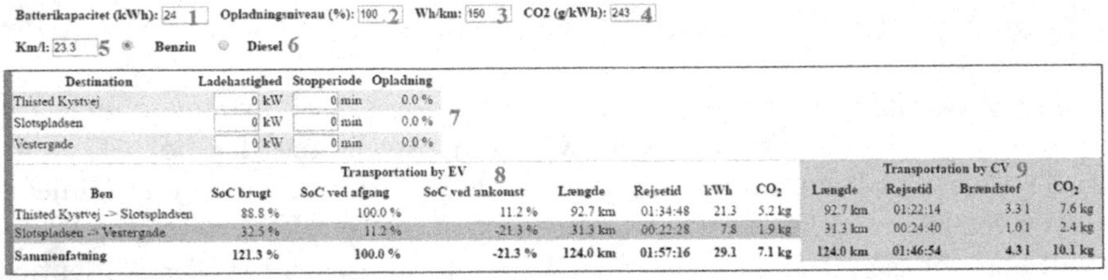

Batterikapacitet (kWh): 24 [1]   Opladningsniveau (%): 100 [2]   Wh/km: 150 [3]   CO2 (g/kWh): 243 [4]

Km/l: 23.3 [5] ● Benzin   ○ Diesel [6]

Destination	Ladehastighed	Stopperiode	Opladning
Thisted Kystvej	0 kW	0 min	0.0 %
Slotspladsen	0 kW	0 min	0.0 % [7]
Vestergade	0 kW	0 min	0.0 %

Ben	Transportation by EV [8]							Transportation by CV [9]			
	SoC brugt	SoC ved afgang	SoC ved ankomst	Længde	Rejsetid	kWh	CO$_2$	Længde	Rejsetid	Brændstof	CO$_2$
Thisted Kystvej -> Slotspladsen	88.8 %	100.0 %	11.2 %	92.7 km	01:34:48	21.3	5.2 kg	92.7 km	01:22:14	3.31	7.6 kg
Slotspladsen -> Vestergade	32.5 %	11.2 %	-21.3 %	31.3 km	00:22:28	7.8	1.9 kg	31.3 km	00:24:40	1.01	2.4 kg
Sammenfatning	121.3 %	100.0 %	-21.3 %	124.0 km	01:57:16	29.1	7.1 kg	124.0 km	01:46:54	4.31	10.1 kg

Fig. 3. Details for EV and CV trip

4   Demonstration

Figure 1 shows the main window of ELVIS. The system uses OpenStreetMap, annotated with routes and pin-markers. The pin-markers show the locations where the user wants to start, where the user wants to stops on a route, and where to end. For these demonstrations the user has chosen to compare the popular EV Nissan Leaf [6] with the CV Citroën Cactus [7]. The user has found the energy consumption for these two vehicles from their webpages. The Nissan Leaf 2017 comes with two different battery capacities, a 24 kWh and a 30 kWh model both with mileage of 150 Wh/km [6]. The Citroën Cactus comes in a gas version with a mileage of 23.3 km/l and a diesel version with a mileage of 32.3 km/l. Further, the user finds the average CO$_2$ emission per kWh in Denmark in 2016, which is 243 g/kWh with (64% from renewable sources in 2016) [8]. The user can enter the information found into ELVIS as shown in Fig. 3.

In the first scenario we assume that a user want to compare the two configurations of the EV and CV on the route shown in Fig. 1. At (1) the user enters the battery capacity of the Nissan Leaf (24 kWh), at (2) the State of Charge (SoC) when the user leaves home (100% is fully charged), at (3) the average Wh/km is entered (150 Wh/km), at (4) the CO$_2$ emission per KWh is entered (243 g/kWh), at (5) the user picks the mileage of the Citroën Cactus (23.3 km/l), and at (6) the user picks gas as the energy source of the CV. With this information ELVIS computes the energy consumption, CO$_2$ emissions and travel time for each leg the EV at (8) in Fig. 3 and for the CV as shown at (9). The second leg is highlighted in red because this leg cannot be completed without charging the EV. The user then has the possibility to pick a charging point, charging speed, and a charging time period at (7) in Fig. 3.

Using the parameters that can be controlled by the user and shown in Fig. 3 we will demonstrate how high the mileage (in km/l) of an CV must be to have a lower CO$_2$ emission on a route than an EV. Second we will demonstrate the effect of having a larger battery capacity, e.g., a 30 kWh Nissan Leaf, would enable longer routes without the need for charging.

As shown in Fig. 3 the route plotted in Fig. 1 cannot be completed with the 24 kWh Nissan Leaf without charging. To see the effect of charging at stops during the route the user can at (7) in Fig. 3 enter information on charging effect and period. If the user charges with 20 kW (1–99 kW) for 30 min (1–1000 min)

Batterikapacitet (kWh): 24		Opladningsniveau (%): 100		Wh/km: 150		CO2 (g/kWh): 243	
Km/l: 23.3	◉ Benzin	◎ Diesel					

Destination	Ladehastighed	Stopperiode	Opladning				
Thisted Kystvej	0 kW	0 min	0.0 %				
Slotspladsen	20 kW	30 min	41.7 %				
Vestergade	0 kW	0 min	0.0 %				

Ben	Transportation by EV								Transportation by CV			
	SoC brugt	SoC ved afgang	SoC ved ankomst	Længde	Rejsetid	kWh	CO₂	Længde	Rejsetid	Brændstof	CO₂
Thisted Kystvej -> Slotspladsen	88.8 %	100.0 %	11.2 %	92.7 km	01:34:48	21.3	5.2 kg	92.7 km	01:22:14	3.31	7.6 kg
Slotspladsen -> Vestergade	32.5 %	52.8 %	20.3 %	31.3 km	00:22:28	7.8	1.9 kg	31.3 km	00:24:40	1.01	2.4 kg
Sammenfatning	121.3 %	100.0 %	20.3 %	124.0 km	01:57:16	29.1	7.1 kg	124.0 km	01:46:54	4.31	10.1 kg

Fig. 4. The effect of charging

at the stop at Slotspladsen the effect on the EV can be seen in Fig. 4. Now none of the legs are high-lighted in red indicating that the route can be driven with the 24 kWh Nissan Leaf. To see the effect of charging in a realistic context, we will use a number of real-world routes used by the home-care sector in Denmark. Here nurses are visiting a larger number of customers (i.e., stops) on a route. The route used in the home-care sector can be quite long and charging is therefore in practice a necessity.

5 Summary

We proposed the ELVIS system for making a direct comparison between EVs and CVs based on very large sets of real-world GPS and CAN bus data from both types of vehicles. ELVIS makes it possible to study the effect of different energy consumption when driving and charging patterns on a route. The user can in a web interface change a number of parameters such as mileage of a CV. As future work, we plan to study and compare the energy consumption from more types of EVs and CVs.

Acknowledgment. We thank www.flexdanmark.dk, www.r2ptracking.dk, www.ens.dk for access to data.

References

1. Guo, C., et al.: EcoMark 2.0: empowering eco-routing with vehicular environmental models and actual vehicle fuel consumption data. GeoInformatica **19**(3), 567–599 (2014)
2. Froehlich, J., et al.: UbiGreen: investigating a mobile tool for tracking and supporting green transportation habits. In: SIGCHI, pp. 1043–1052 (2009)
3. Manzoni, V., et al.: Transportation mode identification and real-time Co₂ emission estimation using smartphones. SENSEable City Lab (2010)
4. Andersen, O., et al.: An open-source based ITS platform. In: MDM, vol. 2, pp. 27–32 (2013)
5. Krogh, B.B., Andersen, O., Lewis-Kelham, E., Torp, K.: CO₂NNIE: personalized fuel consumption and CO₂ emissions. ACM SIGSPATIAL. **4**, 92 (2015)
6. Nissan. https://www.nissan-cdn.net/content/dam/Nissan/gb/brochures/Nissan_Leaf_UK.pdf

7. Citroen. www.citroen.co.uk/new-cars-and-vans/citroen-range/citroen-c4-cactus
8. Energinet.dk: Environmental impact statements for electricity. http://www.energinet.dk/EN/KLIMA-OG-MILJOE/Miljoedeklarationer/Sider/Miljoedeklarering-af-1-kWh-el.aspx

Visualization of Range-Constrained Optimal Density Clustering of Trajectories

Muhammed Mas-Ud Hussain[1]([⊠]), Goce Trajcevski[1], Kazi Ashik Islam[2], and Mohammed Eunus Ali[2]

[1] Department of Electrical Engineering and Computer Science, Northwestern University, Evanston, IL 60208, USA
mas-ud@u.northwestern.edu, goce@eecs.northwestern.edu
[2] Department of Computer Science and Engineering, Bangladesh University of Engineering and Technology, Dhaka, Bangladesh
1205007.kai@ugrad.cse.buet.ac.bd, eunus@cse.buet.ac.bd

Abstract. We present a system for efficient detection, continuous maintenance and visualization of range-constrained optimal density clusters of moving objects trajectories, a.k.a. Continuous Maximizing Range Sum (Co-MaxRS) queries. Co-MaxRS is useful in any domain involving continuous detection of "most interesting" regions involving mobile entities (e.g., traffic monitoring, environmental tracking, etc.). Traditional MaxRS finds a location of a given rectangle R which maximizes the sum of the weighted-points (objects) in its interior. Since moving objects continuously change their locations, the MaxRS at a particular time instant need not be a solution at another time instant. Our system solves two important problems: (1) Efficiently computing Co-MaxRS answer-set; and (2) Visualizing the results. This demo will present the implementation of our efficient pruning schemes and compact data structures, and illustrate the end-user tools for specifying the parameters and selecting datasets for Co-MaxRS, along with visualization of the optimal locations.

1 Introduction

Advances in miniaturization of GPS and sensor-equipped (position-aware) devices, along with progresses in networking and communications, have resulted in the generation of large quantities of (*location, time*) data, (O(Exabyte)) [1]. Movement of people, animals, vehicles, birds, etc. are continuously captured by GPS loggers, and as such, the efficient management of mobility data is at the core of many applications of high societal relevance. Often, such data are being scrutinized by clustering [2,3], mining, information retrieval [4], and visualization techniques [5,6] to detect patterns of interest among the trajectories.

This work explores efficient processing and visualization of the spatio-temporal extension of a particular type of spatial query – the, so called, Maximizing Range-Sum (MaxRS) query, which can be described as follows. Given

M. Mas-Ud Hussain and G. Trajcevski—Research supported by NSF grants III 1213038 and CNS 1646107, ONR grant N00014-14-10215 and HERE grant 30046005.

M. Gertz et al. (Eds.): SSTD 2017, LNCS 10411, pp. 427–432, 2017.
DOI: 10.1007/978-3-319-64367-0_29

428   M. Mas-Ud Hussain et al.

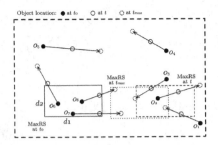

Fig. 1. MaxRS at three different time instants $(t_0 < t < t_{max})$.

a set O of weighted objects and a rectangle R (dimension $d_1 \times d_2$), MaxRS retrieves a location of R that maximizes the sum of the weights of all the objects in its interior. MaxRS, and its variants for scalability, approximate solutions, etc., have been addressed in spatial settings [7,8]. However, many applications involve moving objects and the *continuous* variant of MaxRS is needed – Fig. 1 shows how the locations of the instantaneous MaxRS solutions vary over time for mobile objects.

Co-MaxRS detects and maintains "most interesting" regions over moving objects trajectories, i.e., highest-density clusters given the range R and time-period $T = [t_0, t_{max}]$, which is essential for: environmental tracking (e.g., optimizing a range-bounded continuous monitoring of a herd of animals); – traffic monitoring (e.g., detecting ranges with densest traffic between noon and 6PM); – video-games (e.g., determining a position of maximal coverage in dynamic scenarios). Works on trajectories clustering [2,3] did not track the most dense (based on user-defined criteria) clusters over time. Moreover, proper visualization of Co-MaxRS answer-set for different time-periods and ranges would be handy for focusing on specific optimal regions when analyzing large trajectory data. Recent works on continuous visualization of spatial data have different settings from us: – [5] enables visualizing DBSCAN-based clusters for trajectories (no fixed range or optimal clustering considered) and [6] deals with static spatial events of mobile entities (i.e., does not consider the trajectories themselves).

We provide a system for: (1) Efficiently maintaining Co-MaxRS answer-set for a given range and time-period; and (2) Providing a user-friendly GUI to enable visualization of the answer-set. In [9], we introduced useful pruning schemes and compact data structures (plus TPR*-tree indexing) to enable cost-effective processing of Co-MaxRS queries. This demo paper presents the system which implements the techniques proposed in [9] over several real-world datasets obtained from CRAWDAD (http://crawdad.org) from various domains, along with the GUI for users to specify the desired parameters and even provide their own dataset, and modules for visualizing the Co-MaxRS answer-set.

2   System Design and Demo Details

We now describe the main components of the system architecture and how they interact with each other, and proceed with details of the demo.

Software Architecture: Our final system is a web-based application with interactive and user-friendly interface for both PC and mobile devices, which is implemented using HTML, CSS, and JavaScript (Node.js was used for the server-side programming). We employ *Responsive Web Design* principles in building the website by using CSS media queries, @media rules, and fluid grids—thus, making it suitable to work on every device and screen size. The core Co-MaxRS algorithm is implemented using C++. The software architecture of our system, illustrated in Fig. 2, is organized in the following main categories:

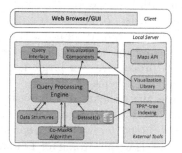

Fig. 2. Software architecture.

- *Query Interface:* The users can select each of the required parameters using the interface by specifying: (1) $\mathbb{F}$ – the query area (depends on the underlying dataset), set dynamically via the zoom-in (+) or out (−) tool; (2) R – specifying d_1 and d_2 values; and (3) T – the time period, specifying t_0 and t_{max}. Users can browse through solutions for each T by sliding the time bar. Additionally, a user will also be able to upload their own dataset, as long as it follows a prefixed format – plain file containing tuples of (*object-id, trajectory-id, latitude, longitude, time, weight (optional, default=1)*).
- *Visualization Components:* We used Google Maps JavaScript API [10] to display a map with respective pins for each object at a particular time instant and the current MaxRS solution. These pins and results will change accordingly as the user drags the time bar. Also, we used various JavaScript visualization tools (such as [11]) to enable a different view of the result – trajectories of the Co-MaxRS solutions in 3D spatio-temporal settings.
- *Co-MaxRS Algorithm:* Even for small object movements, the optimal location of the query rectangle can change while objects participating in the MaxRS solution stay the same. Instead of maintaining a centroid-location (equivalently, a region) as a Co-MaxRS solution, we maintain a list of objects that are located in the interior of the optimal rectangle placement – the

Co-MaxRS answer-set becoming a sequence (*list-of-objects, time-period*). For the example in Fig. 1, Co-MaxRS answer-set is: $\{((o_6, o_7, o_8), [t_0, t_1)), ((o_1, o_2, o_3), [t, t_{max})), ((o_1, o_3, o_7, o_8), [t_{max}, t_{max}))\}$. We identified criteria (i.e., critical times) when a particular MaxRS solution is no longer valid, or a new MaxRS solution emerges. The basic algorithm uses KDS (*Kinetic Data Structures*) – maintaining a priority queue of the critical events and their occurrence time, and recomputing MaxRS solutions at each event in order. Recomputing MaxRS is costly, so we devised efficient pruning schemes to: (1) Eliminate the recomputation altogether at certain qualifying critical events; and (2) Reduce the number of objects that need to be considered when recomputation is unavoidable.

- *Data Structures and Indexing:* We maintain a list for storing each object $o_i \in O$, with its current trajectory, weight, and other necessary information. KDS maintains an event queue, where the events are sorted according to the time-value. Each critical event consists of the related objects, and the occurrence time. Additionally, we also maintained an adjacency matrix to track locality of objects – which is important for smooth processing of our pruning schemes. Moreover, to ensure faster processing over large datasets, we used an existing spatio-temporal data indexing scheme, TPR*-tree (via a C++ library) [12].
- *Datasets:* To demonstrate the benefits of our system in different domains, we use several datasets: (1) GPS traces from 500 taxis collected over 30 days in San Francisco Bay area (http://crawdad.org/epfl/mobility/); (2) (*location, time*) information for 370 taxis within Rome for 1 month with sample-period of 7 s (http://crawdad.org/roma/taxi/); (3) Human mobility data, in [13], where researchers at Microsoft collected data from 182 users in a period of over five years, with 17,621 trajectories; and (4) A small animal movement dataset from (http://crawdad.org/princeton/zebranet/). While the first two datasets are great for demonstrating scalability and traffic-monitoring aspect of our system, the latter two can be used to show applications in human mobility tracing and animal tracking processes. Although all of these datasets had different formats, we converted them into the same format – tuples of (*object-id, trajectory-id, latitude, longitude, time, weight*) values. This enabled our system to handle similarly formatted user-provided data as well.

Demo Specifications: The setup of our demo will consist of a laptop running the web-based application via a web browser. The application is hosted on our server operating at Northwestern University, and will be accessed via a public URL. The demonstration will have three main parts with the following steps:

P1: The first part will have following three main phases:

Phase 1: Specification of the required parameters in the GUI (cf. Fig. 3). This phase will show: (a) Selection of appropriate R, $\mathbb{F}$, and T; and (b) Relation between R and $\mathbb{F}$, i.e., how our system dynamically changes R proportionally to $\mathbb{F}$ given an initial value. $\mathbb{F}$ can be dynamically set by the user ($\pm$ tool).

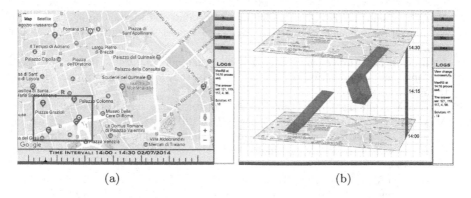

(a)   (b)

Fig. 3. (a) The main GUI with map and time slider; (b) Answer-set 3D view over T.

Phase 2: Visualization of the Co-MaxRS result at a certain point of time t using the user-provided parameters over a map (constrained to $\mathbb{F}$). The locations (linearly interpolated at t) of the objects will be displayed via pins on the map, and the solution, i.e., an optimal placement of R and the objects in its interior will be distinctly marked (see Fig. 3a).

Phase 3: Visualization of the whole answer-set for the time-period T in spatio-temporal (3D) settings (cf. Fig. 3b). Users will be able to analyze how the trajectories of the optimal clusters are evolving over space and time (parallelopipeds).

P2: In the second and most important part of the demo, we will exhibit the benefits of using this tool in analyzing large trajectories data. This part will show: (a) Selection of the area of focus ($\mathbb{F}$) by zooming-in or out; (b) Change the value of t by sliding through per unit time (depends on the data set) within T. We will demonstrate how a user can start from a large $\mathbb{F}$ and relatively larger R, and then continuously refine their region of interest.

P3: The last part of the demo will quickly show steps of **P1** and **P2** for different datasets, emphasizing how the tool can be useful in different domains. Additionally, the steps of uploading a custom dataset will be demonstrated.

References

1. Manyika, J., Chui, M., Brown, B., Bughin, J., Dobbs, R., Roxburgh, C., Byers, A.H.: Big data: The next frontier for innovation, competition, and productivity
2. Nanni, M., Pedreschi, D.: Time-focused clustering of trajectories of moving objects. J. Intell. Inf. Syst. **27**(3), 267–289 (2006)
3. Andrienko, G., Andrienko, N., Rinzivillo, S., Nanni, M., Pedreschi, D., Giannotti, F.: Interactive visual clustering of large collections of trajectories. In: IEEE Symposium on Visual Analytics Science and Technology (2009)
4. Zheng, Y.: Trajectory data mining: an overview. ACM Trans. Intell. Syst. Technol. (TIST) **6**(3), 29 (2015)

5. Shen, Y., Zhao, L., Fan, J.: Analysis and visualization for hot spot based route recommendation using short-dated taxi GPS traces. Information **6**(2), 134–151 (2015)
6. Andrienko, N., Andrienko, G., Fuchs, G., Rinzivillo, S., Betz, H.D.: Detection, tracking, and visualization of spatial event clusters for real time monitoring. In: IEEE DSAA (2015)
7. Choi, D.W., Chung, C.W., Tao, Y.: Maximizing range sum in external memory. ACM Trans. Database Syst. **39**(3), 21:1–21:44 (2014)
8. Nandy, S.C., Bhattacharya, B.B.: A unified algorithm for finding maximum and minimum object enclosing rectangles and cuboids. Comput. Math. Appl. **29**(8), 45–61 (1995)
9. Hussain, M.M., Islam, K.A., Trajcevski, G., Ali, M.E.: Towards efficient maintenance of continuous MaxRS query for trajectories. In: 20th EDBT (2017)
10. Google Maps API. https://developers.google.com/maps/
11. Vis.js: JavaScript visualization library. https://visjs.org/
12. Tao, Y., Papadias, D., Sun, J.: The TPR*-tree: an optimized spatio-temporal access method for predictive queries. In: VLDB (2003)
13. Zheng, Y., Zhang, L., Xie, X., Ma, W.Y.: Mining interesting locations and travel sequences from GPS trajectories. In: ACM World Wide Web (2009)

STAVIS 2.0: Mining Spatial Trajectories via Motifs

Crystal Chen[1(✉)], Arnold P. Boedihardjo[1], Brian S. Jenkins[1],
Charlotte L. Ellison[1], Jessica Lin[2], Pavel Senin[3], and Tim Oates[4]

[1] U.S. Army Engineer Research and Development Center, Alexandria, USA
{crystal.chen,arnold.p.boedihardjo,brian.s.jenkins,
charlotte.l.ellison}@usace.army.mil
[2] George Mason University, Fairfax, USA
jessica@gmu.edu
[3] Los Alamos National Lab, Los Alamos, USA
psenin@lanl.gov
[4] University of Maryland Baltimore County, Baltimore, USA
oates@cs.umbc.edu

Abstract. The increase in available spatial trajectory data has led to a massive amount of geo-positioned data that can be exploited to improve understanding of human behavior. However, the noisy nature and massive size of the data make it difficult to extract meaningful trajectory features. In this work, a context-free grammar representation of spatial trajectories is employed to discover frequent segments or motifs within trajectories. Additionally, a set of basis motifs is developed that defines all movement characteristics among a set of trajectories, which can be used to evaluate patterns within a trajectory (intra-trajectory) and between multiple trajectories (inter-trajectory). The approach is realized and demonstrable through the Symbolic Trajectory Analysis and VIsualization System (STAVIS) 2.0, which performs grammar inference on spatial trajectories, mines motifs, and discovers various pattern sets through motif-based analysis.

Keywords: Spatial trajectory · Motif discovery · Grammar induction · Activity recognition

1 Introduction

The emergence of advanced and embedded sensors has enabled the scientific community to study spatial trajectories in a myriad of domains from animal movements to individual or population human mobility. Spatial trajectories provide an important set of lenses by which one can discover relationships between various groups of species or people and improve understanding of their physical environments. Spatial trajectories are captured from a multitude of data sources, such as Global Positioning System (GPS) traces recorded in vehicle navigation devices, video feeds of animal flock migrations, and online text posts of users sharing their locations and events. Spatial trajectories can be analyzed to discover correlations among themselves and other spatial events to help generate insights into their causal factors and linkages.

© Springer International Publishing AG 2017
M. Gertz et al. (Eds.): SSTD 2017, LNCS 10411, pp. 433–439, 2017.
DOI: 10.1007/978-3-319-64367-0_30

This work studies the problem of developing a novel, compact, and scalable signature representation of large and noisy trajectory data. The signature can serve as an effective surrogate on which existing analytical tasks can be performed. Issues that arise due to voluminous and noisy data are addressed by transforming the multidimensional spatiotemporal (ST) data into a symbolic representation and generating motif-based signatures via grammar induction. The transformation into symbolic sequences effectively compresses the data and enables discovery of informative patterns via applications of statistical language processing algorithms.

To support the analysis of spatial trajectory relationships, we have significantly augmented our prior work [1] and developed the Symbolic Trajectory Analysis and VIsualization System (STAVIS) 2.0. The original STAVIS was the first analytical system that employed motif-based signature representation to support querying, processing, and visualization of trajectories. Different from the previous work, STAVIS 2.0 provides new analytical capabilities for inter-trajectory (i.e., between multiple trajectories) pattern discovery whereas the original STAVIS primarily focused on deriving intra-trajectory (i.e., between segments of a trajectory) patterns. Furthermore, the previous work only considered explicit geographic features of the trajectories and did not consider their shape or directional movements. Trajectory shapes can add significantly deeper understanding of the trajectory's semantic activities and intents. Hence, new techniques were developed for STAVIS 2.0 that support both geographic and shape features in its analytical approach. These extensions enable STAVIS 2.0 to effectively discover patterns and relationships between co-occurring and historical trajectories via their shapes and locations. The contributions of STAVIS 2.0 are summarized below:

1. Development of a new analytical model to flexibly and more effectively support associative operations of motif-based trajectory features.
2. Conceptualization of basis motifs to enable correspondence between motif segments originating from different trajectories.
3. New visualizations that allow more efficient discovery and verification of inter-trajectory patterns.

The remaining sections are organized as follows: Sect. 2 describes the technical approach of STAVIS 2.0 and Sect. 3 gives the demonstration plan of the system.

2   Technical Approach

STAVIS 2.0 was developed to support an end-to-end analysis workflow through a web-based platform. STAVIS 2.0 includes a spatial database that stores recorded trajectory traces composed of the following features: track ID, spatial point coordinates, and times. Although a trajectory is a continuous function mapping from the time domain to the spatial domain, in practice trajectories are recorded as discrete ST point samples (e.g., latitude, longitude, time). As a result, the database represents a trajectory as an ST point set. The following subsections provide a description of the STAVIS 2.0 model of computation (Sect. 2.1) and processing architecture (Sects. 2.2, 2.3 and 2.4).

2.1   Model of Computation

Let us define a spatial trajectory s as a time indexed and ordered set of spatial points sampled from a curve generated by an object traversing a two dimensional (or three dimensional) geographic space. Next, define two mappings, $s \xrightarrow{g} s_g$ and $s \xrightarrow{r} s_r$, where s_g is the space-filling curve (SFC) visit order (e.g., Hilbert SFC) and s_r is the relative direction (e.g., angular degree) between adjacent spatial points in s (details of these mappings are provided in Sect. 2.3). For a given pair of s_g and s_r, two sets of motifs m_{s_g} and m_{s_r} are generated using grammar induction, respectively, where m_s is a set of motif segments [1] for trajectory s.

Based on the above trajectory transformations and their corresponding motifs, the mode of analysis can be modeled as a subset of all pair-wise *feature* associations of s and m_s. Specifically, the following feature analyses are supported: (s_g, m_{s_g}), (s_g, m_{s_r}), (s_r, m_{s_g}), (s_r, m_{s_r}), (m_{s_g}, m_{s_g}), (m_{s_r}, m_{s_r}), and (m_{s_g}, m_{s_r}). For example, (m_{s_g}, m_{s_r}) denotes the analysis between geographic explicit (i.e., SFC visit order) and relative direction feature motif segments of trajectory s. Because STAVIS 2.0 supports analysis of multiple trajectories, each feature type listed above can refer to segments from different trajectories such as $(m_{s_g^1}, m_{s_r^2})$ for trajectories s^1 and s^2. For ease of exposition, the remainder of the paper applies all pair-wise feature associations to a collection of trajectories (i.e., inter-trajectory analysis) unless otherwise noted or made obvious from the context.

To support all of the feature analyses above, STAVIS 2.0 integrates three primary functions: *ST filtering*, *trajectory transformation*, and *signature analysis and visualization*, depicted in Fig. 1. In the first step, *ST filtering*, the trajectories of interest are obtained from the database by performing an ST query. During *trajectory transformation*, the spatial components of ST filtering results are projected to a one dimensional space via a space-filling curve or its relative direction (shape) and transformed into symbols via Symbolic Aggregate ApproXimation (SAX) [2]. In *signature analysis and visualization*, the SAX representations are used to generate motifs via the mSE-QUITUR algorithm [1] and perform pattern analysis and visualization.

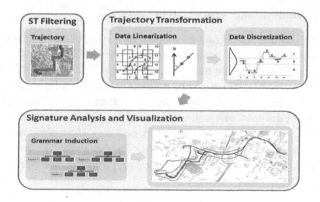

Fig. 1. Workflow of STAVIS 2.0.

2.2 Spatiotemporal (ST) Filtering

Within this component, the spatial area and time period of interest is specified as an ST query. Let $t_i = \{s_{i,1}..s_{i,m}\}$ be trajectory i where $s_{i,j}$ is the jth time-indexed spatial point and query Q is defined as a spatiotemporal cube (2D spatial points and 1D time interval) or hypercube (3D spatial points and 1D time interval). Let trajectory segment $SEG(t_i, s_{i,k}, s_{i,l}) = \{s_{i,p}|k \le p \le l\}$ where $k < l$ and trajectory touch points be the set $TP(t_i, Q) = \{s_{i,p}|s_{i,p} \in \partial Q \forall p\}$. Then a bounded trajectory intersection query is $BTI = \{SEG(t_i, firsttouch(TP(t_i, Q)), lasttouch(TP(t_i, Q))) \forall t_i \in TI\}$, i.e., the continuous segments in TI bounded by the points that first entered Q (*firsttouch*) and last exited Q (*lasttouch*). The obtained trajectory segments are then forwarded to the trajectory transformation module to be mapped into symbolic form.

2.3 Trajectory Transformation

A spatial trajectory is mapped into (a) linearized geographic locations s_g, i.e., geographic explicit, and (b) relative directions s_r where both maps transform the trajectory into one dimensional data indexed by time. For the linearized geographic locations, the Hilbert SFC is employed due to its superior ability to preserve local distances. The relative direction trajectory representation captures the shape-based or directional movements without regard to spatial locations. The relative direction is a calculation of the angles between a north-south line of the globe and a line that connects the current trajectory point to the previous point.

The next step is to represent the transformed trajectories as symbolic sequences using SAX [2]. SAX allows dimension reduction, which not only improves computational complexity, but also removes noise. SAX also allows distance measures to be defined in the symbolic space that *lower bound* corresponding distance measures defined in the original space. This lower-bounding property allows for improved efficiency of analytical operations, while producing comparable or even identical results to the original data.

2.4 Signature Analysis and Visualization

The signature analysis and visualization component is responsible for (a) generating the grammar induction (GI)-based signatures of the SAX transformed trajectories, (b) invoking the pattern mining tasks on the signatures, and (c) visualizing the mined results. STAVIS 2.0 utilizes mSEQUITUR [1] to produce grammar rules of a trajectory. The grammar rules are then processed to obtain the motifs m_s. To enable comparison between motifs from different originating trajectories, grammar induction is performed on a concatenated set of trajectories. Different ordering strategies, such as temporal ordering or random selection, can be applied for concatenation. The resulting grammar rules are then processed for extraction of a *basis motif* set that serves as the final representation of the multiple trajectories.

STAVIS 2.0 visualizations help analysts view the mined results (e.g., motifs) within all available contexts of the data pipeline. The analyst can interactively select results that are mapped to the signature rules, SAX time series representation, and trajectories on a spatial map. Correlating the results interactively to these different stages of the data processing pipeline enhances data provenance, reduces cognitive overhead, and increases understanding of the patterns. The system supports an iterative visualization workflow that allows the user to alter parameters at an arbitrary stage. For example, if after generating the trajectory motifs, the resolution of the space-filling curve or SAX alphabet need to be adjusted, then the parameters for rescaling the transformation can be directly modified without visiting the *ST filtering* step.

3 Demonstration Plan

STAVIS 2.0 will be demonstrated to showcase the applications of the various feature association tasks described in Sect. 2.1. Combinations of these features can help uncover frequency of pattern occurrence within and among trajectories, potential influence between locations and movements, as well as identification of anomalous locations and behaviors.

Prediction Task: The combinations of linearized location (s_g) or relative direction (s_r) trajectories with discovered motifs aid in exposing common trends and potential relationships between locations and how moving objects interact in those environments. These discovered patterns can be applied to support trajectory prediction, interpolation, or route recommendation. Consider segments where a location motif occurs within a trajectory, then (s_g, m_{s_g}) describes the frequency of spatial motif occurrence within trajectories, exposing commonly visited areas and sequences of area visits. Similarly, considering locations where a relative direction motif occurs within a trajectory, (s_g, m_{s_r}) depicts potentially critical movement archetypes at various regions. When a relative direction based trajectory is associated with linearized location motifs, (s_r, m_{s_g}), it can illuminate how location features may impact the movement behavior of objects. Additionally, associating a relative direction based trajectory with relative direction motifs, (s_r, m_{s_r}) highlights common trends in movement behavior among moving objects. Understanding common locations, movement trends, and potential relationships between locations and movements can improve the accuracy of prediction, interpolation, and recommendation applications.

Anomaly Detection Task: Aside from improved understanding of the frequent patterns within trajectories, associations discovered among location motifs (m_{s_g}) and relative direction motifs (m_{s_r}) can reveal anomalous behaviors or regions. Considering location motifs with others of the same motif type, (m_{s_g}, m_{s_g}) shows places or groups of places that are common among users. Associating different relative direction motifs, (m_{s_r}, m_{s_r}) exposes trends in movements among different sets of moving objects. Finally, considering geographic explicit motifs with relative direction motifs, (m_{s_g}, m_{s_r}) shows connections between common locations and common movement archetypes. If the temporal indices for each motif occurrence are compared, trajectories with points of motif correspondence may demonstrate trajectory similarity, while temporal indices where there are no corresponding associated patterns might signify the existence of an anomaly.

Case Example: The following showcases examples of multiple trajectory analysis for trend discovery and anomaly detection using relative direction motifs (s_r, m_{s_r}) and location motifs (m_{s_g}, m_{s_g}), respectively. Two real-world trajectories were concatenated which were derived from separate users tracking their runs during the Dublin, Ireland marathon in two separate years [3, 4]. For purposes of computational efficiency, the GPS trips were down sampled during preprocessing.

In the first case, Fig. 2(a) shows the relative direction time series representation of the concatenated trajectory. The points on the time series show motif locations. Figure 2(b) shows that a certain West-to-East movement is a common trend in both races at various locations. These discovered common movement archetypes can be applied towards developing signatures or constraints for classification or movement prediction.

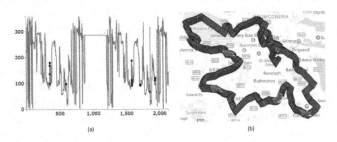

(a)   (b)

Fig. 2. Visualization of all instances of one discovered motif from a relative direction based trajectory. (a) shows the relative direction time series visualization with points indicating the starting locations of each motif occurrence. (b) shows a geographic representation of the trajectory in red, and all occurrences of one grammar rule in blue. (Color figure online)

In the second case example, the location motifs are considered. Figure 3(a) shows the original trajectories, which exactly overlap everywhere except in the region highlighted by the dashed blue box. This anomalous area corresponds to a starting point route change between the two marathon courses [5]. Figure 3(b) and (c) show the results of our analysis within STAVIS 2.0. The system discovers common location motifs throughout the race route and can highlight infrequent common location visits. These infrequent motif co-occurrences can be used to more robustly identify anomalies in noisy trajectories.

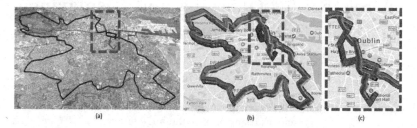

(a)   (b)   (c)

Fig. 3. Results from case study performed on two Dublin marathon trajectories. (a) shows the original trajectories, one in red and one in blue. The trajectories overlap everywhere except in the area highlighted with the dashed blue box. (b) shows a geographic representation of the trajectories in red and the discovered grammar rules in blue. There are no corresponding spatial motif occurrences within the area highlighted by the dashed box, demonstrating the existence of a spatial anomaly. A zoomed-in view of the discovered anomaly is shown in (c). (Color figure online)

References

1. Oates, T., Boedihardjo, A., Lin, J., Chen, C., Frankenstein, S., Gandhi, S.: Motif discovery in spatial trajectories using grammar inference. In: Proceedings of the 22nd ACM International Conference on Information and Knowledge Management, San Francisco (2013)
2. Lin, J., Keogh, E., Lonardi, S., Chiu, B.: A symbolic representation of time series, with implications for streaming algorithms. In: Proceedings of the 8th ACM SIGMOD Workshop on Research Issues in Data Mining and Knowledge Discovery, San Diego (2003)
3. GPSies, GPSies (2017). http://www.gpsies.com/map.do?fileId=aocrnjfdmbrdgcah. Accessed 23 Mar 2017
4. GPSies, GPSies (2017). http://www.gpsies.com/map.do?fileId=xrqmujejpmpxaneo. Accessed 23 Mar 2017
5. Google. Map data: SIO, NOAA, U.S. Navy, NGA, GEBCO, Google Earth Imagery, Google (2016)

A System for Querying and Displaying Typed Intervals

Jianqiu Xu$^{(\boxtimes)}$ and Junxiu Liang

Nanjing University of Aeronautics and Astronautics, Nanjing, China
{jianqiu,liangjunxiu}@nuaa.edu.cn

Abstract. We consider that a database stores a set of typed intervals, each of which defines start and end points, a weight and a type. Typed intervals enrich the data representation and support applications involving diverse intervals. In this demo, we introduce a system that is developed to not only support efficient query processing on typed intervals but also well visualize the index structure and query results. This benefits analyzing the data structure and understanding the data distribution. In particular, complex queries require a graphical interface to help justifying the correctness of answers and visualizing the query results.

1 Introduction

The interval data representing axis-parallel line segments have been widely used in spatial and temporal databases. In some applications, interval data are associated with types corresponding to different parameters. For example, the website "www.booking.com" provides various hotels for tourists such as five stars, apartments and motels, and their room rates change over time. The system needs to (i) efficiently manage a large amount of typed intervals representing rate per night for each hotel and (ii) display interesting query results in a clear way.

Querying typed intervals is proposed in [8] with a preliminary solution to answer top-k point and interval queries. In this demo, we introduce a system that is able to efficiently answer a range of queries on typed intervals and provide a graphical interface to help analyzing the data, structures and query results. To help understand the problem, consider an example in Fig. 1. The database stores a list of typed intervals. There are three types in total: {A, B, C}. Each interval is associated with a type and a weight.

We build a system that provides the data representation for typed intervals and supports three top-k queries: point, interval and continuous. Consider the top-k point query, the system returns k intervals at a point that have the largest weights and contain a particular type, e.g., Q_1 in Fig. 1. In addition to a point, an interval query is supported. Continuous queries are complex because the system returns top-k typed intervals at each time instant. Results may change at certain points. In the example, Q_2 returns $<([5.5, 7], o_7), ([7, 9.5], o_3)>$. The problem to be solved lies in two aspects: (i) efficiently indexing and querying typed intervals because the query combines different predicates; (ii) graphically displaying the index structure and the traversal path in the tree, and animating the query results.

© Springer International Publishing AG 2017
M. Gertz et al. (Eds.): SSTD 2017, LNCS 10411, pp. 440–445, 2017.
DOI: 10.1007/978-3-319-64367-0_31

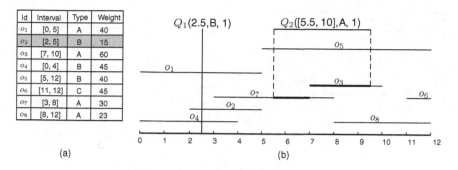

Fig. 1. A database of typed intervals

Existing works mainly focus on processing standard intervals (without the type) such as intersecting [6,9], stabbing [3] and joins [4]. Our work differs from them in two ways. On one hand, typed intervals enrich the data representation and standard intervals is in fact a special form of typed intervals. So far little attention has been paid to process diverse intervals. On the other hand, we aim to not only support efficient query processing but also develop a tool that graphically displays data structures, query results and the execution procedure. This benefits analyzing and understanding built-in structures in the system and how the structure is accessed and traversed. The index structure is complex for large datasets and thus well knowing the structure shape requires a visualization tool. Furthermore, the correctness of query results can also be easily justified.

To provide the query optimization, we make use of four existing indexes and develop two new structures based on standard interval tree [5]. This provides a comprehensive framework to compare the performance of different indexes. In the system, query results are displayed in a graphical interface by mapping intervals into spatial points and lines. Index structures are visualized by transforming into forms that can be graphically displayed and show the structure shape. This helps users well understand the structure and data distribution, in particular, is useful for a large tree including thousands of nodes and even more. A visualization tool benefits the analysis instead of just imaging the structure in mind or checking some parameters by performing queries. We provide the animation for continuous queries by transforming intervals into moving points such that one can view how results change at certain places. The visualization tool is general in terms of supporting not only typed intervals but also standard intervals.

In the demonstration, we display viewing the index structure built on one million intervals at different scales and the traversal path in the tree for point and interval queries. We animate the results of continuous queries by making use of moving points.

2   The Framework

We provide provide a flexible way for representing interval data that supports both typed intervals and standard intervals (without any type). One can scale

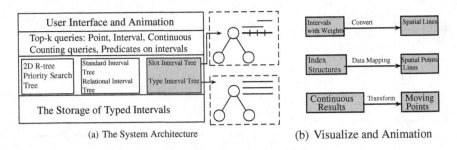

(a) The System Architecture   (b) Visualize and Animation

Fig. 2. An Overview

the number of types to simulate different dataset. An extreme case is that all intervals have the same type, leading to uniform type intervals. To manage typed intervals, the developed system includes four components: data storage, indexes, query algorithms and visualization, as depicted in Fig. 2(a). We use a relational interface to integrate standard intervals, types and weights into one framework. The relation schema is of the form

TypeInterval(Id:*int*, Range:*interval*, Type: *int*, Weight: *int*) in which a composite data type for interval data is implemented and embedded as an attribute. Several index structures are built including traditional structures such as priority search tree and the interval tree, and new structures such as slot interval tree and type interval tree by extending existing structures. The slot interval tree partitions the data space of each node into a set of slots and distributes intervals into slots. In each slot, we maintain two tables for intervals containing and overlapping the slot, respectively. The type interval tree maintains a set of interval lists in each node in which one list stores intervals with the same type. Algorithms for top-k queries are implemented by using different indexes to compare the performance and test the correctness. Auxiliary structures are developed to maintain continuous results. We maintain a binary tree in which each node corresponds to an interval and stores large weight intervals.

The system also supports counting queries that return the number of intervals intersecting a query point or an interval. Predicates can be integrated to return particular intervals, e.g., the length is smaller than a threshold. In some applications, very long intervals that span the overall data space may not be interesting. Updating indexes are also supported in the system in order to manage both historical and incoming data.

Query results are displayed in a graphical interface by transforming the data into the visualization form. Transforms are developed for interval data, indexes and continuous query results, see Fig. 2(b). Top-k queries return k intervals with the largest weights. To display intervals as well as weights in an integrated way, we convert the data into spatial lines by defining the x coordinates as interval start and end points, and the y coordinates as weights. Index structures are displayed by mapping tree nodes into spatial points and the creating lines to connect parent and child nodes. Since nodes are located at different levels, the y coordinates are used to represent node levels. Such values should be set appropriately such that users are easy to distinguish the nodes at different levels. In particular, given a large amount of data, the index will contain a lot of nodes. Graphically displaying the structure in a clear way benefits the analysis such as whether the tree is correctly built and how the data is distributed in the index.

3   Demonstration

The implementation is developed in an extensible database system SECONDO [7] and program in C/C++ and Java. Both synthetic and real datasets are used. We develop a tool to generate typed intervals in different settings. The start point of an interval is randomly chosen within the domain, and the length is a stochastic value between the defined range, e.g., [1, 1000]. Types and weights are randomly generated within their domains. One can scale the number of intervals, the number of types and interval lengths to produce datasets with different distributions. This provides comprehensive datasets to test the system in terms of efficiency and effectiveness. Queries are executed by calling functions in which users specify parameters following the system defined form. Different operators are implemented to report the final or intermediate result.

Visualize index structures. We build an interval tree on 1,000,000 typed intervals. The index contains 9,726 nodes and has the height 15, as displayed in Fig. 3. We visualize tree nodes by zooming out and zooming in to view the structure at different scales. One can see the structure in a global view and look at some branches in detail. Given a query, one traverses the tree from top to bottom level and the search path can be recorded and later displayed to help analyzing the query procedure. After inserting or deleting a node, the system visualizes the structure for users to understand the update and test the correctness. For example, a rotation is performed to keep the tree balanced and the parent-child relationship is changed among rotated nodes.

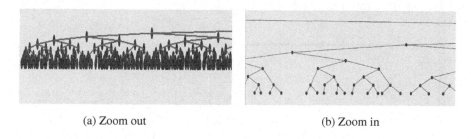

(a) Zoom out   (b) Zoom in

Fig. 3. Display large tree nodes

Locate the node and track the traversal path. Given an interval (or a set of intervals), we can find the node in which the interval is located, as illustrated in Fig. 4(a). We can also find the nodes storing intervals with particular types.

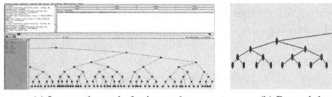 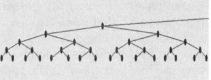

(a) Locate the node for intervals   (b) Record the traversal path

Fig. 4. Track the query

To analyze the query procedure, one may want to keep record of the traversal path in the tree and graphically display the result. For a point query, one node is accessed at each level, and for an interval query, the traversal path will split at a point, as demonstrated in Fig. 4(b). Graphically displaying the path helps understanding the execution procedure and debugging the program, in particular, when a large tree is built.

Animate continuous top-k queries. We use real datasets (bus card records) from a data company DataTang [2] including 4,575, 970 typed intervals. A sample data can be found at [1]. Each interval represents the number of passengers in a bus during a time interval. The type is the bus route and the weight is the number of passengers. We show both texts and spatial lines for intervals. Intervals are colored according to types to helps users find certain intervals among large datasets. For continuous top-k queries, e.g., *"top-k buses with the maximum number of passengers during [7:00am,8:00am]"*, reported intervals usually change at some places. To clearly view how the data changes over the query, we create k moving points to represent the result. The location of a moving point corresponds to a valid range (x-coordinate) and the weight (y-coordinate). Results changing over time are simulated by intervals appearing and disappearing at different points, This results in moving points in the graphical interface such that users are easy to figure out the place where the result is updated, as illustrated in Fig. 5.

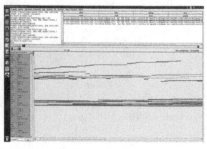 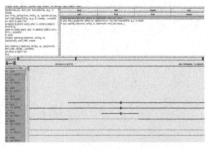

(a) Buses with different numbers of passengers   (b) Animate continuous queries (k=3)

Fig. 5. Visualize results

Acknowledgment. This work is supported by NSFC under grant numbers 61300052 and 61373015.

References

1. http://dbgroup.nuaa.edu.cn/jianqiu/
2. http://factory.datatang.com/en/
3. Arge, L., Vitter, J.S.: Optimal external memory interval management. SIAM J. Comput. **32**(6), 1488–1508 (2003)
4. Dignös, A., Böhlen, M.H., Gamper, J.: Overlap interval partition join. In: ACM SIGMOD, pp. 1459–1470 (2014)
5. Edelsbrunner, H.: Dynamic data structures for orthogonal intersection queries. Tech. Univ. Graz, Graz, Austria, Technical report (1980)
6. Enderle, J., Schneider, N., Seidl, T.: Efficiently processing queries on interval-and-value tuples in relational databases. In: VLDB, pp. 385–396 (2005)
7. Güting, R.H., Behr, T., Düntgen, C.: SECONDO: a platform for moving objects database research and for publishing and integrating research implementations. IEEE Data Eng. Bull. **33**(2), 56–63 (2010)
8. Xu, J., Lu, H., Yao, B.: Indexing and querying a large database of typed intervals. In: EDBT, pp. 658–659 (2016)
9. Kriegel, H.P., Pötke, M., Seidl, T.: Managing intervals efficiently in object-relational databases. In: VLDB, pp. 407–418 (2000)

MVSC-Bench: A Tool to Benchmark Classification Methods for Multivariate Spatiotemporal Data

Siddhant Kulkarni[✉] and Farnoush Banaei-Kashani

Department of Computer Science and Engineering,
University of Colorado, Denver, CO, USA
{siddhant.kulkarni,
farnoush.banaei-kashani}@ucdenver.edu

Abstract. Applications focusing on analysis of multivariate spatiotemporal series (MVS) have proliferated over the past decade. Researchers in a wide array of domains ranging from action recognition to sports analytics have come forward with novel methods to classify this type of data, but well-defined benchmarks for comparative evaluation of the MVS classification methods are non-existent. We present MVSC-Bench, to target this gap.

Keywords: Benchmarking · Multivariate data · Spatiotemporal data · Classification

1 Introduction

A *Multivariate Spatiotemporal Series (MVS)* dataset captures trajectories of several moving objects in a given space and timeframe. Each spatiotemporal source corresponds to one object, where objects can be birds in a flock, players in a sports team, vehicles in a transportation network or joints in human body. Figure 1 is an example of MVS dataset capturing human gait, where trajectories of same joints are visualized. Classification of MVS datasets has numerous knowledge discovery use cases such as gait identification [6], mobility analysis [2] and action recognition [1] from gait MVS as shown in Fig. 1, traffic behavior analysis from traffic MVS and team tactic analysis from sports MVS. Several methods have been proposed for MVS classification, including solutions based on feature extraction [6], time series analysis [10], episode mining [11], etc. Despite the vast research interest in MVS classification (MVSC), reliable and generic benchmarks as well as tools to evaluate the effectiveness of new methods are non-existent. In turn, this results in confusion with respect to the transferability of approaches across domains, and therefore difficulty in choosing the appropriate methodology.

To fill this gap, in this paper we propose the demonstration of a benchmarking and evaluation tool, dubbed MVSC-Bench, which addresses three needs:

1. A benchmark for comparative analysis of the performance of existing MVS classification methods.

© Springer International Publishing AG 2017
M. Gertz et al. (Eds.): SSTD 2017, LNCS 10411, pp. 446–451, 2017.
DOI: 10.1007/978-3-319-64367-0_32

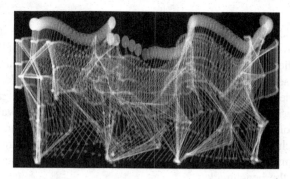

Fig. 1. MVS dataset capturing joint trajectories of human gait

2. A standard for MVS data format as well as unified interface for MVS classification methods
3. A tool to evaluate performance of new MVS classification methods versus existing methods.

To the best of our knowledge, currently MVSC-Bench is the first and only tool available for benchmarking MVS classification methods. Design and development of MVSC-Bench attempts to tackle several challenges such as defining and creating an interface to allow additional methods to be integrated as well as defining a generic data format that will allow different types of MVS datasets to be evaluated.

The remainder of this paper is organized as follows. Section 2 presents the benchmark we have defined for MVS classification including datasets, benchmark experiments, and baseline classification methods. Section 3 describes MVSC-Bench system we developed to implement our benchmark and the challenges we addressed toward this end. Section 4 elaborates on our demonstration plan to showcase functionalities of the MVSC-Bench system.

2   Benchmark

Our benchmark is defined to evaluate performance of the MVS classification methods based on a variety of parameters including:

- Number of MVS datasets (e.g., human gaits) for training and testing
- Number of instances considered for each MVS dataset (e.g., number of gait instances for each human subject)
- Number of objects considered in each MVS data instance (e.g., number of joints considered for each gait MVS dataset)
- Any other method specific parameters (e.g., features to be extracted, type of similarity measure to be used, minimum support, etc.)

Users of MVSC-Bench can use these parameters to define a thorough and extensive set of benchmarking experiments to evaluate performance of various existing (and new) MVS classification methods. With MVSC-Bench we have also implemented two baseline methods for MVS classification to serve as reference for comparison (see Sect. 3 for more details on baseline methods). The benchmark also defines standard interface that allows "plugging" in new methods to be compared with the baseline and other methods. The methods are compared based on the following measures: Accuracy, Time to train, Time to test and Memory required to save learning model. Moreover, our benchmark includes two sample MVS datasets, one capturing human gaits of over 140 humans with 4 or 5 instances of MVS data captured for each human walking across the view of a Microsoft Kinect [6] and the other one which records location of players in a soccer game which spans over 40 min capturing various spatial features of each player [8, 9]. Our benchmark also defines a standard format that allows incorporating other MVS datasets in the benchmark.

3   System Description

MVSC-Bench is the demonstration system we have developed to implement our MVS classification benchmark. In this section, we briefly present components of the MVSC-Bench system. A video demonstration of MVSC-Bench is available online [12].

3.1   User Interface

MVSC-Bench supports two user interfaces: Graphical User Interface (GUI) and command line interface. With GUI (Fig. 2), users interactively modify the parameters for experimentation. The command line interface is for users who wish to create jobs to execute several experiments using established configuration files. Moreover, this mode allows users to customize the configuration file to include a new method or dataset.

3.2   Baseline Methods

There are two baseline methods that have been embedded into the implementation of MVSC-Bench. The first approach is based on Feature Extraction [3–6] and it focuses on extracting certain features from the data that will represent the multivariate spatiotemporal series for classification. These features are a generalization of the features specified in [6]. Once these features are extracted from given data, they are used to implement nearest neighbor classification.

The second approach is based on the idea of MVS Time Series Similarity analysis where entries for each individual variate is considered as a time series. The similarity between two MVS datasets is then evaluated by measuring similarity based on corresponding variates using Dynamic Time Warping (DTW) or Euclidean Distance and aggregating the pair-wise similarity of variates to compare similarity of MVS datasets.

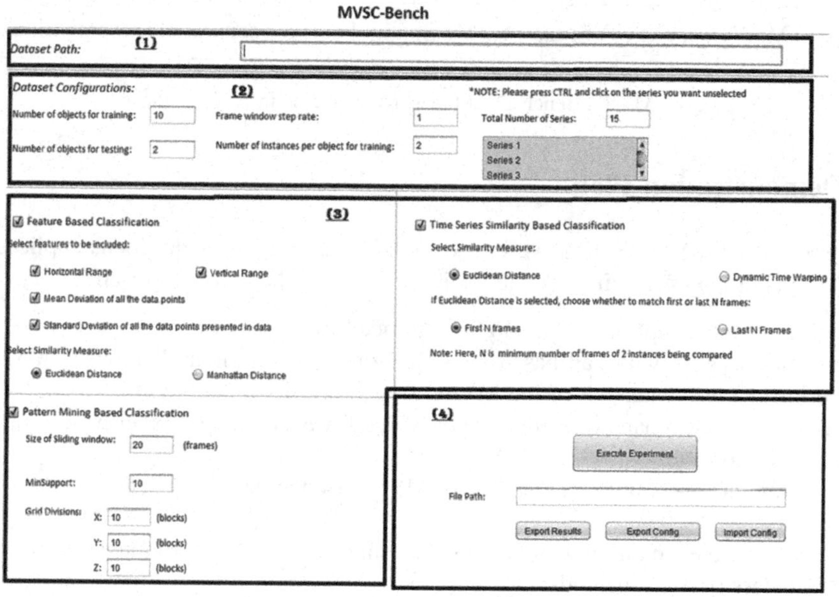

Fig. 2. MVSC-Bench GUI Sections: (1) Dataset path, (2) Generic Benchmarking parameters, (3) Approach specific configuration, and (4) Export/Import section

3.3   Standard Plug-in Interface for New Methods

MVSC-Bench defines an "IApproachImplementation" interface which defines methods, their parameters and their return values that must be strictly followed by any method that is to be integrated with MVSC-Bench. This interface defines methods for training (where the new method should build the learning model), testing (where the learning model should be applied to one MVS dataset at a time) as well as two miscellaneous methods to retrieve the name of the approach and to write any information for the specific approach to a file.

3.4   Standard Data Format for New Dataset Integration

Finally, MVSC-Bench also defines a standard format that allows the integration of new MVS datasets into the benchmark. The format defines a folder structure where each MVS dataset (e.g. gait data for each individual) is stored in respective folders with the name of the folder indicating the class represented by the dataset. Each such folder may contain one or more text files that represent different instances of data. Figure 3 presents the format for an entry for each object of each instance of data stored. Use of this format will allow integration of most MVS datasets into MVSC-Bench regardless of how many spatial features are to be used. Moreover, the proposed format allows for efficient retrieval of MVS datasets that are often humongous.

Entry<M>Name;<X>;<Y>;<Z>;<Other Spatial Features separated by semicolon>

Fig. 3. MVSC-Bench Data Format for entries in each instance

4 Demonstration Plan

To demonstrate MVSC-Bench, we intend to use gait analysis as the focus application for demonstration. We will walk the audience through the following steps:

1. Explain the concept of MVS, MVSC and need of MVSC-Bench.
2. Present the users with gait based identification [2, 6] as an application of MVSC.
3. Present the user interface of the MVSC-Bench and its functionality.
4. Describe the command line interface provided for execution of experiments using a configuration file.
5. Elaborate the programming interface of MVSC-Bench defines to accommodate new methods.
6. Walk the audience through the process of using MVSC-Bench to integrate two new MVS classification methods.
7. Execute sample experiment and (while that executes) present the data format and how they can leverage it to add more gravity to their comparative study.
8. Present and elaborate on the contents of the results.

The pattern based classification method extracts frequent patterns from MVS data. Each gait MVS dataset is then represented by the patterns derived from the dataset. Finally, in analogy with document classification, we consider each pattern as a word representing a gait dataset, and use TF-IDF [7] as similarity measure to classify MVS datasets.

References

1. Raheja, J.L., Minhas, M., Prashanth, D., Shah, T., Chaudhary, A.: Robust gesture recognition using Kinect: a comparison between DTW and HMM. Optik Int. J. Light Electron Optics **126**(11–12), 1098–1104 (2015)
2. Kashani, F.B., Medioni, G., Nguyen, K., Nocera, L., Shahabi, C., Wang, R., Blanco, C.E., Chen, Y.-A., Chung, Y.-C., Fisher, B., Mulroy, S., Requejo, P., Winstein, C.: Monitoring mobility disorders at home using 3D visual sensors and mobile sensors. In: Proceedings of the 4th Conference on Wireless Health (WH 2013). ACM, New York (2013)
3. Gianaria, E., Grangetto, M., Lucenteforte, M., Balossino, N.: Human classification using gait features. In: Cantoni, V., Dimov, D., Tistarelli, M. (eds.) Biometric Authentication, BIOMET 2014, vol. 8897. Springer, Cham (2014)
4. Sinha, A., Chakravarty, K.: Pose based person identification using kinect. In: 2013 IEEE International Conference on Systems, Man, and Cybernetics, Manchester, pp. 497–503 (2013)

5. Araujo, R.M., Graña, G., Andersson, V.: Towards skeleton biometric identification using the microsoft kinect sensor. In: Proceedings of the 28th Annual ACM Symposium on Applied Computing (SAC 2013). ACM, New York, pp. 21–26 (2013)
6. Andersson, V.O., Araujo, R.M.: Person identification using anthropometric and gait data from kinect sensor. In: Proceedings of the 29th AAAI Conference (2015)
7. Ramos, J.: Using TF-IDF to Determine Word Relevance in Document Queries (1999)
8. Pettersen, S.A., Johansen, D., Johansen, H., Berg-Johansen, V., Gaddam, V.R., Mortensen, A., Langseth, R., Griwodz, C., Stensland, H.K., Halvorsen, P.: Soccer video and player position dataset. In: Proceedings of the 5th ACM Multimedia Systems Conference (MMSys 2014). ACM, New York, pp. 18–23 (2014)
9. Yu, S., Tan, T., Huang, K., Jia, K., Wu, X.: A study on gait-based gender classification. IEEE Trans. Image Process. **18**(8), 1905–1910 (2009)
10. Morse, M.D., Patel, J.M.: An efficient and accurate method for evaluating time series similarity. In: Proceedings of the 2007 ACM SIGMOD International Conference on Management of Data (SIGMOD 2007). ACM, New York, pp. 569–580 (2007)
11. Han, J., Dong, G., Yin, Y.: Efficient mining of partial periodic patterns in time series database. In: Proceedings 15th International Conference on Data Engineering (Cat. No. 99CB36337), Sydney, NSW, pp. 106–115 (1999)
12. Kulkarni, S.: siddhantkulkarni/MVSClassification. GitHub (2017). https://github.com/siddhantkulkarni/MVSClassification. Accessed 26 Mar 2017

Author Index

Printed in the United States
By Bookmasters